Mechanics of Biomaterials

Teaching mechanical and structural biomaterials concepts for successful medical implant design, this self-contained text provides a complete grounding for students and newcomers to the field. Divided into three sections – Materials, Mechanics, and Case Studies – it begins with a review of sterilization, biocompatibility, and foreign body response before presenting the fundamental structures of synthetic biomaterials and natural tissues. Mechanical behavior of materials is then discussed in depth, covering elastic deformation, viscoelasticity and time-dependent behavior, multiaxial loading and complex stress states, yielding and failure theories, and fracture mechanics. The final section on clinical aspects of medical devices provides crucial information on FDA regulatory issues and presents case studies in four key clinical areas: orthopedics, cardiovascular devices, dentistry, and soft tissue implants. Each chapter ends with a list of topical questions, making this an ideal course textbook for senior undergraduate and graduate students, and also a self-study tool for engineers, scientists, and clinicians.

Lisa A. Pruitt is the Lawrence Talbot Chair of Engineering at the University of California, Berkeley and also serves as an Adjunct Professor in the Department of Orthopedic Surgery at the University of California, San Francisco. She recently served as the Associate Dean of Lifelong Learning and Outreach Education in the College of Engineering and has received numerous awards including the Presidential Award for Excellence in Science, Mathematics, and Engineering Mentoring (2004) and the Graduate Student Instructor Mentor Award from UC Berkeley (2009).

Ayyana M. Chakravartula received her Ph.D. in Mechanical Engineering from the University of California, Berkeley in 2005. She currently works at Exponent, Inc. in Menlo Park, California in its Mechanics and Materials practice. She has worked as a research scientist at the Cambridge Polymer Group in Boston, MA, and has served as an Adjunct Lecturer at Boston University. She has mentored numerous students, interns, and research assistants in her graduate and post-graduate career.

CAMBRIDGE TEXTS IN BIOMEDICAL ENGINEERING

Series Editors
W. Mark Saltzman, Yale University
Shu Chien, University of California, San Diego

Series Advisors
William Hendee, *Medical College of Wisconsin*
Roger Kamm, *Massachusetts Institute of Technology*
Robert Malkin, *Duke University*
Alison Noble, *Oxford University*
Bernhard Palsson, *University of California, San Diego*
Nicholas Peppas, *University of Texas at Austin*
Michael Sefton, *University of Toronto*
George Truskey, *Duke University*
Cheng Zhu, *Georgia Institute of Technology*

Cambridge Texts in Biomedical Engineering provides a forum for high-quality accessible textbooks targeted at undergraduate and graduate courses in biomedical engineering. It covers a broad range of biomedical engineering topics from introductory texts to advanced topics including, but not limited to, biomechanics, physiology, biomedical instrumentation, imaging, signals and systems, cell engineering, and bioinformatics. The series blends theory and practice, aimed primarily at biomedical engineering students. It also suits broader courses in engineering, the life sciences, and medicine.

Mechanics of Biomaterials is the textbook I have been waiting for. This comprehensive work synthesizes the science and engineering of biomaterials that has developed over the past three decades into a highly useful textbook for training students in two of the senior undergraduate/first-year graduate student courses I teach: Advanced Biomechanics, and Biomaterials and Medical Devices. In fact, as I reviewed this work it felt like I was reviewing my own lecture notes developed over 20 years. The work combines materials science, mechanics and medical device design and analysis in a seemless and thorough manner incorporating many critical studies from the literature into a clear and comprehensive work.

Pruitt and Chakravartula have succeeded in developing an outstanding text and reference book that should be required reading for all who aspire to design, develop and evaluate medical devices.

Jeremy L. Gilbert, *Syracuse University*

The authors have written a detailed yet easy-to-read book that can be used by materials scientists and biomedical engineers, from both the budding biomedical engineering student to the seasoned medical device designer. It combines the fundamentals of plastics, metals, and ceramics behavior with the required properties for the often challenging loading and environmental conditions found in the body. I particularly liked Pruitt and Chakravartula's technique of introducing a detailed discussion of the theoretical explanation of a particular material class's response to a loading environment, and then providing a real-life case study demonstrating how the theoretical response translates to clinical performance.

The book is rich in practical examples of biomaterials used in permanent implants currently on the market. Sufficient historical information is provided on implant successes and failures to appreciate the challenges for material and design selection in the areas of both hard and soft tissue replacement.

Stephen Spiegelberg, *Cambridge Polymer Group, Inc., MA, USA*

Mechanics of Biomaterials: Fundamental Principles for Implant Design provides a much needed comprehensive resource for engineers, physicians, and implant designers at every level of training and practice. The book includes a historical background which outlines the engineering basis of traditional implant designs, and interactions of materials, biology, and mechanics resulting in clinical success or failure of these devices. Each chapter contains a detailed description of the engineering principles which are critical to understand the mechanical behavior of biomaterials and implants in vivo. The scope of the text covers orthopaedics, cardiovascular devices, dental, and soft tissue implants, and should help considerably in our efforts to improve the function and durability of biomaterials and implants used in clinical practice.

Michael Ries, *University of California, San Francisco*

Mechanics of Biomaterials

Fundamental Principles for Implant Design

Lisa A. Pruitt
University of California, Berkeley

Ayyana M. Chakravartula
Exponent, Inc., Menlo Park, California

CAMBRIDGE
UNIVERSITY PRESS

University Printing House, Cambridge CB2 8BS, United Kingdom

One Liberty Plaza, 20th Floor, New York, NY 10006, USA

477 Williamstown Road, Port Melbourne, VIC 3207, Australia

4843/24, 2nd Floor, Ansari Road, Daryaganj, Delhi - 110002, India

79 Anson Road, #06-04/06, Singapore 079906

Cambridge University Press is part of the University of Cambridge.

It furthers the University's mission by disseminating knowledge in the pursuit of education, learning and research at the highest international levels of excellence.

www.cambridge.org
Information on this title: www.cambridge.org/9780521762212

© L. A. Pruitt and A. M. Chakravartula 2011

This publication is in copyright. Subject to statutory exception
and to the provisions of relevant collective licensing agreements,
no reproduction of any part may take place without the written
permission of Cambridge University Press.

First published 2011

A catalogue record for this publication is available from the British Library

Library of Congress Cataloging in Publication data
Pruitt, Lisa A.
Mechanics of biomaterials : fundamental principles for implant design / Lisa A. Pruitt, Ayyana M. Chakravartula.
    p. cm. – (Cambridge texts in biomedical engineering)
Includes bibliographical references and index.
ISBN 978-0-521-76221-2
1. Biomedical materials.   2. Prosthesis – Design and construction.   I. Chakravartula, Ayyana M.   II. Title.   III. Series.
R857.M3P78   2011
610.284 – dc23   2011019855

ISBN 978-0-521-76221-2 Hardback

Cambridge University Press has no responsibility for the persistence or accuracy of URLs for external or third-party internet websites referred to in this publication, and does not guarantee that any content on such websites is, or will remain, accurate or appropriate.

Contents

Symbols	*page* xi
Prologue	xiv

Part I **Materials** 1

1 Biocompatibility, sterilization, and materials selection for implant design 3

1.1	Historical perspective and overview	3
1.2	Learning objectives	4
1.3	Successful device performance and implant design	4
1.4	Biocompatibility	7
1.5	Sterility	8
1.6	Regulatory issues	9
1.7	Structural requirements	10
1.8	Classifying biomaterials	13
1.9	Structure–property relationships	16
1.10	Attributes and limitations of synthetic biomaterials	17
1.11	Case study: deterioration of orthopedic-grade UHMWPE due to ionizing radiation	20
1.12	Summary	22
1.13	Problems for consideration	23
1.14	References	23

2 Metals for medical implants 26

2.1	Historical perspective and overview	26
2.2	Learning objectives	27
2.3	Bonding and crystal structure	28
2.4	Interstitial sites	32
2.5	Crystallographic planes and directions	34
2.6	Theoretical shear strength	36
2.7	Imperfections in metals and alloys	38
2.8	Metal processing	42
2.9	Corrosion processes	53

	2.10	Metals in medical implants	60
	2.11	Case study: corrosion in modular orthopedic implants	64
	2.12	Summary	67
	2.13	Problems for consideration	67
	2.14	References	68

3 Ceramics   70

3.1	Historical perspective and overview	70
3.2	Learning objectives	71
3.3	Bonding and crystal structure	71
3.4	Mechanical behavior of ceramics	75
3.5	Processing of ceramics	82
3.6	Ceramics in medical implants	85
3.7	Case study: the use of coral as a bone substitute	87
3.8	Summary	89
3.9	Problems for consideration	89
3.10	References	90

4 Polymers   92

4.1	Historical perspective and overview	92
4.2	Learning objectives	94
4.3	Bonding and crystal structure	95
4.4	Molecular weight distribution in polymers	106
4.5	Mechanical behavior of polymers	109
4.6	Polymer processing	112
4.7	Polymers in medical implants	113
4.8	Case study: resorbable sutures and suture anchors	124
4.9	Summary	125
4.10	Problems for consideration	126
4.11	References	127

5 Mechanical behavior of structural tissues   129

5.1	Historical perspective and overview	129
5.2	Learning objectives	131
5.3	Building blocks of tissues	131
5.4	Load-bearing tissues	136
5.5	Case study: creating a scaffold for tissue engineering	156
5.6	Summary	158
5.7	Problems for consideration	158
5.8	References	159
5.9	Bibliography	163

Part II Mechanics   165

6 Elasticity   167
    6.1 Overview   167
    6.2 Learning objectives   169
    6.3 Stress and strain   169
    6.4 Bending stresses and beam theory   193
    6.5 Composites   200
    6.6 Case study: modifying material and cross-section to reduce bone absorption   203
    6.7 Summary   205
    6.8 Problems for consideration   206
    6.9 References   207
    6.10 Bibliography   207

7 Viscoelasticity   208
    7.1 Overview   208
    7.2 Learning objectives   209
    7.3 Introduction to viscoelasticity   209
    7.4 Linear viscoelastic networks   214
    7.5 Frequency domain analysis   227
    7.6 Time-temperature equivalence   233
    7.7 Nonlinear viscoelasticity   235
    7.8 Case study: creep behavior of UHMWPE used in total joint replacements   237
    7.9 Summary   238
    7.10 Problems for consideration   238
    7.11 References   239

8 Failure theories   241
    8.1 Overview   241
    8.2 Learning objectives   244
    8.3 Yield surfaces   244
    8.4 Maximum shear stress (Tresca yield criterion)   245
    8.5 Maximum distortional energy (von Mises yield criterion)   249
    8.6 Predicting yield in multiaxial loading conditions   255
    8.7 Modified yield criteria   261
    8.8 Maximum normal stress failure theory   265
    8.9 Notches and stress concentrations   266
    8.10 Failure mechanisms in structural biomaterials   269
    8.11 Case study: stress distribution in a total joint replacement   276
    8.12 Summary   279

8.13	Problems for consideration	280
8.14	References	281

9 Fracture mechanics 283

9.1	Overview	283
9.2	Learning objectives	284
9.3	Linear elastic fracture mechanics (LEFM)	285
9.4	Modified methods in LEFM	299
9.5	Elastic-plastic fracture mechanics (EPFM)	303
9.6	Time-dependent fracture mechanics (TDFM)	318
9.7	Intrinsic and extrinsic fracture processes	320
9.8	Fracture mechanisms in structural materials	322
9.9	Case study: fracture of highly crosslinked acetabular liners	324
9.10	Summary	325
9.11	Problems for consideration	326
9.12	References	327

10 Fatigue 329

10.1	Overview	329
10.2	Learning objectives	331
10.3	Fatigue terminology	331
10.4	Total life philosophy	334
10.5	Strain-based loading	343
10.6	Marin factors	346
10.7	Defect-tolerant philosophy	348
10.8	Case study: fatigue fractures in trapezoidal hip stems	362
10.9	Summary	364
10.10	Problems for consideration	364
10.11	References	366

11 Friction, lubrication, and wear 369

11.1	Overview	369
11.2	Learning objectives	370
11.3	Bulk and surface properties	371
11.4	Friction	373
11.5	Surface contact mechanics	375
11.6	Lubrication	377
11.7	Wear	381
11.8	Surface contact in biomaterials	386
11.9	Friction and wear test methods	387
11.10	Design factors	389
11.11	Case study: the use of composites in total joint replacements	390
11.12	Summary	391

11.13	Problems for consideration	391
11.14	References	391

Part III Case studies — 395

12 Regulatory affairs and testing — 397

12.1	Historical perspective and overview	397
12.2	Learning objectives	398
12.3	FDA legislative history	398
12.4	Medical device definitions and classifications	401
12.5	CDRH organization	404
12.6	Anatomy of a testing standard	408
12.7	Development of testing standards	409
12.8	International regulatory bodies	410
12.9	Case study: examining a 510(k) approval	411
12.10	Summary	413
12.11	Problems for consideration	413
12.12	References	414

13 Orthopedics — 416

13.1	Historical perspective and overview	416
13.2	Learning objectives	421
13.3	Total joint replacements	421
13.4	Total hip arthroplasty	422
13.5	Total knee arthroplasty	435
13.6	Fracture fixation	446
13.7	Spinal implants	453
13.8	Engineering challenges and design constraints of orthopedic implants	462
13.9	Case studies	462
13.10	Summary	470
13.11	Looking forward in orthopedic implants	470
13.12	Problems for consideration	470
13.13	References	471

14 Cardiovascular devices — 477

14.1	Historical perspective and overview	477
14.2	Learning objectives	478
14.3	Cardiovascular anatomy	479
14.4	Load-bearing devices	483
14.5	Case studies	496
14.6	Looking forward	500

14.7 Summary 500
14.8 Problems for consideration 501
14.9 References 501

15 Oral and maxillofacial devices 505
15.1 Overview 505
15.2 Learning objectives 507
15.3 Oral and maxillofacial anatomy 507
15.4 Dental implants 510
15.5 Temporomandibular joint replacements 532
15.6 Case studies 540
15.7 Looking forward 544
15.8 Summary 546
15.9 Problems for consideration 546
15.10 References 547

16 Soft tissue replacements 560
16.1 Historical perspective and overview 560
16.2 Learning objectives 563
16.3 Sutures 565
16.4 Synthetic ligament 570
16.5 Artificial skin 575
16.6 Ophthalmic implants 578
16.7 Cosmetic implants 584
16.8 Case studies 587
16.9 Looking forward 588
16.10 Summary 589
16.11 Problems for consideration 590
16.12 References 590

Epilogue 595
Appendix A. Selected topics from mechanics of materials 597
Appendix B. Table of material properties of engineering biomaterials and tissues 600
Appendix C. Teaching methodologies in biomaterials 611
Glossary 620
Index 645

Symbols

Roman letters

a	crack length
a, b, c	characteristic lengths in crystal system
a_T	shift factor for time-temperature superposition
A	area
A, B, C	reciprocal of Miller indices
b	Burgers vector, Basquin exponent
B	thickness
c	maximum distance from neutral axis in beam theory
C	degrees Celsius, Marin factor, material constant in Paris Equation
d	atomic diameter, grain diameter, diameter, incremental fatigue damage
D	total fatigue damage
D_e	Deborah number
E	elastic modulus, energy
F	force, function of, degrees Fahrenheit
g	weight coefficient for generalized Maxwell model
G	shear modulus, energy release rate
h	height, lubricant film thickness, weight coefficient for generalized Kelvin model
h, k, l	Miller indices
H	enthalpy
i	imaginary number defined as $i^2 = -1$
I	area moment of inertia
J	stress invariants, energy release rate
k	stiffness, Archard's coefficient
K	stress intensity factor, degrees Kelvin
l	lamellar thickness, object length
L	load
m	material constant in Paris Equation
M	molecular weight, moment

n	vector normal to plane, number of polymer chains, number of cycles
N	number of fatigue cycles, total number of polymer chains
p	pressure, hydrostatic pressure
P	force
r	interatomic separation, distance from crack tip, radius
R	crystallographic direction vector, radius, resistance to crack growth, roughness
s	deviatoric part of stress tensor
S	stiffness, strength, compliance, stress
t	time, direction of dislocation line, thickness
T	temperature, force
u, v, w	directional vector components
U	energy, bond energy, strain energy
v	viscosity
V	volume, wear volume
w	strain energy density
W	width, work
x	separation distance, degree of crystallinity, distance traveled
x, y, z	spatial coordinates
Y	geometric flaw factor

Greek letters

α	thermal expansion coefficient
α, β, γ	characteristic angles in crystal system
γ	surface energy, shear strain
δ	crack opening displacement, phase shift
ε	strain
η	viscosity
θ	angle
κ	curvature, bulk modulus
λ	mean free spacing of particles
μ	pressure sensitivity coefficient, coefficient of friction
ν	Poisson's ratio, frequency
π	pi
ρ	bond density, atomic packing density, dislocation density, mass density, radius of curvature, length of plastic zone
σ	stress
τ	shear stress, shear strength, relaxation time
ω	rotational velocity, frequency

Subscripts

0	equilibrium separation (r_0)
a	amorphous (ρ_a), average (R_a)
c	critical (K_{IC}), crystalline (ρ_c)
f	fracture (σ_f), failure (S_y)
f	final (l_f)
g	glass transition (T_g)
i	initial (l_i)
I, II, III	fracture mode (K_I)
K	relating to the Kelvin model (ε_K)
m	melt (T_m), mean (σ_m)
max	maximum (σ_{max})
min	minimum (σ_{min})
M	relating to the Maxwell model (ε_M)
n	number average (M_n)
N	normal (F_N)
o	yield (σ_o)
p	plastic zone (r_p)
S	shear (F_S)
SCC	stress corrosion cracking (K_{ISCC})
SLS	relating to the Standard Linear Solid model (ε_{SLS})
th	theoretical (σ_{th})
w	weight average (M_w)
y	yield (σ_y)

Prologue

Mechanics of Biomaterials: Fundamental Principles for Implant Design provides the requisite engineering principles needed for the design of load-bearing medical implants with the intention of successfully employing synthetic materials to restore structural function in biological systems. This textbook makes available a collection of relevant case studies in the areas of orthopedics, cardiology, dentistry, and soft tissue reconstruction and elucidates the functional requirements of medical implants in the context of the specific restorative nature of the device. Each chapter opens with an exploratory question related to the chapter content in order to facilitate inquiry-based learning. Subsequently, a general overview, learning objectives, worked examples, clinical case studies, and problems for consideration are provided for each topic.

The organization of the book is designed to be self-contained, such that a student can be trained to competently engineer and design with biomaterials using the book as a standalone text, while practicing engineers or clinicians working with medical devices may use its content frequently in their careers as a guide and reference. The book comprises three basic sections: (i) overview of the materials science of biological materials and their engineered replacements; (ii) mechanical behavior of materials and structural properties requisite for implants; and (iii) clinical aspects of medical device design.

The first section of this book begins with a synopsis of medical devices and the fundamental issues pertaining to biocompatibility, sterilization, and design of medical implants. Engineered biomaterials including metals, ceramics, polymers, and composites are examined. The first segment of the book concludes with a description of structural tissues. In these chapters, the mechanical properties of common synthetic and natural materials are discussed within the context of their structure-property relationships.

The second portion of the book addresses mechanical behavior of materials and provides the framework for how these properties are important in load-bearing medical implants. This section commences with a review of elastic deformation, constitutive behavior of biomaterials, multiaxial loading, and complex stress states in the body. Elastic behavior is followed by time-dependent behavior and failure theories. The topics of fracture mechanics, fatigue, and tribology and the concomitant design concerns round out this section of the text. Key example problems are carried through successive chapters in this section in order to demonstrate the layered critical thinking that accompanies engineering design decisions.

The final part of the textbook is concerned with the clinical applications of medical devices. This section opens with a broad overview of the FDA, medical device classifications, and regulatory information for implants. Chapters devoted to orthopedics, cardiology, dentistry, and soft tissue repair that include embedded case studies as well as the current challenges associated with implant design in these specialized fields complete the third section of the book.

Appendices that provide the ASTM testing protocols used for medical devices in the FDA regulatory process, structural equations, and teaching strategies that may be employed in the classroom for this multidisciplinary topic appear at the end of the textbook to serve as references for engineers, researchers, clinicians, and faculty working in this field.

The compilation of this textbook is the result of the synergistic efforts of many people. The fundamental technical content has been developed over the past decade through a mezzanine undergraduate-graduate course entitled *Structural Aspects of Biomaterials*. This is a technical elective course designed for students in the fields of bioengineering, materials science, and mechanical engineering and has been evolving since its inception at UC Berkeley in 1998.

This book would not have been possible without the hundreds of dedicated undergraduates and graduate students who enrolled in this course over the past decade. Their continued feedback in course content and format has been invaluable to its development. Doctoral research students in the Medical Polymer Group at UC Berkeley have served as the graduate instructors for this course. They have relentlessly offered their time, pedagogy, and expertise in the implementation of this highly successful course. Specifically, we thank Sara Atwood, Dezba Coughlin, Donna Ebenstein, Jevan Furmanski, Jove Graham, Shikha Gupta, Sheryl Kane, Catherine Klapperich, Cheng Li, and Eli Patten.

For more than a decade we have collaborated with the Lawrence Hall of Science in the development of an annual course project entitled *Body by Design* that has focused on teaching mechanical behavior and medical device technology to elementary school and middle school children in the surrounding communities. We are most grateful to Barbara Ando, Craig Hansen, JohnMichael Selzer, Brooke Smith, and Gretchen Walker for their inspiring commitment to k-12 engineering education. Elements of our outreach teaching and active learning strategies are provided in the Appendix of the textbook. In the educational domain we would also like to acknowledge Rebecca Brent and Richard Felder (National Effective Teaching Institute) for sharing many great practices in pedagogy including methods for teaching to audiences with diverse learning styles, activities for active learning, and formats for inquiry-based learning.

The clinical content of this book has been facilitated by a number of ongoing collaborations in the field of biomaterials science and the medical device research industry. We are most appreciative to Tamara Alliston (University of California, San Francisco), Paul Ashby (Lawrence Berkeley National Labs), Mehdi Balooch (Lawrence Livermore National Labs), Anuj Bellare (Brigham and Women's Hospital), Alessandro Bistolfi (Centro Traumatologico Ortopedico, Torino), Luigi Costa (University of Torino), Avram Ediden (Kyphon), Alistair Elfick (University of Edinburgh), Seth Greenwald (Cleveland

Clinic), Steve Gunther (private practice), Bob Hastings (DePuy Orthopaedics, Inc.), Paul Hansma (University of California, Santa Barbara), Tony Keaveny (University of California, Berkeley), Karen King (University of California, San Francisco), Steve Kurtz (Exponent, Inc.), Steve Kaplan (4th State, Inc.), Kyriakos Komvopoulos (University of California, Berkeley), Jeff Lotz (University of California, San Francisco), Sally Marshall (University of California, San Francisco), Corey Maas (University of California, San Francisco), Tom Norris (San Francisco Shoulders and Elbows), Alan Pelton (Nitinol Devices & Components), Christian Puttlitz (Colorado State University), Michael Ries (University of California, San Francisco), Clare Rimnac (Case Western Reserve University), Robert Ritchie (University of California, Berkeley), David Saloner (University of California, San Francisco), Stephen Spiegelberg (Cambridge Polymer Group, Inc.), Subra Suresh (Massachusetts Institute of Technology), Linda Wang (private practice), Tim Wright (Hospital for Special Surgery), and Thomas Wyrobek (Hysitron, Inc.).

We would like to express gratitude to Dr. Jevan Furmanski for his technical input and contributions to this textbook, especially on the topics of tissues, viscoelasticity, and fracture mechanics. Additionally we thank Dr. Shikha Gupta for her contribution on dental materials. We are indebted to the numerous students who provided the high-quality technical illustrations for this book. We are most thankful to AJ Almaguer, Perry Johnson, Matt Kury, Mimi Lan, David Lari, Amir Mehdizadeh, Helen Wu, as well as Kayla and Eli Patten. We are most grateful for the funding from the Lawrence Talbot Chaired Professorship in Engineering that provided financial support for the technical illustration of this book.

We are deeply indebted to our friends and colleagues who served as our technical editors. In particular we thank Sara Atwood (Elizabethtown College), Bhaskararao Chakravartula, Neela Chakravartula (32BJ Benefit Fund), Dezba Coughlin (University of California, San Francisco), Donna Ebenstein (Bucknell University), Jove Graham (Geisinger Center for Health Research), Shikha Gupta (Food and Drug Administration), Cheng Li (Trivascular, Inc), Elise Morgan (Boston University), Gerry Pruitt (Northrop Grummen), Greg Supple (Hospital of the University of Pennsylvania), and Jingli Wang (PRD USA).

Finally and most important, we thank our families for their unwavering support. This book is dedicated to our families. We are most grateful to Raiden and Will, Savanna and Ric, and Juan and JJ for their relentless support in this process. Thank you.

Part I **Materials**

1 Biocompatibility, sterilization, and materials selection for implant design

Inquiry

All medical implants must be sterilized to ensure no bacterial contamination to the patient. How would you sterilize a total hip replacement comprising a titanium stem, a cobalt-chromium alloy head, and an ultra-high molecular weight polyethylene acetabular shell? Could the same method be employed for all three materials? How do you ensure that there is no degradation to the material or its structural properties? What factors would you need to consider in the optimization of this problem?

The inquiry posed above represents a realistic challenge that one might face in the field of orthopedic biomaterials. At a minimum, one would want to know the **sterilization** methods available for medical implants and which materials they best serve. For example, steam or autoclaving work well for sterilization of metals and ceramics but are generally unsuitable for polymers due to the lower melting and distortion temperatures of medical plastics. Also, one needs to consider whether there are any changes in the mechanical properties or if any time-dependent changes are expected owing to the sterilization method employed; for example, gamma radiation is known to leave behind free radicals (unpaired electrons) and these free radicals are highly reactive with elements such as oxygen that may be present or may diffuse into the implant material. In certain polymer materials such as ultra-high molecular weight polyethylene, **gamma radiation** can result in oxidation-induced embrittlement (shelf aging) that can severely degrade its wear and fracture properties. The case study presented at the end of this chapter addresses this issue.

1.1 Historical perspective and overview

Designing medical implants is a complex process, and this textbook aims to provide insight into the material, mechanical, and clinical factors that affect implant design and performance. The goal of this book is to integrate all aspects of implant design including clinical issues, structural requirements, materials selection, and processing treatments.

Historically, medical implant designs were driven solely by the need to restore function to a patient. Early **medical devices** utilized the skill of the resident surgeon and materials available at that time; materials such as ivory, bone, and wood were the first materials utilized to replace lost or damaged limbs. As time passed, implant designs utilized available metals such as gold, silver, or amalgams in facial reconstruction or dentistry; natural materials such as cat gut for sutures, porcine valves as heart valves; various steels in orthopedic implants; and polymers in soft tissue repair (Park and Bronzino, 2003; Ratner et al., 1996). It has only been in the last 50–60 years that engineering materials have been widely utilized for medical implants; the majority of material development used in medical devices has occurred in the last 50 years and has been accompanied by the growing research field of biomaterials science. Figure 1.1 shows the evolution of design-material combinations used for total hip repair in the last century.

1.2 Learning objectives

This introductory chapter provides a broad overview of biomaterials used in medical implants. The basic factors contributing to medical device design are presented. Issues of biocompatibility, sterility, and basic structure of biomaterials used in implants are addressed. The benefits and limitations of each material class are discussed. At the completion of this chapter, the student will be able to:

1. name the key factors that contribute to successful device performance
2. explain biocompatibility
3. define sterility and recommend sterilization schemes for various biomaterials
4. classify medical devices according to FDA regulatory requirements
5. identify mechanical properties used in describing structural requirements
6. describe the classification schemes for biomaterials
7. illustrate the materials selection process
8. discuss structure-property relationships associated with bonding mechanisms
9. list limitations and benefits of each class of synthetic material including composites used in medical devices
10. elucidate the role of the design process used in medical devices
11. describe a clinical case example involving sterilization and medical implants

1.3 Successful device performance and implant design

Successful medical device design brings together multifactorial challenges and synergistic solutions that build from diverse fields including engineering, manufacturing, biology, and clinical medicine. A schematic illustration of the variables that factor into the long-term success of a medical device is provided in Figure 1.2.

The *clinical issues* are paramount in the design process and in the long-term performance of the medical device. The primary requirement of any implantable device is that it is **biocompatible**; the implant must be able to restore function without adverse reaction or chronic inflammatory response. Combinations of materials selection (Chapters 2–5), loading (Chapters 6–11), and design (Chapters 13–16) can cause *in vivo* degradation.

1.3 Successful device performance and implant design

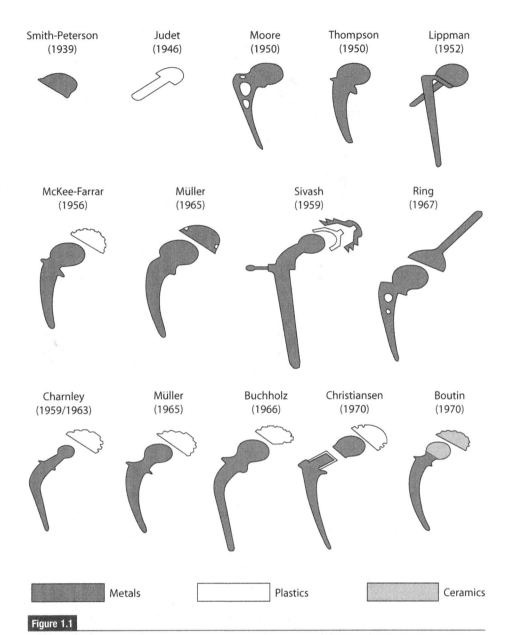

Figure 1.1

An example of material and design evolution in total hip replacements. (After Park and Bronzino, 2003.)

Moreover, the patient's specific immunological response and body environment also affect the long-term integrity and performance of the implant. The role of a good surgeon cannot be underestimated: a good surgeon provides appropriate assessment of the patient needs and assures optimized surgical placement of the implant. Patient factors also include health, anatomy, weight, and physical activity levels as these contribute directly to the structural requirements of the implant.

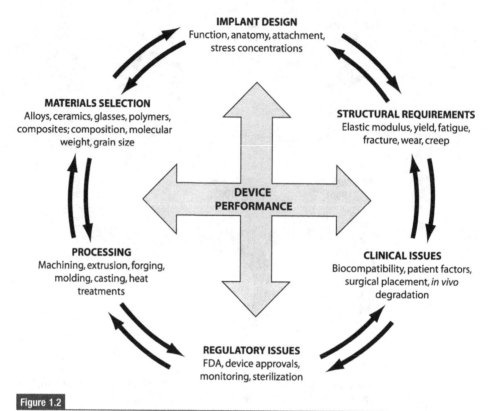

Figure 1.2

The multifactorial factors contributing to medical device performance.

The *structural requirements* of a medical device are typically found through an assessment of the expected **physiological stresses** on the implant. These stresses will vary depending upon the patient's anatomy, weight, and physical activity. For example, activities such as running can amplify the forces in a hip joint by nearly a factor of ten. Analysis of these stresses is key in making certain that the appropriate **elastic modulus** (Chapter 6), **yield strength** (Chapter 8), and other important material properties such as **creep** (Chapter 7), **fracture** (Chapter 9), **fatigue** (Chapter 10), and **wear** (Chapter 11) resistance are known. Simulator studies that mimic the physiological stresses are generally employed in the design process and help to assess the integrity of the implant in a laboratory environment. Such tests are often necessary in obtaining regulatory (FDA) approval of the medical implant.

Implant design involves the actual creation of the blueprints for the implant and calls out necessary materials to be utilized as well as geometric requirements such as component **tolerance**, **sphericity**, **surface finish**, and **notches**. The design of the implant directly incorporates the necessary geometry for anatomical constraints and desired function of the device.

The *materials selection* is first addressed by analyzing the structural requirements of the implant. The choice of a specific **alloy** (Chapter 2), **ceramic** (Chapter 3), **polymer** (Chapter 4), or **composite** is founded upon the knowledge of the structural requirements and function of the implant. Appropriate choice of material is necessary to make sure that appropriate material properties are provided. Within the materials selection process it is also important to understand the role of variations in microstructure, molecular weight, and composition as these directly contribute to the material properties and can be affected by the processing conditions of the implant.

Processing of the implant includes taking the raw material and bringing it into component form. Such processes include **machining**, **casting**, **molding**, **sintering**, **extrusion**, **forging**, and other manufacturing methods. Additionally the cleaning and sterilization protocols must be specified. Good manufacturing practice is necessary to ensure that the implant is void of defects and that specified tolerances and surface finish of the implant are achieved. Moreover, processing can directly alter the material properties and hence it is important to understand the critical interplay of such variables.

Regulatory issues include the FDA approval and monitoring of the implant (Chapter 12). The approval process often requires biocompatibility analysis of the material, simulated loading conditions in the implant, and clinical trials. This aspect of the design process is reliant upon all the other variables in this multifactorial challenge of medical device design.

1.4 Biocompatibility

Biocompatibility is the primary requirement of any material used in the body. Unless designed to degrade *in vivo*, the material must offer long-term resistance to biological attack. Biocompatibility is a multifaceted issue in that both the composition and size scale of the biomaterial can dictate the cellular or inflammatory response (Black, 1999; Bronzino and Yong, 2003). Materials that are considered biocompatible in bulk form can trigger an inflammatory response if the material becomes small enough to be ingested by an inflammatory cell such as a **macrophage** (Howie *et al.*, 1993). The generation of debris associated with mechanical loading or corrosion of the implant can result in an acute **inflammatory response** and premature failure of the implant.

A standard definition of biocompatibility is "the ability of a biomaterial to perform its desired function with respect to a medical therapy, without eliciting any undesirable local or systemic effects in the recipient or beneficiary of that therapy, but generating the most appropriate beneficial cellular or tissue response in that specific situation, and optimizing the clinically relevant performance of that therapy" (Williams, 2008). Consequently, the total number of biocompatible materials that are suitable for use in medical devices is quite limited.

Assessment of biocompatibility is quite complex as variations in immune response, activity level, and overall health of individual patients can be considerable. *In vivo* degradation of implants is often nucleated by the coupled effect of mechanical loading

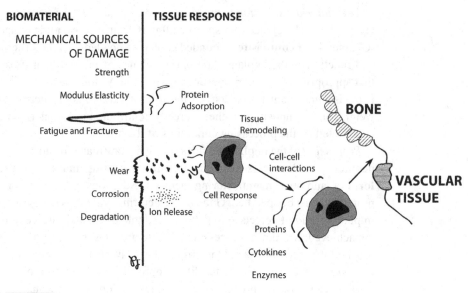

Figure 1.3

Illustration showing typical interactions at the interface between the biological system and a mechanically damaged synthetic biomaterial used in implants. (After Ratner et al., 1996.)

and environment. Complications often arise because of corrosion, wear, fatigue, or fracture of the biomaterial (Figure 1.3). For this reason, the biocompatibility of a material often is not known until the material is used in its intended clinical environment. Accordingly, the historical and clinical performance of a material is crucial in assessing whether that material is suitable in a given device or application.

1.5 Sterility

Sterilization involves the elimination of bacterial contamination through a mechanism of **DNA** disablement. Sterility is defined as less than one in one million surviving bacterial spores in the medical device prior to implantation. There are several sterilization options available and these include autoclaving, irradiation, ethylene oxide gas, and gas plasma.

Autoclaving works by subjecting devices or materials to high-pressure steam at temperatures on the order of 121°C in order to destroy bacterial contamination. Autoclaving is highly accessible and is often available in hospitals and surgical units. These systems are commonly employed to sterilize surgical tools. Because of the temperature of the steam, autoclaving is generally not used with polymeric systems.

The gamma sterilization process uses high-energy photons that are emitted from a Cobalt 60 isotope source to produce ionization (electron disruptions) throughout the medical device (Bruck and Mueller, 1988). In living cells, electron disruptions result in damage to the DNA and other cellular structures. These photon-induced changes at the molecular level cause the death of the organism or render the organism incapable of reproduction. The gamma process is deeply penetrating and has been employed in many devices and materials including the orthopedic-grade polyethylenes used in total joint replacements (Kurtz *et al.*, 1999). However, **irradiation** can result in **crosslinking** or chain **scission** in polymers and leave behind free radicals that can lead to long-term **oxidation** (Premnath *et al.*, 1996). Similarly, the **electron beam** method accelerates electrons to very high speeds in order to increase their energy and penetrate products to achieve sterility by damaging the DNA strands of the microorganisms (Bruck and Mueller, 1988). This method does not penetrate as deeply as gamma irradiation and is better suited for low-density, uniformly packaged materials.

Ethylene oxide (EtO) **gas** sterilization is a chemical process that utilizes a combination of gas concentration, humidity, temperature, and time to render the material sterile. Ethylene oxide is an alkylating agent that disrupts the DNA of microorganisms and prevents them from reproducing. Ethylene oxide sterilization is considered a low-temperature method and is commonly employed in a variety of materials including many polymers such as orthopedic-grade polyethylene (Ries *et al.*, 1996). Materials sterilized with EtO are typically encased in final breathable packaging as an aeration process completes the sterilization cycle. **Gas plasma** is a low-temperature, sterilization method that relies upon ionized gas for deactivation of biological organisms on surfaces of devices or implants. Low-temperature hydrogen peroxide gas plasma is accomplished at temperatures lower than 50°C. This method is commonly employed in polymeric materials that are susceptible to irradiation damage (Goldman and Pruitt, 1998). The attributes and limitations of the primary sterilization methods are summarized in Table 1.1.

In general, materials that can withstand high temperatures, such as metals and ceramics, can employ any of these methods for sterilization. Polymers, because of their low melt temperatures, require low-temperature methods such as gas plasma, ethylene oxide gas, or irradiation. However, in using irradiation for polymeric materials it is extremely important to be aware of the radiation chemistry of the specific polymer system (Birkinshaw *et al.*, 1988; Dole *et al.*, 1958; Pruitt, 2003). Specifically, the irradiation process can result in a chain scission or crosslinking mechanism that can alter the physical structure, mechanical properties, and long-term stability of the polymeric implant. A case study examining the effect of gamma irradiation sterilization on medical grade ultra-high molecular weight polyethylene used in total joint replacements is presented at the end of this chapter.

1.6 Regulatory issues

Regulatory aspects of medical implants in the United States are governed by the **Food and Drug Administration** (FDA) and play an important role in the medical device development process (Figure 1.2). The FDA classifies implants based on risk to the patient in the event of a device failure. **Class I** implants are low-risk devices such as

Table 1.1 Summary of primary sterilization methods employed in biomaterials

Sterilization Type	Mechanism	Benefits	Drawbacks	Applications
Autoclaving	High-pressure steam (121°C) disables DNA	• Efficient • Easily accessible	• High temperature	• Metals • Ceramics
Gamma irradiation	Radiation disables DNA	• Efficient • Penetrating	• Radiation damage	• Metals • Ceramics • Polymers
E-beam irradiation	Accelerated electrons disable DNA	• Efficient • Surface treatment	• Radiation damage • Limited penetration	• Metals • Ceramics • Polymers
Ethylene oxide gas	Alkylating agent disables DNA	• No radiation damage • Surface treatment	• Requires extra time for outgassing • Requires special packaging	• Metals • Ceramics • Polymers
Gas plasma	Plasma chemistry disables DNA	• Low temperature • No radiation damage • Surface treatment	• Limited penetration • Requires special packaging	• Metals • Ceramics • Polymers

bandages; **Class II** implants are moderate-risk devices such as total joint replacements; and **Class III** implants are high-risk implants such as heart valves or pacemakers. Generally, devices that are life-supporting or life-sustaining must undergo the most rigorous FDA approval process before they can be marketed. When the implant or device is approved for marketing, it is still subject to further analysis. For example, clinical performance is a true marker of an implant's performance; clinical trials and retrieval studies provide invaluable insight into adverse reactions or complex problems that may not have been predicted in the laboratory. The FDA analyzes the performance of many devices annually to monitor safety and good manufacturing practice, and it also facilitates the actions of device recalls if necessary.

1.7 Structural requirements

The structural requirements of medical devices can vary widely. For example, some loads in the body are very high as is the case for many orthopedic and dental implants. These loads are then resolved into stresses in the medical device and vary depending on geometry and mode of loading of the implant. For example, in total knee reconstruction the two bearing surfaces of the **condyles** are highly non-conforming and the contact stresses are high, while the ball-and-socket joint of the hip is highly conforming and the contact stresses are low. Variations in anatomical positioning and joint function result in a need for different material properties in the hip versus the knee. In the former, wear resistance is paramount because the surface area of contact is high, while in the latter fatigue resistance is critical owing to the high cyclic contact stresses. This difference makes it unlikely that one material choice would be suitable for both applications.

Most medical implants require very specific properties. For example, an arterial graft needs to offer flexibility, **anisotropic** behavior, and **compliance** that match that of adjacent tissue. A balloon **angioplasty** catheter needs to be tractable over a guide wire and must be stiff enough to prevent kinking yet flexible enough to navigate the vasculature. Sutures need to provide high tensile strength, and for the case of **resorbable** sutures this strength must decrease over time in a controlled manner. In a femoral stem of a hip replacement or other such applications, the biomaterial should offer compliance match to the adjacent bone in order to prevent stress shielding or loss of bone caused by lack of loading. For such reasons, the structural properties of the biomaterials are extremely important for the long-term success or performance of the implant. One such property that is extremely important in implant design is the elastic modulus of the material as this plays a key role in determining the geometric stiffness of the implant and load transfer to adjacent tissues.

A simple way to estimate the elastic modulus is through the assessment of interatomic force potential. The general form of the bond energy, $U(r)$, as a function of atomic separation, r, is shown in Figure 1.4. The equilibrium separation distance is denoted as r_o. If two atoms are displaced by an amount $r - r_o$, then the force, F, that resists this deformation is proportional to the separation distance for small displacements in both tension and compression.

This illustration given in Figure 1.4 also demonstrates that the force for the separation of atoms is given by the following relationship:

$$F = \frac{dU}{dr}. \tag{1.1}$$

Similarly, the stiffness of the bond is given as

$$S = \frac{dF}{dr} = \frac{d^2U}{dr^2}. \tag{1.2}$$

When the displacement between bonds is small, the bond stiffness behaves in a linear elastic fashion and is given as:

$$S_o = \left(\frac{d^2U}{dr^2}\right)_{r_o}. \tag{1.3}$$

The above relationship provides the physical foundation for elastic modulus. For small displacements (small strains), S_o is constant and represents the spring constant of the bond. One can envision that the material is held together by springs with a spring constant equal to S_o as shown in Figure 1.4. This assumes a very simple arrangement

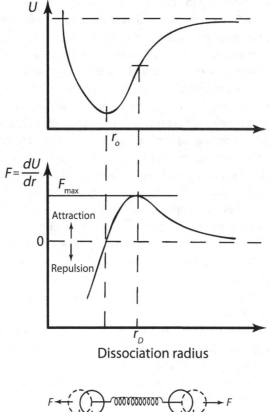

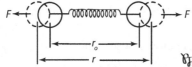

Figure 1.4

Schematic plot of the general form of the bond energy, $U(r)$, as a function of interatomic separation, r, and the force for the separation of atoms.

of atoms within the crystal but provides a basis for elastic modulus that captures small deformation, or linear elastic behavior. The force between the atoms can then be rewritten as:

$$F = S_o(r - r_o). \tag{1.4}$$

The uniaxial stress, σ, represents the total force that acts upon a unit area if two planes are displaced by an amount $(r - r_o)$.

This stress can be written as:

$$\sigma = \rho S_o(r - r_o) \tag{1.5}$$

where ρ is the bond density (bonds/unit area). The average area per atom is defined as $1/r_o^2$.

The elastic strain, ε, is defined as $(r - r_o)/r_o$ and thus Equation (1.5) can be rewritten as:

$$\sigma = \frac{S_o}{r_o}\varepsilon. \qquad (1.6)$$

Thus the elastic modulus is given as:

$$E = \frac{S_o}{r_o}. \qquad (1.7)$$

As each material class comprises different bonding types, a range of elastic moduli is associated with each material class. Covalently bonded materials, such as diamond, offer the greatest elastic modulus; covalent-ionic ceramic structures offer high elastic modulus; metallic bonding offers intermediate modulus; and polymers offer lower elastic modulus due to the mixture of covalent bonds and van der Waals forces. Composites and tissues offer a range of elastic moduli depending on specific constituents and orientation of such elements. Thus each material class offers unique mechanical properties that are fundamentally linked to its basic structure and bonding.

1.8 Classifying biomaterials

In classifying biomaterials, it is customary to categorize the materials by property, clinical application, or basic structure. For example, biomaterials can be described as hard or soft. In such a classification, polymers, **vascular** tissues, skin, and **cartilage** would be classified as soft; while metals, ceramics, bone, **dentin**, and **enamel** would be considered hard. Another classification scheme is the materials' use or clinical application. These categories typically include surgical (staples, sutures, scalpels, etc.); orthopedic (**total joint replacements**, **spinal rods**, **fracture plates**, **bone grafts**, etc.); cardiovascular (**heart valves**, balloons, **pacemakers**, **catheters**, **grafts**, **stents**, etc.); dental (**fillings**, **crowns**, **temporomandibular joint** replacements, **orthodontics**, etc.); soft tissue (wound healing, reconstructive and **augmentation**, **ophthalmologic**, etc.); and diagnostics (implantable sensors, drug delivery). The three synthetic material categories based on structure, bonding, and inherent properties include metals, ceramics, and polymers.

Biomaterials are used for the replacement of and restoration of function to tissues or constituents of a living being that may have deteriorated through disease, aging, or trauma (Black, 1999; Park and Lakes; 1992; Park and Bronzino, 2003; Ratner *et al.*, 1996). Biological tissues (Chapter 5) are complex hierarchical materials that are linked closely to their biological, chemical, and mechanical environments. Engineering materials (Chapters 2–4) used in the body are generally much simpler in their molecular structure and mechanical properties. With appropriate understanding of the clinical and structural needs of the application, medical devices can be designed to provide long-term success *in vivo* using synthetic biomaterials. Over the past several decades vast advances have been made in the design of medical devices for load-bearing applications

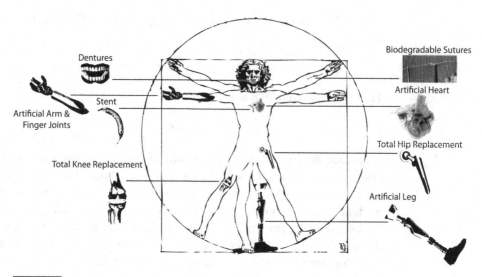

Figure 1.5

Illustration of typical modern medical implants utilized in the body.

in the body. Examples of structural implants used successfully in the body include hip, shoulder, and knee replacements; fracture fixation devices; sutures and suture anchors; dental implants; vascular grafts, heart valves, pacemakers, balloon catheters, and stents. Figure 1.5 provides an illustration of several types of implants currently utilized in the body.

In selecting materials for devices and implants, the primary goal is to restore function or integrity to the tissue or biological component. For example, in an arthritic joint the function of the implant is to restore smooth articulation to the ends of bones. Typically, degraded or osteoarthritic cartilage and underlying bone are removed through a surgical process and a total joint replacement composed of synthetic materials is implanted to restore function to the joint (Figure 1.6). The mechanical requirements on such an implant can be demanding, with cyclic contact stresses approaching or surpassing the elastic limit of the material. For such reasons, in addition to biocompatibility, the materials used in implants must offer suitable structural properties for the given application including corrosion resistance, wear resistance, fracture resistance, toughness, fatigue strength, and compliance match to surrounding tissue. Thus, as shown in Figure 1.2, an intricate relationship that includes implant design, materials selection, manufacturing, sterilization protocols, patient health, and clinical environment will ultimately dictate the longevity and success of the implant (American Academy of Orthopaedic Surgeons, 2007).

The majority of materials used in load-bearing medical devices are synthetic biomaterials such as metals, ceramics, and polymers. These well-characterized materials are known as "engineering materials." The choice of man-made materials over biological materials is due not only to the excellent mechanical properties they offer but also

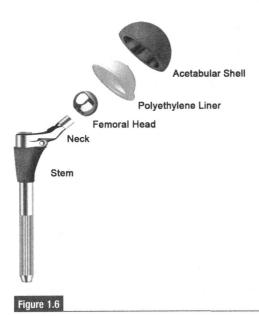

Figure 1.6

Total hip replacement utilizing a titanium alloy for the femoral stem with a beaded coating in the proximal region of the stem, a cobalt-chromium steel alloy for the femoral head, and an UHMWPE acetabular cup with a porous-coated acetabular shell often made of titanium or cobalt-chromium alloys.

the superior quality control available in terms of material purity and high component tolerances during manufacturing. Furthermore, many biological materials are prone to immunorejection or normal resorption processes found in the body. To the lay person, it may seem unusual that degraded articular cartilage at the end of the long bones in the hip or knee joint is replaced with a bearing couple made of metal and plastic (Figure 1.6). However, the combination of a cobalt-chromium alloy (metal) and an ultra-high molecular weight polyethylene (plastic) implant system offers restored articulation, relief of pain, and structural integrity for nearly two decades of use in the body (American Academy of Orthopaedic Surgeons, 2007). In comparison, the ability to regenerate cartilage *in vivo* or to utilize cartilage repair systems remains limited without comparable clinical performance to the synthetic systems. Thus, until genetics or tissue engineering pathways provide a viable alternative, the use of engineering materials in implant design remains central to the medical device community.

The choice of an engineering material for an implant design is typically founded on the need for certain material properties such as elastic modulus (stiffness), yield strength (resistance to permanent deformation), or fracture toughness (resistance to fracture in the presence of a flaw) to ensure structural integrity of the device. The primary difference between the basic classifications of structural materials and their physical properties originates at the molecular bonding level. The physical processes responsible for the attractive interactions between atoms and molecules in these materials include strong interactive forces such as **covalent**, **ionic**, and **metallic bonding**; and weak interactions

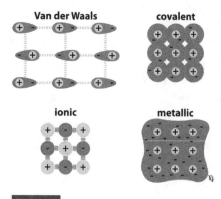

Figure 1.7

Bonding mechanisms utilized in metals (metallic), ceramics (covalent-ionic), and polymers (covalent-van der Waals).

such as **van der Waals forces** or **hydrogen bonding**. These bonding types form the essential building blocks of metals, ceramics, and polymers and they are schematically illustrated in Figure 1.7.

1.9 Structure-property relationships

The nature and structural arrangement of chemical bonds within a material dictate many of its basic properties. For example, the elastic modulus (E) is a material property that is directly linked to structure and bonding, and which describes the correlation between applied force (stress) and resulting displacement (strain) for a given material (as shown in Figure 1.4). For most engineering metals or ceramics, E is a material constant for a given chemical composition. However, for polymer systems E can be highly dependent on molecular variables such as crystallinity, molecular weight, and entanglement density; and external factors such as strain rate or temperature. Elastic deformation correlates to the displacement of the chemical bonds between atoms. In ceramics and metals this is limited to bond stretching, but in polymers, bond rotations of side groups and chain sliding are also enabled.

A material with a high density of strong bonds in closely packed atomic planes will have a high elastic modulus, while a material with weaker bonds that are less densely packed will have a substantially lower elastic modulus. For example, ceramics utilize close packing of strongly bonded (covalent or ionic) atoms and they have high elastic modulus values. An example of a ceramic crystal lattice utilizing a blend of anions and cations in a Halite structure is shown in Figure 1.8(a). Similarly, metals make use of metallic bonding and closely packed planes of atoms. An example of a face-centered-cubic crystal lattice common to ductile metals such as aluminum or copper is shown in Figure 1.8(b). In contrast, polymers employ covalent bonds along the backbone of individual chains but only weak van der Waals forces between chains, and this results

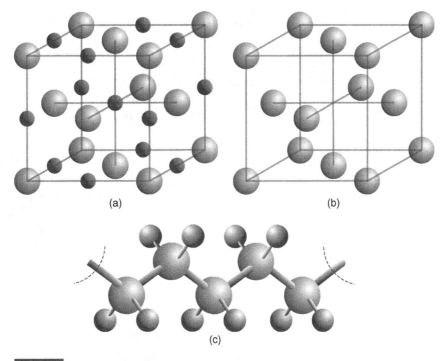

Figure 1.8

Illustration showing typical molecular structures found in (a) ceramics, (b) metals, and (c) polymers.

in low elastic modulus values. An example of a linear chain structure as is found in polyethylene is shown in Figure 1.8(c). These different material structures result in a large range of elastic modulus.

A comparison of elastic modulus values among the engineering and biological materials is provided in Figure 1.9. It is noteworthy that the engineering polymers provide the greatest overlap of modulus with both hard and soft tissues.

1.10 Attributes and limitations of synthetic biomaterials

Metals offer many important mechanical properties for load-bearing medical devices including strength, ductility, and toughness. Alloys have the properties of metals but consist of two or more elements of which only one needs to be a metal (such as iron and carbon to make steel). Typically, alloys can have different properties than the individual elements, and alloying one metal with another generally enhances engineering properties such as strength or toughness. **Intermetallics** contain two or more metallic elements blended with a non-metallic element and offer properties such as hardness that fall into the range spanned by metals and ceramics. A general limitation of using metals in long-term implants is that they are prone to corrosion and may be susceptible to **stress**

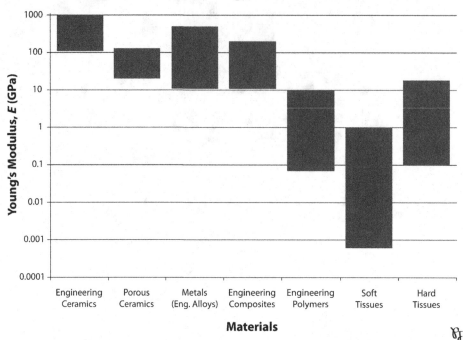

Figure 1.9

Plot showing relative values of elastic modulus for engineering and biological materials.

corrosion cracking, **fretting** (loss of material and abrasion due to rubbing), or **galvanic corrosion**. An interesting aspect of corrosion in metals is that in certain environments the metal or alloy may fare quite well, but changes in pH, local chemistry, stress state, or coupling with another material can swiftly turn a stable system into a highly corrosive one. Titanium, for example, is used in orthopedic hip stems because of its excellent fatigue strength and its ability to minimize stress shielding and subsequent bone loss around the implant. Titanium forms an outer oxide layer that protects it from corrosion; however, this TiO_2 layer lacks stability and is readily worn off under mechanical contact. In situations where fretting or rubbing can occur, the oxide is continually lost and reformed resulting in third-body particulate and material loss (wear) of the underlying titanium. Thus, unless special surface treatments are utilized with titanium alloys, they do not fare well under sliding contact applications such as bearing surfaces in articulating joints.

Ceramics generally have high **compressive strength**, hardness, and wear resistance. These properties make them useful as hard bearings in applications such as the femoral head in total hip replacements or in heart valves. However, ceramics have poor **tensile strength** and are inherently brittle owing to their low fracture toughness. Such limitations render ceramics susceptible to fracture when stress concentrations are present. When designing with ceramic materials it is important to consider the fracture toughness, critical flaw size, and the expected stress states of the implant. One philosophy used when designing with materials with low tolerances to cracks or defects is the defect-tolerant approach that utilizes linear elastic fracture mechanics to predict crack growth and fracture in the component.

Polymers as a class of materials are known to be resilient (capable of storing and releasing elastic energy) and offer good modulus match to many biological materials. Moreover, the molecular variables such as the backbone chemistry, molecular weight, crosslinking, entanglement density, and crystallinity can be tailored to create specific mechanical properties required for a device. Polymers are generally known to be **viscoelastic**, that is, they have both solid and fluid characteristics (Chapter 7). Thus care must be taken in designing with polymers to account for creep (time-dependent strain) and stress relaxation (time-dependent stress decay). Additionally, many polymers readily absorb fluid in aqueous environments, and while this can be beneficial in facilitating the bioresorbability of certain polymers such as **polyglycolic acids** (PGA) used in sutures, it can cause complications with unwanted biodegradation in other polymeric materials such as **polymethlymethacrylate** (PMMA) used in bone cement. Polymeric biomaterials also face unique demands when utilized in load-bearing medical devices in that the mechanical stresses in which they function often put them at risk for yield, fatigue, wear, creep, and fracture.

A composite system typically utilizes fibers or particles bound to a matrix in order to tailor mechanical properties such as tensile strength and elastic modulus. Composite systems are generally anisotropic, that is, the material properties are dependent on direction within the material (Chapter 6). An illustration of a unidirectional composite is given in Figure 1.10. This figure depicts two loading scenarios for the composite that illustrate the basis for (a) an upper bound elastic modulus and (b) a lower bound elastic modulus.

If the composite is loaded in the direction of the fiber, then an upper bound elastic modulus is achieved. In this case, the displacement in the fiber and the matrix are assumed to remain equal and the upper bound elastic modulus, E_{upper}, takes the form:

$$E_{\text{upper}} = V_f E_f + V_m E_m \tag{1.8}$$

where V_f is the volume fraction of fibers, V_m is the volume fraction of matrix, E_f is the elastic modulus of the fiber, and E_m is the elastic modulus of the matrix. If the composite is loaded orthogonal to the fibers then the stresses in the fibers and matrix are assumed equal, and the lower bound elastic modulus is obtained:

$$E_{\text{lower}} = \frac{E_f E_m}{V_f E_f + V_m E_m}. \tag{1.9}$$

Composites can have matrices made of polymers, metals, or ceramics. Similarly, the fibers and particle reinforcements may be from any of the three material classes. Polymer composites such as carbon fiber reinforced **polyetheretherketone** (PEEK) can be used

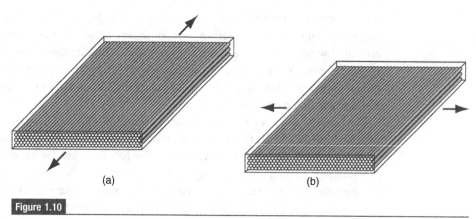

Figure 1.10

Schematic of two loading options for a unidirectional fiber composite that illustrates the basis for the (a) upper bound elastic modulus and (b) lower bound elastic modulus.

in orthopedic applications where modulus match and load transfer to adjacent bone tissue are needed. Similarly, they may be employed in spinal cages or orthopedic bone plates where high flexural strength is required. A concern when designing with composites is that lack of interfacial strength between the matrix and fibers can result in low fracture toughness and may also render the material susceptible to particulate debris generation. Such failure modes can severely degrade the biocompatibility and structural integrity of the system. Modern systems take great care to eliminate such complications. Natural tissues, in general, are composite structures with different constituents and a range of hierarchical structures.

1.11 Case study: deterioration of orthopedic-grade UHMWPE due to ionizing radiation

One of the most significant challenges in orthopedics has been the poor oxidation resistance of **ultra-high molecular weight polyethylene** (UHMWPE) concomitant with ionizing radiation sterilization (Birkinshaw *et al.*, 1989). This degradation is associated with osteolysis (bone loss around the implant) due to the poor wear resistance of oxidized UHMWPE polymer used as the bearing surface in total joint replacements (Jasty, 1993). Until about 1995, UHMWPE was typically sterilized with a nominal dose of 25–40 kGy of gamma radiation in the presence of air. By 1998, all of the major manufacturers had shifted their sterilization method away from gamma irradiation in air (Kurtz, 2009).

Gamma irradiation, like other high-energy photons, generates free radicals in polymers through homolytic bond cleavage. Radicals generated in the crystalline regions of the polymer have long lifetimes, allowing them to diffuse into the amorphous regions of the polymer and undergo oxidative reactions when oxygen is available. This time-dependent free radical reaction mechanism poses serious concern for the structural

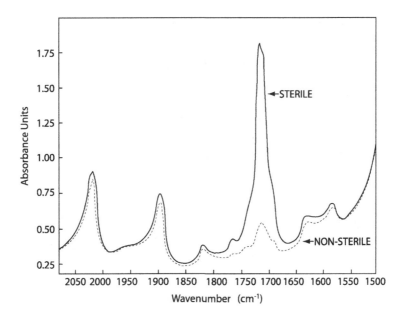

Figure 1.11

FTIR spectra showing oxidation of an UHMWPE component that has shelf aged for 5 years following gamma radiation in air.

degradation of UHMWPE (Premnath *et al.*, 1996). Oxidation results in a decrease in molecular weight, an increase in crystallinity and density due to chain scission (Goldman *et al.*, 1996) and loss of mechanical properties (Kurtz *et al.*, 1999; Kurtz, 2009). Moreover, oxidation results in severe degradation of fatigue and fracture properties of the UHMWPE (Baker *et al.*, 2000; Pruitt and Ranganathan, 1995).

Fourier transform infrared (FTIR) spectroscopy is commonly employed to assess oxidation in UHMWPE and other polymers. This is done by measuring the area under the carbonyl peak (representing oxidative species) and normalizing with the methylene peak (representing non-oxidized bonding of the polymer). Figure 1.11 depicts an FTIR spectrum showing oxidation of an UHMWPE tibial (knee) component that has shelf-aged for 5 years following gamma radiation in air. There is a clear increase in the carbonyl ketone peak at 1720 cm^{-1} indicating oxidation of the UHMWPE (Goldman *et al.*, 1996). Figure 1.12 shows the concomitant loss in fatigue fracture resistance (Pruitt and Ranganathan, 1995). Evidence of most significant levels of oxidation occurred beneath the articulating surface where contact stresses are highest. This subsurface peak in oxidation is due to the coupled effect of free radical distribution and oxygen diffusion. The coupled effect of oxidation and contact stresses is thought to be linked to delamination wear of UHMWPE as illustrated in Figure 1.13 (Sutula *et al.*, 1995).

Today UHMWPE components are sterilized with EtO or gas plasma methods. Radiation is now only employed when it is coupled with a thermal treatment to eliminate free

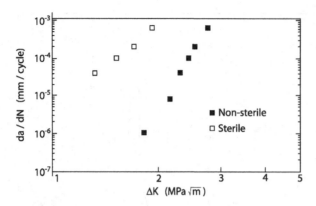

Figure 1.12

Plot of crack velocity as a function of stress intensity showing embrittlement of UHMWPE component following gamma radiation sterilization.

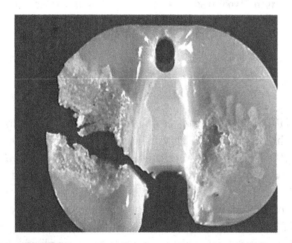

Figure 1.13

Delamination in a tibial component of a total knee replacement that has oxidized due to gamma irradiation sterilization in an air environment.

radicals and is often utilized to crosslink UHMWPE components for improved wear resistance (Kurtz, 2009).

1.12 Summary

This chapter provided a broad introduction to biocompatibility, sterilization protocols, and biomaterials selection for medical devices. Factors concerning medical device performance were addressed and include issues related to materials selection, structural requirements, processing, clinical factors, regulatory affairs, and device design. A query and case study on the topic of sterilization of medical grade polymeric implants was offered. The primary objective of Chapter 1 was to provide the reader with fundamental skills in biomaterials science. These skills include the ability to describe the key factors that contribute to successful device performance, define biocompatibility and recommend appropriate sterilization schemes for various biomaterials, explain the three FDA classifications of medical devices, discuss the limitations and benefits of each class of synthetic material used in implants, and illustrate the fundamental structure-property relationships in structural tissues and the associated synthetic biomaterial classes.

1.13 Problems for consideration

Checking for understanding

1. Describe the primary factors that contribute to medical device performance.
2. Discuss the variables that factor into the biocompatibility of a material or a medical implant. Why is size scale important?
3. Evaluate the sterilization methods currently available for synthetic biomaterials. What are the benefits and limitations of each process?
4. How would you sterilize a vascular graft consisting of a porous polymer? A modular hip implant with a titanium stem and a CoCr head? A pyrolytic carbon heart valve?
5. Discuss the attributes and limitations of metals, ceramics, polymers, and composites.

For further exploration

6. Derive the Equation (1.8) for upper bound modulus for a unidirectional composite. State all assumptions.
7. Consider a unidirectional composite that has 50 vol. % carbon fibers in a polyetheretherketone (PEEK) resin matrix. The modulus of the carbon fiber is 100 GPa and that of the resin is 3 GPa. What is the predicted value of the upper bound modulus? What is the ratio of the upper bound modulus to lower bound modulus for this material?

1.14 References

American Academy of Orthopaedic Surgeons (2007). Osteolysis and implant wear: biological, biomedical engineering and surgical principles. *Journal of the AAOS*, 16(1), S1–S128.

Baker, D., Hastings, R., and Pruitt, L. (2000). Compression and tension fatigue resistance of medical grade UHMWPE: the effect morphology, sterilization, aging and temperature. *Polymer*, 41, 795–808.

Birkinshaw, C., Buggy, M., and Dally, S. (1988). Mechanism of aging in irradiated polymers. *Polymer Degradation and Stability*, 22, 285–94.

Birkinshaw, C., Buggy, M., Dally, S., and O'Neill, M. (1989). The effect of gamma radiation in the physical structure and mechanical properties of ultrahigh molecular weight polyethylene. *Journal of Applied Polymer Science*, 38, 1967–73.

Black, J. (1999). *Biological Performance of Materials: Fundamentals of Biocompatibility*. New York: Marcel Dekker.

Bronzino, J.D., and Yong, J.Y. (2003). *Biomaterials*. Boca Raton, FL: CRC Press.

Bruck, S.D., and Mueller, E.P. (1988). Radiation sterilization of polymeric implant materials. *Journal of Biomedical Materials Research*, 22, 133–44.

Dole, M., Milner, D.C., and Williams, T.F. (1958). Irradiation of polyethylene. II. Kinetics of unsaturation effects. *Journal of the American Chemical Society*, 80, 1580–88.

Goldman, M., and Pruitt, L. (1998). A comparison of the effects of gamma radiation and plasma sterilization on the molecular structure, fatigue resistance, and wear behavior of UHMWPE. *Journal of Biomedical Materials Research*, 40(3), 378–84.

Goldman, M., Ranganathan, R., Gronsky, R., and Pruitt, L. (1996). The effects of gamma radiation sterilization and aging on the structure and morphology of medical grade ultra high molecular weight polyethylene. *Polymer*, 37(14) 2909–13.

Howie, D.W., Haynes, D.R., Rogers, S.D., McGee, M.A., and Pearcy, M.J. (1993). The response to particulate debris. *Clinical Orthopedics: North America*, 24(4) 571–81.

Jasty, M. (1993) Clinical reviews: particulate debris and failure of total hip replacements. *Journal of Applied Biomaterials*, 4, 273–76.

Kurtz, S.M. (2009). Packaging and sterilization of UHMWPE. In *UHMWPE Biomaterials Handbook*, 2d edn., ed. S.M. Kurtz. London: Elsevier, pp. 21–30.

Kurtz, S.M., Muratoglu, O.K., Evans, M., and Edidin, A.A. (1999) Advances in the processing, sterilization, and crosslinking of ultra-high molecular weight polyethylene for total joint arthroplasty. *Biomaterials*, 20, 1659–88.

Park, J.B., and Bronzino, J.D. (2003). *Biomaterials: Principles and Applications*. Boca Raton: CRC Press.

Park, J.B., and Lakes, R.S. (1992). *Biomaterials: An Introduction*. New York: Plenum Press.

Premnath, V., Harris, W.H., Jasty, M., and Merrill, E.W. (1996). Gamma sterilization of UHMWPE articular implants: an analysis of the oxidation problem. *Biomaterials*, 17, 1741–53.

Pruitt, L. (2003). Radiation effects on medical polymers and on their mechanical properties. In *Advances in Polymer Science: Radiation Effects*, ed. H.H. Kausch. Heidelberg: Springer-Verlag, pp. 63–93.

Pruitt, L., and Ranganathan, R. (1995). Effect of sterilization on the structure and fatigue resistance of medical grade UHMWPE. *Materials Science and Engineering*, C3, 91–93.

Ratner, B.D., Hoffman, A.S., Schoen, F.J., and Lemons, J.E. (1996). *Biomaterials Science: An Introduction to Materials in Medicine*. New York: Academic Press.

Ries, M.D., Weaver, K., and Beals, N. (1996). Safety and efficacy of ethylene oxide sterilized polyethylene in total knee arthroplasty. *Clinical Orthopedics*, 331, 159–63.

Sutula, L.C., Collier, J.P., Saum, K.A., Currier, B.H., Currier, J.H., Sanford, W.M., Mayor, M.B., Wooding, R.E., Sperling, D.K., Williams, I.R., Karprazak, D.J., and Surprenant, V.A. (1995). Impact of gamma sterilization on clinical performance of polyethylene in the hip. *Clinical Orthopaedics and Related Research*, 319, 28–40.

Williams, D.F. (2008). On the mechanisms of biocompatibility. *Biomaterials*, 29(20), 2941–53.

2 Metals for medical implants

Inquiry

All metals, with the exception of high noble metals, are susceptible to corrosion. This problem is exacerbated in the body, where implants must function in aqueous environments under complex loading and often utilize designs that contain inherent stress concentrations and material discontinuities. How would you design a modular hip system composed of metal components that minimizes the likelihood for fretting and crevice corrosion?

The inquiry posed above represents a realistic challenge that one might face in the field of orthopedic biomaterials. At a minimum one would want to minimize the galvanic potential of the alloys involved in the implant design. Fretting is the loss of material owing to the rubbing or contact of components; to minimize this type of surface degradation it is necessary to utilize designs that optimize component tolerances for device fixation. Crevice corrosion occurs in metal components that contain fissures, notches, or other geometries that facilitate local changes in the environment. This type of corrosion is further enabled in the presence of elevated stresses or stress concentrations within the component. To minimize fretting and crevice corrosion in a modular hip it is necessary to diminish geometric gaps at junctions such as the Morse Taper (where the stem and acetabular head join) and to keep the stem and neck at reasonable lengths in order to reduce bending stresses at the stem-neck juncture. The case study presented at the end of this chapter addresses this issue.

2.1 Historical perspective and overview

Metals have been utilized in medical implants for several hundred years; in fact gold has been used as dental material from as early as 500 BC and noble metals have been employed in structural dentistry for use in fillings, crowns, bridges, and dentures since the 15th century (Williams, 1990). Noble alloys contain varying percentages of gold, palladium, silver, or platinum. **High noble metals** are metal systems that contain more than 60% noble metals (gold, palladium, or platinum) with at least 40% of the metal being composed of gold. It is known that gold was used in the 16th century to repair cleft palates; gold, bronze, and iron evolved as suture wires over the 17th–18th centuries; and steel plates were commonly employed as internal fixation devices by the early 19th century (Aramany, 1971; Williams *et al.*, 1973). By the mid-1920s, stainless steels were introduced as implant materials and a decade later cobalt-chromium alloys were utilized in the body to provide enhanced corrosion resistance (Park and Bronzino, 2003). Titanium found its use as a surgical material shortly thereafter (Leventhal, 1951). Much of what motivated the development of metals as implant materials was the need for both strength and corrosion resistance in load-bearing devices.

The materials science of medical alloys including their chemical nature, bonding, and crystal structure is paramount to the performance of the metal in the body for medical device applications. As described below, all metals utilize metallic bonding and this results in a number of material behaviors such as conductivity of electrons that can render this class of materials highly susceptible to corrosion in the body. The nature of the chemistry, such as nickel content, may affect patient sensitivity and immunological response *in vivo*. Moreover, the crystal structure that describes how atoms are organized in space and the concomitant microstructure that captures factors such as grain size or orientation affects the mechanical properties such as yield strength (the stress at which the onset of permanent deformation occurs), **strain to failure** (the amount of deformation that can be sustained before the material fractures), and **ultimate strength**. The mechanical properties in combination with the **corrosion** resistance of the alloys ultimately determine the success of the material in a medical device application. For this reason, as will be seen below, only a small number of metal systems are utilized as medical alloys. It is well known that mechanical properties can be tailored with appropriate chemical formulation, processing, and heat treatment; and for these reasons these pathways are discussed in the first section of this chapter.

This chapter provides a broad overview of metals and alloys used in medical implants. The basic structure, bonding, material properties as well as mechanisms for strengthening are reviewed. Methods of processing and heat treatments are examined for metal systems. Issues of medical device design with metals or alloys and appropriate material selection for specific types of implants are addressed.

2.2 Learning objectives

This chapter provides an overview of metals used in medical implants. The basic bonding, structure, and inherent properties of metals and alloys are reviewed. The key objectives of this chapter are that at the completion of this chapter the student will be able to:

1. explain the principles of metallic bonding mechanisms
2. illustrate the primary crystal structures, interstitial sites, and crystallographic planes of structural metals
3. describe and illustrate the types of imperfections that exist in metal structures
4. explain the fundamental strengthening mechanisms used in medical grade metals and alloys
5. depict the primary processing methods used in the manufacture of metal components
6. advocate design strategies which minimize corrosion in metallic components
7. recommend appropriate types of metals for specific biomedical applications

Table 2.1 Seven general crystal systems

Crystal system	Bravais lattice	Lengths	Angles
Triclinic	simple	$a \neq b \neq c$	$\alpha \neq \beta \neq \gamma$
Monoclinic	simple base-centered	$a \neq b \neq c$	$\alpha = \gamma = 90° \neq \beta$
Orthorhombic	simple base-centered body-centered face-centered	$a \neq b \neq c$	$\alpha = \beta = \gamma = 90°$
Tetragonal	simple body-centered	$a = b \neq c$	$\alpha = \beta = \gamma = 90°$
Cubic	simple body-centered face-centered	$a = b = c$	$\alpha = \beta = \gamma = 90°$
Trigonal	simple	$a = b = c$	$\alpha = \beta = \gamma \neq 90°$
Hexagonal	simple	$a = b \neq c$	$\alpha = \beta = 90°; \gamma = 120°$

2.3 Bonding and crystal structure

Metals and alloys establish their basic building blocks through metallic bonding. **Metallic bonding** involves the sharing of free electrons among positively charged metal ions. That is, each atom contributes its valence electrons to the formation of a "sea of electrons" that is shared by the solid and flows freely among the constituent atoms (Figure 1.7). The bond itself is the result of electrostatic attraction between conduction electrons and the metallic ions within metals, which are arranged in a three-dimensional crystalline lattice. In metallic bonding all positive ions are equivalent and this facilitates ductile behavior when a bond is broken and reformed in an adjacent position. Since metallic bonds, unlike covalent bonds, can easily be reformed when broken, atoms can move past each other with ease when metals are deformed. This motion, also known as **slip**, is the primary plastic deformation mode when the atoms are loaded in shear. This deformation mechanism also enables metals to have good malleability.

Metals and alloys are generally classified as crystalline materials. **Crystalline solids** utilize a repeated pattern of unit cells. These cells are defined by a specific and uniquely symmetric arrangement of atoms in space, as described in Table 2.1. These geometric patterns or crystalline lattices provide both short- and long-range order to the material. There are seven classifications of crystal systems: triclinic, monoclinic, orthorhombic, tetragonal, cubic, trigonal (rhombohedral), and hexagonal. The 14 associated **Bravais space lattices** are schematically illustrated in Figure 2.1.

2.3 Bonding and crystal structure

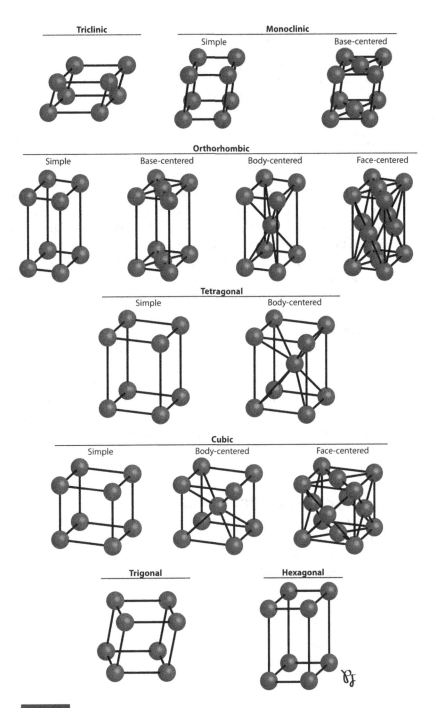

Figure 2.1

Illustration of the 14 Bravais space lattices associated with the seven crystal systems.

Figure 2.2

Planar section of close-packed spheres.

Metallic bonding enables the atoms to be geometrically modeled as hard spheres that are arranged in a configuration of closest packing (Figure 2.2). For close-packed structures, such as the face-centered cubic (**FCC**) and hexagonal closest packed (**HCP**) lattices, the atoms will pack in a hexagonal manner (in planar section) in order to achieve maximum packing efficiency. These planes are known as close-packed planes, and the **coordination number** (defined as the number of nearest neighbors) of atoms in this two-dimensional plane is six.

This close-packed layer is denoted as layer A. The next layer of close-packed atoms can sit over one of two sets of valleys created in layer A. The layer that sits over one set of valleys is termed layer B, and the layer that sits over the complementary valleys is termed layer C. Thus there are two ways in which a closest packed structure can be created. A structure created by layering $ABCABC$ will yield a FCC structure, while a structure created by layering $ABABAB$ will create a HCP structure as shown in Figure 2.3. The coordination number for both of these lattice types is 12 and the packing efficiency (the percentage of space in the volume of the single crystal that is occupied by the metal atoms) is 74% (See Example 2.1).

Example 2.1   **Packing density of FCC and HCP structures**

The FCC lattice comprises close-packed planes that are stacked in the sequence $ABCABC$ (Figure 2.3). The packing density can be determined from basic geometry of the structure. In the FCC lattice, there are four atoms per unit cell (each corner atom shares an eighth of itself and each face atom shares one half) and the volume of this cell is given as:

$$V = a^3.$$

As can be seen in Figure 2.1, the atomic diameter is geometrically given as

$$d = \frac{a\sqrt{2}}{2}.$$

The volume occupied by the atoms is then

$$V' = \frac{\pi a^3 \sqrt{2}}{6}$$

2.3 Bonding and crystal structure

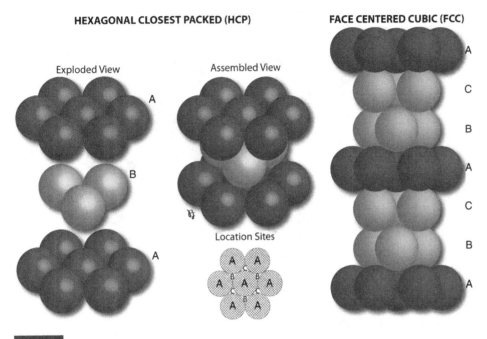

Figure 2.3

Section revealing layering of close-packed planes. Stacking of planes in sequence *ABABAB* to create the HCP lattice and the *ABCABC* sequence to yield the FCC lattice.

and the packing density is thus

$$\rho = \frac{V'}{V} = \frac{\pi\sqrt{2}}{6} = 0.742 = 74\%.$$

The same packing efficiency is found for the hexagonal close-packed structure. However, in the HCP structure, the atom spacing along the c axis is not independent of the atomic spacing, a, in the basal plane (Table 2.1, Figure 2.1). In order to determine the idealized c/a ratio for this structure, one can imagine a tetrahedron whose base comprises three atoms in the basal plane and one atom mid-plane along the c axis. Using basic geometry of a tetrahedron, one finds that the idealized c/a ratio for the HCP structure is given as $c = 2a\sqrt{(2/3)} = 1.633a$.

In the HCP lattice, there are two atoms/cell: eight atoms in the corners that each contributes one-eighth and one atom/cell at the mid-plane position (Figure 2.1). The volume of the HCP unit cell is given as:

$$V = a\left(\frac{a\sqrt{3}}{2}\right)c = a\left(\frac{a\sqrt{3}}{2}\right)2a\frac{\sqrt{2}}{\sqrt{3}} = a^3\sqrt{2}.$$

The atomic diameter is given as

$$d = a.$$

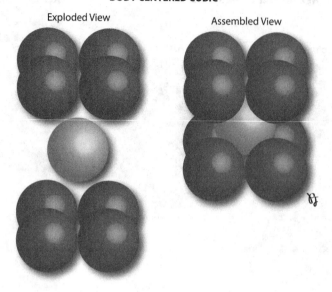

Figure 2.4

Section revealing layering of closest packed planes to create the BCC lattice.

The volume occupied by the atoms is then

$$V' = \frac{2\pi a^3}{6} = \frac{\pi a^3}{3}$$

and the packing density is thus

$$\rho = \frac{V'}{V} = \frac{\pi}{3\sqrt{2}} = 0.742 = 74\%.$$

Another common arrangement of atoms in metallic materials is the body-centered cubic (**BCC**) crystal structure in which the planes are also stacked in the *ABABAB* arrangement but the layering is such that the atoms in a given plane are not in direct contact (Figure 2.4). It should be noted that in the BCC structure, the A and B planes are not closest packed planes; however, a close-packed plane can be achieved on the diagonal plane in which the atoms actually touch. The BCC structure has a coordination number of 8 and a packing efficiency of 68%.

2.4 Interstitial sites

Interstitial sites are the spaces between packing of atoms in the FCC, HCP, and BCC structures. In metals, there may be two or more types of atoms that are combined to

INTERSTITIAL SITES

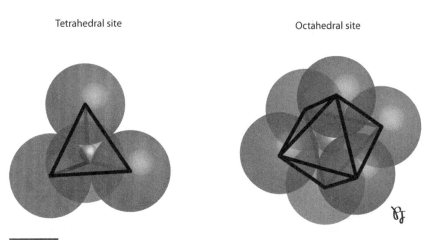

Figure 2.5

Schematic illustration showing the tetrahedral site and octahedral site in the close-packed crystal lattice.

form the lattice structure. This commonly occurs in alloys where two or more elements are used to create the crystal structure, such as carbon in iron to form steel, or aluminum and vanadium in titanium to form a titanium alloy. The resulting crystalline structure generally utilizes the basic hexagonal or cubic lattice but accommodates the additional element in one of the tetrahedral or octahedral interstitial sites. Or, if the atoms of the alloying element are near equal size, they may substitute for a lattice position. The addition of such atoms may cause distortion to the crystal lattice depending on the difference in atomic radii. For this reason, these atoms are often treated as point defects in metals. In general, it is the distortion to the crystalline lattice due to the additional element that makes most alloys stronger than single constituent metals.

The simplest of the interstitial sites is the **tetrahedral site**. The three atoms in the closest packed plane and an atom sitting in a valley above them create it. The coordination number for the tetrahedral site is 4 and this space is found in the inside corners in the FCC lattice. The maximum size atom that can be accommodated in this site has a radius that is 22.5% the size of the lattice atoms. This radius ratio of 0.225 is based on the geometry of the tetrahedron. The **octahedral site** is contained within an octahedron with a coordination number of 6. The octahedral void space is larger and the radius ratio for this interstitial atom is 0.414. The octahedral sites are found in the center and halfway along the edges of the FCC lattice. Figure 2.5 schematically illustrates the tetrahedral and octahedral interstitial sites for the close-packed structures.

Similar sites can be found in the BCC structure, but the tetrahedron and octagon are distorted because of the close-packed planes (rather than closest packed planes). In the BCC lattice the tetrahedral site is actually larger than the octahedral site, and the radius ratios are 0.291 and 0.154, respectively.

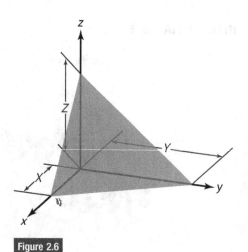

Figure 2.6

Schematic illustration depicting an arbitrary plane that intersects the x, y, and z axes at points X, Y, and Z, respectively.

2.5 Crystallographic planes and directions

Certain planes and directions play a pivotal role in the plastic deformation within the crystal systems. For this reason it is useful to have a quantitative system that describes planes and directions within the lattice structure. The crystal lattices themselves can be oriented in a coordinate system such that one atom is centered at the intersection of the x, y, and z axes. In such a system, an arbitrary plane can be imagined that intersects the x, y, and z axes at points X, Y, and Z, respectively (Figure 2.6). These intercepts can also be described in terms of the lattice parameters, a, b, and c (Table 2.1):

$$X = Aa, \ Y = Bb, \ Z = Cc \tag{2.1}$$

where A, B, C denote the distance from the point of origin to the intercept points. The convention to describe crystallographic planes is denoted by the reciprocals of these values, $1/A$, $1/B$, $1/C$, and these are known as **Miller indices**. These indices are given as h, k, and l.

In a cubic system, the close-packed plane intercepts x, y, and z, such that $A = B = C = 1$ (Figure 2.7). Thus, for this plane the Miller indices are given as $h = 1/1$, $k = 1/1$, and $l = 1/1$. The close-packed plane for the face=centered cubic system is denoted as (111). The family of such planes that are equivalent is denoted as {111}. The brackets denote a family of planes and represent equivalent planes in a crystal system. For example, the {100} family of planes has six equivalent planes denoting the faces of the cube: (100), (010), (001), ($\bar{1}$00), (0$\bar{1}$0), and (00$\bar{1}$).

The vector components within the lattice are used to describe the crystallographic direction, denoted by u, v, and w, using the lattice coefficients, a, b, and c:

$$\vec{R}_x = u\vec{a}, \ \vec{R}_y = v\vec{b}, \ \vec{R}_z = w\vec{c} \tag{2.2}$$

2.5 Crystallographic planes and directions

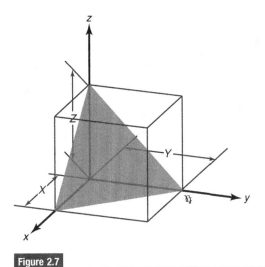

Figure 2.7

Schematic illustration depicting the (111) plane that intersects the x, y, and z axes at points $X = 1$, $Y = 1$, and $Z = 1$, respectively.

and the direction vector is given as:

$$\vec{R} = u\vec{a} + v\vec{b} + w\vec{c}. \tag{2.3}$$

In an orthogonal system where $\alpha = \beta = \gamma = 90°$,

$$\left|\vec{R}\right|^2 = u^2 |\vec{a}|^2 + v^2 \left|\vec{b}\right|^2 + w^2 |\vec{c}|^2. \tag{2.4}$$

For the body centered atom in a BCC lattice, the position is given as $u,v,w = \tfrac{1}{2}, \tfrac{1}{2}, \tfrac{1}{2}$, and the direction is denoted as $[uvw] = [111]/2$. This is an equivalent direction to [111]. Like the planes, a family of directions is denoted as $<uvw>$. The family of directions $<111>$ has eight equivalent directions: [111], [$\bar{1}$11], [1$\bar{1}$1], [11$\bar{1}$], [$\bar{1}\bar{1}$1], [1$\bar{1}\bar{1}$], [$\bar{1}$1$\bar{1}$], and [$\bar{1}\bar{1}\bar{1}$]. These planes and directions are extremely important in defining the slip systems for the various crystal lattices. The close-packed direction in a BCC lattice is $<111>$, while the close-packed direction in an FCC lattice is $<110>$. A schematic illustration of the (110) plane is shown in Figure 2.8. These closest packed planes and directions for each crystal system are the **slip systems** for that structure. These slip systems (described below) play an important role in plastic deformation and **hardenability** (the ability to have the yield strength increased through mechanical work) of the various metals.

In general, when metals are plastically deformed, the atoms slide along close-packed directions on close-packed planes. This deformation is generally accommodated in a mode of shear and is generally the most efficient way to break and reform metal bonds. This atomic movement is denoted as slip and facilitates both ductility and hardenability in metals. The most readily activated slip systems for a given crystal system are denoted as the primary slip systems. A summary of the primary slip systems for the FCC, HCP, and BCC lattices are provided in Table 2.2 (Richman, 1974).

Table 2.2 Slip systems in metals

Crystal structure	Slip plane	Slip direction	Number of slip systems
FCC	{111}	<110>	12 (primary)
HCP	{001}	<110>	3 (primary)
BCC (primary)	{110}	<111>	12 (primary)
BCC (secondary)	{112},{123}	<111>	36 (secondary)

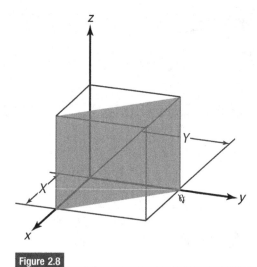

Figure 2.8

Schematic illustration depicting the (110) plane that intersects the x and y axes at points $X = 1$ and $Y = 1$. The plane does not intersect the z axis.

2.6 Theoretical shear strength

The theoretical shear strength of a metal is based on deformation of a perfect crystal and is a measure of the stress needed to slip one atom over another atom when ordered in close-packed planes (Table 2.2). It is also referred to as the **yield strength**, or stress at which plastic deformation begins. In the process of slip, planes of atoms move past each other. An illustration of displacement of atoms across the crystal planes is given in Figure 2.9. As the atoms are displaced, the potential energy is maximized when the atoms have been displaced by half their transit distance and can be modeled as a sinusoidal function (Equation. 2.5).

The applied stress needed to overcome the resistance of the crystalline lattice is termed τ_{max} and the stress that represents the theoretical strength is assumed to take the form of a sinusoidal stress-displacement curve:

$$\tau = \tau_{max} \sin \frac{2\pi x}{b} \tag{2.5}$$

2.6 Theoretical shear strength

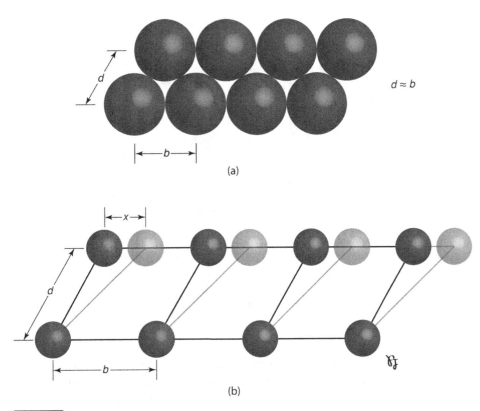

Figure 2.9

Schematic illustration showing (a) two adjacent planes of close-packed atoms with an interplanar distance d and atomic distance b, and (b) displacement of upper plane of atoms by an amount x as one plane is sheared (slipped) past the other.

At small values of strain, the shear stress, τ, and shear strain, γ, are related through the shear modulus, G, as $\tau = G\gamma$, and hence $d\tau/d\gamma = G$. Taking the derivative of the stress-displacement equation yields:

$$\frac{d\tau}{dx} = \frac{2\pi \tau_{\max}}{b} \cos \frac{2\pi x}{b} \qquad (2.6)$$

and

$$\left(\frac{d\tau}{dx}\right)_{x=0} = \frac{2\pi \tau_{\max}}{b}. \qquad (2.7)$$

When the displacement is small, the shear strain is given as x/a where a is the spacing on the slip plane and $d\tau/dx = d\tau/d\gamma \cdot d\gamma/dx$,

$$\left(\frac{d\tau}{d\gamma}\right)_{x=0} = \frac{2\pi a}{b} \tau_{\max} \qquad (2.8)$$

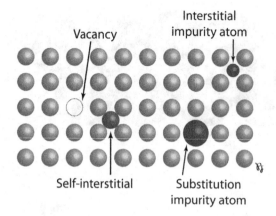

Figure 2.10

Schematic illustration showing various types of point defects including vacancies, self-interstitials, interstitial impurity atoms, and substitution impurity atoms.

which is equated to the shear modulus to yield the maximum shear stress:

$$\tau_{max} = \frac{Gb}{2\pi a}. \tag{2.9}$$

The expression for the theoretical strength of a metal predicts a yield strength value on the order of $G/2\pi$. However, none of the commercial alloys or metals have yield strength values on this order. The reason for the vast discrepancy is the presence of imperfections and their ability to help facilitate slip mechanisms in the crystal structure.

2.7 Imperfections in metals and alloys

2.7.1 Point defects

The long-range order found in the crystalline lattice can be readily disturbed by defects or imperfections. The first-order imperfections are known as point defects and include vacancies, self-interstitials, interstitial impurity atoms, and substitution impurity atoms (Figure 2.10). The vacancy is an unoccupied atom position in the crystalline lattice. This vacant lattice site results in a slight distortion of the adjacent atoms as they tend to displace toward the empty space. Another type of point defect is the self-interstitial atom that is accommodated in the space between regular atom positions in the lattice. The interstitial defect can also involve atoms of another element. If the atom is small relative to the lattice atoms, then a greater concentration of the impurity atom can be obtained in the interstitial sites (this is the case in steel where the concentration of interstitial carbon in iron can reach a few percent). If the atom is nearly the same size as the lattice element, this will result in a substitution of the impurity atom for a lattice atom. The self-interstitial point defect requires much more energy and results in a much greater

2.7 Imperfections in metals and alloys

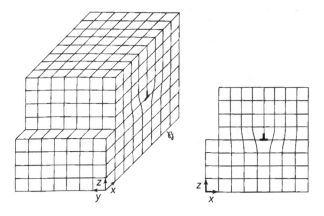

Figure 2.11

Schematic illustration showing an edge dislocation visualized as an extra half plane of atoms in a crystal lattice.

distortion than that resulting from the presence of a vacancy. For this reason there are much fewer self-interstitial defects than vacancies present in the lattice. Vacancies play an important role in diffusion, deformation mechanisms, chemical reactions, creep, and corrosion mechanisms in metal systems.

2.7.2 Line defects

The second-order imperfections are termed line defects and include **dislocations** of edge, screw, and mixed character (dislocation loops). The edge dislocation is easiest to visualize as an extra half plane of atoms into an otherwise ideal crystal lattice (Figure 2.11). The consequence of the extra half plane of atoms is that it deforms the crystal lattice and results in elastic displacement, strain, and stress fields around the defect. The Burgers vector, b, is the displacement vector needed to reconcile the perfect lattice and defected lattice. If b is orthogonal to the direction of the dislocation line, t, then the planar imperfection is termed an **edge dislocation**. If the Burgers vector is parallel to the dislocation, then the imperfection is a **screw dislocation** (Figure 2.12). The dislocation line can be imagined as the boundary between the area of the material that has slipped (movement of atoms from one position to an adjacent position through the action of shear force) and the undeformed crystal. The Burgers vector quantifies the amount of slip that occurs as the dislocation moves through the crystal (this movement is incremental and can be visualized conceptually to be similar to the movement of an inchworm). The slip plane that captures the motion of the dislocation is the plane that contains both b and t and is defined by the normal to this plane ($b \times t$). A full treatment of dislocation theory and its role in plastic deformation of metals can be found elsewhere (Hirthe and Lothe, 1982; Hull and Bacon, 2001; Read, 1953; and Weertman and Weertman, 1992).

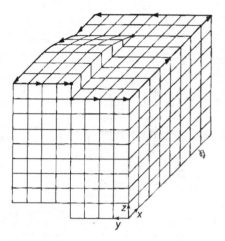

Figure 2.12

Schematic illustration showing a screw dislocation in a crystal lattice.

The energy associated with the existence of a dislocation is proportional to the square of its Burgers vector, $U \propto b^2$, and thus it is energetically preferable to have dislocations with the smallest possible Burgers vector. For this reason, slip tends to occur in the direction of closest packing and on the closest packed planes (Table 2.2). It should be noted that since BCC crystals are less efficiently packed, it is harder to achieve conditions of single crystal slip while FCC metals such as copper and gold easily flow under stress. The number of slip systems activated lays a fundamental role in **strain hardening** and **cold working** of metals and alloys (discussed below). However, as stated above, the BCC systems have secondary slip systems (36) available to them so these can help facilitate hardening mechanisms when the metal is polycrystalline (Table 2.2).

Dislocations play a pivotal role in the shear processes in metals. As a result, metals yield or plastically deform when a critical resolved shear stress is achieved on certain slip planes in the crystal (Hertzberg, 1989). This is schematically illustrated in Figure 2.13.

Using Figure 2.13, one can geometrically determine that the resolved shear stress is given as:

$$\tau_{rss} = \frac{F}{A} \cos\phi \cos\lambda \qquad (2.10)$$

where $\cos\phi \cos\lambda$ is the orientation factor. If there are a limited number of slip systems available then the orientation of the slip system plays a key role in determining the necessary force, F, for activating yield. The fact that dislocations have energy fields means that they will interact with each other in a way that can restrict motion (Figure 2.14). For example, edge dislocations that are parallel and separated by a distance defined by x and y in Cartesian coordinates (denoted as configuration A in Figure 2.14) will repel each other at short distances and will be attracted over long distances of separation. On the other hand, two equal yet opposite edge dislocations (configuration B) will be attracted

2.7 Imperfections in metals and alloys

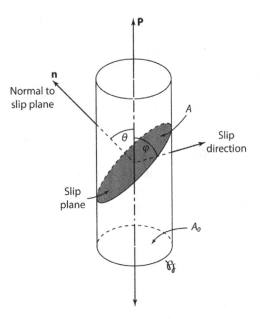

Figure 2.13

Schematic illustration showing shear plane resulting in a critical resolved shear stress in the crystal.

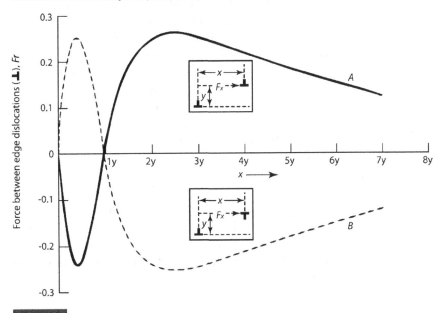

Figure 2.14

Illustration showing forces between two edge dislocations in different configurations (A and B). In configuration A, the two edge dislocations are parallel (both are positive) and separated by a distance defined by x and y. In this scenario, the two dislocations will repel each other at short distances and lead to dislocation pile-ups and strain hardening. On the other hand, two equal yet opposite edge dislocations (configuration B) will be attracted and annihilate each other in close contact (the extra half plane above and below the slip plane will restore crystalline structure).

and tend to annihilate each other in close contact (the extra half plane above and below the slip plane will restore crystalline structure). Interactions of dislocations of same orientation will result in **dislocation pile-ups** and contribute to strain hardening, while dislocations that are opposite in orientation will be attracted at short distances and tend to eliminate (annihilate) each other.

Dislocation force fields can be caused by the intersection of dislocations themselves, grain boundaries, precipitates, and inclusions. These dislocation pile-ups can result in a need for additional stress for continued slip, and this results in strain hardening in many metals. Because metals tend to deform readily in shear due to the slip processes, the onset of plastic deformation is critically linked to the resolved shear stresses. In fact, even in the mode of tensile loading, the deformation of a polycrystalline metal tends to occur in a shear mode. For this reason, the criterion for yielding in metals is based on shear stresses rather than normal stresses.

2.7.3 Planar defects

Planar defects are known as third-order imperfections and include defects such as grain boundaries. The grain boundary is the border within the metal that separates regions that differ in crystal orientation. The regions of single crystal orientation are termed grains. Metals are polycrystalline materials unless they are processed in a very specialized manner to create a single crystal structure or random (amorphous) structure. Though metals are called crystalline solids due to their lattice structure, they are more precisely termed polycrystalline solids with many grains that represent a crystal lattice of a specific orientation. The degree of orientation mismatch between grains can be small or large. Often the grain boundary mismatch is modeled with an array of parallel edge dislocations as the grain boundary is an area of localized energy and resistance to deformation (Figure 2.15). The grain boundary can serve to impede dislocation motion and can be a source of dislocation pile-up and strain hardening. This mechanism is depicted in Figure 2.16. For this reason, making the grain size of the metal or alloy smaller through processing can strengthen metals. The grain boundary serves as a nucleation site for dislocation pile-ups resulting in an increased strain field. Consequently, creating more grain boundaries in a given volume of material through reducing grain size results in a higher yield strength (Figure 2.17). However, this higher-energy region of grain boundaries also has an increased amount of interstitial space which can be a source of degradation that increases the local material reactivity, and can also allow a second phase or element of an alloy to segregate and corrode.

2.8 Metal processing

Processing, heat treatments, and mechanical operations are used to alter the structure and properties of metals and alloys. An interesting aspect of heat-treating alloys is that the yield and post-yield processes can be greatly altered, but if the basic chemistry

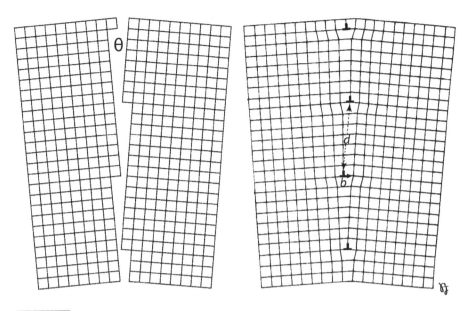

Figure 2.15

Illustration showing grain boundary modeled as an array of edge dislocations.

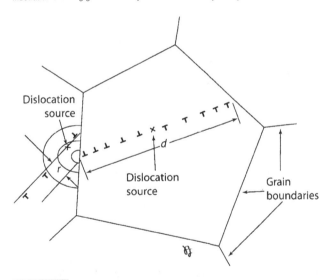

Figure 2.16

Illustration showing that the grain boundary can serve to impede dislocation motion and can be a source of dislocation pile-up and strain hardening.

and bonding remains the same then the elastic modulus remains unaltered. Figure 2.18 schematically illustrates how the material properties such as yield stress, fracture stress, ductility, and toughness of an alloy of one composition can be tailored with heat treatment.

44 Metals for medical implants

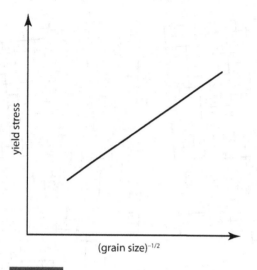

Figure 2.17

Plot of yield stress as a function of grain size. As the grain size becomes smaller, the yield stress increases.

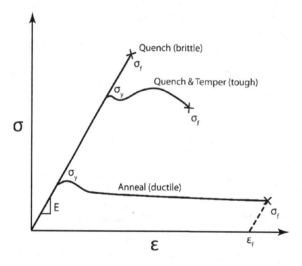

Figure 2.18

Schematic illustration showing how the material properties such as yield stress, fracture stress, ductility, and toughness of an alloy of one composition can be tailored with heat treatment.

Thus, steels as a class of metals have near identical elastic modulus values regardless of heat treatment and concomitant differences in yield or post-yield material properties. The same is true of modulus values for alloy classes used in medical implants, such as stainless steel, titanium alloys, and cobalt-chromium alloys. The ability to tailor material properties through heat treatment enhances manufacturability of metals. They can be

Table 2.3 Strengthening mechanisms in metals

Type of method	Works through	Example
Deformation	Strain-hardening Reduction in grain size	Cold-working/Strain-hardening Extrusion Drawing Rolling (hot)
Solid-solution strengthening	Residual strain field Geometric mismatch Modulus mismatch	Alloying Interstitial solid solution or substitutional solid solution strengthening
Second-phase particles	Creation of dislocation loop	Age-hardening Precipitation-hardening

treated to be soft for manufacturing operations (forging) and then hardened in their finished shape with little distortion to provide a strong component.

2.8.1 Processing for improved material properties

There are several mechanisms or processes that can be used to strengthen metals. Table 2.3 provides a summary of three different strengthening mechanisms used in metals: (i) deformation (cold-working, strain-hardening, extrusion, drawing, rolling); (ii) lattice substitution (alloying, solid-solution strengthening); and (iii) second-phase particles (age-hardening, precipitation-hardening).

In strain-hardening, line defects (dislocations) are added to the material through deformation. These added dislocations cause lattice strain that impedes the motion of other dislocations, thus improving the material's strength. Deformation can also assist in altering the orientation of grains within the polycrystalline structure in mechanisms such as drawing, extrusion, or cold-rolling. Slip is then more difficult due to the grain orientation mismatch.

The principal mechanism underlying the ability of a metal or alloy to be hardened through mechanical work is founded in the strain-hardening processes that are activated as dislocations interact with each other on activated slip planes. Consequently, the basic lattice structure and number of slip systems available in the alloy or metal affects the strain-hardening ability of that material. For example, polycrystalline alloys with the FCC or BCC crystal structure typically have a higher rate of work hardening than HCP alloys. The cubic structures have more available slip systems (Table 2.2) and the interaction of a greater number of dislocations results in more pile-ups and an increased stress for continued deformation. The shear stress required for continued deformation can be estimated from the following structure-property relationship:

$$\tau = \tau_o + \alpha G b \sqrt{\rho} \qquad (2.11)$$

where τ_o is the shear stress in the absence of work-hardening, α is a material coefficient with a value of about 0.5, G is the shear modulus, b is the Burgers vector, and ρ is the dislocation density.

A simple demonstration of strain-hardening can be achieved with a metal paper clip or metal coat hanger where little resistance to deformation is found when bending the metal for the first time but the process of unbending requires substantially more force. This strengthening of the metal is achieved through a process termed **cold-working**. Cold-working utilizes the complex interactions and force fields between dislocations and takes advantage of altering grain orientations such that slip is more difficult (Suresh, 1998). As a result, more stress is required for continued deformation as the material is worked. As an example, cold-rolled brass is strengthened by continual rolling to reduced thicknesses. A 50% reduction in thickness through cold-working can nearly double the strength. Significant changes in grain orientation and texture can occur with cold-working, unlike in hot-working where the grain size remains unaltered as grains recrystallize simultaneously with permanent deformation and little change of force is needed to obtain deformation. Thus, the hardness and yield strength can be substantially increased with cold-working. The trade-off to this strengthening is a loss of ductility. Many metals become embrittled and may crack if cold-worked too much. The embrittlement can cause manufacturing difficulty but can be tempered with intermittent **annealing** (heating below the melt temperature) to aid in recrystallization and relieve stresses. Extrusion and wire drawing are examples of other manufacturing processes that involve strain-hardening.

Paradoxically, the primary mechanisms utilized to strengthen metals involve the use of imperfections, which are conventionally thought of as weak points. One such process, as mentioned above, is cold-working. A second is reduction in grain size. Most metals are polycrystalline and are made up of grains whose orientations do not match. This orientation mismatch serves to discourage dislocation motion. The grain boundary also serves as a nucleation site for dislocation pile-ups resulting in an increased strain field. Consequently, creating more grain boundaries in a given volume of material through reducing grain size results in a higher yield strength. This mechanism is schematically illustrated in Figure 2.17. The equation relating yield strength to grain size is premised on the fact that at a grain boundary the adjacent grain may not be oriented favorably for slip and dislocations may pile up at the grain boundary. The Hall-Petch equation captures this structure-property relationship and takes the form:

$$\sigma_y = \sigma_i + \frac{k}{\sqrt{d}} \quad (2.12)$$

where σ_y is the yield strength of the material, σ_i is the friction stress inherent to the lattice, k is a material constant, and d is the grain diameter. This relationship holds for most pure metals and alloys. Following the Hall-Petch relationship above, it is evident that larger grains will help facilitate the working process. This relationship holds for most pure metals and alloys. An example of a method that can be used to reduce grain size is **hot rolling** that is performed above the **recrystallization** temperature for the alloy and results in the nucleation of a fine equiaxed grain structure. If the rolling temperature is below the recrystallization temperature, then it is termed **cold rolling**. Hot rolling increases the yield strength by reducing the grain size, while cold rolling increases the yield strength through mechanical work and of dislocation interactions.

The term "wrought" is used in conjunction with a specification of a steel alloy whose mechanical properties can be controlled through composition and heat treatment. Wrought steel generally undergoes two process steps. First, it is either poured into ingots or sand cast. Then, the metal is reheated and hot rolled into the finished, wrought form. **Wrought steel** can be subsequently heat treated to improve machinability or to adjust mechanical properties. Wrought cobalt-chromium–based steels are commonly used in orthopedic bearings due to their exceptional mechanical properties. Note that this is different than wrought iron which is characterized by its very low carbon content.

Alloying a metal system can also be used to strengthen a metal by **solid-solution strengthening**. This mechanism works through the addition of an element that has a geometric mismatch to the lattice atoms. Depending on geometric mismatch between the alloying element and the lattice atoms, either a **substitutional solid solution** or an **interstitial solid solution** can form. In both of these cases the overall crystal structure is essentially unchanged, but the local strain fields around the substitution atoms can interact with dislocation strain fields and potentially impede dislocation motion. The elastic interaction with a dislocation depends on a few structural factors. The first structural factor is solute atom size. Substitutional solid-solution strengthening, also known as **lattice substitution**, occurs when the solute atom is large enough that it can replace solvent atoms in their crystalline lattice positions (FCC, BCC, or HCP sites); the size mismatch between elements generally varies by less than 15% and must be of the same crystalline lattice type (Hume-Rothery, 1960). The gold-silver binary system used in dentistry is an example of a substitutional solid solution. If the solute atom is much smaller than the solvent (lattice) atoms, then an interstitial solid solution forms (carbon in iron is an example of an interstitial solid solution). Elements such as carbon, oxygen, nitrogen, and hydrogen are used to facilitate this type of strengthening mechanism. A large mismatch in size between the lattice and solute atoms will result in a greater elastic distortion and hence more resistance to dislocation motion. The site that the solute atom occupies is also important. A solute atom with tetrahedral distortion results in a nonsymmetric stress field that can interact with both screw and edge dislocations. Another structural factor is if the solute atom locally alters the shear modulus of the material. The self-energy of a dislocation is proportional to the shear modulus and therefore the dislocations will be attracted to the solute atom if the shear modulus is locally reduced, but repelled if the shear modulus is locally increased.

Precipitation-hardening is also one of the most effective ways of strengthening an alloy. In order to accomplish this type of strengthening mechanism the alloy must exhibit a decreasing solubility with decreasing temperature such that precipitates can be formed in the matrix with appropriate heat treatment. A **supersaturated solid solution** is created by heating the alloy to an elevated temperature and then quenching the system to take advantage of the decreasing solubility. The supersaturated solid solution is

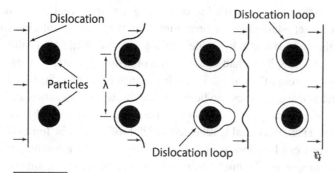

Figure 2.19

Schematic illustration depicting the strengthening achieved with second-phase particles or precipitates.

thermodynamically unstable and will nucleate **precipitates** upon moderate heating or extended time (**age-hardening**). These precipitates help improve the yield strength of the alloy. Precipitates cause local lattice strains in the host lattice and help impede dislocation movement. These precipitates or second-phase particles are modeled as dispersed particles in the matrix. If the particles are strong enough that dislocations cannot penetrate them, then dislocation motion will arrest upon meeting the precipitate and begin to bend around the particles. An externally applied stress will force the dislocations to pass between the particles, but they will leave behind a constrained dislocation loop around the particles (Figure 2.19). The shear stress necessary to force the dislocations between the particles is given by the following structure-property relationship:

$$\tau = \frac{2T}{\lambda b} \tag{2.13}$$

where T is the force needed to overcome the line tension of the dislocation, b is the Burgers vector, and λ is the mean free spacing of the particles. The primary benefit of this strengthening mechanism is achieved when the precipitates are coherent with the lattice and elastic distortion fields are maximized (Suresh, 1998).

This second-phase particle strengthening mechanism can be achieved with powder metallurgy methods or through simple heat treatments commonly referred to as age-hardening. In alloy systems with a decreasing solubility with decreasing temperature, such as aluminum alloyed with copper, or titanium alloys alloyed with aluminum and vanadium, the alloy can be heated to dissolve the solute atoms and then quenched to create a supersaturated solid solution. This system can then be annealed to cause precipitation of a second phase and to effectively harden the alloy. Similarly, chromium precipitates can help strengthen stainless steel alloys that are unable to be strengthened through traditional heat treatment owing to the large amount of chromium and nickel used for increased corrosion resistance.

Heat treatments include thermal processes such as annealing, quenching, and tempering. **Annealing** is a treatment in which the metal or alloy is heated below its melting point but at a sufficiently high temperature to activate both diffusion processes for annihilation of dislocations and for the reformation of grains (a process known as recrystallization)

(Richman, 1974). Thermal treatments can be used to control grain size, which as discussed above can be used to alter strength, and can also serve to relieve stresses in the material. **Quenching** is the term used to describe very rapid cooling and is accomplished by cooling the metal rapidly enough to create a metastable structure. Quenching can be used to create non-equilibrium systems such as supersaturated solid solutions and amorphous metals. **Tempering** is a thermal treatment that provides sufficient energy to facilitate precipitation, grain growth, and relief of residual stresses.

Thermal, mechanical, or combined processes can be used to tailor the desired component geometry, microstructures, and material properties. For example, heating and quenching can be used to harden steels. Such adaptability is facilitated by multiphase systems and solubility that decreases with decreasing temperature. Such heat treatability can be observed in the phase diagram that plots equilibrium phases as a function of temperature and composition.

Example 2.2 **Heat treatment of low-carbon steel**

Steel is an excellent example of a material whose microstructure and mechanical properties can be controlled through heat treatment. Figure 2.20 provides the phase diagram for steel that is heat-treatable and shows the associated microstructure development through the eutectoid reaction. The eutectoid reaction describes the phase transformation of one solid into two different solids via a decomposition reaction. It can be seen from Figure 2.20 that the gamma (γ) phase of iron (austenite) is stable above the eutectoid temperature (723°C). In the Fe-C system, there is an eutectoid point at approximately 0.8wt% C, 723°C. Austenite (also known as the γ phase) has an FCC structure and a greater solubility of carbon than the BCC α phase. As steel is slowly cooled through the eutectoid temperature, it undergoes a eutectoid reaction ($\gamma \Rightarrow \alpha + Fe_3C$) and precipitates iron carbide. The resulting two-phase microstructure is called pearlite and has a low yield strength but much ductility. Slow cooling results in a fine laminar microstructure called *pearlite*, which can be tailored to have a wide range of strength and ductility characteristics. If the system is quenched from the austenite phase, the carbon is confined in the non-equilibrium structure and distortion of the BCC lattice occurs. This rapid cooling results in a brittle microstructure known as *martensite* that has a body-centered tetragonal (BCT) structure. Martensite has high strength and hardness due to residual stresses but almost no ductility. In fact, excessive quenching can result in rupturing of the component due to the magnitude of the residual stress. When martensite is tempered, i.e., heated to increase carbon diffusion, the carbon is precipitated to provide a strong yet tough structure. Tempering is commonly used after a quenching process to restore stability to the structure and to toughen the metal or alloy system. Tempering is performed well below the melting point but provides sufficient thermal energy to enable precipitation of the trapped carbon. In steels, the tempered martensite results in a new microstructure that includes a mixture of martensite and one of the above microstructures, and provides both strength and energetic toughness (which captures how much energy the material can absorb prior to

50 Metals for medical implants

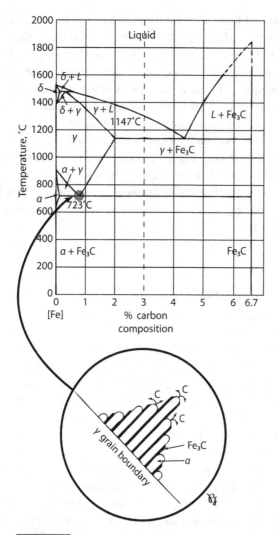

Figure 2.20

Phase diagram for Fe-C steel that is heat-treatable and shows the associated microstructure development through the eutectoid reaction. (After Richman, 1974.)

fracture). Figure 2.21 shows the time-temperature-transformation (TTT) plot for steel, which is used to tailor heat treatments of steel for desired microstructures and concomitant mechanical properties. A **TTT** plot is helpful to engineers as it enables one to determine what microstructure or phase such as austenite (A), Ferrite (F), cementite (C), or martensite (M) will transform (**T**) if an alloy is held at temperature (**T**) for a certain amount of time (**T**). The mechanical properties concomitant with variations in microstructure are schematically illustrated in Figure 2.18.

2.8 Metal processing

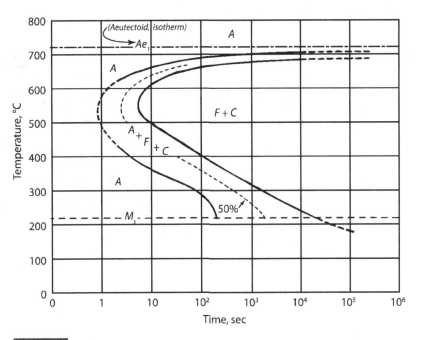

Figure 2.21

Time-temperature-transformation (TTT) plot for steel used to tailor heat treatments to obtain desired microstructures.

2.8.2 Processing for shape-forming

Standard metal processing can be separated into four categories: **forming**, machining, casting, and other. Forming is the process of working metal that keeps a constant cross-section. Machining involves the removal of material to create a final shape or finish. Casting uses molds to create a specific shape. Finally, welding, powder processing, and rapid prototyping fall under the category of "other."

The bulk deformation processes such as rolling, drawing, extrusion, and forging are schematically illustrated in Figure 2.22. Drawing is used to reduce or change the diameter of a wire or rod by pulling the wire or metal rod through one die or a sequence of drawing dies. It is a useful manufacturing process for making wires such as those found in wire leads of pacemakers. Extrusion, while similar in concept to drawing, typically utilizes initial bar stock material that is pushed through a die of desired cross-section. In extrusion, unlike drawing, the material experiences only compression and shear stresses rather than tensile stresses, and can thus be utilized on less ductile materials. Extrusion can be done either hot or cold, and is often characterized as hot or cold working. Rolling is reduction of thickness by movement through rollers, and can be performed hot or cold, as described in the thermo-mechanical section above. Forging involves cold working of the component through a hammer, press, or stamping or other mechanical operations, and typically involves high compressive stresses to the component.

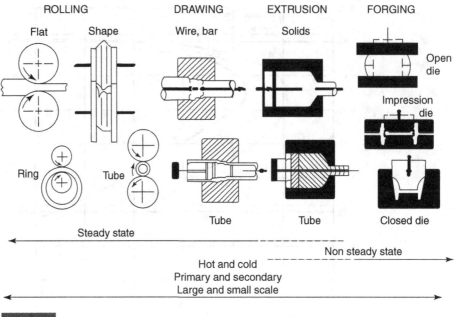

Figure 2.22

Illustration of bulk deformation processes such as rolling, drawing, extrusion, and forging.

Machining encompasses a number of operations to finish a component into a desired shape or surface finish and may include lathes, mills, and drills as a few examples. Figure 2.23 schematically illustrates the processes involved in machining a hip stem: (1) start with bulk material, (2) utilize a multiaxis mill to remove excess material and to achieve desired shape, (3) control tooth path for milling operations and desired final geometry, (4) use heat treatment and surface polishing, and (5) finish with porous or powder coating (to facilitate device integration with bone).

Casting can utilize an expendable or non-expendable mold. An example of the former is investment casting or lost-wax casting where the wax prototype to be molded is encased with a very high melting temperature material (Figure 2.24). Upon heating, the wax is melted or "lost." The refractory encasement can be used to mold alloys into net-shaped products with complex geometries, and are commonly employed in dental applications. Non-expendable molds are typically made of steel with a refractory coating on the inner wall to lengthen the life of the mold and to enhance release of the molded component. Die casting with steel molds is a good example of this method. Multiple molds can be processed at once and thus the process is highly automated.

The intricacy of molds and the batch size typically dictates the manufacturing run cost. That is, the more parts that are made, the more economical the casting process becomes and vice versa. For example, a hip prosthesis that is custom-made in a one-of-a-kind compression mold will be far more costly than one machined from an extruded

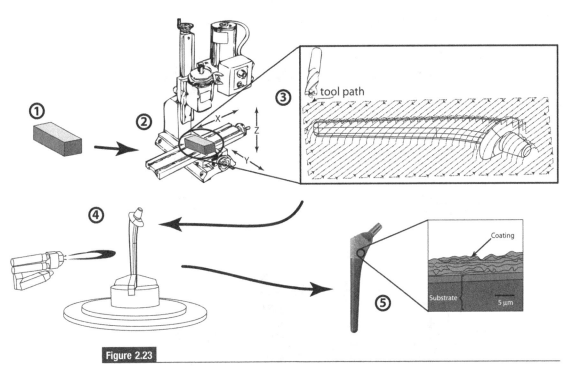

Figure 2.23

Processes involved in machining a hip stem: (1) start with bulk material, (2) utilize a multiaxis mill to remove excess material and to achieve desired shape, (3) control tool path for milling operations and desired final geometry, (3) heat treatment and polishing, and (4) porous or powder coating (to facilitate device integration with bone).

rod or made in a large batch process. Such costs can dictate the manufacturing method utilized in an implant.

2.9 Corrosion processes

One of the restrictions of using metallic materials in implant applications is that they are susceptible to chemical attack or corrosion. Most metals are intrinsically metastable and unless they are noble metals, oxidation is a preferable energy state. This limitation is owed to the basic bonding structure in metallic bonding where the "sea" of free electrons surrounds the positive ions. These positively charged ions are not only attracted to the electrons within the metal itself but are also attracted to negatively charged particles outside of the metal, such as those found within the aqueous environment of the body (Park and Bronzino, 2003).

The degradation of metallic implants in the body is cause for concern not only for structural integrity of the device but also the systemic response *in vivo*. The loss of material in the form of corrosion debris can result in a chronic inflammatory response and premature failure of the device. In addition, the reactivity of

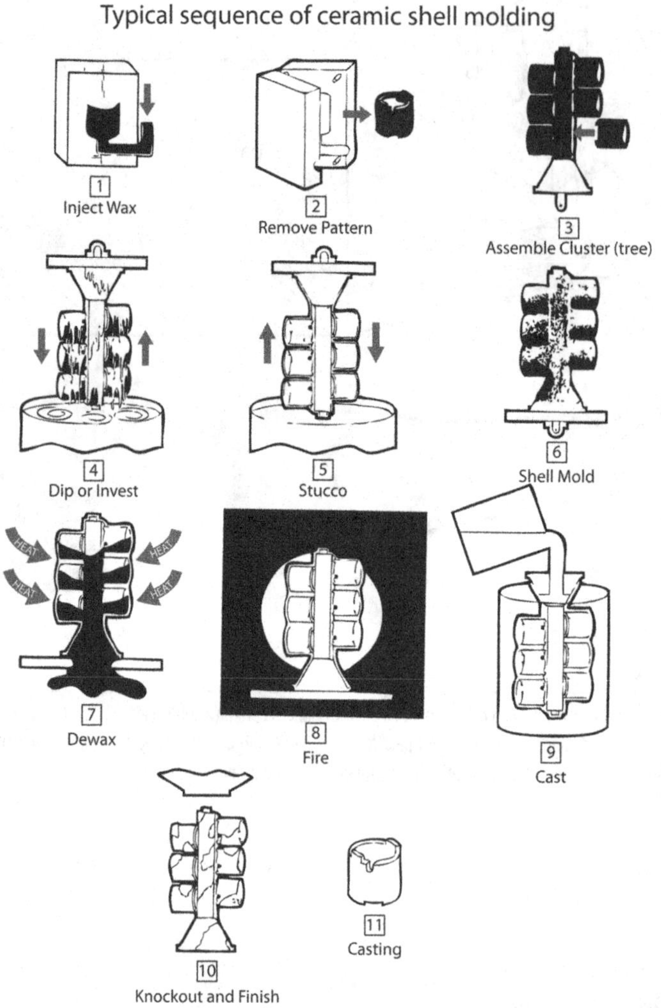

Typical sequence of ceramic shell molding

1. Inject Wax
2. Remove Pattern
3. Assemble Cluster (tree)
4. Dip or Invest
5. Stucco
6. Shell Mold
7. Dewax
8. Fire
9. Cast
10. Knockout and Finish
11. Casting

Figure 2.24

Illustration of investment casting where the wax prototype to be molded is encased with a very high melting temperature material. Such methods can be used for prototyping and custom implant development.

many metals and release of metal ions such as nickel, chromium, and titanium is speculated to trigger both inflammatory response and carcinogenic pathways in the body.

Corrosion processes are generally the result of metals reacting with an aqueous environment of salts, acids, or bases. In metal implants, designing for corrosion resistance can be quite challenging as such reactions are exacerbated by geometry, composition, galvanic coupling, and stresses (Fontana, 1967). The basic underlying reaction that occurs during corrosion is the increase of the valence state. That is, the loss of electrons of the metal atom to form an ion, as expressed by the equation: $M = M^{n+} + ne^-$. In general, two basic reactions occur when metals corrode. An oxidative reaction occurs when electrons are lost from the metal. This causes the metal to dissolve as an ion or to become a combined state such as an oxide. The reduction reaction consumes the electrons lost by the anodic reaction. The material undergoing the reduction reaction is called the cathode and that undergoing oxidation is known as the anode. As a general rule, the anode corrodes and the cathode does not.

Metals used in the body should be intrinsically resistant to oxidative or chemical attack or able to utilize protective mechanisms such as **passivation** or surface treatments. Two essential characteristics determine whether metal corrosion will occur and the extent of the degradation. The first of these is the thermodynamic driving force that drives the corrosion (oxidation and reduction) reactions. The thermodynamic driving forces that cause corrosion correspond to the energy required or released during a reaction. A scale of corrodibility is often presented in qualitative form and is commonly referred to as the galvanic series (Table 2.4). This series is somewhat subjective as local environments or couplings can alter the relative scale. However, it serves as a general point of reference and provides a relative list of more corrosion-resistant materials (cathodic) versus those more susceptible to basic corrosion processes (anodic). For comparison, materials such as pyrolytic carbon and diamond, which are highly inert materials in the body, would be ranked with the high noble metals.

The second characteristic that determines the corrosion resistance of a material is the presence of kinetic barriers that prevent or limit corrosion. This protective process is known as passivation and often utilizes the formation of a metal-oxide passive film on the metal surface. Such barriers generally prevent the migration of metal ions and electrons across the metal-solution interface. Most orthopedic alloys rely on the formation of a passive oxide film to prevent corrosion from taking place. Passive films are normally non-porous and must be continuous over the underlying metal. To be successful, films should have an atomic structure that limits the migration of ions and electrons across the metal oxide-solution interface. Moreover, these passivation layers must be able to withstand mechanical stressing or abrasion that accompany most orthopedic implants due to contact stresses, articulation, and fretting between components such as screws and plates. The kinetic barriers to corrosion are related to factors that impede or prevent corrosion reactions from taking place, whereas the chemical driving force determines whether corrosion will take place under certain conditions.

Corrosion in metallic systems can occur through a number of pathways including uniform attack, galvanic corrosion, intergranular corrosion, crevice corrosion, and stress corrosion cracking. Each of these is cause for clinical concern in medical implants. **Uniform attack** occurs when the whole surface of the metal is corroded in an evenly distributed manner (Figure 2.25).

Table 2.4 Galvanic series with noble metals at top and more anodic metals toward bottom

Metals
Platinum
Palladium
Gold
Silver
Titanium
CoCr alloys
Stainless Steel (316 passive)
Stainless Steel (304 passive)
Stainless Steel (316 active)
Molybdenum
Brass
Stainless Steel (304 active)
Chromium
Nickel
Steel
Aluminum
Zinc

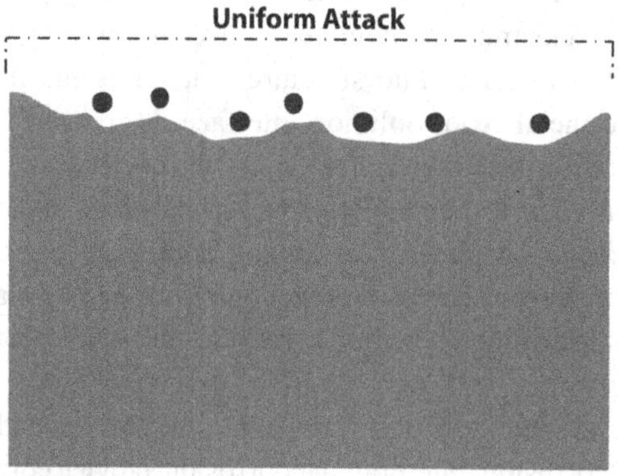

Figure 2.25

Illustration of uniform attack mode of corrosion.

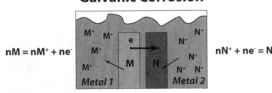

Figure 2.26

Schematic of the galvanic corrosion process.

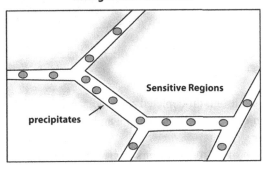

Figure 2.27

Graphic of the intergranular corrosion mechanism.

Galvanic corrosion occurs when two materials of dissimilar electrochemical potential are placed in contact in the presence of an electrolyte in such a way that an electrochemical cell is created (Figure 2.26). The electrolyte provides the transfer medium for ions between the cathodic and anodic metal and facilitates corrosion of the anodic metal. Such scenarios occur in modular implants where two materials are coupled together to achieve desired structural properties. An example is a modular metallic hip implant where a titanium alloy may be used for the femoral stem to provide fatigue resistance, while a cobalt-chromium alloy may be used in the femoral head to provide wear resistance.

Intergranular corrosion, as its name suggests, is corrosion that occurs along grain boundaries (Figure 2.27). Recall from above that grain boundaries are high-energy surfaces that have sufficient free volume to be susceptible to chemical or oxidative attack. Thus, while grains can be a source of strength they can also be a source of weakness in materials. This is an isolated form of corrosion and typically occurs when there is a local differential in alloy chemistry between the grain and the grain boundary. It often occurs in materials that are inherently known to be corrosion resistant. An example is found in stainless steel, where chromium is added to facilitate passivation. The addition of this element can result in precipitation of chromium carbide along grain

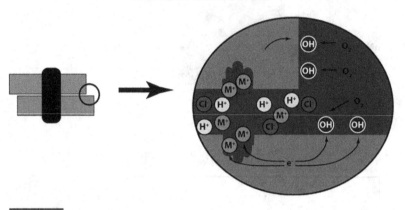

Figure 2.28

Schematic illustrating the mechanism of crevice corrosion.

boundaries and can cause chromium-depleted zones adjacent to the boundary. This local deficiency in chromium causes a galvanic couple and can lead to regions of corrosion along the grain boundary.

Another localized form of corrosion occurs in **crevice corrosion** (Figure 2.28). This process sets up a local behavior much like a galvanic cell, except that with galvanic corrosion there are two dissimilar materials and one environment, while in crevice corrosion there are two dissimilar environments but only one material. For example, the conditions for crevice corrosion can be satisfied by placing a tightly wound fishing line or a snug-fitting elastic band around a stainless steel component and then placing this system in salinated water. This corrosion mechanism has interesting consequences when one thinks of using stainless steel components such as screws or fasteners in the body. Similar conditions are often set up at the junction between the femoral stem and head, in a region known as the Morse taper, and this mismatch can result in corrosion at the junction. It is important to note that a passivated surface that is shown to be highly corrosion resistant in physiologically relevant tests can nevertheless degrade rapidly in the environments that facilitate crevice corrosion.

A subset of crevice corrosion is **pitting** and concomitant stress corrosion cracking. Pits may preferentially grow deep from the surface (becoming more like crevices), and thus serve as initiators for cracks. This mechanism is illustrated in Figure 2.29. In the mechanism shown, the pit serves as the site of environmental mismatch and is often found in components with internal tensile stresses. Once the flaw or crack grows to a critical size, no further corrosion is necessary for component failure. The condition for fast fracture is satisfied when the stress intensity (the coupling of the flaw size and stress) reaches the critical value for the material. The basic form of the stress intensity is given as:

$$K = \sigma\sqrt{\pi a} \qquad (2.14)$$

Pitting Corrosion

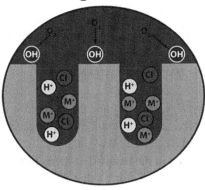

Figure 2.29

Schematic diagram showing the mechanism of pitting corrosion.

where σ is the applied stress and a is the flaw size. When the flaw size reaches a critical value, a_c, then the conditions for fracture are met. Under plane strain conditions, this value of stress intensity is a material property known as the fracture toughness, K_{IC} (presented in detailed form in Chapter 9, *Fracture Mechanics*). Stress corrosion cracking is of critical concern in device components as it occurs at stress intensity levels that are below the fracture toughness of the material. The value for the onset of crack propagation under corrosive conditions is known as K_{ISCC} and this value is generally lower than the fracture toughness of the material. Since $K_{ISCC} < K_{IC}$, the conditions of coupled stress and corrosive environment can lead to unpredicted component failure. An additional complication is that developing laboratory tests to accurately predict K_{ISCC} for the physiological conditions expected *in vivo* is quite difficult. The following is an example of stress corrosion cracking behavior in a medical grade alloy.

Example 2.3 **Stress corrosion cracking**

Stainless steel alloys are viable material candidates for fracture fixation devices but are susceptible to stress corrosion cracking. The following stress corrosion data were obtained for a medical grade stainless steel alloy subjected to a range of physiological loads in a saline solution used to simulate the local body environment:

Stress (MPa)	Crack length, a (mm)	Crack growth rate, da/dt (m/s)
28	5	8×10^{-10}
28	10	26×10^{-9}
56	5	8×10^{-7}
56	7.5	8×10^{-7}

This alloy will be used in a fracture fixation plate and the form of its stress intensity factor is $K = \sigma\sqrt{(\pi a)}$. The fracture toughness, K_{IC}, for this alloy is 25 MPa$\sqrt{m}$. How long would a fracture fixation device made of this alloy last if an initial flaw 5 mm in length was present, and the component was subjected to an operational stress of 60 MPa? Is this an acceptable time period if the alloy is used in a fracture fixation plate where the time necessary for tissue healing is 12 weeks?

Solution The stress intensity takes the form, $K = \sigma\sqrt{(\pi a)}$, and the far-field stress is given as $\sigma = 60$ MPa. For an initial crack size, and $a_i = 0.005$ m, one finds the initial stress intensity using $K = \sigma\sqrt{(\pi a)}$ to find $K_i = 7.52$ MPa$\sqrt{m}$.

The stresses and crack lengths given in the first two columns of the table above are used to calculate values of stress intensity, K. For a value of $K_i = 7.52$ MPa$\sqrt{m}$, the crack velocity is $da/dt = 8 \times 10^{-7}$ m/s. We use $K_{IC} = 25$ MPa$\sqrt{m}$ and $\sigma = 60$ MPa to find the critical crack length, $a_c = 0.055$ m. Knowing that
$$\frac{da}{dt} = \frac{a_c - a_i}{\Delta t} = 8 \times 10^{-7} \text{ m/s},$$ one can solve for $\Delta t = 17.4$ hours.

This is not an acceptable time period for this fracture fixation plate, or many other scenarios where the device may be left in the patient indefinitely and is structurally necessary for several weeks or months.

2.10 Metals in medical implants

The structural properties of metals and alloys make them suitable candidates for many load-bearing applications. However, while there are numerous metals or alloys that are suitable engineering materials, there are only a few metal systems that are appropriate for medical implants. The primary limitation in material selection is due to the necessity for biocompatibility in medical devices. Few metals can be utilized in the body without adverse reaction or chronic inflammatory response. Furthermore, as discussed above, the corrosion resistance of metals is paramount when making materials selection considerations for the design of an implant. In general, all metals are somewhat susceptible to corrosion; however, an inadvertent coupling of dissimilar materials, a design that utilizes stress concentrations, or mechanical fretting can result in corrosion processes that lead to premature failure of a metallic component. In designing components made with metals it is necessary to consider the material selection as a multifactorial problem that factors in structure, geometry, stresses, and biological environment.

Medical alloys span a variety of uses in the body including but not limited to fracture fixation plates, surgical tools, total joint replacements, dental implants, pacemaker components, and stents. Candidate systems include stainless steels, cobalt-chromium alloys, titanium and titanium alloys, as well as zirconium alloys. In general, the composition of these alloys facilitates biocompatibility, corrosion resistance, and appropriate mechanical properties for use in structural or load-bearing devices. Table 2.5 provides

Table 2.5 Comparison of elastic modulus, yield strength, and tensile strength for stainless steels, CoCr alloys, and Ti alloys

Material	ASTM	Thermal treatment	Elastic modulus (GPa)	Yield strength (MPa)	Tensile strength (MPa)	Applications
Stainless Steels	F745	Annealed	190	220	480	Surgical implants
	F55–56,	Annealed	190	331	586	Bone screws
	F138–139	Cold-worked (30%)	190	792	930	Fracture fixation
		Cold-forged	190	1200	1351	
CoCr Alloys	F75	As-cast/Annealed	210	450–517	655–890	Knee bearing
	F799	Hot isostatic press	250	840	1300	Hip Implants
	F90	Hot-forged	210	900–1200	1400–1600	Hip implants
	F562	Annealed	210	450–650	950–1200	Orthopedic implants
		Cold-worked (40%)	210			
		Hot-forged	232	1600	1900	Fixation
		Cold-worked, aged	232	965–1000	1200	Intervertebral disc
				1500	1795	Orthopedics
Titanium Alloys	F136	Forged annealed	116	896	965	Dental implants
	F67	Forged, heat-treated	116	1034	1103	Hip stem
		Cold-worked (30%)	110	485	760	Dental implants

a comparison of elastic modulus, yield strength, and tensile strength for the most commonly used medical alloys. It should be noted that not all of these alloys are cleared for all device applications. For example, cast cobalt-chromium alloys are typically used in the hip, while the wrought form is commonly employed in the knee.

Stainless steels are utilized in medical implants primarily for their corrosion resistance; additionally they are relatively inexpensive and easy to manufacture. The most common stainless steel alloy used in medical devices is designated 316L where the "L" designates low carbon content. **ASTM** International (formerly the American Society for Testing and Materials) specifies the composition of this alloy as follows (in standard specifications ASTM F138 and F139): 0.03% (max) carbon, 2% (max) manganese, 0.03% (max) phosphorous, 0.75% (max) silicon, 17–20% chromium, 12–14% nickel, 2–4% molybdenum, and the balance is iron. The amount of carbon in this alloy is limited to 0.03% to minimize corrosion. Similarly, chromium is added to facilitate passivation and to improve corrosion resistance. This alloy is known as an austenitic stainless steel because the addition of nickel stabilizes the austenite (FCC) phase. Because of its composition, this alloy cannot be strengthened with heat treatment. However, stainless steel is highly amenable to strengthening by cold-working. The elastic modulus for stainless steel is 190 GPa but a wide range of yield and tensile properties can be obtained by varying the degree of cold work. For example, an annealed form of the alloy may have a yield strength of 330 MPa, while a cold-forged alloy may be on the order of 1200 MPa.

While stainless steel is generally known for its corrosion resistance, it remains susceptible to stress corrosion cracking and crevice corrosion, as shown in Example 2.2. For this reason, most stainless steel alloys must be utilized with great scrutiny and are often limited to structural devices where strength is not necessary for extended lengths of time.

Cobalt-chromium alloys are utilized in device applications where high strength and fatigue resistance are important. These alloys have exceptional corrosion resistance and can be utilized in highly loaded implants such as total hip or knee replacements. These alloys are generally cast or wrought. The cast form of this alloy is specified by ASTM (ASTM F75) with the following composition: 27–30% chromium, 5–7% molybdenum, 2.5% (max) nickel, 0.75% (max) iron, 0.35% (max) carbon, 1.0% (max) silicon, and 1.0% (max) manganese. The two primary elements of this alloy, cobalt and chromium, form a solid solution of up to 65% cobalt. The cobalt is utilized to limit grain size and hence improve strength of the alloy. As can be seen in Table 2.5, the yield strengths can be as high as 1500 MPa for a cold worked and aged condition with a concomitant tensile strength of nearly 1800 MPa. This is substantially higher that the maximum yield (1200 MPa) and tensile strength (1500 MPa) for the cold forged stainless steels. The primary benefit over stainless steel, however, is the improved corrosion resistance obtained with the additional chromium. This can be seen in the galvanic series (Table 2.4).

The cast form of the CoCr alloy has a few limitations. The first of these is that non-uniform cooling can result in chromium-depleted zones at grain boundaries that can facilitate intergranular corrosion. The solidification process can also result in large grains, which can reduce strength, and casting defects that can facilitate crack inception. The presence of flaws can make the cast form of this alloy more prone to fracture or fatigue failures. For this reason, powder metallurgy methods have been developed to improve the properties of this alloy. Hot isostatic pressing (HIP), which utilizes high temperatures and pressures, is utilized to compact and sinter a powder form of this alloy. Forging is then used to achieve the desired geometry. This process results in improved strength (due to a reduced grain size) and a slight increase in elastic modulus because of compaction.

The ASTM specifications for wrought CoCr alloys used in medical implants are: F799, F90, and F562. The first of these is essentially the same composition as that specified by F75, but the mechanical properties are greatly improved by hot forging after casting. Mechanical work performed on this alloy can double the yield and ultimate tensile strength of the alloy while maintaining the elastic modulus. The structural improvement is achieved through strain hardening of the grains and a transformation of the FCC lattice into a HCP structure. The F90 alloy utilizes the addition of tungsten and nickel for processing. These element additions also facilitate strengthening processes and improve tensile behavior of the alloy. The F562 alloy is a multiple phase alloy that is 33–37% nickel, 29–38.8% cobalt, 19–21% chromium, 9–10.5% molybdenum, 1% titanium (max), 0.15% (max) silicon, 0.15% (max) manganese, 0.01% (max) sulphur, and 1% iron. This alloy can be strengthened through a number of methods including solid solution strengthening, cold-working, and precipitation hardening to yield a wide range of tensile properties.

The elastic modulus of the CoCr alloys remains largely unchanged (220–234 GPa) while the yield strengths (240–1585 MPa) and tensile strengths (655–1795 MPa) can vary immensely with heat treatment, alloying, and hardening mechanisms utilized. These alloys offer excellent corrosion resistance and mechanical properties and are most suitable for long-term, high-strength implants such as total joint replacement systems. The cast form is often used in the hip while the wrought form is typically used in the knee. Like all materials, there are certain limitations of these alloys that need to be considered in the design process. While these alloys are generally quite resistant to corrosion, the processing can render the material susceptible to intergranular corrosion, and the use of varying composition or different alloy systems together in modular devices can render them susceptible to galvanic corrosion. Additionally, the high elastic modulus of this alloy in comparison to titanium can make this material more prone to complications with stress shielding (loss of bone due to lack of stress transfer to surrounding bone tissue) unless geometric designs are utilized that reduce component stiffness. The consequence of stress shielding is bone loss, implant loosening, and potential for premature failure of the device.

Titanium and titanium alloys offer exceptional strength with a substantially lower density and elastic modulus than CoCr and stainless steel alloys. Titanium and its alloys are highly resistant to corrosion and facilitate long-term biological stability. Titanium is virtually unique in that it has a natural propensity for integration with bone. It is thought that the structure of the titanium oxide surface layer interfaces well with the structure of bone and that there is a natural propensity for bone attachment and integration to this surface. The **osseointegration** at the bone-implant interface enables excellent fixation for titanium systems used in orthopedic and dental implants (Davies and Baldan, 1997). Further, the lower elastic modulus (at 110 GPa, nearly half that of cobalt chromium or stainless steel) reduces the complication of stress shielding that can occur with stainless steels and cobalt–chromium-based alloys.

Pure titanium exists as a HCP structure (α phase) below 882°C and as a BCC structure (β phase) above this temperature. The use of alloying elements enables the microstructure to be tailored. Aluminum is used to help stabilize the α phase while vanadium is used to help stabilize the β phase. Heat treatment of the alloyed structure enables a precipitation hardening mechanism to be utilized for additional strength. The most notable titanium alloy is processed utilizing the addition of aluminum (5.5–6.5%) and vanadium (3.5–4.5%) to yield Ti-6Al-4V. In addition to corrosion resistance, this alloy is noted for its exceptional fatigue strength, yield strength, and moderate elastic modulus. Titanium alloys are commonly used in orthopedic and dental implants because they offer exceptional corrosion resistance, osseointegration, and material properties.

Another well-known titanium alloy utilizes 50% (atomic) titanium and 50% (atomic) nickel and is commonly known as **Nitinol**. This Ni-Ti alloy derives its name from its Ni-Ti composition and its initial discovery at the Naval Ordinance Laboratory (NOL). This alloy exhibits both superelastic (extreme elastic shape recovery) and shape memory

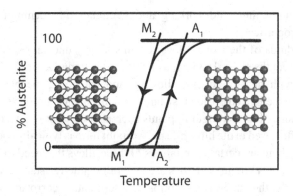

Figure 2.30
Diagram showing the martensite-austenite phase transformation that contributes to shape recovery through thermal and mechanical means. (After Pelton *et al.*, 2003.)

(thermal shape recovery) behavior. These unique traits enable the alloy to undergo large recoverable deformations and to regain an original shape after plastic deformation if energy is added to the system. At high temperatures the alloy is stable as an ordered cubic phase also known as the austenitic (parent) phase. At low temperatures the stable structure is a non-cubic martensitic (daughter) phase (Figure 2.30).

The shape memory behavior is attributed to a diffusionless martensitic phase transformation that is also thermoelastic due to the ordering in the parent and daughter phases (Pelton *et al.*, 2003). The superelastic traits enable the alloy to undergo large elastic deformation (strain) without a need for increasing stress. This alloy is capable of large recoverable strain on the order of 8–11% (as compared to 0.2% recoverable strain for many metals) due to the recovery of the stress-induced martensitic transformation on loading and the spontaneous reversion upon unloading as well as through linear elasticity of the stress-induced martensite phase. Nitinol has a restorative force that is at least an order of magnitude greater than stainless steel. These unique mechanical characteristics make this alloy an exceptional material for medical devices where elastic deployment, kink resistance, and fatigue resistance are important. However, this alloy is expensive to manufacture, particularly for parts with large cross-sections, and thus the device must warrant the cost. Nitinol has been successfully employed in vascular stents, guide wires, and surgical tools (Pelton *et al.*, 2004).

2.11 Case study: corrosion in modular orthopedic implants

In orthopedic implants there has been an increasing trend toward designs that include metal-on-metal conical taper connections to assist in modularity; thus the coupled effects of environment, material, geometry, stress, and motion are critical to the success of the device. Modularity offers several benefits, including the pairing of mismatched sizes and materials by the surgeon to best accommodate the specific physiological and anatomical

2.11 Case study: corrosion in modular orthopedic implants

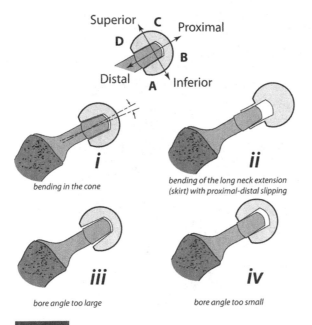

Figure 2.31

Schematic picture of the Morse taper and various scenarios of loading in the superior-inferior and proximal-distal orientations that can lead to enhanced stresses and corrosion conditions in a modular implant. (After Jacobs *et al.*, 1998.)

needs of a patient without exorbitant manufacturing cost. Additionally, if part of the component needs revision at a later date this can be accomplished with a partial revision rather than full revision. Most modular implants, such as the junction of the femoral stem and acetabular cup in a hip replacement, use a conical press-fit fixation known as a Morse taper, which is illustrated in Figure 2.31.

Modular implants generally utilize a tapered shank that fits into a socket of the corresponding taper, and facilitates alignment and frictional fixation. Under normal loading, the fixation achieved between the two surfaces is stable. However, if bending occurs in the cone and is accompanied by **proximal-distal** slipping, then fretting at the junction can lead to a gap at the interface. Similarly, a gap can be created if the bore angle is too large or small. Contamination of this junction increases the likelihood for premature failure of the device. The degradation of this junction in modular implants is evidenced by particulate corrosion and wear products in tissue surrounding the implant (Agins *et al.*, 1988; Jacobs *et al.*, 1998). Such biological inflammatory response may ultimately lead to bone loss around the implant or the leaching of metal ions into the immune system (Black *et al.*, 1989).

Retrieval studies have shown that severe corrosive attack can occur in the crevices formed at the Morse tapers of hip implants used clinically (Collier *et al.*, 1992). Gilbert *et al.* reported that up to 35% of retrieved total hip implants showed signs of moderate-to-severe corrosive attack in the **Morse taper** junction of the femoral head and neck

Figure 2.32

Scanning electron micrograph of Morse taper cone showing burnishing, fretting, and intergranular attack.

connection (Gilbert *et al.*, 1993). Such corrosive degradation has been found in components comprising Ti-6Al-4V alloy stems and cobalt chrome alloy femoral heads as well as in systems using only cobalt chrome alloys. It is believed that the corrosion process is the combined result of stress, motion, and crevice geometry at the taper connection (Gilbert and Jacobs, 1997). In such cases, it is thought that the contact stresses in the implant result in continual breakdown and abrasion of the passivation film resulting in a more negative (anodic) surface potential. Further, the chemistry of the crevice solution is altered by the continual breakdown and re-passivation of the surface and can result in a local loss of oxygen in the crevice and an accompanying decrease in pH (Gilbert and Jacobs, 1997). Thus, conditions where stresses are unusually high due to poor taper fit or geometric misalignment in modular components can result in crevice corrosion of the junction. This clinical condition may result in corrosive attack in the taper region and premature fracture of the component.

Figure 2.32 shows typical corrosion processes that can occur in a modular total hip replacement. This figure shows burnishing and fretting at the Morse taper junction and concomitant intergranular attack of the CoCr alloy in the stem. This implant utilized a 12 mm extension of the modular neck (known as a skirt) that enabled the surgeon to match the patient anatomy. However, this added neck length results in greater bending stresses and it is believed it was this geometric fracture that set up the initial conditions of crevice corrosion. Upon retrieval, the surgeon noted a large amount of white fluid with black particulate in the hip joint. The surgeon noted that there was a substantial amount of corrosion at the Morse taper and that it had a burnished appearance. The geometric gap facilitated by the extended neck stresses resulted in localized fretting and burnishing and intergranular attack of the cobalt alloy. The resulting particular debris resulted in biological failure of the implant after only six months *in vivo*.

2.12 Summary

Metals and alloys are crystalline materials that offer excellent mechanical properties such as toughness, strength, and ductility. These properties are intrinsically linked to the structure of the material. The elastic modulus is proportional to the density and strength of the metallic bonds. The theoretical strength of a metal is directly proportional to the elastic modulus but is rarely obtained due to inherent defects such as vacancies, dislocations, and grain boundaries. Paradoxically these same defects are used as strengthening mechanisms in metals. The yield strength is increased as grain boundary density and dislocation density increase. A limitation of metals is their susceptibility to corrosion. This restraint must be considered when designing implants as corrosion can degrade biocompatibility and cause premature failure of devices. Metals and alloys are commonly employed as fracture fixation devices, components in total joint replacements, guide wires, and stents.

2.13 Problems for consideration

Checking for understanding

1. Describe and schematically illustrate the strengthening mechanisms utilized in metals alloys. Why do some metal strains harden and others do not?
2. Discuss three alloys used in medical devices and an appropriate application for each.
3. What are good design philosophies to minimize corrosion at taper junctions in orthopedic implants? Why is crevice corrosion so problematic for modular implants?

For further exploration

4. Identify the likely corrosion mechanisms at the indicated locations in each of these implant systems: (a) hip stem made of titanium coupled with a cobalt chromium head, (b) Nitinol stent, (c) fracture fixation plate made of stainless steel.
5. A temporary fracture fixation implant made of stainless steel was utilized with cobalt-chromium screws. The implant failed catastrophically due to a combination of environmental and mechanical factors. Speculate on what these were, and provide schematic illustrations diagrams to support your descriptions.

2.14 References

Agins, H.J., Alcock, N.W., Bansal, M., Salvati, E.A., Wilson, P.D., Jr., Pellicci, P.M., and Bullough, P.G. (1988). Metallic wear in failed titanium-alloy total hip replacements. A histological and quantitative analysis. *Journal of Bone and Joint Surgery*, 70-A, 347–56.

Aramany, M.A. (1971). A history of prosthetic management of cleft palate. *Cleft Palate Journal* 8, 415–30.

Beddoes, A., Bibby, M.J., Bibby, M., and Biddy, M.J. (1999). *Principles of Metal Manufacturing Processes*. New York: Elsevier.

Black, J., Skipor, A., Jacobs, J., Urban, R.M., and Galante, J.O. (1989). Release of metal ions from titanium-base alloy total hip replacement prostheses. *Transactions of the Orthopedics Research Society*, 14, 50.

Collier, J.P., Surprenant, V.A., Jensen, R.E., Mayor, M.B., and Surprenant, H.P. (1992). Corrosion between the components of modular femoral hip prostheses. *Journal of Bone and Joint Surgery*, 74B(4), 511–17.

Davies, J., and Baldan, N. (1997). Scanning electron microscopy of the bone-bioactive implant interface. *Journal of Biomedical Materials Research*, 36, 429–40.

Fontana, G. (1967). *Corrosion Engineering*. New York: McGraw Hill.

Gilbert, J.L., Buckley, C.A., and Jacobs, J.J. (1993). In vivo corrosion of modular hip prosthesis components in mixed and similar metal combinations. The effect of crevice, stress, motion and alloy coupling. *Journal of Biomedical Materials Research*, 27, 1533–44.

Gilbert, J.L., and Jacobs, J.J. (1997). The mechanical and electrochemical processes associated with taper fretting crevice corrosion: a review. In *Modularity of Orthopedic Implants*, ed. D. E. Marlowe, J.E. Parr, and M.B. Mayor. West Conshohocken, PA: American Society for Testing and Materials, Special Technical Publication 1301, pp. 45–59.

Hertzberg, R.W. (1989). *Deformation and Fracture of Engineering Materials*. New York: John Wiley & Sons.

Hirthe, J.P., and Lothe, J. (1982). *Theory of Dislocations*. New York: John Wiley & Sons.

Hull, D., and Bacon, D.J. (2001). *Introduction to Dislocations*. London: Butterworth-Heinemann.

Hume-Rothery, W. (1960). *Atomic Theory for Students of Metallurgy*. London: Institute of Metals.

Jacobs, J.J., Skipor, A.K., Doom, P.F., Campbell, P., Schmalzried, T.P., Black, J., and Amstutz, H.C. (1996). Cobalt and chromium concentrations in patients with metal on metal total hip replacements. *Clinical Orthopedics*, 329(Suppl.), 256–63.

Jacobs, J., Gilbert, J.L., and Urban, R.M. (1998). Current concepts review – Corrosion of Metal Orthopaedic Implants. *Journal of Bone and Joint Surgery (American)*, 80, 268–82.

Leventhal, G.B. (1951). Titanium, A Metal for Surgery. *Journal of Bone and Joint Surgery (American)*, 33, 473–74.

Park, J.B., and Bronzino, J.D. (2003). *Biomaterials: Principals and Applications*. New York: CRC Press.

Pelton, A.R., Duerig, T., and Stockel, J. (2004). A guide to shape memory and superelasticity in Nitinol medical devices. *Minimally Invasive Therapy and Applied Technology*, 13(4), 218–21.

Pelton, A.R., Russell, S.M., and DiCello, J. (2003). The physical metallurgy of Nitinol for medical applications. *Journal of Minerals, Metals and Materials*, 55(5), 33–37.

Read, W.T. (1953). *Dislocations in Crystals*. New York: McGraw-Hill.

Richman, M.H. (1974). *An Introduction to the Science of Metals*. Lexington, MA: Ginn Publishing.

Suresh, S. (1998). *Fatigue of Materials*, 2d edn. Cambridge: Cambridge University Press.

Weertman, J., and Weertman, J.R. (1992). *Elementary Dislocation Theory*. Oxford: Oxford University Press.

Williams, D.F. (1990). *Concise Encyclopedia of Medical and Dental Materials*. Oxford: Pergamon Press.

Williams, D.F., Roaf, R., and Maisels, D.O. (1973). *Implants in Surgery*. London: W.B. Saunders.

3 Ceramics

Inquiry

How would you create a bone graft for the supporting structure of a dental implant if the patient has periodontal disease accompanied by bone loss in the jaw and also suffers from osteoporosis (porous bone)?

The above inquiry presents a realistic challenge that one might face in the field of dentistry. A dental implant must be integrated into the underlying bone in order to have the necessary structural support needed for function. A patient who has periodontal disease accompanied by bone loss in the jaw and who also suffers from osteoporosis (porous bone) is unlikely to have bone tissue that can be used as a structural graft. Traditionally bone grafts are obtained from elsewhere in the patient's body such as ribs, pelvis, or skull. If the patient has had severe trauma or other disease such as osteoporosis, there may be insufficient quantity or quality of bone available. In such cases, viable alternatives such as coral may be necessary. The case study presented at the end of this chapter examines coral as a bone substitute.

3.1 Historical perspective and overview

Structural materials, such as **plaster of Paris** (loosely categorized as a resorbable ceramic), were first employed as a bone substitute in the late 1800s (Pelter, 1959). Even today this material is used as a structural support for fractured bones. Plaster of Paris, which is made of calcium sulphate hemihydrate ($CaSO_4 \cdot H_2O$), is also used in radiotherapy to make immobilization casts for patients and in dentistry for modeling of oral tissues. Only within the last 50 years have medical ceramics been incorporated into load-bearing applications. The use of ceramics in structural applications has been restricted due to their inherent susceptibility to fracture and sensitivity to flaws (Kingery, 1976). For this reason, ceramics have had limited use in medical implants in which any tensile stresses were expected. Modern manufacturing methods, including sintering and high-pressure compaction, have enabled these materials to be produced with fewer defects and better mechanical properties. As a result of such technological development, ceramics are now utilized in a variety of applications including heart valves (More and Silver, 1990), bone substitutes (Bajpai, 1990), dental implants (Hulbert *et al.*, 1987), femoral heads (Oonishi, 1992), middle ear ossicles (Grote, 1987), and bone screws (Zimmermann *et al.*, 1991). Additionally, ceramics can be utilized in inert, active, and resorbable forms.

Inert biomaterials are broadly defined as nonreactive and do not cause tissue response while implanted in the body. The basic inert ceramics include alumina (Hench, 1991), zirconia (Barinov and Baschenko, 1992), diamond-like carbon and pyrolytic carbon (Bokros, 1972), and dense hydroxyapatite (Whitehead *et al.*, 1993). Active materials, on the other hand, encourage interaction with surrounding tissue and may stimulate new tissue growth at the implant-tissue interface. Bioactive ceramics include amorphous silicate glasses such as Bioglass (Ducheyne, 1985) and polycrystalline ceramic glasses such as Ceravital (Hench, 1993). Resorbable or **biodegradable** materials are typically incorporated into the adjacent tissue and may be fully or partially metabolized by the body. The biodegradable or resorbable ceramics include calcium phosphate (Parks and Lake, 1992), **hydroxyapatite** (Bajpai and Fuchs, 1985), and **corals** (Wolford, 1987). The development of ceramics as implant materials was motivated by the need for specific mechanical properties such as compressive strength and wear resistance as well as tailored bioreactivity. One of the greatest challenges for designing with ceramic materials for use in load-bearing implants is their susceptibility to fracture. The improvement of processing as well as the improvement of fracture toughness with various toughening mechanisms has enabled ceramic materials to be utilized in load-bearing devices such as the femoral head in total hip replacements and in safety-critical components such as synthetic heart valves.

3.2 Learning objectives

This chapter provides an overview of ceramics used in medical implants. The key objectives of this chapter are that, at the completion of the chapter the student should be able to:

1. describe the bonding mechanisms and basic structure of ceramic materials
2. discuss the meaning of theoretical cohesive strength
3. calculate a critical flaw size for a ceramic material
4. portray the basic toughening mechanisms used in structural ceramics
5. delineate common processing techniques for ceramic biomaterials
6. explain the range of bioreactivity in the basic ceramic biomaterials
7. evaluate appropriate types of ceramics for specific biomedical applications

3.3 Bonding and crystal structure

Ceramics, in general, contain both metal and non-metal constituents and have a range of bonding from highly covalent to highly ionic. Covalent bonds are achieved when atoms share outer valence electrons (Figure 1.7) and this tendency is greatest when the atoms have similar **electronegativity**. Covalent bonds have fixed coordination with bond strengths that help facilitate high melting temperatures in covalently bonded ceramic materials. A high degree of covalent bonding is found in diamond and silicon structures. Ionic bonds are created by the donation and acceptance of one or more valence electrons. **Cations** donate electrons and carry a positive charge, while **anions** accept electrons and carry a negative charge. Ionic bonds are strong, have preferential bond directions, and enable high melting points in ceramic structures. In general, ceramic materials comprising ionic bonding utilize the packing of anions around cations to maintain charge neutrality in crystalline structures. That is, each ion has a certain number of nearest neighbors of opposite charge to balance charges between anions and cations. Ionic bonds are prevalent in many ceramics and include crystalline ceramics such as alumina (Al_2O_3) and zirconia (ZrO_2) as well as various ceramic glasses. The primary engineering attributes of ceramics used in medical implants are compressive strength, non-reactive surface properties, wear resistance, and customizable **bioreactivity**.

Ceramics have complex crystal structures that generally utilize a number of octahedral and tetrahedral interstitial sites to maintain charge neutrality. There are several factors that contribute to the basic lattice structure of ceramic materials. For example, the bonding type, such as covalent or ionic, can impose restrictions on the geometric packing of nearest neighbors. Furthermore, the relative size of atoms dictates the structural arrangement needed to maximize attractive forces. This is especially true for ceramics that utilize ionic bonding. Figure 3.1 provides an example of some of the common crystal structures found in structural ceramics including diamond, alumina, and sodium chloride.

There is a strong resemblance between crystalline ceramics and metal structures that utilize interstitial sites. For example, the sodium chloride (NaCl) structure can be thought of as an FCC lattice of sodium atoms with all of its octahedral sites filled with chlorine atoms. Similarly, the diamond cubic lattice utilizes the FCC lattice with all of its tetrahedral sites filled. In comparison, the structure of a silica glass is less ordered, and termed amorphous, but still maintains charge neutrality between cations and anions (Figure 3.2).

Pauling's rules of chemistry are useful in understanding how atoms are arranged in space to maximize packing efficiency and to maintain charge neutrality in ceramics (Pauling, 1960). These rules state:

(i) *A coordinated polyhedron of anions is formed about each cation with the cation-anion distance equaling the sum of their characteristic packing radii and the coordination polyhedron being determined by the radius ratio.* In this model, anions and cations are treated as hard spheres that are packed together closely such that the spheres are touching. The number of large spheres (anions) that can fit around a small sphere (cation) is called the anion coordination number and depends on the relative sizes of the small and large sphere (the radius ratio).

3.3 Bonding and crystal structure

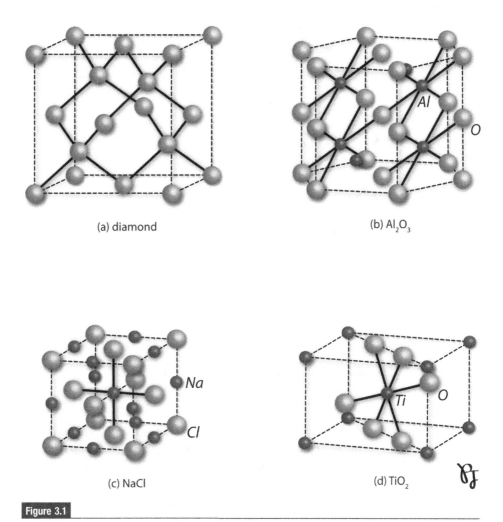

(a) diamond (b) Al$_2$O$_3$

(c) NaCl (d) TiO$_2$

Figure 3.1

Schematic of common crystal structures found in structural ceramics including diamond, alumina, and sodium chloride.

(ii) *In a stable crystal structure, the total strength of the valence bonds that reach an anion from all the neighboring cations is equal to the charge of the anion. The valence charge divided by the coordination number, called the electrostatic valency, is a measure of the strength of the bonds.* This means that a cation charge is balanced by neighboring anions. For example, in NaCl, each Cl atom (−1) is balanced by six sodium neighbors that each contributes 1/6 of its charge (+1).

(iii) *The existence of edges, and particularly of faces, common to two anion polyhedra in a coordinated structure decreases its stability. This effect is large for cations with high valency and small coordination number, and is especially large when the radius ratio approaches the lower limit of stability of the polyhedron.* This

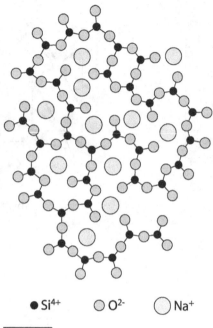

Figure 3.2

Schematic illustration of the organization of cations and anions in silica glass.

means that when a crystalline material is highly ionic in character, the structure of the crystal lattice minimizes its energy by maximizing the distance between cations. This optimization is achieved if the coordination polyhedra share only corners, rather than edges or faces.

(iv) *In a crystal containing different cations, those of high valency and small coordination number tend not to share polyhedral elements with each other.* This means that if a structure contains a highly charged cation that has a small number of anions around it, then the bond strengths between this cation and anions will be strong and the anion will be unlikely to have more than one bond to that same cation. *The number of essentially different kinds of constituents in a crystal tends to be small, because characteristically, there are only a few types of contrasting cation and anion sites.* This is found in many ceramic systems that have only two constituents such as Al_2O_3, Si_3N_4, and SiC.

Pauling's rules are highly useful in understanding how ceramic structures organize themselves. These simple rules pertaining primarily to charge neutrality (valency balance) and geometric size differences in atoms predict where cations and anions can pack in space. Thus, while ceramics may have complex chemistry and crystal structures, the

3.4 Mechanical behavior of ceramics

3.4.1 Theoretical cohesive strength

The **theoretical cohesive strength** is a measure of the stress needed to cause bond rupture in close-packed planes. The applied stress needed to overcome the bond strength of the crystalline lattice is termed the theoretical cohesive strength, σ_{th}. Like the theoretical shear strength for a metal, the stress that represents the theoretical cohesive strength is based on a sinusoidal stress-displacement curve where the half period is given as $\lambda/2$ but loading is normal to the plane rather than acting in shear (Figure 3.3).

For small displacements, the theoretical strength takes the form:

$$\sigma_{th} = \frac{E\lambda}{2\pi x_o} \quad (3.1)$$

where x_o is the equilibrium atomic separation. Assuming that x_o is on the order of $\lambda/2$, the theoretical value of strength is predicted to be:

$$\sigma_{th} = \frac{E}{\pi}. \quad (3.2)$$

The above equations predict that the theoretical cohesive strength should be very high given the typical range of elastic modulus values (28 GPa– 1000 GPa) found in ceramics. However, very few ceramics come close to obtaining such values experimentally. In particular, the tensile strength, $\sigma_{f,tensile}$, of most ceramic structures is quite low and very rarely exceeds a few hundred MPa; while the compressive strength, $\sigma_{f,compressive}$, can be an order of magnitude greater and may have values reaching several thousand MPa (Table 3.1). The reason for the enormous discrepancy between the theoretical strength and the actual tensile strength is the presence of imperfections. The presence of flaws in ceramics facilitates fracture at much lower tensile stresses due to the inherently low fracture toughness (K_{IC}) of these materials (see discussion below). Table 3.1 shows a comparison of the theoretical strength, tensile strength, compressive strength, and elastic modulus for alumina, zirconia, and pyrolytic carbon.

3.4.2 Fracture behavior and toughening mechanisms

The structure and mechanical behavior of ceramics can be drastically altered by the presence of defects. These imperfections, like those described for metals, may exist at a number of length scales and may include vacancies, impurity atoms, dislocations, grain boundaries, or cracks. While dislocations can exist in ceramic structures,

Table 3.1 Comparison of mechanical properties of structural medical ceramics

Ceramic	$\sigma_{th} = E/\pi$ (GPa)	E (GPa)	$\sigma_{f,tensile}$ (MPa)	$\sigma_{f,compressive}$ (MPa)	K_{IC} (MPa√m)
Alumina	127	380–400	260	4500	3–5
Zirconia	66	190–207	248	2500	8–10
Pyrolytic carbon	9.5	28–30	207	500	0.5–1

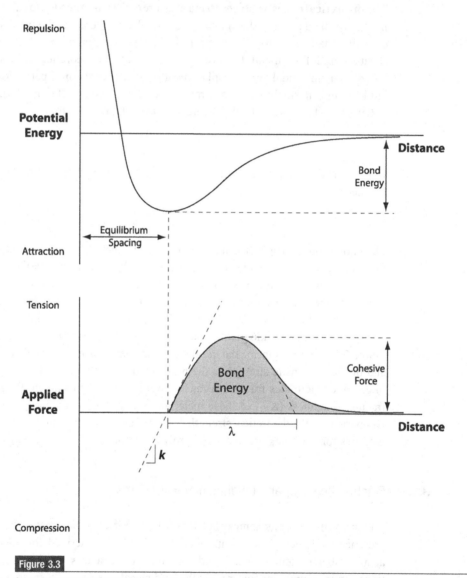

Figure 3.3
Schematic illustration of loading of crystal atoms that gives rise to the theoretical strength of a ceramic.

3.4 Mechanical behavior of ceramics

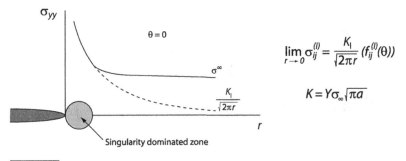

Figure 3.4

Illustration showing the stress intensity factor, K, and the concomitant stresses that develop at the crack tip.

the required energy to move a dislocation is quite large due to nature of the bonding and the need to maintain charge neutrality in the lattice (Pauling's rules). Therefore, plastic deformation rarely occurs in ceramic structures at low temperatures because dislocations are so difficult to move through the complex structures. Ceramics are highly sensitive to tensile normal stresses that facilitate opening modes of crack growth and provide the energy necessary to cleave atomic bonds and create fracture surfaces. Consequently, the onset of failure in ceramics is usually premised on a critical normal stress or critical flaw size rather than yielding criteria based on a shear mechanism.

When a flaw or crack in the ceramic is subjected to a tensile stress, the local stress field at the crack tip is greatly enhanced. The stresses at the crack tip can be calculated using linear elastic fracture mechanics (the details are presented in Chapter 9, *Fracture Mechanics*). The peak intensity of these stresses occurs at the crack tip. Further, near-tip stresses are maximized under opening (tensile) modes of loading; this is termed Mode I loading. In contrast, it is believed that under compression loading the flaws close and do not give rise to a stress intensity. When the critical combination of stress and flaw size occurs, the conditions for fracture can be readily met in ceramic structures. The parameter that quantifies these effects is known as the stress intensity factor and is schematically illustrated in Figure 3.4. The stress intensity factor couples the effects of stress and flaw size in the material, and is given as $K = Y\sigma\sqrt{\pi a}$ where Y is the geometric factor for the flaw and component, σ is the applied stress, and a is the flaw size. The geometric factor, Y, is dictated by the geometry of the flaw and its length; it is often interchangeable with the forms $F(\alpha), f_{ij}(\theta)$, or $F(a/W)$ where a is the crack length and W is the uncracked ligament length of the specimen or component. The critical value of the stress intensity for a material is known as the **fracture toughness**, K_{IC}. Ceramics as a class of materials have low values of fracture toughness in comparison to metal systems that can undergo plastic deformation through shear mechanisms.

The fracture toughness of a material is characterized experimentally using notched specimens. This fracture toughness can be used to determine the **critical flaw size**

Figure 3.5
Bileaflet heart valve design.

in an implant that is subjected to known physiological stresses. The critical flaw size determined from fracture toughness is given as:

$$a_c = \frac{K_{IC}^2}{\sigma^2 F(\alpha)^2 \pi} \tag{3.3}$$

and will be derived fully in Chapter 9 (*Fracture Mechanics*). The critical crack length is often used as an inspection parameter for quality control in safety-critical medical devices (see Example 3.1). Care must be taken to evaluate flaws in ceramic components to ensure safe performance of the device, especially when the implant is utilized in safety-critical applications such as heart valves (Figure 3.5).

Example 3.1 **Critical crack length in a ceramic heart valve**

The critical flaw size of a component is a useful parameter in medical device design. Consider a bileaflet heart valve (Figure 3.5) that is known to have inherent stress concentrations. The form of stress intensity factor for the flaw is given as $K = \sigma F(\alpha)\sqrt{\pi a}$ and the expected *in vivo* stresses are on the order of 100 MPa. Consider a bileaflet valve made of pyrolytic carbon that has a fracture toughness of 0.5 MPa$\sqrt{m}$. What is the critical crack length? How does this compare to a heart valve made of zirconia ceramic that has a fracture toughness of 10 MPa$\sqrt{m}$?

Solution The form of the critical crack length is given as:

$$a_c = \frac{K_{IC}^2}{\sigma^2 F(\alpha)^2 \pi}$$

Where K_{IC} is the fracture toughness, $F(\alpha)$ is the geometric factor for the flaw, and σ is the physiological stress. Assuming a geometric factor for the flaw to be on the order of

3.4 Mechanical behavior of ceramics

1, and substituting $K_{IC} = 0.5$ MPa$\sqrt{m}$ and $\sigma = 100$ MPa, we find the critical flaw size for pyrolytic carbon:

$$a_c = \frac{(0.5 \text{ MPa}\sqrt{m})^2}{(100 \text{ MPa})^2 \pi} = 7.96 \times 10^{-6} \text{m}.$$

For comparison, the critical flaw size for a toughened zirconia ceramic with a fracture toughness of 10 MPa$\sqrt{m}$ is:

$$a_c = \frac{(10 \text{ MPa}\sqrt{m})^2}{(100 \text{ MPa})^2 \pi} = 3.2 \times 10^{-3} \text{m}.$$

This simple example indicates how the fracture toughness of a material influences the critical flaw size for all other factors being equivalent. An interesting observation is that due to the square root relationship between fracture toughness and crack length there is a 20-fold increase in the fracture toughness of zirconia as compared to that of pyrolytic carbon, yet there is a 400-fold difference in the critical crack length. Also of interest is that most bileaflet valve designs utilize the pyrolytic carbon owing to its exceptional hemocompatibility, and this further supports the multifactorial aspects of medical device design.

There are several mechanisms or processes that can be used to toughen ceramics, and these are schematically illustrated in Figure 3.6. These processes are theoretically

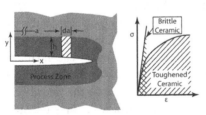

(a) Process zone formed by growing crack

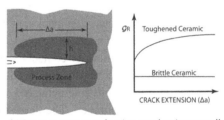

(b) Process zone toughening mechanism usually results in a rising R curve

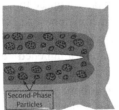

(c) Nonlinear deformation of second-phase particles

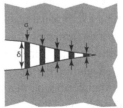

(d) Fiber bridging mechanism for ceramic toughening

Figure 3.6

Schematic illustration of toughening mechanisms utilized in ceramics. (After Ritchie *et al.*, 2000.)

Table 3.2 Toughening mechanisms in ceramics

Mechanism	Structure-property relationship	Systems
Phase transformation	Phase transformation ahead of the crack tip results in a volume change that is constrained by surrounding material and results in compressive stresses that retard crack growth.	Stabilized Zirconia (ZrO_2)
Process zone microcracking	Use of numerous cracks to dissipate energy in front of the crack tip as energy is used to create surfaces	Alumina (Al_2O_3)
Second-phase additions	Second-phase particles capable of deformation are responsible for the energy dissipation	Hydroxyapatite composites
Crack bridging	Advancing crack leaves second-phase particles or fibers intact behind the crack tip	Bone

developed in Chapter 9; however, a general description is provided here and summarized in Table 3.2.

Examples of common toughening mechanisms utilized in ceramic structures include controlled **microcracking**, **phase transformations**, **second-phase addition**, and **crack bridging**. These schemes are utilized by many structural ceramics used in medical implants. Alumina is sometimes employed in bearing surfaces and is used as the femoral head of a total hip replacement. In such an application it is critical for the component to withstand high contact stresses and to be resistant to both wear and fracture. The toughening mechanism employed in alumina is microcracking. In this toughening mechanism, numerous microcracks are created within a process zone ahead of a crack or stress concentration. The development of each microcrack requires the utilization of surface energy and the creation of multiple microcracks requires a substantial increase in the amount of work required for crack advance. Zirconia is also commonly used as a bearing material in total hip replacements and incorporates a phase toughening mechanism whereby a tetragonal phase is converted to a monoclinic phase under elevated stresses. Phase transformation toughening employs a stress-induced transformation of crystal structure to retard crack growth. In a transformation toughened ceramic, an applied stress magnified by the stress concentration such as that found at a crack tip causes a phase transformation that includes an increase in volume in the deformed phase. This volume change is constrained by the surrounding material that has not transformed, and results in associated compressive stresses at the crack tip that retard the growth of the flaw and extend the lifetime of the component. Second-phase additions to ceramics can also provide a toughening mechanism. In this case, the second phase becomes a source for energy dissipation ahead of the crack or stress concentration and increases

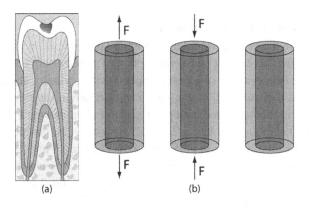

Figure 3.7

(a) Schematic of a tooth with a cavity in the upper enamel, and (b) an illustration a cylindrical plug of filling material within the enamel that is subjected to a tensile force, compressive force, and no force.

the stress required for crack advance. This mechanism can be found in **hydroxyapatite** composites used as bone substitutes. Crack bridging utilizes a mechanism by which the crack has traction behind its crack tip and this is facilitated by the presence of microstructural elements such as grains, second phases, or fibers that span the crack wake. This mechanism is found in bone and is highly successful in toughening the tissue structure.

3.4.3 Thermal stresses

Most implants are utilized in the body such that the temperature remains near constant at approximately 37°C. An exception to this, however, is the thermal variation experienced by dental implants. Thermal stresses or strains arise in a material due to thermal expansion as the material undergoes a temperature change. The basic one-dimensional form of the thermal strains associated with a temperature differential is given as:

$$\varepsilon = \alpha \Delta T \quad (3.4)$$

and similarly the thermal stress is given as:

$$\sigma = E\alpha \Delta T \quad (3.5)$$

where α is the **thermal expansion coefficient**. The thermal stresses depend on both the thermal expansion coefficient, which is a material property that gives a normalized measure of atomic displacement owing to a change in temperature, and the elastic modulus of the material. The thermal expansion difference between an implant and the adjacent tissue can give rise to thermal stresses sufficient to result in premature failure of a component if not appropriately designed. The interplay between thermal expansion coefficient and elastic modulus is important in dental implants that are used to fill cavities in the tooth enamel (Figure 3.7; Example 3.2).

Example 3.2 **Thermal stresses in dental implants**

Dental implants can readily experience a temperature differential on the order of 50°C in normal daily function (consider the consumption of a cold or hot beverage). Consider a very simple example of a dental implant that is cylindrical in shape (Figure 3.7). If this implant is 1.0 mm in radius and 4.0 mm in length, what is the expected volumetric expansion for an implant made of alumina with a thermal expansion coefficient, $\alpha_{\text{alumina}} = 8.5 \times 10^{-6}$°C? How does this compare to an implant made with a resin with a coefficient of thermal expansion equal to $\alpha_{\text{resin}} = 85 \times 10^{-6}$°C? Assume a temperature change of 50°C. The thermal expansion coefficient for enamel is $\alpha_{\text{enamel}} = 8.0 \times 10^{-6}$°C. The modulus for the ceramic is 400 GPa and for the resin is 2.5 GPa.

Solution The one-dimensional strain is given in Equation (3.4). If the implant is assumed to deform isotropically (the same in all dimensions) then the volumetric thermal expansion can be estimated by first approximation as:

$$\alpha_v = 3\alpha = \frac{\Delta V}{V_o \Delta T}.$$

Accordingly, the change in volume is found as:

$$\Delta V_{\text{Al}_2\text{O}_3} = V_o 3 (\alpha_{\text{Al}_2\text{O}_3}) \Delta T = \pi (1 \text{ mm})^2 (4 \text{ mm})^3 (8.5 \times 10^{-6} - 8 \times 10^{-6})(50)$$
$$= 9.42 \times 10^{-4} \text{ mm}^3$$

and

$$\Delta V_{\text{Resin}} = V_o 3 (\alpha_{\text{Resin}}) \Delta T = \pi (1 \text{ mm})^2 (4 \text{ mm})^3 (85 \times 10^{-6} - 8 \times 10^{-6})(50)$$
$$= 0.15 \text{ mm}^3.$$

Correspondingly, the one-dimensional force owing to this thermal expansion is approximated as:

$$F = EA(\Delta \varepsilon) = E(\Delta T)(\Delta \alpha) \pi D h$$
$$F_{\text{Al}_2\text{O}_3} = 251 \text{ N}$$
$$F_{\text{Resin}} = 250 \text{ N}$$

An interesting outcome is that the force is nearly the same even though the volume expansion for the ceramic is three orders of magnitude lower than the resin. This is due to the large difference in elastic modulus between the resin and the ceramic.

3.5 Processing of ceramics

One challenge in designing with ceramic materials in load-bearing devices is control of manufacturing. Any residual stresses, flaws, or stress concentrations can lead to a

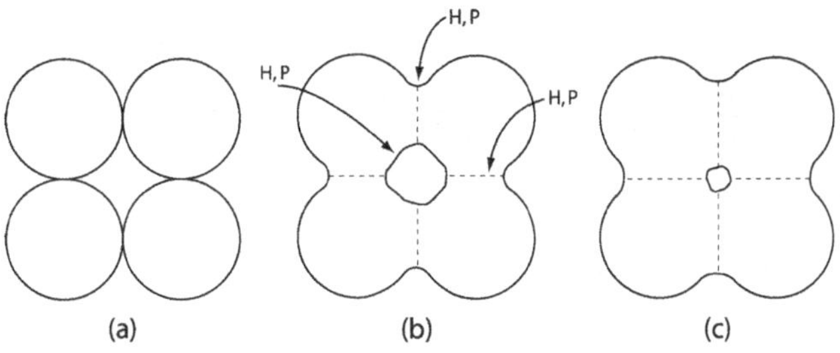

3.5 Processing of ceramics

Figure 3.8

Illustration of the sintering process in which (a) initial particles are (b) subjected to heat (H) and pressure (P) as denoted by the arrows in order to activate diffusion processes, and (c) densification into a solid structure.

complication with fracture owing to the low fracture toughness of this material. The most common method used for making compacted ceramic structures is **sintering** (Figure 3.8). In this method, powders of a material are placed in a mold and heated below the melting temperature until all the particles adhere to one another.

Prior to heating, the packed powder is termed a green body. This green body form is subsequently heated to utilize diffusion processes and to solidify the structure. The sintering mechanism is driven by surface tension, which drives the reduction in pore space. Low porosity in these structures can usually be achieved with appropriate use of pressure, temperature, and time. This method enables very complex geometries to be generated and, while it is similar in concept to direct compression molding methods in metals, it is far more adaptable as numerous modifications are possible. This method is also used in **powder metallurgy** where at least one element of the alloy exists in a liquid state (termed liquid phase sintering), and in polymer processing (termed compression molding) for highly viscous polymers or high-temperature polymers. Similarly, **injection molding** and **emulsion** methods can be used for ceramics used for hard tissue replacement. **Hydrophilic** ceramics can be molded in the hydrated "clay" form and then fired to achieve its final, hardened state. **Rapid prototyping** can also be used to develop complex geometries composed of ceramics. This method utilizes basic powder metallurgy methods coupled with a low-pressure injection molding process or an **inkjet method** in which the structure is built layer by layer to achieve the final geometry. **Glasses**, also known as **amorphous** ceramics, are generally formed using molding or blowing methods, while the ceramic is in molten form. Extrusion or drawing from the melt is used to create glass fibers.

Pyrolytic carbon is typically in the form of a coating and its processing is different than that employed for most other materials. Fine-grained isotropic carbon is typically used as the base substrate for the pyrolytic carbon coating. The process of deposition utilizes severe temperature conditions and for this reason the thermal expansion coefficient of the underlying material and the coating must match closely, otherwise thermal stresses will result in flaws and susceptibility to fracture. The coating is performed in a vaporized chamber at elevated temperatures much like processes employed in chemical

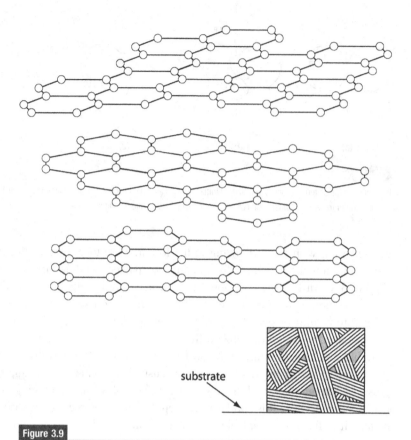

Figure 3.9

Structural arrangement of carbon in pyrolytic carbon.

vapor deposition of diamond-like carbon (DLC) coatings. Pyrolytic carbon belongs to the family of **turbostratic carbon** (crystalline form of carbon where the basal planes have slipped relative to each other resulting in a non-ideal planar spacing) that has similar structure to graphite (Figure 3.9). In graphite, carbon atoms that are covalently bonded in hexagonal arrays are layered and held together by weak interlayer bonding. This results in decreased shear strength and also facilitates easy layer transfer (lubricity). In pyrolytic carbon, the layers are disordered and distorted, which results in a stronger system with enhanced durability.

Resorbable (biodegradable) ceramics are based primarily on calcium phosphate structures and utilize wet precipitation techniques to yield hydroxyapatite structures that are readily dissolved and remodeled by the body. Bioactive ceramics rely on porosity to aid in tissue integration. For this reason, methods of processing are employed that facilitate porosity and include foaming methods, gel casting, and slip casting (Park and Bronzino, 2003).

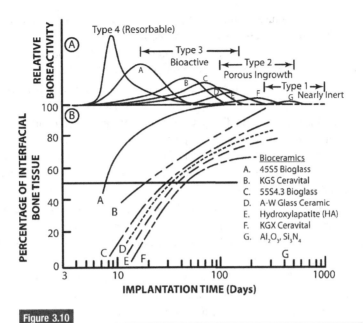

Figure 3.10

Schematic illustration showing the full spectrum of (A) bioreactivity and (B) interfacial bone tissue growth (adapted from Ratner et al., 1996).

3.6 Ceramics in medical implants

The properties of ceramics and glasses make them ideal candidates for many load-bearing applications where hardness, wear resistance, and corrosion resistance are needed but where tensile stresses are mitigated. As with metals, the biocompatibility of ceramics will dictate their suitability for medical devices. One benefit of this class of materials is that the **bioreactivity** can be tailored from inert to fully resorbable depending on the system chosen. The drawback of using ceramics is that the fracture resistance is inherently low and thus designing for appropriate stress levels is critical for the safe design of the implant. Medical ceramics are used in a wide range of medical implants including restorative dentistry, orthopedics, and cardiovascular devices.

Ceramics are generally classified as **inert**, **bioactive**, or **resorbable** (Figure 3.10(A)). Inert ceramics facilitate long-term structural stability and biocompatibility and are used in applications such as acetabular liners, heart valves, dental crowns, or reconstruction in the eye. Bioactive ceramics are used to facilitate tissue ingrowth and are often in the form of coatings used with metal implants (Figure 3.10(B)). Ceramics that are bioresorbable provide temporary structure and are utilized in bone repair, for filling tissue space associated with tumor removal, device removal, and in drug delivery devices (Ratner et al., 1996). Figure 3.10 illustrates the spectrum of bioreactivity and tissue integration that is available in ceramic systems ranging from fully resorbable to inert.

Inert medical ceramics include alumina (Al_2O_3), zirconia (ZrO_2), and pyrolytic carbon. The mechanical properties of these ceramics are given in Table 3.1. Alumina (α-phase) is a polycrystalline ceramic material that utilizes hexagonal close packing of oxygen ions with two-thirds of the octahedral sites filled with aluminum ions. Alumina makes a good bearing material and is a suitable articulating surface in acetabular cups and femoral heads for total hip replacements. Zirconia also has exceptional wear resistance and finds use as the femoral head in total hip replacements. Pyrolytic carbon is part of the carbon family with properties residing between those of graphite and diamond, but it is commonly classified as an inert ceramic. Pyrolytic carbon is highly hemocompatible and is successfully utilized in heart valves, although it has a low value fracture toughness (Ritchie, 1996).

Biodegradable ceramics are commonly used as bone substitutes and are capable of resorption or controlled degradation in the body. Factors that determine the degradation rate include chemical susceptibility, degree of crystallinity, amount of media available, and the surface area to volume ratio of the material. Biodegradable ceramics are generally composed of calcium phosphate or some derivate of this structure and can be crystallized into salts of hydroxyapatite. Hydroxyapatite is chemically similar to the mineral component of bones and hard tissues. This mineral supports bone ingrowth (**osseointegration**) when used in orthopedic, dental, and maxillofacial applications. The chemical nature of hydroxyapatite lends itself to substitution, meaning that it is not uncommon for non-stoichiometric hydroxyapatites to exist. The most common substitutions involve carbonate, fluoride, and chloride substitutions for hydroxyl groups, while defects can also exist resulting in deficient hydroxyapatites. The apatite form of calcium phosphate is quite similar in structure to the mineral phase of bone and dental tissue that comprises hexagonal prisms of $[Ca_{10}(PO_4)_6(OH)_2]$ and this similarity of structure facilitates integration with mineral components of bone and teeth. The optimum Ca:P ratio is 10:6 but the structure can vary widely depending on process conditions. Accordingly, there is a large variation in mechanical properties with elastic modulus ranging from 4–117 GPa depending on Ca:P ratio and density (Parks and Lake, 1992). However, this range appropriately spans the moduli values expected for bone, dentin, and enamel. More importantly, this material offers exceptional biocompatibility and the ability to directly bond to hard tissues.

Another bioresorbable ceramic is coral, which exists in crystalline form with a primary constituent of calcium carbonate. This naturally porous structure closely matches the microstructure of bone and can be processed to serve as an effective bone substitute. The porosity of the coral structure dictates many of the mechanical properties with elastic modulus ranging from 8–100 GPa. Further, the calcium carbonate component gradually dissolves in the body, making this an ideal resorbable material for repair of bone tissue or defects (see Case Study).

Another category of medical ceramics is the bioactive ceramics that are typically used in the form of coatings and include bioactive glasses and glass ceramics. With time in the body these materials form a biologically active carbonated hydroxyapatite that enables a bond to form with mineralized tissues (Hench, 1991). Hench recognized

Table 3.3 Summary of medical ceramics and their applications

System	Structure	Bioreactivity	Applications
Alumina	Polycrystalline ceramic	Inert	Orthopedic bearings; dental implants; maxillofacial prosthetics
Zirconia	Polycrystalline ceramic	Inert	Orthopedic bearings, maxillofacial prosthetics
Pyrolytic carbon	Polycrystalline ceramic	Inert	Heart valves
Carbonated hydroxyapatite	Polycrystalline ceramic	Bioactive	Coatings for tissue ingrowth
Silica glass	Glass	Bioactive	Dental implants; percutaneous tissue devices; periodontal; orthopedic fixation
Coral	Polycrystalline ceramic	Biodegradable	Bone grafts; bone substitutes
Calcium phosphate	Polycrystalline ceramic	Biodegradable	Bone grafts; bone substitutes

that specific compositional ranges (those containing less than 60% SiO_2, with high content of CaO and Na_2O, and high ratio of CaO/P_2O_5) facilitated good interfacial bond strength with bone. This integration with surrounding tissues provides good fixation to the implants that utilize such coatings. Bioglasses have a composition that comprises 45% SiO_2, 14.5–24.5% CaO, 24.5% Na_2O, 6% P_2O_5, and balance is CaF_2. Ceravital has a composition that comprises 38–46% SiO_2, 20–33% CaO, 14–26% $Ca(PO_3)_2$, 3–5% MgO, 4–5% Na_2O, 6% P_2O_5, and balance is Al_2O_3 and TiO_2. These materials are commonly employed as coatings in orthopedic and dental implants. A summary of the most commonly used medical ceramics and their applications is given in Table 3.3.

3.7 Case study: the use of coral as a bone substitute

In the past few decades, sea **coral** has been studied as a bone substitute and has found use in surgical procedures requiring bone grafts (Demers *et al.*, 2002). The need for bone grafts is widespread; loss of bone may be due to disease, trauma, or congenital defects. For example, **periodontal disease** is often accompanied by depleted bone in the jaw and afflicts millions of people each year. Traditionally bone grafts are obtained from elsewhere in the patient's body, such as ribs, pelvis, or skull. If the patient has had severe trauma or other disease, there may be insufficient quantity or quality available. Taking bone from elsewhere in the body also usually requires an additional incision,

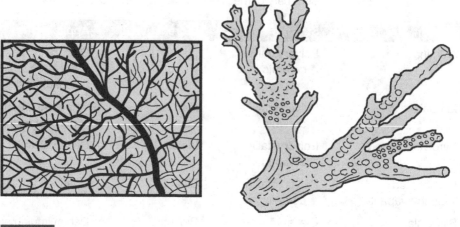

Figure 3.11
Basic structure of sea coral used as a bone substitute.

surgical procedure, or risk of infection. Moreover, while bone grafts from bone banks are available, there is associated risk with immunological complications. The porosity and architecture of any bone substitute is essential to its success, as these structural elements are key for sustained bone growth and osseointegration.

Coral consists of calcium carbonate and is a hard mineral deposit built up by **coral polyps**. Figure 3.11 shows the basic structure of sea coral. In its native form, the calcium carbonate is inappropriate for grafting as it is highly resorbable in the body. However, this structure can be converted with addition of phosphates, heat, and water to a **hydroxylapatite** (Miyazaki et al., 2009). This calcium phosphate structure is the major mineral component of mineralized tissue and teeth. The evolution in the body involves incursion of blood vessels, **vascularization**, resorption of coral by **osteoclasts**, **neoformation** of bone (**osteoblasts**), and remodeling of the bone architecture.

The physiology of successful bone grafting relies on three basic processes known as **osteogenesis, osteoinduction**, and **osteoconduction**. In osteogenesis, the cellular elements of the bone substitute synthesize new bone at the graft site. When new bone is generated by recruitment of bone-building cells known as osteoblasts from the host, the process is known as osteoinduction. This process is aided by the presence of growth factors such as bone morphogenic proteins (BMP) within the graft. Osteoconduction is a term used to describe integration of blood vessels and the creation of bone into the porous architecture.

Coral offers many attributes for bone grafting owing to its excellent osteoconductive properties. Its biological activity closely mirrors that of bone and facilitates gradual resorption of the coral substitute by bone remodeling cells known as osteoclasts and generation of new bone in its place with the activation of osteoblasts. The open pore architecture (150–500 μm) of coral enables blood cells and bone marrow cells to participate in the regeneration of bone. Further, the coral offers structural properties that span

Table 3.4 Structural properties of coral used as bone substitutes (www.biocoral.com)

Coral porosity	Compressive fracture stress (MPa)	Elastic modulus (GPa)
Dense	395	100
20%	110	25
50%	25	8

that of cortical bone or trabecular bone depending on porosity (which can range from 20–50%). The mechanical properties of various corals, depending on their porosity, are given in Table 3.4. Due to its highly biocompatible properties, full resorption of biocoral with complete bone integration is usually complete within nine months.

Converted coral has many applications as a bone substitute. In orthopedics, it is used as a substitute for **trabecular** and **cortical bone**. Coral provides a viable alternative to **autogenous** bone grafts. Coral serves as a viable alternative where there is osteoporotic bone and offers structural integrity in fractures associated with bone demineralization. In dental applications, coral is used for the reconstruction and filling of bone defects. It is utilized in periodontal, endodontic, and implant surgeries. Coral is also used in craniomaxillofacial surgery and is used in facial reconstruction. This coral-derived bone substitute has shown widespread clinical success in biomaterial applications (Demers et al., 2002).

3.8 Summary

Ceramics and glasses utilize ionic and covalent bonding to create vastly different structures with unique properties such as wear and corrosion resistance, bioactivity, and controlled resorption in the body. The primary design concern when utilizing ceramics in load-bearing implants is their susceptibility to fracture. Care must be taken to fully analyze the presence of flaws, cracks, or stress concentrations in these materials as they can lead to premature failures of the medical device. Ceramics are commonly employed as bearing materials in total joint replacements, heart valves, orbital reconstruction, dental crowns, coatings for bone fixation and tissue integration, and resorbable constructs for bone defect repair. Looking forward in the field of bioceramics, one may envision more biomimetic structures that make use of inherent and hierarchical toughening mechanisms that are available to healthy bone tissue.

3.9 Problems for consideration

Checking for understanding

1. Describe the essential features of ionic and covalent bonding.
2. Describe bioreactivity as it relates to inert, bioactive, and resorbable ceramics. Give an example of each.

3. Describe and schematically illustrate the toughening mechanisms utilized in ceramics.
4. Explain why fracture is problematic for modular implants that utilize ceramics.

For further exploration

5. How would you design a ceramic component to minimize its likelihood for fracture?

3.10 References

Bajpai, P.K. (1990). Ceramic amino acid composites for repairing traumatized hard tissues. In *Handbook of Bioactive Ceramics, Volume II*, ed. T. Yamamuro, L.L. Hench, and J. Wilson-Hench. Boca Raton: CRC Press, pp. 255–70.

Bajpai, P.K., and Fuchs, C.M. (1985). Development of a Hydroxypatite bone grout. In *Proceedings of the First Scientific Session of the Academy of Surgical Research*, ed. C.W. Hall. New York: Pergamon Press, pp. 50–54.

Barinov, S.M., and Baschenko, Y.V. (1992). Applications of ceramic composites as implants. In *Bioceramics and the Human Body*, ed. A. Ravaglioli and A. Krajewski. London: Elsevier Applied Science, pp. 206–10.

Barrenblatt, D.S. (1962). The mathematical theory of equilibrium cracks in brittle solids. In *Advances in Applied Mechanics*. New York: Academic Press, pp. 55–129.

Bokros, G.I. (1972). Deposition structure and properties of pyrolytic carbon. *Chemistry and Physics of Carbon*, 5, 70–81.

Demers, C., Hamdy, C.R., Corsi, K., Chellat, F., Tabrizian, M., and Yahlis, L. (2002). Natural coral as a bone substitute. *Biomedical Engineering*, 12(1), 15–35.

Ducheyne, P. (1985). Bioglass coatings and bioglass composites as implant materials. *Journal of Biomedical Materials Research*, 19 (3), 273–91.

Ducheyne, P., and McGuckin, J.F. (1990). Composite bioactive ceramic-metal materials. In *Handbook of Bioactive Ceramics*, ed. T. Yamamuro, L.L. Hench, and J. Wilson. Boca Rotan: CRC Press, pp. 75–86.

Dugdale, D.S. (1960). Yielding of steel shots containing slits. *Journal of Mechanics and Physics of Solids*, 8(2), 100–108.

Griffith, A.A. (1920). The phenomena of rupture and flow in solids. *Philosophical Transactions A*, 221, 163–98.

Grote, J.J. (1987). Reconstruction of the ossicular chin with hydroxyapatite prostheses. *American Journal Otolaryngology*, 8, 396–401.

Hench, L.L. (1991). Bioceramics: From concept to clinic. *Journal of the American Ceramic Society*, 74, 1487–1510.

Hench, L.L. (1993). Bioceramics: From concept to clinic. *American Ceramic Society Bulletin*, 72, 93–98.

Hulbert, S.F., Bokros, J.C., Hench, L.L., Wilson, J., and Heimke, G. (1987). Ceramics in clinical applications: past, present, and future. In *High Tech Ceramics*, ed. P. Vincezini. Amsterdam: Elsevier Press, pp. 189–213.

3.10 References

Kingery, W.D. (1976). *Introduction to Ceramics*. New York: John Wiley.

Miyazaki, M., Tsumura, H., Wang, J.C., and Alanay, A. (2009). An update on bone substitutes for spinal fusion. *European Spine Journal*, 18: 783–99.

More, R.B., and Silver, M.D. (1990). Pyrolitic carbon prosthetic heart valve occluder wear: in vitro results for the Bjork-Shiley prosthesis. *Journal of Applied Biomaterials*, 1, 267–78.

Oonishi, H. (1992). Bioceramic in orthopaedic surgery – our clinical experiences. In *Bioceramics*, vol. 3, ed. J.E. Hulbert and S.F. Hulbert. Terre Haute, IN: Rose Hulman Inst. Technology, pp. 31–42.

Park, J.B., and Bronzino, J.D. (2003). *Biomaterials: Principles and Applications*. Boca Raton: CRC Press.

Parks, J.B., and Lake, R.S. (1992). *Biomaterials: An Introduction*. New York: Plenum Press.

Pauling, L. (1960). *Nature of the Chemical Bond*. Ithaca, NY: Cornell University Press.

Pelter, L.F. (1959). The use of plaster of Paris to fill large defects in bone. *American Journal of Surgery*, 97, 311–15.

Ratner, B.D., Hoffman, A.S., Schoen, F.J., and Lemons, J.E. (1996). *Biomaterials Science: An Introduction to Materials in Medicine*. New York: Academic Press.

Ritchie, R.O. (1996). Fatigue and fracture of pyrolytic carbon: a damage-tolerant approach to structural integrity and life prediction in ceramic heart valve prostheses. *Journal of Heart Valve Disease*, 5 (Suppl. I), S9–S31.

Whitehead, R.Y., Lacefield, W.R., and Lucas, L.C. (1993). Structure and integrity of a plasma sprayed hydroxyapatite coating on titanium. *Journal of Biomedical Materials*, 27, 1501–7.

Wolford, L.M., Wardrop, R.W., and Hartog, J.M. (1987). Coralline porous hydroxylapatite as a bone graft substitute in orthognathic surgery. *Journal of Oral Maxillofacial Surgery*, 45, 1034–42.

Zimmermann, M.C., Alexander, H., Parsons, J.R., and Bajpai, P.K. (1991). The design and analysis of laminated degradable composite bone plates for fracture fixation. In *High Text Textiles*, ed. T.L. Vigo and A.F. Turback. Washington, DC: American Chemical Society (Symposium 457), pp. 132–48.

4 Polymers

Inquiry

How would you design a polymeric suture anchor so that it offered strength for 6 months yet would be resorbed by the body within a year?

The above inquiry presents a realistic challenge that one might face in the field of tissue repair. **Suture anchors** are fixation devices used for the attachment of tendons and ligaments to bone. These devices are made up of the anchor itself, which is inserted into the bone using a screw mechanism or an interference fit and is often made of biodegradable polymers that dissolve in the body over time. The anchor also is equipped with an opening at its end and a securing suture is attached to the anchor through the eyelet of the anchor. The implant must be integrated into the underlying bone in order to provide structural support until the bone is able to support the necessary loads. Typically, a **bioresorbable** suture anchor is designed to provide structural integrity for up to 6 months and is often fully resorbed by the body within 1–2 years. The rate at which the polymer degrades is controlled primarily through its backbone chemistry, molecular weight, and crystallinity. The case study presented at the end of this chapter examines bioresorbable polymers as sutures and suture anchors.

4.1 Historical perspective and overview

Polymers have been utilized in medical implants for nearly 80 years. **Polymethylmethacrylate (PMMA)** was first used in the body in the 1930s and was initially chosen for its biocompatibility, stiffness, and optical properties. Today, PMMA is widely used as a medical implant in such applications as bone cement for dental and orthopedic applications (Kühn, 2000). In the decades following their introduction into the implant field, the use of polymers expanded in the body with a diverse range of applications including blood pumps, heart-lung machines, vascular grafts, angioplasty catheters, dental and orthopedic implants, and mammary prosthetics (ASM, 2003). Materials such as **silicones, polyurethanes**, and adhesives were employed; however, there were no stringent device regulations in place and many biomaterials fared poorly in the body. Although medical devices were regulated by the Food and Drug Administration (FDA), the 1938 Federal Food, Drug and Cosmetic Act did not require premarket approval for medical devices (this is discussed further in Chapter 6). In the 1960s and 1970s, developments

Table 4.1 Examples of polymers in load-bearing applications in medical devices

Application	Devices	Polymers	Performance requirements
Cardiovascular	Balloons	Nylon/polyester polyester/LDPE	Rupture resistance, flexibility, friction
	Catheters	LLDPE/HDPE	Compliance, tractability
	Grafts	e-PTFE	Tissue integration
Soft Tissues	Suture anchors	PLDLA/PLA PLLA/PEEK	Resilience, strength, compliance
	Sutures	Polyester/PLLA/PGA	Tensile strength
	Breast implants	Silicone	Burst strength
Dental	Crown/filling	Acrylic resins	Resistance to wear, fatigue, fracture
	Cements	PMMA	Adhesive strength, interface toughness
Orthopedics	Joint replacements	UHMWPE	Resistance to fracture, wear, and fatigue
	Spinal implants	Polyurethanes	Resistance to fracture, wear, and fatigue
	Tendon-ligament	HDPE, e-PTFE	Tensile strength, resistance to creep
	Bone cement	PMMA/PS	Interface fracture toughness
	Spinal fusion	PEEK	Resistance to fracture, wear, and fatigue

in heart valve technology, orthopedics, sutures, dental materials, and soft tissue reconstruction prevailed. Early heart valve cages made of silicones fared poorly due to lipid adsorption, polyurethane foams used on the surface of mammary prosthetics for better adhesion were susceptible to chronic inflammatory response, and the use of **polytetrafluoroethylene (PTFE)** as a bearing surface in both dental and orthopedic implants was complicated by high wear rates and premature device failures (Feinerman and Piecuch, 1993; Zardenta *et al.*, 1996). Starting in 1976, medical devices were subject to pre-market approval by the FDA. Throughout the decades that followed, the implementation of regulatory constraints on medical implants, certain material applications, such as silicone breast implants and PTFE used in dental implants, were removed (silicone was temporarily banned by the FDA and later reapproved for use in breast implants) from the biomaterials market, while other polymers such as **ultra-high molecular weight polyethylene (UHMWPE)** used in total joint replacements became the mainstay of the orthopedics industry. Much of the past two decades has focused on the improvement of polymers known to perform well in the body. A summary of current uses for polymers in orthopedics, dentistry, cardiovascular, and soft tissue implants is given in Table 4.1.

Polymers are widely used in the cardiovascular, soft tissue, dental, and orthopedic device industry with diverse applications in vascular grafts, balloon catheters, stent coatings, orthopedic bearings, screws, suture anchors, bone cement, sutures, and soft tissue reconstruction. In orthopedics, UHMWPE has served as the gold standard as a bearing material in total joint replacements since the early sixties (Kurtz, 2009). **High density polyethylene (HDPE)** is used in tendon reconstruction and catheter tubes. Ligaments can be reconstructed with braided forms of **expanded polytetrafluoroethylene (e-PTFE)** that offer both strength and stability (Bolton and Bruchman, 1983; Hanff *et al.*, 1992; Paavolainen *et al.*, 1993). This same polymer, e-PTFE is most successful in cardiovascular applications (Campbell *et al.*, 1975; Watanabe, 1984) where its porous structure offers tissue integration and opportunity for neovascularization (growth of new blood vessels). E-PTFE is also used in facial reconstruction where the open pore structure facilitates tissue ingrowth and fixation of the implant (Catanese *et al.*, 1999; Greene *et al.*, 1997; Maas *et al.*, 1993). **Nylon** and **polyester** polymers have good tensile strength and are often employed in applications such as sutures (Lawrence and Davis, 2005) and in cardiovascular devices such as balloon catheters (Wheatley *et al.*, 2007). Polyurethanes have shown excellent promise as bearing materials and are being utilized in orthopedic applications such as the spine (Geary *et al.*, 2008). Bioresorbable polymers, such as **polyglycolic acid (PGA)** and **polylactic acid (PLA)** and their derivatives, are used where structural properties are temporarily needed and where tissue integration and resorption are required in a medical device. Degradable polymers are used in applications such as resorbable sutures (Amass *et al.*, 1998), suture anchors (Stanford, 2001), bone screws (Bailey *et al.*, 2006), and drug delivery devices (Arosio *et al.*, 2008). Polymeric coatings can also be used to obtain desired surface properties, such as the use of a fluorinated surface treatment for improved lubricity (Klapperich *et al.*, 2001) or for improved biocompatibility of a device (Ratner *et al.*, 1996).

One primary advantage of polymeric materials is that their macromolecular structure provides many structural properties that can be utilized in the body. The elastic modulus and strength of a polymer can be tailored through chemistry and processing to provide values that are bounded by those of hard and soft biological materials. This feature enables polymers to be utilized in a way that optimizes compliance match with many tissues and enables appropriate load transfer between the implant and the adjacent tissue. Further, polymers, like ceramics, offer a range of bioreactivity that spans from inert to fully resorbable.

4.2 Learning objectives

This chapter provides an overview of polymers used in medical implants. The basic bonding, structure, and inherent properties of polymeric biomaterials are reviewed. The key learning objectives are that at the completion of this chapter the student will be able to:

1. describe the bonding mechanisms and basic structure of polymers
2. portray the range of bioreactivity in the basic polymeric biomaterials
3. specify the effect of molecular variables on the mechanical behavior of polymers
4. discuss the methods utilized for polymer processing
5. recommend appropriate types of polymers for specific biomedical applications

4.3 Bonding and crystal structure

Polymeric materials utilize covalent bonds and secondary interactions to establish their basic structures. Polymers consist of repeated units termed "mers." Homopolymers are created when one mer or monomer (A) reacts with itself to yield a chain comprising mers (-A-A-A-A-A-A-). Co-polymers utilize two monomers (A and B) and these monomers many react with themselves or with each other (-A-B-A-B-A-B- or -A-A-B-B-A-A-B-B-). The way in which molecules pack together in a polymer is as important in dictating its properties as is its chemical constitution. Polymers differ primarily from metals and ceramics in that these repeated "mers" are typically in the form of chains or macromolecules rather than lattice structures. The macromolecules tend to utilize covalent bonds along the backbone of the chain but only use weak secondary forces such as hydrogen bonds or van der Waals forces for cohesion between chains. The foundation element in most polymers is carbon that utilizes sp^3 hybridization to facilitate four bonds and serves as the basis for covalent bonding along the backbone structure (Figure 4.1). The covalent bonding along the backbone provides strength to the molecules, while the secondary van der Waals attractive forces between molecules contribute to the basic structural properties but also facilitate low melting points and ease of processing. Polymer chains can be folded into ordered structures or they may be randomly grouped together. If the chains are sufficiently long, they may become entangled and the mechanical drag or friction in moving these chains past each other (owing to the secondary forces) can offer further structural integrity to the polymer system.

The structural properties of polymers depend on many variables, including how their chains are organized in space. The chains can order themselves in three dimensions (network), two dimensions (planar), or in one dimension (chains). A schematic illustration of the different spatial configurations found in polymers is shown in Figure 4.2. Most three-dimensional polymer networks are highly crosslinked (covalently bonded between chains) but have no long-range order or lattice structure. Such materials have little or no crystallinity and offer relatively high hardness, strength, and modulus but provide little ductility or strain to failure. Two-dimensional systems have covalent bonding and ordering in the planar layer but only weak van der Waals forces between layers. The two-dimensional structure provides a lower hardness and low shear strength that facilitates molecular transfer under shear loading (such attributes are necessary when writing with a pencil or in the generation of solid lubricants). One-dimensional systems (chains) are the most prevalent architectural group in polymers. These systems utilize

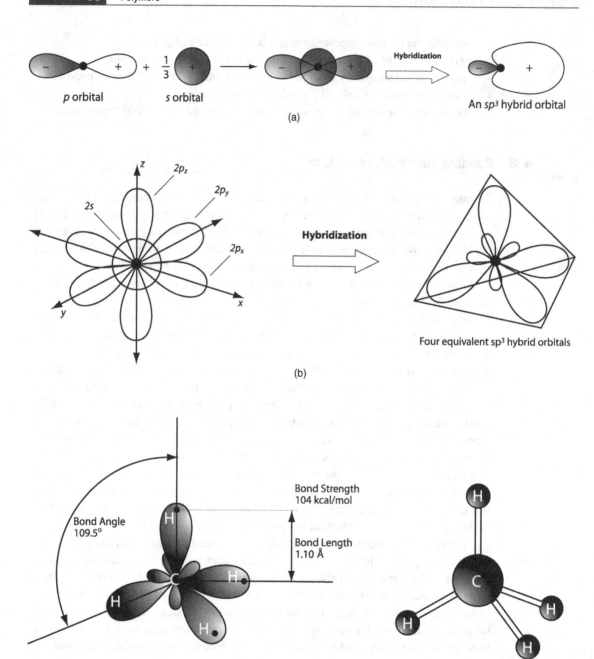

Figure 4.1

(a) The development of an sp^3 hybrid orbital, (b) hybridization to create four equivalent sp^3 hybrid orbitals, and (c) sp^3 hybridization in carbon resulting in four equivalent bonds.

4.3 Bonding and crystal structure

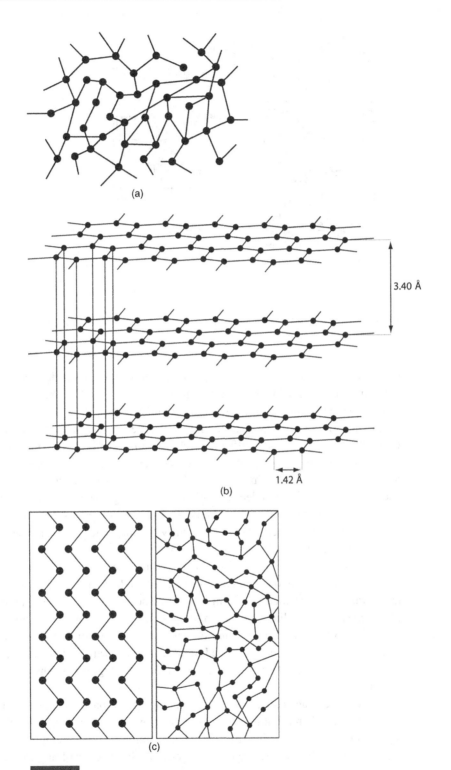

Figure 4.2

Schematic of a (a) three-dimensional network polymer, (b) planar two-dimensional polymer, (c) one-dimensional chain structure that is ordered and found in crystalline structures (left) and randomly organized as is found in amorphous systems (right).

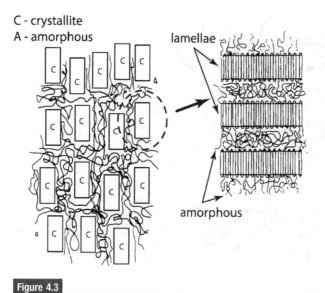

Figure 4.3

Schematic illustration showing a semi-crystalline polymer with crystalline domains (lamellae) comprising folded chains and amorphous domains with randomly organized chains.

covalent bonding on the backbone of polymer chains but weak van der Waals interactions between chains and achieve structural integrity through packing and entanglements of chains. These systems may be crystalline or amorphous.

Another structural factor contributing to the mechanical properties of a polymer is **crystallinity**. It is extremely difficult to achieve full crystallinity in a polymer, and therefore polymers are generally classified as semi-crystalline or **amorphous**. The ability to pack into crystalline domains is usually the consequence of the chain architecture, backbone chemistry, or molecular weight. If the long chains pack regularly, side-by-side, they tend to form crystalline domains as shown in Figure 4.2(c). If the long chain molecules have large side groups or extensive chain length, they may become irregularly tangled and they will tend to be amorphous (no long-range order). An amorphous polymer lacks order throughout its structure while a semi-crystalline polymer has both crystalline and amorphous domains. An illustration depicting the amorphous and crystalline domains in an *idealized* semi-crystalline polymer is provided in Figure 4.3. The idealized semi-crystalline polymer comprises uniform stacking of crystalline domains, while real systems have crystalline domains that are randomly organized within the amorphous matrix. In the crystalline **lamellae** the chains are folded back and forth with order and symmetry, while in the adjacent amorphous regions the chains are entangled but have no order.

Molecular characteristics such as branching of the backbone, bulky side groups, chain length, and crosslinking (covalent bonds between chains) can limit the ability of a polymer to form crystalline structures. For example, chain branching prevents regular organization of the backbone chain and generally limits crystallinity to less than 30%

Table 4.2 Molecular and structural property range in polyethylene polymer systems

Polyethylene	Molecular weight (g/mol)	Crystallinity (%)	Density (g/cm^3)	Applications
LDPE *branched structure*	30,000–50,000	25–35	0.910–0.925	Medical packaging
LLDPE *lamellar, light branching*	30,000–50,000	35–45	0.925–0.935	Angioplasty catheters
MDPE *spherulitic structure*	60,000–100,000	50–60	0.926–0.940	Facial implants
HDPE *spherulitic structure*	200,000–500,000	70–90	0.941–0.980	Tendons; catheters
UHMPWE *lamellae structure*	4–6 million	45–60	0.925–0.935	Orthopedic bearings

in the polymer, as is the case with low-density polyethylene (LDPE). Similarly, bulky side groups can prevent packing of chains with any long-range order and may result in fully amorphous structures, as is the case with polystyrene (PS). Another factor that may impede long-range ordering is **molecular weight** or chain length. If chains become extremely long, as in very high molecular weight polymers, they may become more prone to entanglement rather than crystallization. An interesting result of such factors is that polymers with the same chemistry can vary greatly in molecular weight, crystallinity, and density along with concomitant mechanical properties. Table 4.2 shows the range of molecular and structural properties in polyethylene polymer systems.

Process time, temperature, and pressure conditions enable a broad range of crystallinity, molecular weight, and microstructures to be obtained in polymers. **Ziegler-Natta catalysts** provide steric hindrance to the growing chain during polymerization and are used to maintain linearity of the polymer chains. Without this type of catalyst, the polymer branches off the main chain backbone as the polymerization process continues and its ability of the polymer to order its chains efficiently in space (crystallize) is quite limited. The branches minimize the ability of the polymer chains to pack with order and this is reflected in both its lower crystallinity and density. Linear chains facilitate efficient packing of molecules and thus the system has higher crystallinity (up to 90%) and density. The crystallinity is facilitated through chain folding during chain growth. The linear chains can minimize their energy with appropriate fold lengths. In many highly linear systems, the crystalline domains tend to grow from a nucleation site and form crystalline domains termed **spherulites** (Figure 4.4). Spherulites are structurally similar to grains in that they represent the limit of a specific orientation of crystal growth and serve as a boundary that can resist deformation and contribute to the strength of the polymer.

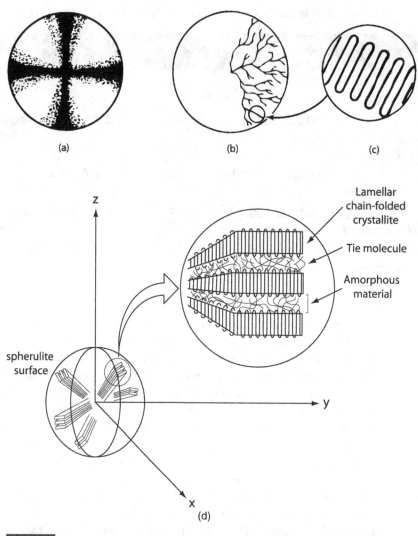

Figure 4.4

Illustration of spherulite development: (a) birefringence develops in the structure, (b) polymer chains nucleate from a central site, (c) chains fold into crystalline domains, and (d) fully formed spherulite with embedded crystalline and amorphous domains. (After Rosen, 1993.)

If the polymerization is allowed to continue so that the linear chains get very long, the molecular weights can become very large. The very high molecular weight has an interesting effect on the structure, and due to the greater propensity for entanglement of very long chains, the ability to crystallize is much more limited. The chains can fold back and forth regularly or irregularly. The latter is characterized as the switchboard model and is depicted in Figure 4.5.

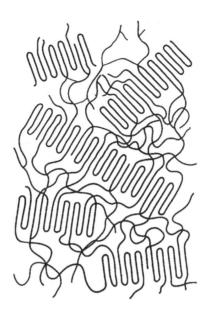

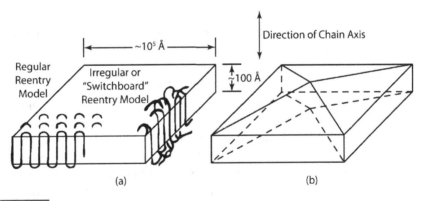

Figure 4.5

Illustration of crystalline lamellae. Chains can fold (a) irregularly or regularly at the crystalline amorphous interface and lamella tend to be only on the order of 100 Angstroms for polyethylene crystals; (b) chains oriented along the main chain axis in the short axis of the lamellae plate. (After Rosen, 1993.)

The crystallinity of a polymer system directly affects its mechanical properties and hence it is important to understand the nature of crystallite development in polymer systems. There are several factors that indicate whether a polymer will be likely to have a high degree of crystallinity. These include linear and symmetric chains, small side groups, lower molecular weight (smaller chain lengths pack more efficiently), and slow cooling from the melt (enables crystal formation). The crystallinity of a polymer can be

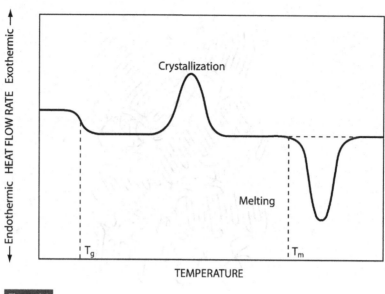

Figure 4.6

Typical differential scanning calorimetry (DSC) scan showing heat flow rate as a function of temperature.

directly correlated to the density of the polymer structure through the following basic relationship:

$$\rho = \rho_a x + \rho_c (1 - x) \tag{4.1}$$

where ρ is the overall polymer density, ρ_a is the density of amorphous region, ρ_c is density of the idealized crystal lattice for the polymer system, and x is the degree of crystallinity. Because of the relationship between density and crystallinity, a density gradient column with calibrated floats can be utilized to measure the density of a polymer and to provide a first estimate of the crystallinity. It is more common, however, to utilize **differential scanning calorimetry (DSC)** to measure the crystallinity of a polymer. In this technique, a small sample (usually 5 mg or less) of polymer is heated from room temperature at a heating rate of 5–10°C/min through the melting point or flow temperature of the polymer (Figure 4.6). An endothermic reaction occurs as the sample absorbs heat to melt the crystals. The peak of the melting curve represents the melting temperature of the polymer. Melting curves can be affected by defects in lamellae, constrained chains in amorphous regions, and other factors such as lamellae size distribution. The degree of crystallinity is equal to the ratio of the area under the thermographs and the heat of fusion of a perfect crystal, $x = \Delta H / \Delta H_f$.

Long polymer chains are accommodated through chain folding and are added to the crystalline domain by folding along the edge (Young, 1991). The thickness of the polymer crystals, which is a measure of fold length, is nearly a constant for a given polymer crystallized from a given solvent and temperature. In general, if the crystal is annealed at a higher temperature, it will refold to the thickness of the new characteristic temperature, and as the annealing temperature is increased, the lamellae thickens. The relationship between the melting temperature of the polymer and the lamellae thickness is given as:

$$T_m = T_m^o \left(1 - \frac{2\gamma}{\Delta H_f l^*}\right) \quad (4.2)$$

where T_m is the melting point under the ambient conditions, T_m^o is the equilibrium melting temperature, ΔH_f is the heat of fusion, γ is the surface energy of the fold, and l^* is the lamellae thickness (Young, 1991).

In contrast to semi-crystalline polymers, an amorphous polymer comprises randomly ordered or entangled chains throughout its structure. A large side group (-R) on the vinyl polymers (-CH_2-CH-R-) or other steric hindrance on the backbone chain can prevent crystallization. In a case such as polystyrene, where R is a benzene ring, the result is a fully amorphous (randomly organized) polymer structure. The characterizing feature of an amorphous polymer or the amorphous phase of a semi-crystalline polymer is the **glass transition temperature**, T_g. This is the temperature where the polymer transitions from glassy behavior to rubbery behavior and represents the transition between large chain motions and small chain motions in a polymer so that above T_g large chain mobility and sliding are readily available, whereas below T_g chain mobility and sliding are quite limited. T_g is a second-order thermodynamic parameter of the material and it is typically measured using dynamic testing techniques or a thermal method such as DSC. T_g is denoted by an inflection point in enthalpy as a function of temperature or in the specific volume as a function of temperature. The specific volume is the inverse of density and is a function of pressure, temperature, and cooling rate. At a slow cooling rate, polymers have more time to position themselves together and accordingly the "free volume" (the volume not occupied by chains) is larger than if the cooling rate is high. Below T_g the motion of polymer chains is severely limited. Changes in free volume or specific density are much greater above T_g. A schematic illustration depicting both T_g (second-order transition seen as an inflection) and T_m (first-order transition observed as a step change) in a semi-crystalline polymer is provided in Figure 4.7(a). For comparison, the T_g characterization plot for an amorphous polymer is shown in Figure 4.7(b).

If polymers have a high degree of chain branching, entanglements, or large side groups, then the mobility of chains is limited or restricted, and T_g will be much higher than for other polymer systems. Polymers with high glass transition temperature (on the order of $100°C$ or greater) are known as **glassy polymers**. If the polymer chains are linear and without large side groups, then chain mobility is easier and T_g is typically quite low. Below T_g, amorphous polymers exhibit little deformation. This is useful in applications where the component needs to have a high elastic modulus and little deformability, such as polymethylmethacrylates in acrylic bone cements. Glassy polymers tend to offer high strength, elastic modulus, and optical clarity (in the absence of additives or scattering

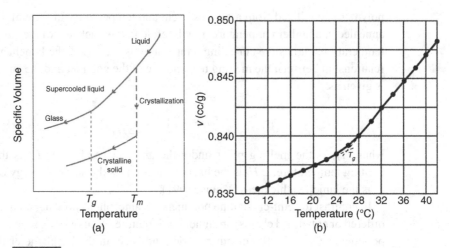

Figure 4.7

Illustration of specific volume as a function of temperature for (a) a semi-crystalline polymer exhibiting both T_g as well as T_m and (b) an amorphous polymer where the inflection denotes T_g. (After Rosen, 1993.)

particles) but offer limited toughness or strain to failure. Such materials may be used in certain applications where a high degree of resilience or toughness is not required but where optical clarity is needed, such as in their use in ophthalmologic implants.

Rubbery polymers, on the other hand, exhibit large recoverable deformations and tend to have very low glass transition temperatures. A classic rubber material is a lightly crosslinked long chain polymer structure that is well above its T_g. Large motions are available to the chains at elevated temperatures and this facilitates large recoverable deformation as the crosslinks (and equivalently hard phases below T_g) provide an intrinsic memory throughout the chain extension. Large reversible deformations are often utilized in the design of vascular grafts that are reconstructing elastin-rich vessels.

The degree of crosslinking has an interesting effect on glass transition temperature of a polymer. At low levels of crosslinking, the covalent ties between backbone chains facilitate recoverable deformation under large extension and the T_g of the system is low as chain coiling and uncoiling is not hindered. However, with extensive crosslinking, the extensive deformation is prevented by the covalent bonds between molecular chains and the system behaves like a glassy polymer with a high T_g.

The molecular weight, like crosslinking, affects the glass transition of a polymer system. This effect can be predicted from the Fox and Flory equation (Flory, 1953):

$$T_g = T_g^\infty - \frac{K}{M_n} \qquad (4.3)$$

where T_g^∞ is the glass transition temperature at infinite molecular weight, K is a material constant for the polymer, and M_n is the molecular weight of the polymer. The equation predicts that increasing molecular weight increases the glass transition temperature of the polymer. Extensive crosslinking has the same effect as an infinite molecular weight

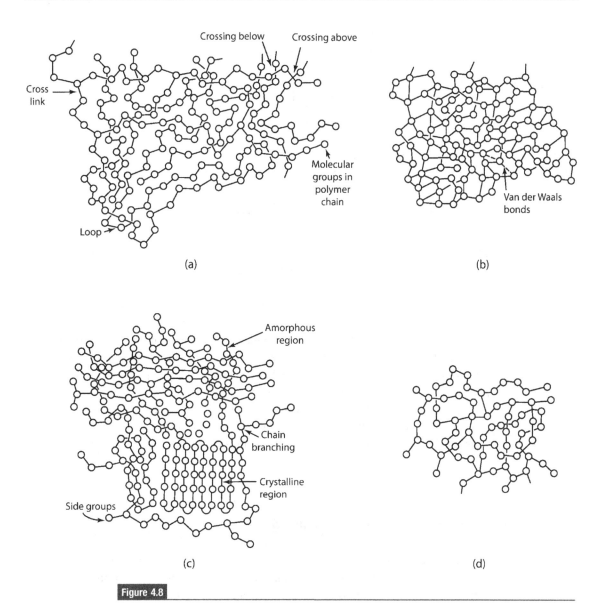

Figure 4.8

Schematic showing a (a) lightly crosslinked rubber above T_g with capacity for large, recoverable extensions; (b) lightly crosslinked rubber below T_g with limited amount of deformation; (c) semi-crystalline polymer with an amorphous phase above its T_g and capable of some recoverable extension before the onset of permanent deformation (crystallite deformation); (d) fully crosslinked network structure incapable of large elastic deformation.

and hence the glass transition temperature for a fully crosslinked network is T_g^∞. The aforementioned molecular effects are schematically illustrated in Figure 4.8.

Many of the semi-crystalline polymers used in medical devices have glass transition temperatures that are well below body temperature, which provides polymer structures

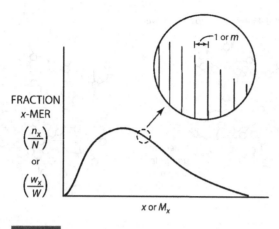

Figure 4.9

Illustration of molecule size distribution in a typical polymer as a result of synthesis reactions.

that are resilient and tough. Toughness is necessary where high loads and impacts are expected, such as orthopedic bearing surfaces, and resilience is needed in applications such as suture anchors. The coupling of both crystalline and amorphous components generally facilitates toughness where strength (or resistance to plastic deformation) is offered by the crystalline component and elastic extensibility is provided by the amorphous phase above its glass transition temperature. Thus, semi-crystalline polymers are used in applications where high stresses and strains are expected. Amorphous polymers are utilized above their glass transition temperature when large extensibility (low strength) is needed and below their glass transition temperature when resistance to deformation and strength is required.

4.4 Molecular weight distribution in polymers

Polymers are generally processed using either a **condensation reaction** in which a water molecule is released as part of the reaction or through an **addition reaction** in which free radicals are used in the initiation and propagation stages of chain growth (Young, 1991). The result of basic polymer synthesis is that the polymer structure contains many chains and hence there is a distribution of chain lengths depending on individual termination sequences in polymerization. Figure 4.9 schematically illustrates a representative molecule size distribution in a typical polymer.

Because of this variability in chain size, the molecular weight is typically averaged and can be done on a number or weight basis. The **number average molecular weight**, M_n, is an average over the number of molecules of given size:

$$M_n = \frac{\sum n_i M_i}{\sum n_i} \tag{4.4}$$

where n_i is the number of moles and M_i is the molecular weight of mer i and is known as the first moment of molecular weight. Similarly, the **weight average molecular weight**, M_w, is based on the weight distribution of chains:

$$M_w = \frac{\sum w_i M_i}{\sum w_i} = \frac{\sum n_i M_i^2}{\sum n_i M_i} \qquad (4.5)$$

where w_i is the weight fraction and M_i is the molecular weight of mer i. The weight average molecular weight is known as the second moment of molecular weight and is always greater than or equal to M_n. The **polydisperity index** (PDI) is the ratio of M_w/M_n, and is always greater than or equal to one. The molecular weight distribution of a polymer is generally determined using **gel permeation chromatography** (GPC). This technique utilizes a solvent to enable the polymer chains to go into solution and to take on a hydrodynamic volume or molecule size in solution. The molecules are then passed through porous media, usually through a series of packed columns and separated from one another based on their differences in size and elution time (time to pass through the packed columns). The smaller molecules are able to visit more of the porous pathways and take longer to elute from the column than the larger molecules that are limited to a smaller number of pores due to their size. GPC separates and quantifies the molecules by size or elution time and is the most widely accepted method for determining polymer molecular weight distribution and polydispersity index.

The distribution of molecular weight and molecule size for a given chemistry makes polymers vastly different than metals and ceramics; while this limits the ability to generally classify properties based solely on chemistry of the polymer, it enables a wide range of properties to be tailored through processing. The increased length of polymer chains improves strength but reduces the mobility of the chains and can limit the methods by which a polymer is processed. For example, a very high molecular weight polymer may not be able to be injection molded if the **melt viscosity** is too high and will be limited to a sintering type process coupled with molding, machining, or extrusion. The melt viscosity η scales with molecular weight as:

$$\eta = M_w^{3.2}. \qquad (4.6)$$

The chain length or molecular weight also affects the chain entanglement density. As the molecules become larger, they have a greater likelihood for entanglement of chains. These entanglements can act as physical crosslinks but are different in that the entanglements will dissipate at elevated temperatures or sufficient length scales of loading; crosslinks, on the other hand, are often created with energetic or chemical means and result in permanent covalent bonds between chains (Figure 4.10). With sufficient crosslink density, the molecular weight between crosslinks is decreased and the nonlinear elastic deformation (rubber elasticity) or time-dependent strain (viscoelasticity) is severely limited.

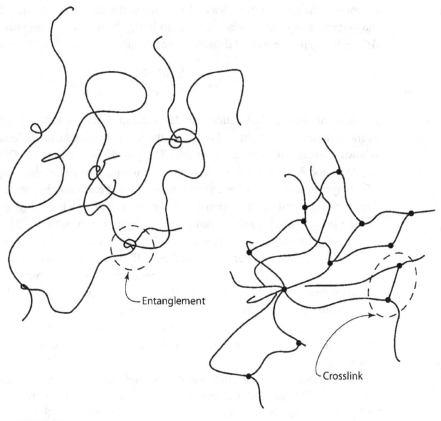

Figure 4.10

Schematic illustration showing the difference between physical entanglements and covalently bonded crosslinks.

Example 4.1 **Determination of molecular weight**

Consider two linear polymers, such as MDPE and HDPE, which are combined to create one polymer. Assume both polymers are monodisperse, that is, PDI = 1, and that the molecular weight of the MDPE is 100,000 g/mol and that of the HDPE is 400,000 g/mol. Consider a mixture that is one part by weight MDPE and two parts by weight HDPE. Determine the weight and number average molecular weight and PDI for this polymer mixture.

Solution The number of moles of MDPE and HDPE is given as:

$$n_{\text{MDPE}} = \frac{1}{100{,}000} = 1E-5$$

$$n_{\text{HDPE}} = \frac{2}{400{,}000} = 0.5E-5$$

The number average molecular weight is given as:

$$M_n = \frac{\sum n_i M_i}{\sum n_i} = \frac{(1 \times 10^{-5})(10^5) + (0.5 \times 10^{-5})(4 \times 10^5)}{(1 \times 10^{-5} + 0.5 \times 10^5)} = 2E5 \text{ g/mol}$$

The weight average molecular weight is given as:

$$M_w = \sum \left(\frac{w_i}{W}\right) M_i = \frac{1}{3}(1 \times 10^5) + \frac{2}{3}(4 \times 10^5) = 3E5 \text{ g/mol}$$

The polydispersity index (PDI) of the mixture is

$$\text{PDI} = \frac{M_w}{M_n} = \frac{300{,}000}{200{,}000} = 1.5$$

4.5 Mechanical behavior of polymers

The deformation of polymers is complicated by the role of various molecular variables such as backbone chemistry, molecular weight, crystallinity, crosslinking, and chain entanglement density. The mechanical behavior is also strongly affected by both temperature and strain rate. Polymers deform elastically under small stresses by displacement of bond lengths and angles (including rotation) from their minimum energy arrangement of chains. This is similar to an elastic deformation of a crystalline metal or ceramic, except that in addition to bond stretching there is also elastic deformation that is accommodated by angular changes and rotational changes of the polymer chains. This results in elastic strains being much greater for semi-crystalline polymers and amorphous polymers above their glass transition temperatures than for other crystalline materials. Typical limits of elastic strains are as high as 2% for polymers but are limited to 0.2% for metals and less than 0.1% for ceramics. In the rubbery state, the full coiling and uncoiling effect can be activated much like the deformation of an elastic band. This enables large amounts of elastic deformation to occur (up to several hundred percent strain in some elastomers) and elastic strain energy to be stored when the polymer is above its glass transition temperature. Figure 4.11 schematically illustrates the stress-strain behavior of several representative polymers: a semi-crystalline polymer, an amorphous polymer above and below its glass transition temperature, a rubbery polymer above its glass transition temperature, and a fully crosslinked network.

The yield stress and elastic modulus of a polymer generally increase with rate of deformation or the rate at which strain is imposed on the polymer system (**strain rate**) and decrease with increasing temperature. This is due to the viscoelastic nature of most polymers. **Viscoelasticity** refers to the ability of a polymer to exhibit both elastic (solid) and viscous (fluid) behavior (this topic is discussed in detail in the chapter allocated to viscoelasticity). The **Eyring model** captures the strain rate and temperature dependence of a polymer and is theoretically founded in chemistry and quantum mechanics (Young, 1991). If a chemical reaction, diffusion, or corrosion is causing degradation leading to

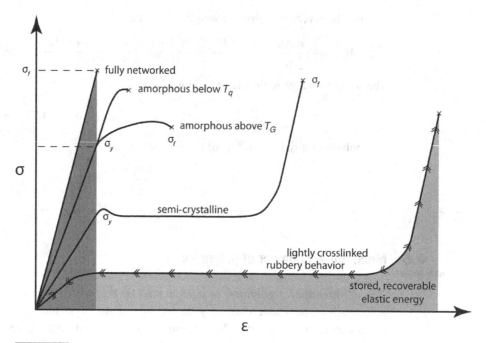

Figure 4.11

Stress-strain behavior of a semi-crystalline polymer, an amorphous polymer above and below its T_g, a rubbery polymer above its T_g, and a fully crosslinked network.

failure, the Eyring model describes how the rate of degradation varies with stress or, equivalently, how time to failure varies with stress. The model includes temperature and can be expanded to include other relevant stresses. This model takes the form of an Arrhenius equation and utilizes a molecular model to correlate yield stress with strain rate and temperature:

$$\frac{\sigma_y}{T} = \frac{2}{V^*} \left| \frac{\Delta H}{T} + 2.3 R \log\left(\frac{d\dot{\varepsilon}_y}{d\dot{\varepsilon}_o}\right) \right| \qquad (4.7)$$

where σ_y is the yield stress of the polymer, ΔH represents the enthalpy required to move a mole of polymer segment over its activation barrier to move from one position to another in the solid, $\dot{\varepsilon}_y$ is the strain rate at yield, $\dot{\varepsilon}_0$ is a constant for the material, and V^* is the activation volume. The **activation volume** is broadly defined as the rate of decrease of activation enthalpy with respect to flow (yield) stress at fixed temperature. The activation volume affects the rate-controlling mechanisms in the plastic deformation of the crystalline component of the structure. The premise of this model is that the segments of the polymer chain must overcome activation barriers to move from one position to another in the structure (Figure 4.12).

The application of a stress reduces the activation barrier in the direction of the applied stress, enabling **strain** (the deformation of the polymer with respect to its initial or current dimensions) to occur. The model assumes that the imposed strain rate is

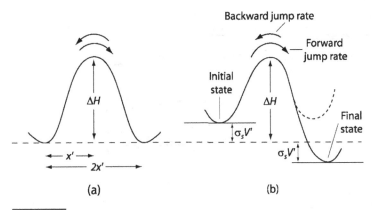

Figure 4.12

Eyring model of solid flow. (a) Before stress is applied, the polymer segments are separated by an enthalpy barrier (ΔH) and a distance $2x'$. (b) Once a stress, σ_s, is applied there is a forward jump with a barrier of $\Delta H - \sigma_s V'$ and a backward jump barrier of $\Delta H + \sigma_s V'$ (V' is free volume of the polymer). This results in a preferential jump in the direction of the applied stress.

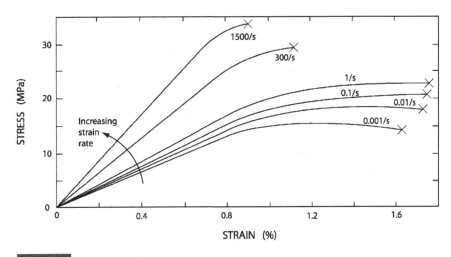

Figure 4.13

Effect of increasing strain rate on the stress-strain behavior of an amorphous PVA polymer (above its glass transition temperature) (After Rosen, 1993.)

proportional to the rate of preferred segmental jumping in the direction of the applied stress. There is also evidence that the dislocation initiation rates in polymers are the limiting shear strain rates in the crystalline phase. The equation predicts that (σ_y/T) increases proportionately with log ($d\dot{\varepsilon}_y$), and that (σ_y/T) increases with decreasing temperature at constant strain rates. That is, the yield stress increases with increasing strain rate and decreasing temperature. Similarly, the elastic modulus increases with increasing strain rate (Figure 4.13). These effects, however, are less substantial below the glass transition temperature of the polymer.

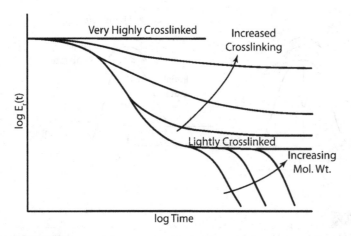

Figure 4.14

Schematic illustration showing the effect of molecular variables on the time-temperature behavior of polymer systems. (After Young, 1991.)

The extent of this time-dependent behavior depends on many molecular variables such as crystallinity, molecular weight, entanglement density, and degree of crosslinking. Figure 4.14 schematically illustrates such effects and shows that as the molecular weight, crystallinity, and crosslinking density increase, the polymer becomes more solid-like and less susceptible to viscous mechanisms. Specifically, as a polymer becomes fully crosslinked, it behaves like an elastic solid with an elastic modulus that is not time dependent and a yield strength that is independent of strain rate. A lightly crosslinked system will behave like a rubber system and will demonstrate an upper bound modulus (glassy) below its glass transition temperature or over very short length scales and a lower bound modulus (rubbery) above its glass transition or over very long length scales. The molecular weight affects the onset of flow and at infinitely high values the system behaves like a fully crosslinked structure.

In general, the following molecular variables can be utilized to strengthen a polymer system: stronger bonding within the backbone structure, larger side groups on the polymer backbone chain, enhanced forces between chains such as polar effects or hydrogen bonding, increased molecular weight, greater crystallinity, and higher density of crosslinks or entanglements.

4.6 Polymer processing

The manufacturability of polymers can be divided into two subclasses distinguished by structure: **thermosetting polymers** and **thermoplastics**. Three-dimensional amorphous polymers that are highly crosslinked are generally classified as thermosetting polymers. Thermosetting polymers are polymerized through a chemical reaction resulting in crosslinking of the structure into one large three-dimensional molecular network. Once the chemical reaction or polymerization is complete, the polymer is a hard, infusible, insoluble material, which cannot be melted or molded without destroying the molecule. A common analogy is an egg: when an egg is polymerized, the proteins are crosslinked and the egg turns white but it cannot be re-melted. Two-part epoxy systems are examples of thermosetting plastics. A resin and hardener (both in a viscous state) are mixed and within several minutes, the polymerization is complete and a hard epoxy plastic is the result. Most one-dimensional polymers are typically classified as thermoplastics. In these systems, there are covalent bonds on the polymer backbone but only weak van der Waals forces between molecules. At elevated temperature, it is easy to melt these bonds and have molecular chains readily slide past one another. A thermoplastic, under the application of appropriate heat, can be melted into a liquid state. In general, thermoplastics are easier to process and can be manufactured by a range of methods including injection molding, extrusion, compression molding, and blow molding (Rosen, 1993). These manufacturing processes are depicted in Figure 4.15.

4.7 Polymers in medical implants

Medical polymers are used in a wide range of medical devices including restorative dentistry, soft-tissue reconstruction, orthopedic implants, and cardiovascular structures. The structural properties of polymers make them ideal candidates for many biomaterial applications where compliance match, lubricity, and resilience are needed but where long-term time-dependent deformation can be tolerated. The viscoelastic nature of polymers can be a design challenge but can also be useful in offering time-dependent properties such as recoverable strains that mimic certain tissues such as bone or cartilage. As with the other synthetic biomaterials, the success of the polymer as a medical material is dependent on its biocompatibility. Polymers, like ceramics, range in **bioreactivity** and can be fully inert, bioactive, or **bioresorbable**.

Inert medical polymers include **polyethylenes**, **polymethylmethacrylates**, **polypropylenes**, **polyesters**, **polyurethanes**, **polyetheretherketones**, and **polyamides**. Inert polymers are used where long-term structural stability and biocompatibility are needed and are utilized in applications such as bearing surfaces in hip, knee, shoulder, or spine implants; vascular grafts or catheters; fillings and resins for teeth; nose, chin, and cheek implants; ocular implants and non-resorbable sutures. Bioactive polymers encourage tissue interaction or ingrowth. Polymers that are porous by design such as **expanded polytetrafluoroethylene** can facilitate tissue ingrowth and enable long-term stability. Bioresorbable polymers are used where structural properties are needed only temporarily and may be utilized in sutures, suture anchors, bone grafts, tissue engineering, and drug delivery devices. The bioresorbable polymers include **polylactic acid, polyglycolic acid**, and derivatives thereof. A summary of commonly used medical polymers along

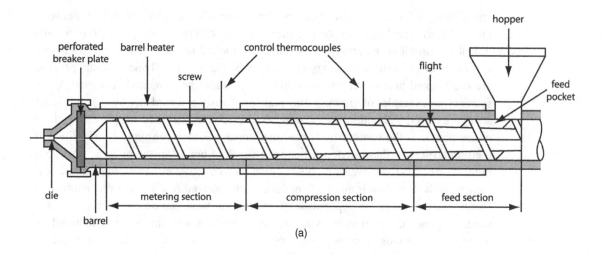

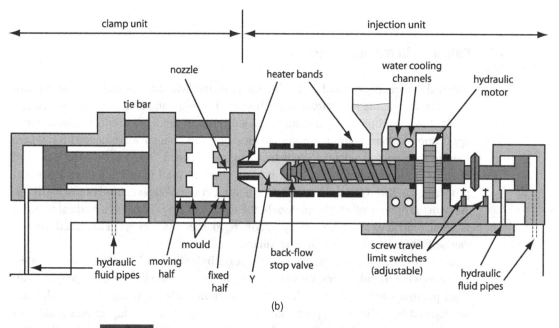

Figure 4.15

Example of common manufacturing processes utilized in polymeric materials: (a) extrusion, (b) injection molding, (c) compression molding, and (d) blow molding. (After Rosen, 1993.)

4.7 Polymers in medical implants

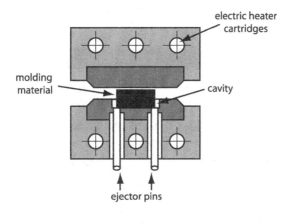

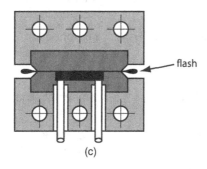

(c)

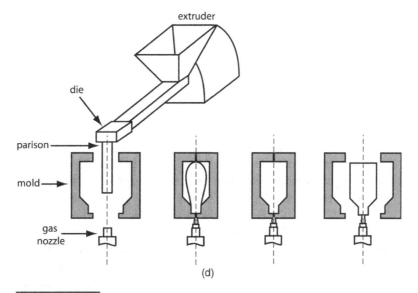

(d)

Figure 4.15 (*cont.*)

Table 4.3 Characteristic linkages, morphology and thermal properties of medical polymers

Polymer	Linkage	Morphology	T_g (C)	T_m (C)
Polyethylene (PE)	-(C$_2$H$_4$)-	semi-crystalline	−80	125–135
Polypropylene (PP)	-(CH$_2$CH CH$_3$)-	semi-crystalline	−10	125–167
Polytetrafluoroethylene (PTFE)	-(C$_2$F$_4$)-	semi-crystalline	−70	340
Polyester (PET)	-(RC=OOR')-	semi-crystalline	50–70	250–265
Polymethylmethacrylate (PMMA)	-(C$_5$O$_2$H$_8$)-	amorphous	118	–
Polyurethane (PU)	-(R$_1$-O-C=0NR$_2$R$_3$)-	amorphous-semi-crystalline	-80–140	240
Polyamide (Nylon)	-(NHCO(CH2)$_4$CONH(CH$_2$)$_6$)-	amorphous or semi-crystalline	45	190–350
Silicone	-(OSiR$_2$)-	amorphous	−127	300
PEEK	-C$_6$H$_4$-O-C$_6$H$_4$-O-C$_6$H$_4$-CO-	semi-crystalline	140	340
Polylactic acid (PLA, PLLA, PDLA)	-(OCH (CH3)C(=O))-	amorphous-semi-crystalline	50–80	175

with their primary mer linkage, reaction process, morphology, melting temperature, and glass transition temperature are provided in Table 4.3.

4.7.1 Polyethylenes

Polyethylenes are commonly employed as biomaterials and depending on their properties can be employed as catheter tubes, facial implants, artificial tendons, or bearing components in total joint replacements. The polyethylene polymer is made through an addition polymerization reaction that utilizes an ethylene monomer (-C$_2$H$_4$-) repeated along the chain and is generally a semi-crystalline polymer with a T_g of −80°C and a T_m of 125–135°C. The branched, low molecular weight form of polyethylene is known as low-density polyethylene (LDPE). If the chains of this low molecular weight polyethylene are linearly maintained with catalysts during polymerization, the result is a linear low-density polyethylene (LLDPE). High-density polyethylene (HDPE) is a linear polyethylene with chains that have little or no branching along extended lengths of chain. Figure 4.16 provides a schematic illustration of the structural differences between the chains in the linear and branched systems. As can be seen in Table 4.2, the molecular weight of polyethylene can vary from 30,000–6,000,000 g/mol, density

4.7 Polymers in medical implants

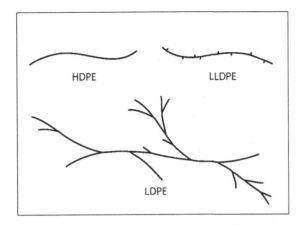

Figure 4.16

Illustration of the structural differences owing to the degree of chain branching in LDPE, LLDPE, and HDPE.

can range from 0.91–0.98 g/cc, and crystallinity can span 30–90%. For this reason, polyethylene is generally subcategorized as low-density polyethylene (LDPE), medium-density polyethylene (MDPE), high-density polyethylene (HDPE), or ultra-high molecular weight polyethylene (UHMWPE). It is imperative that the details of the processing, structure, and mechanical properties are well-characterized when using polyethylene in medical implants.

Polyethylenes are used broadly in orthopedic applications and soft tissue reconstruction. The most widely used and studied polyethylene material is medical grade **ultra-high molecular weight polyethylene (UHMWPE)** and typically has a molecular weight of about 4–6 million g/mol, an elastic modulus on the order of 1 GPa, and crystallinities of about 50–60%. In UHMWPE, the polymer chains fold themselves into crystalline lamellae (plate-like structures) dispersed within amorphous phases. Within the crystalline lamellae, the polyethylene chains tend to orient with hexagonal symmetry along the chain axis (c-axis) as is shown in Figure 4.17. The unit lattice of polyethylene can be altered with change in pressure-temperature conditions by using an alternative stable phase in the processing.

UHMWPE has exceptional **energetic toughness** (ability to absorb energy) owing to its very high molecular weight, crystallinity, and chain entanglement density. Additionally, its low **surface energy** and **coefficient of friction** enable the polymer to perform well as a bearing material against CoCr and ceramic counter bearings in hips, knees, shoulders, and other joint replacements. Decades of polymer research have been devoted to the characterization of microstructure, processing, sterilization, oxidation, and mechanical behavior of UHMWPE used in total joint replacements of this (Kurtz, 2009). A typical total hip replacement utilizing an UHMWPE acetabular cup that articulates against a CoCr femoral head is shown in Figure 4.18. Such implant designs can have survivorship rates of up to 90% after two decades of *in vivo* service.

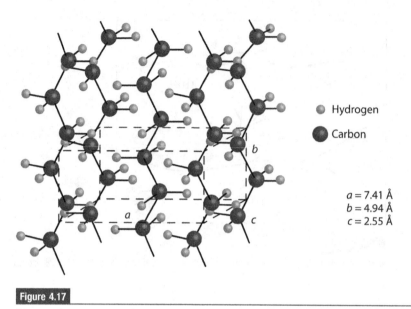

Figure 4.17

Schematic illustration showing the symmetry achieved in packing linear polyethylene chains with their backbone axes aligned.

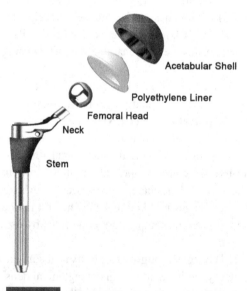

Figure 4.18

Technical illustration showing the use of polyethylene (UHMWPE) as the acetabular liner in a modular total hip replacement design.

4.7.2 Polymethylmethacrylates

Polymethylmethacrylates (PMMA) are commonly employed in orthopedics and dentistry as the grouting material between bone and a synthetic implant. The polymethylmethacrylate polymer is made through an addition polymerization reaction that utilizes

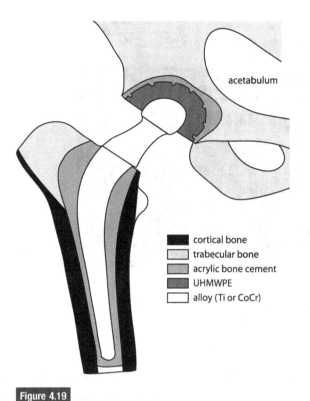

Figure 4.19

Technical illustration showing the use of acrylic bone cement (PMMA) as the load transfer medium and grouting agent between the femoral stem and the adjacent bone tissue.

a monomer of the form -($C_5O_2H_8$)- repeated along the chain. The polymer is generally amorphous with a T_g on the order of 100°C. In general, this polymer is resilient but it has a limited amount of ductility. PMMA has an elastic modulus on the order of 3–4 GPa and its molecular weight can vary between 200,000–700,000 g/mol depending on processing conditions. PMMA can be co-polymerized with other monomers to tailor the properties somewhat, but in general it is known as a glassy polymer. Without any additives this material is known to have excellent transparency. For this reason PMMA is used in ocular implants, although it is best known for its application as a dental or bone cement (Kühn, 2000). An example of PMMA as an orthopedic bone cement providing load transfer between the stem and the adjacent bone in a cemented total hip replacement design is provided in Figure 4.19.

4.7.3 Fluorocarbon polymers

Fluorocarbon polymers are flexible polyolefin polymers that are commonly employed as vascular grafts, facial implants, or artificial tendons. The fluorocarbon polymer is made through an addition polymerization reaction that utilizes a fluoroethylene monomer

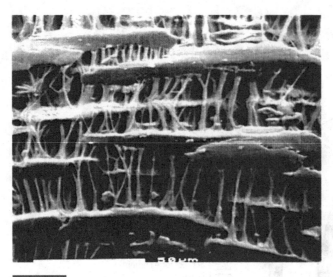

Figure 4.20

Microstructure of expanded polytetrafluoroethylene (e-PTFE) used in facial reconstruction implants. The intermodal distance ranges between 23–32 μm and facilitates tissue ingrowth.

($-C_2F_4-$) repeated along the chain and is generally a semi-crystalline polymer with a T_g of $-70°C$ and a T_m on the order of 300–340°C. These polymers provide exceptional properties such as chemical inertness, mechanical integrity, and lubricity that are owed to the highly electronegative fluorine sheath that protects the carbon backbone from chemical attack and results in a hydrophobic polymer with low surface energy. The primary fluoropolymer used in biomedical applications is predominantly based on expanded polytetrafluoroethylene (e-PTFE). The success of this biomaterial results from its microporous structure that allows bio-integration for fixation and provides structural integrity. The first successful medical applications of e-PTFE initiated with vascular grafts and then evolved to soft tissue reconstruction (Pruitt, 2001). The mechanical properties of e-PTFE depend strongly on the porosity and microstructure of the polymer as well as strain rate (Catanese *et al.*, 1999) However, the predominant success of e-PTFE as a biomaterial is owed to its microporous structure that allows bio-integration for fixation and long-term stability *in vivo* (Greene *et al.*, 1997). An example of the e-PTFE microstructure used in facial reconstruction implants is depicted in Figure 4.20.

4.7.4 Polypropylenes

Polypropylenes (PP) are commonly used in finger joint prostheses, grafts, and sutures (ASM, 2003). These polyolefins are generated through addition polymerization and have a repeat mer of the form -($CH_2CH\ CH_3$)-. The methyl side group can affect the crystallinity of the polymer depending on its arrangement on the backbone. The melting

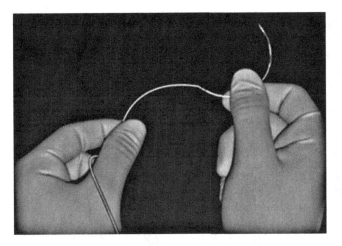

Figure 4.21

Illustration of a non-resorbable polypropylene suture used for wound repair.

temperature spans 125–167°C and its glass transition temperature is on the order of −10°C. Typically, the molecular weight ranges between 200,000–700,000 g/mol and the density is generally bounded between 0.85–0.98 g/cc. This polymer material has an exceptional fatigue life in flexion and it is for this reason that PP is used in small joint reconstruction. The high tensile strength of this polymer also serves as a good material choice for non-resorbable sutures (Figure 4.21). The details of materials used in sutures and the concomitant aspects of suture design are addressed in Chapter 16 (*Soft Tissue Replacements*).

4.7.5 Polyesters and polyamides

Polyesters are utilized in non-resorbable sutures and are also employed as vascular graft materials owing to their **non-thrombogenic** properties. Polyesters have a characteristic ester linkage, -(RC=OOR′)-, in their backbone chain and include materials such as **polyethylene terephthalate (PET)** and **Dacron**® fibers. These systems are generally semi-crystalline. The melting temperature spans 250–260°C and its glass transition temperature is on the order of 50–70°C. These materials are highly flexible and the fabric form enables good tissue integration. Another group of condensation polymers includes the **polyamides** or **nylons**. These materials have a repeat mer with the form -(NHCO(CH2)$_4$CONH(CH$_2$)$_6$)-. Nylons can be amorphous or semi-crystalline depending on process conditions. Polyamides typically have a glass transition temperature on the order of 50°C and a melting range between 200–300°C. Nylons are susceptible to swelling in aqueous solutions but are often employed in short-term medical device applications such as the outer layer of cardiovascular catheters and as the balloon used

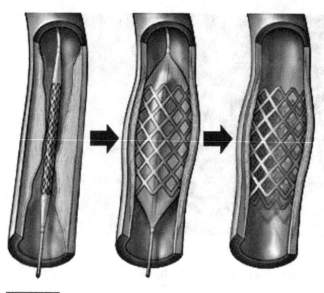

Figure 4.22

Illustration of a nylon balloon used in the deployment of a vascular stent.

for deployment of a stent or the expansion of an occluded artery. Nylons offer strength and stiffness to balloon angioplasty systems; an example of a nylon balloon used in the deployment of a stent is illustrated in Figure 4.22.

4.7.6 Polyurethanes

Polyurethanes are the broadest classification of polymers and are generated through a step-growth polymerization that combines a monomer comprising at least two isocyanate functional groups with another mer containing at least two hydroxyl (alcohol) groups. The urethane links that build the polymer chains have the chemical formulation (R_1-O-C=ON$R_2 R_3$). Polyurethanes can be processed to have hard and soft segments of varying glass transition temperatures and can span a full range of structural properties ranging from elastomeric to glassy (the glass transition temperature can range between $-80°C$ and $140°C$). The structural and mechanical properties are highly dependent upon the backbone chemistry utilized in synthesis and the processing conditions employed in the manufacture of the polymer. As a biomaterial, polyurethanes can be used for a wide range of applications, including bearing materials and soft tissue reconstruction.

4.7.7 Silicones

Silicones are a broad class of polymers also known as siloxanes. These amorphous polymers are composed of -(OSiR_2)- on their backbones, have a glass transition on

Figure 4.23

Image of an early silicone implant design that used polydimethylsiloxane fluid to fill breast implants. If ruptured, these implants could leak the silicone fluid into the body.

the order of $-127°C$, and a melting temperature in the range of $300°C$. Silicones are known best for their elastomeric-viscoelastic properties. A well-known subclassification of silicone is polydimethylsiloxane (PDMS) that was once used as the fluid filler in silicone breast implants, but this technology was terminated due to concerns of leaking (Figure 4.23). Modern silicone breast implants utilize a crosslinked form of silicone that is a soft solid and elastomeric in nature.

4.7.8 PEEK

Polyetheretherketone (PEEK) is used in spinal cages and is also used in suture anchors in soft tissue repair. The backbone structure of PEEK is $(-C_6H_4-O-C_6H_4-O-C_6H_4-CO-)_n$ and this semi-crystalline macromolecule provides strength and resistance to deformation. PEEK has a glass transition temperature of $140°C$ and a melting temperature on the order of $340°C$. Suture anchors must maintain an anchoring point in bone to securely fasten a suture and are often employed in shoulder arthroplasty (Figure 4.24).

4.7.9 Poly(lactic) acids

Poly(lactic acid) (PLA) and its derivatives are bioresorbable polyesters that are designed to break down in the body. These materials are used as bioresorbable sutures, tissue engineering meshes, and temporary fracture fixation components. PLA has a glass transition on the order of $50-80°C$ and a melting temperature of $165-175°C$. The chemical formulation of PLA is $(-O-CH(CH3)-C(=O)-)_n$ and it can be found in amorphous or semi-crystalline form depending on its conformation. PLA has two forms: D-lactide

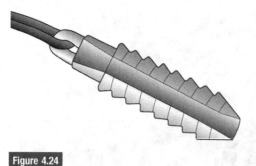

Figure 4.24

Image of a polymeric suture anchor that provides the fixation site for a suture in soft tissue repair.

rotates a plane of polarized light in a clockwise orientation while the L-lactide results in a counterclockwise rotation. An equimolar mixture of enantiomers is optically inactive and is designated by the prefix DL. The polymer degradation in the body is accompanied by a loss in mechanical properties and must be tailored for the specific clinical application. The polymers can be blended or co-polymerized to achieve a tailored resorption rate that is controlled primarily through its molecular weight and crystallinity.

4.8 Case study: resorbable sutures and suture anchors

Suture materials have been utilized for several thousand years (see Chapter 16 for a complete overview of suture design). The first successful suture materials comprised natural materials and included silk, animal gut, hair, cotton, and flax. By the early 1900s, the most successful suture materials were catgut and silk. The former was widely recognized as a resorbable material in the body and was often used in surgical procedures where removing the suture would be burdensome to the patient. Sutures made of nylon (nondegradable) were introduced into surgical medicine in the 1960s and it has only been within the past few decades that bioresorbable polymer sutures with resorption rates tailored through chemistry (hydrophilicity), molecular weight, and crystallinity have been developed. Sutures (Figure 4.21) must offer appropriate strength to the healing tissue for an amount of time necessary for the tissue to regenerate itself or for the wound to be repaired, and this can range from 6 weeks to a year or more. Suture anchors are inserted into bone and provide an attachment site for ligaments, joint capsules, and tendons. Suture anchors are equipped with an eyelet that accommodates a suture (Figure 4.24). In many instances the suture anchor and the sutures are both designed to be resorbed in the body and are often made of PLA, PLLA, or PL,DLA depending on the length of time needed for tissue repair.

Erosion through a process that utilizes the hydrophilic nature of a polymer is the primary resorption mechanism utilized in the body. Bulk erosion occurs where the entire polymeric component is disintegrated through the absorption of water and only a small amount of the component surface is dissolved at a time. When the hydrophilic

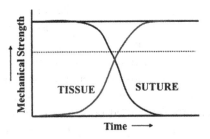

Figure 4.25

Schematic plot of polymer strength and tissue strength as a function of time. The rate of polymer resorption should coincide with tissue repair rate.

resorbable polymer is implanted into the body, the material becomes saturated and begins to break down the polymer. Cracks may form and break the component into small pieces that are continually eroded and absorbed by the body, or the component may slowly dissolve. The timeframe of structural integrity is key for the specific application. The degree of hydrophilicity in the polymer can be used to help tailor the length of time in which the component offers structural integrity in the body. Polymers that are not intrinsically hydrophilic require a chemical pathway such as enzymatic degradation to dissociate the polymer structure into water-soluble chains. Enzymes, which can be found in body fluids, serve as a solubilizing agent to the polymer. The result is that the polymer is reduced into smaller molecular structures that are soluble in water.

PLA and its derivatives can be tailored readily for strength and resorption rates in sutures and their anchors. Ideally the implant tissue system should enable the resorption rate or degradation rate to coincide with that for tissue repair (Figure 4.25). Copolymerization with PLA provides a material that is less hydrophilic and is more resistant to hydrolysis, and this aids in extending the time of structural integrity *in vivo*. Hydrophilic polymers such as polyglycolic acid (PGA) hydrolyze rapidly and are used when the sutures are needed for only a short length of time. PGA-PLA polymers have increased resorption time without a loss in biocompatibility. To obtain a suture anchor system that offers structural integrity for several months, the polymer must resist immediate hydrolysis and have sufficient crystallinity and molecular weight to provide strength to the healing wound. PLLA and PL,DLA systems can be tailored for such needs.

4.9 Summary

Polymers utilize covalent bonding and secondary interactions to create macromolecular structures with a broad range of crystallinity, molecular weight, density, and mechanical properties. They can be tailored to have high toughness, resilience, wear resistance, compliance match to tissue, and controlled resorption in the body. The material properties are intrinsically linked to the basic chemistry and structure of the polymer. A unique aspect of polymers is their viscoelastic nature, demonstrated in their susceptibility to creep and stress relaxation. One of the major challenges in designing with polymeric materials is degradation of the polymer structure and concomitant loss of mechanical integrity (Furmanski and Pruitt, 2009; Pruitt, 2003). This can occur through oxidation and chain scission mechanisms and typically results in embrittlement of the polymer. One source of degradation is the sterilization procedure used prior to implantation (Chapter 1). Another challenge is *in vivo* degradation owing to biological attack or immunological response around the implant (Hughes *et al*., 2003). These factors must be considered when designing implants, as fracture can be life-threatening and may cause premature failure of devices. Polymers are commonly employed as bearing materials in total joint replacements, artificial tendons and ligaments, vascular grafts, catheter balloons, bone fixation, and resorbable constructs for tissue repair. Looking forward, it is likely that future polymers will better integrate tissues, cells, and growth factors that can assist in the tissue repair process.

4.10 Problems for consideration

Checking for understanding

1. Describe the essential features of covalent bonding and van der Waals interactions in polymers.
2. Draw representative polymers depicting 1D, 2D, and 3D structures.
3. A polymer shipment is to be made up by blending three lots of polyethylene, A, B, and C. How much of each lot is needed to make up a shipment of 50,000 kg with a weight average molar mass of 250,000 kg/mol and a polydispersity of 3.65? Let $w_i = W_i / \sum W_i$.

Lot	Mw	PD
A	500,000	2.5
B	250,000	2.0
C	125,000	2.5

For further exploration

4. Provide an example of the types of polymers that will perform well in the following medical implants: an artificial chin, a ligament replacement, a quickly resorbing suture, a bearing surface in a spinal disk replacement, an ocular implant, a vascular graft, and a dental implant.
5. Discuss the benefits and limitations of using polymers in structural medical device applications where long-term cyclic and static loads are expected.

4.11 References

Amass, W., Amass, A., and Tighe, B. (1998). A review of biodegradable polymers: uses, current developments in the synthesis and characterization of biodegradable polyesters, blends of biodegradable polymers and recent advances in biodegradation studies. *Polymer International*, 47(2), 89–144.

American Society of Metals. (2003). *Handbook of Materials for Medical Devices*, ed. J.R. Davis. Materials Park, OH: ASM International.

Arosio, P., Busini, V., Perale, G., Moscatelli, D., and Masi, M. (2008). A new model of resorbable device degradation and drug release. *Polymer International*, 57(7), 912–20.

Bailey, C.A., Kuiper, J.H., and Kelly, C.P. (2006). Biomechanical evaluation of a new composite bioresorbable screw. *Journal of Hand Surgery (British and European Volume)*, 31(2), 208–12.

Bolton, C.W., and Bruchman, B. (1983). Mechanical and biological properties of the GORE-TEX (PTFE) expanded ligament. *Aktuelle Probleme in Chirurgie Orthopadie*, 26, 40–51.

Campbell, C.D., Goldfarb, D., and Roe, R. (1975). A small arterial substitute: expanded microporous polytetrafluoroethylene: patency versus porosity. *Annals of Surgery*, 18, 138–43.

Catanese, J., Cooke, D., Maas, C., and Pruitt, L. (1999). Mechanical properties of medical grade expanded polytetrafluoroethylene: the effects of internodal distance, density, and displacement rate. *Applied Biomaterials*, 48, 187–92.

Feinerman, D.M., and Piecuch, J.F. (1993). Long-term retrospective analysis of twenty-three Proplast-Teflon temporomandibular joint interpositional implants. *International Journal of Oral Maxillofacial Surgery*, 2(1), 11–16.

Flory, P.J. (1953). *Principles of Polymer Chemistry*. Ithaca, NY: Cornell University Press.

Furmanski, J., and Pruitt, L. (2009). Polymeric biomaterials for use in load bearing medical devices: The need for understanding structure-property-design relationships. *Journal of Metals*, September, 14–20.

Geary, C., Birkinshaw, C., and Jones, E. (2008). Characterization of bionate polycarbonate polyurethanes for orthopedic applications. *Journal of Materials Science*: Materials in Medicine (2008) 19, 3355–63.

Greene, D., Pruitt, L., and Maas, C. (1997) Biomechanical effects of e-PTFE implant structure on soft tissue implant stability. *Laryngoscope*, 107, 957–62.

Hanff, G., Dahlin, L.B., and Ludborg, G. (1992). Reconstruction of ligaments with expanded poytetrafluoroethylene. *Scandinavian Journal of Plastic Reconstructive Surgery and Hand Surgery*, 26(1), 43–9.

Hughes, K., Ries, M.D., and Pruitt, L. (2003). Structural degradation of acrylic bone cement due to in vivo and simulated aging. *Journal of Biomedical Materials Research*, 65A, 126–35.

Klapperich, C., Pruitt, L., and Komvopoulos, K. (2001). Chemical and biological characteristics of low-temperature plasma treated ultra-high molecular weight polyethylene for biomedical applications. *Journal of Materials Science: Materials in Medicine*, 12, 549–56.

Kühn, K.D. (2000). *Bone Cements: Up-to-date Comparison of Physical and Chemical Properties of Commercial Materials*. New York: Springer.

Kurtz, S.M. (2009). *The UHMWPE Handbook: Principles and Clinical Applications in Total Joint Replacement*. New York: Elsevier Academic Press.

Lawrence, T., and Davis, R.C. (2005). A biomechanical analysis of suture materials and their influence on a four-strand flexor tendon repair. *Journal of Hand Surgery*, 30(4), 836–41.

Maas, C.S., Gnepp, D.R., and Bumpous, J. (1993). Expanded polytetrafluoroethylene (Gore-Tex) in facial bone augmentation. *Archives of Otolaryngology: Head and Neck Injuries*, 90, 1008–15.

Paavolainen, P., Makisalo, S., Skutnabb, K., and Holmstrom, T. (1993). Biological anchorage of cruciate ligament prosthesis. *Acta Orthopedics Scandanavia*, 64, 323–28.

Pruitt, L. (2001). Fluorocarbon polymers in biomedical engineering. In *Advances in Polymer Science: Encyclopedia of Materials: Science and Technology*, ed. D.F. Williams. Oxford: Elsevier Science Limited.

Pruitt, L. (2003). Radiation effects on medical polymers and on their mechanical properties. In *Advances in Polymer Science: Radiation Effects*, ed. H.H. Kausch. Heidelberg: Springer-Verlag.

Ratner, B.D., Hoffman, A.S., Schoen, F.J., and Lemons, J.E. (1996). *Biomaterials Science: An Introduction to Materials in Medicine*. London: Academic Press.

Rosen, S.L. (1993). *Fundamental Principles of Polymeric Materials, Second Edition*. New York: Wiley.

Stanford, R. (2001). A novel, resorbable suture anchor: Pullout strength from the human cadaver greater tuberosity. *Journal of Shoulder and Elbow Surgery*, 10(3), 286–91.

Watanabe, T. (1984). Experimental study on the influence of porosity on the development of the neointima in e-PTFE grafts. *Journal of Japanese Surgery Society*, 85, 580–91.

Wheatley, G.H., McNutt, R.T., and Diethrich, E.B. (2007). Introduction to thoracic endografting: imaging, guidewires, guiding catheters, and delivery sheaths. *Annals of Thoracic Surgery*, 83(1), 272–78.

Young, R.J. (1991). *Introduction to Polymers*. London: CRC Press.

Zardenta, G., Mukai, H., Marker, V., and Milam, S.B. (1996). Protein interactions with particulate teflon: implications for the foreign body response. *Journal of Oral Maxillofacial Surgery*, 54, 873–78.

5 Mechanical behavior of structural tissues

Jevan Furmanski and Ayyana Chakravartula

Inquiry

How does the arrangement of collagen fibers in tendons assist in their function?

The question above represents a fundamental philosophy of this textbook. All materials, natural tissues included, have mechanical properties that reflect the chemical makeup and physical architecture of their microstructures. Tendons are assemblies of bundled collagen fibers arranged on a long axis that connect muscle to bone. Because tendons regularly experience axial tensile stresses with only very rare instances of other types of loading, this alignment optimizes the strength conferred by the collagen fibers by keeping them oriented in the direction of the greatest stress.

5.1 Historical perspective and overview

The earliest written records of anatomical study date back to 500 BCE, when Alcmaeon of Crotona asserted that the brain is the organ that governs intelligence. Later experiments by Claudius Galen (129–299 CE) were conducted on monkeys because human dissection was forbidden at that time. His theories on medicine (many of them incorrect, such as the concept that blood vessels were filled with grasping fibers that were responsible for blood flow) remained the basis of medical education for the next 1,400 years. Andreas Vesalius (1514–1564) made the next well-known advance in the understanding of human anatomy. Vesalius sought to recreate several of Galen's findings using human cadavers (by that time, human dissection was more accepted) and ultimately disproved many of them. In 1543, Vesalius published his results in *De humani corporis fabrica* (Mow and Huiskes, 2005; O'Malley, 1964), where, among other things, he concluded that it was the heart that caused blood to flow through arteries and veins. More recent developments in the study of natural tissues can be attributed to advances in histological techniques (microscopy), biochemical techniques (sample preparation and storage), and understanding of molecular biology (Piesco, 2002). While growth in these fields has led to a dramatic increase in our understanding of natural tissues, there remain aspects of many tissues that are still not fully understood, particularly regarding the relationship between their microstructure and overall physiological performance.

There has been a long history of employing applied mechanics alongside the investigation of tissues and physiological problems. Leonardo da Vinci was a student of both anatomy and mechanics, and was the first to solve a number of problems in the strength of materials and static structural analysis, though his work in anatomy was largely descriptive and not as applied as his engineering exploits. Galileo, a student of medicine and physics, later formalized many problems of statics, particularly the bending and breaking of beams and columns, and he noted how the weight-bearing bones of animals do not scale proportionally with size, but heavier animals must have stouter bones to account for their increased mass, while hollow bird bones represent a more weight-conservative structure. Galileo also pioneered precise measurement methods in physiological problems; for instance, he measured the human pulse and reported its frequency as the length of an equivalent pendulum. His legacy also lies in the passing on of his methods in philosophy to his contemporaries; one example is William Harvey, who studied at Padua while Galileo was there, and who by 1628 had a working model of blood circulation. Harvey established the one-way operation of heart valves, and measured the capacity of the heart, which led to his deduction of the existence of a yet-undiscovered bridge between the arteries and veins (capillaries) to form a closed circulation system some 45 years before the observation of capillaries was reported. Much later, as engineering mechanics advanced, so did its application to natural problems. Thomas Young, the namesake of the Young's Modulus, was an outstanding student of both mechanics and medicine, whose work encompassed the mechanics of impact, the physics of sight in the eye, and in 1808, a treatise on "Functions of the Heart and Arteries." In 1867, in an early prominent instance of applied mechanics in bone, von Meyer produced his explanation of the architecture of trabecular bone in the proximal femur by examining the lines of principal stress in a "crane" with similar shape and applied loading. Perhaps one of the greatest contributors during the mid-19th century to the burgeoning science of biomechanics was von Helmholtz, who held appointments as a professor of physiology, pathology, and physics at various times, and whom Y. C. Fung nominates as the "Father of Bioengineering" (Fung, 1981). Von Helmholtz made important contributions to the understanding of the operation of the eye, studied acoustics and hearing, measured the velocity of nerve pulse signals, and studied muscle contraction and contractile heat generation.

As engineering mechanics has grown more sophisticated, so have its applications in tissues, a trend that has been accelerating rapidly since the 1960s. Prominent books on biomechanics emerged during the early stages of this most recent period of growth and synthesized much of this progress, notably, J. B. Park's "Biomaterials: An Introduction" (1979), Y. C. Fung's "Biomechanics: Mechanical Properties of Living Tissues (1981), and "Basic Orthopaedic Biomechanics" (1991) edited by V. C. Mow and W. C. Hayes. Research and publication in the mechanical performance of physiological systems and tissues has exploded in subsequent years, as biological and materials research have become increasingly synergistic, to the enrichment of both the medical and scientific communities.

When designing an implant using synthetic biomaterials, it is necessary to know how the materials will respond to different stress states and environments. This response is related to the material's microstructure, as discussed in Chapters 1–4. Thus, it is equally important to understand the constituents and microstructure of the natural tissues being replaced. The tissues of the body are more complex than most synthetic biomaterials and many exhibit a multi-level hierarchical organization. Tissue microstructure is as predictive of the material response of that tissue as the nature of the chemical bonding is predictive of the behavior of metals or ceramics.

In this chapter, the mechanical behavior of natural tissues will be discussed within the framework of their constituents and microstructural organization. The aim is to provide the reader with a broad appreciation for the mechanical performance characteristics and structure-property relationships of the principal load-bearing tissues of the body.

5.2 Learning objectives

The focus of this chapter is on the relationships between a natural tissue's microstructure and its mechanical function. Specific material properties are given wherever possible. By the end of this chapter, students should be able to:

1. describe the microstructure and mechanical properties of the basic building blocks of structural tissues
2. explain the hierarchical organization of dental tissues, bone, tendons, ligaments, cartilage, skin, and blood vessels
3. predict the macroscopic mechanical performance of tissues based on their microstructure and constituents
4. compare and contrast the mechanical performance of various structural tissues

5.3 Building blocks of tissues

Natural tissues are made up of basic structural elements whose physical and chemical structure determine the macroscopic mechanical behavior of the tissues. As in the study of synthetic biomaterials, an understanding of these basic building blocks is essential when developing the structure-property relationships for natural tissues. The two dominant proteins in the body's structural tissues are collagen and elastin. Collagen is one of the most abundant proteins in the body, and is found in tissues as diverse as the strong, resilient tendon connecting muscle to bone and the loose, gel-like vitreous humor in the eye. Elastin is a rubbery protein that allows for elastic stretching and, perhaps more importantly, recovery after large deformations. This enables tissues such as skin and arteries to experience physiological conditions without permanent damage. Some natural tissues, such as bone and enamel, also contain hydroxyapatite, a mineral that is responsible for giving these tissues their hardness, strength, and durability. Familiarity with the roles of collagen, elastin, and hydroxyapatite, both in healthy and in aged or diseased tissues, is crucial when seeking to replicate their performance or to restore their function with synthetic materials.

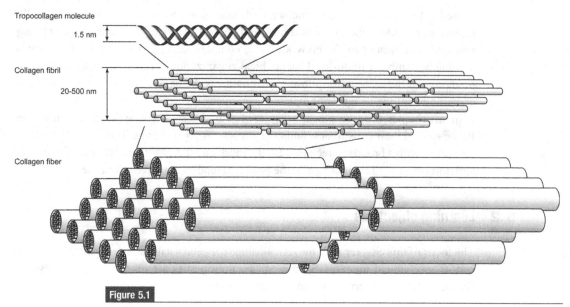

Figure 5.1

Schematic of a collagen fiber showing its hierarchical organization.

5.3.1 Collagen

Collagen fibers appear in a variety of conformations specific to the functional demands of the different tissues of the body. There are more than 20 known types of collagen and they are named according to the order in which they were discovered. All collagen was initially thought to consist of a single protein, until a second type of collagen, Type II, was discovered in 1970. The discovery of the other major collagen types followed shortly (McGrath *et al.*, 2004). The primary variants are Types I, II, and III, which are the rope- or fibril-forming collagens. Type I collagen is found in bone, dentin, ligament, tendon, and blood vessel wall. Type II is the main collagen constituent in articular cartilage and the nucleus pulposus of the intervertebral disc. Type III collagen is found in skin, artery walls, and hollow organ walls (Bhat, 2005). The other types of collagen are also associated with specific structures: Types IV (found in the basement membrane) and VII (found beneath stratified squamous epithelia) form meshes, while Types IX (commonly found in some cartilages) and XII (present in tendons and ligaments) bond to the surface of collagen fibrils, helping to link them to one another or to the extracellular matrix (the structural support of connective tissues) (Champe *et al.*, 2008; Safadi *et al.*, 2009).

Three left-handed helical polypeptide chains combine to form a right-handed superhelix known as a tropocollagen molecule. As shown in Figure 5.1, the tropocollagen molecules are 1.5 nm in diameter and 280 nm long and are stacked quasi-hexagonally in a staggered overlapping fashion, with each molecule indexed ahead of its neighbor by one-fourth of its length. The spaces between the head and tail of neighboring molecules are known as the hole zone, while the lateral spaces between molecules are called pores.

The stacked tropocollagen molecules are assembled into collagen fibrils, which are 20–40 nm in diameter. Finally, these fibrils are bundled together to form collagen fibers, whose diameter ranges from 0.2–1.2 μm (Park and Lakes, 2007).

Collagen requires hydration in order to maintain its elasticity. When dehydrated, it has an elastic modulus on the order of 6 GPa and it becomes very brittle (Vincent, 1990). It is believed that a small amount of water is incorporated directly into the tropocollagen triple helix and interacts strongly with it. Further hydration causes the collagen to swell and gives it a more extensible and plastic character. The elastic modulus of hydrated collagen fibers is approximately 1–2 GPa (Park and Lakes, 2007; Vincent, 1990). Collagen fibers are flexible but inelastic and have high tensile strength (Champe *et al.*, 2008). The tensile strength is reported to be 50–100 MPa while the ultimate strain is 10% (Park and Lakes, 2007). The mechanical properties of tendon are often substituted for the mechanical properties of Type I collagen, due to the high proportion of Type I collagen in this type of tissue (Lucchinetti, 2001).

Collagen molecules are also covalently crosslinked internally between helixes and with neighboring molecules by an enzymatic process that acts on specific sites in the molecules (Bhat, 2005). Crosslinks are also formed by glycosylation reactions between glucose and amino acids in collagen to form non-enzymatic crosslinks; this process can be accelerated in diabetes due to the resulting increased glucose content in the tissue (Lucchinetti, 2001). The degree of crosslinking generally increases with age, with increased crosslinking resulting in a more brittle tissue, though the specific contribution of enzymatic and non-enzymatic crosslinks to the embrittlement process is a topic of ongoing research (Ruppel *et al.*, 2008; Siegmund *et al.*, 2008).

Example 1 **Locations of cartilage variants**

Types I, II, and III are the primary variants of collagen. What are some locations where other types of collagen are consistently found?

Solution Type IV collagen is found in the basement membrane, which is underneath the epithelium or the endothelium (the linings of cavities, organs, and blood vessels). Type V collagen is found in large blood vessels, cornea, and bone (in small amounts). Type VII is found beneath stratified squamous epithelia (layered lining cells, such as skin). Types IX and XI are found in cartilage and Type XII is found in tendons and ligaments.

5.3.2 Elastin

Elastin is a highly crosslinked, and therefore insoluble, protein that is found in elastic collagenous tissues. Elastic fibers are made up of elastin and 10–12 nm diameter fibrillin-based microfibrils (Champe *et al.*, 2008; Kadar, 1989; Kielty *et al.*, 2003). Fibrillin is

134 Mechanical behavior of structural tissues

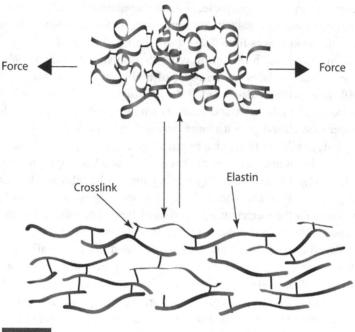

Figure 5.2
Schematic of elastic fibers demonstrating the use of crosslinks for preventing slip of the elastin molecules.

a glycoprotein that forms a scaffold for elastin deposition. The crosslinks in elastin are responsible for the recoil in elastic fibers as portrayed in Figure 5.2. These fibers can be stretched to 1.5 times their normal length without permanent deformation.

Elastin is responsible for the stiffness of many soft tissues when they are relaxed, owing to the near zero elastic modulus of crimped collagen fibers in that condition. Figure 5.3 shows a schematic of the total response of a ligament superimposed over the response of the tissue with either the collagen or elastin enzymatically removed. It can be seen that the two constituents interact essentially in parallel. Removing the elastin may, however, allow the remaining collagenous matrix to flow in a viscous manner. The initial low modulus response, due to the elastin and still-crimped collagen, is termed the toe region. Beyond the heel region (the higher slope portion of the curve), the response is almost entirely due to the collagen constituents and is termed the linear region. When the collagen is selectively digested, the toe region response is nearly unaltered and the reverse is true of the linear region when the elastin is removed.

The mechanical properties of elastic fibers have not been widely studied. Using atomic force microscopy (AFM), Koenders *et al.* have found the bending modulus for hydrated elastic fibers without microfibrils to be 0.3–1.5 MPa. The microfibrils were not found to contribute significantly to the bending modulus of the elastic fibers (Koenders *et al.*, 2009). The Young's modulus for elastic fibers is 0.6 MPa, the tensile strength is 1 MPa, and the ultimate strain is 100% (Park and Lakes, 2007). When compared to collagen

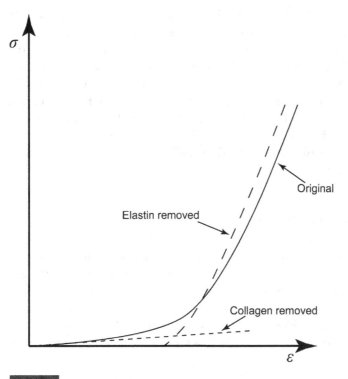

Figure 5.3

The behavior of a ligament in its unaltered state compared to when the elastin or collagen is enzymatically digested, showing the dominance of these constituents in the toe and linear regimes (respectively). (After Park and Lakes, 2007.)

fibers, elastin fibers have much lower moduli and tensile strengths, while having an ultimate strain that is an order of magnitude higher (Park and Lakes, 2007).

5.3.3 Hydroxyapatite

Hydroxyapatite is the mineral responsible for stiffening teeth and bones. It is a variant of calcium phosphate, $Ca_{10}(PO_4)_6(OH)_2$. In the body, hydroxyapatite is not stoichiometric, as it contains carbonate, sodium, potassium, citrate, and other trace elements. The carbonate content of bone is 4–8% and increases with age (Roverri and Palazzo, 2006). In bone, hydroxyapatite forms plate-like crystals that are approximately 4 nm thick, 50 nm wide, and 50 nm deep (Currey, 1998). This geometry creates a large surface area which, when combined with its trace impurities, increases the mineral's solubility. Since the secondary function of hydroxyapatite is to serve as a reservoir for mineral exchange, such a high solubility is desirable in bone (Boskey, 2001).

In enamel, the crystals are much larger, 30 nm × 90 nm in cross-section, with a length estimated to be on the order of millimeters and potentially spanning the entire enamel thickness (Piesco and Simmelink, 2002). These crystals are orders of magnitude larger than the ones found in bone. Because of this size and the fact that enamel is mainly composed of hydroxyapatite, the mechanical properties of these crystals are assumed to be close to that of enamel and are measurable using nanoindentation. Habelitz *et al.* found the elastic modulus of dental enamel to be 87.5 GPa in the direction parallel to the long axis of the hydroxyapatite crystals, and 72.7 GPa in the direction perpendicular to the long axis (Habelitz *et al.*, 2001). Park and Lakes report the elastic modulus of enamel to be 74 GPa without specifying orientation. Synthetic hydroxyapatites used for bone reconstruction have elastic moduli in the range of 40–117 GPa and $v = 0.27$ (Park and Lakes, 2007).

5.4 Load-bearing tissues

The constituents of the primary load-bearing tissues in the body are collagen, elastin, and hydroxyapatite as described in the previous section. The materials given in this section are listed in order roughly from most highly mineralized or high-strength tissues, moving toward the weaker tissues. It is useful to think of the natural tissues as existing along a continuum, with their location determined by the quantities and organization of the basic building blocks. The constituents of these tissues are given in Table 5.1.

5.4.1 Enamel and dentin

Enamel and dentin are structural tissues with properties suited to the severe mechanical demands of mastication, or chewing. The location of these tissues is illustrated in Figure 5.4. A tooth has two sections: the crown, which sits above the gumline, and the root, which sits below the gumline. The outermost surfaces of the crown are covered by enamel, which is an extremely hard and wear-resistant material composed almost entirely of hydroxyapatite, as discussed previously. Dentin makes up the majority of the subsurface of the tooth, apart from the non-structural pulp core, and is less mineralized but substantially more compliant and tougher than enamel. Unlike other structural tissues, enamel and dentin are exposed to a complex and aggressively corrosive environment in combination with high stresses, which can degrade or destroy the tissue.

Enamel is by far the most mineralized tissue in the body, with 96 wt% hydroxyapatite, 1 wt% organic matter, and 3 wt% water (Healy, 1998). As described earlier, enamel consists of hydroxyapatite crystals that form long prisms extending from the surface of the tooth to the dentino-enamel junction. The space between the mineral prisms contains proteins that control the prism geometry and mitigate the intrusion of the environment.

The structure of dentin is more complex than that of enamel. It contains 70 wt% carbonated hydroxyapatite, 20 wt% organic matter (mainly Type I collagen), and 10 wt% water (Healy, 1998). The microstructure of dentin consists of the dentin matrix, dentinal tubules, mineral, and dentinal fluid. The mineral crystals in dentin are smaller

Table 5.1 Representative values for load-bearing human tissue constituents. In some cases, the percentage of collagen and elastin was given as a single number, which is indicated by a shared cell

	Collagen	Elastin	Hydroxyapatite	Water	Other (proteoglycans, cells)
Enamel	<1 wt%[1]		96 wt%	3 wt%	
Dentin	22 wt%, mainly Type I[2]		65 wt%	13 wt%	
	20 wt%[3,4]		70 wt%	10 wt%	
Cortical Bone	40% dry wt, mainly Type I[5]		60% dry wt		
	22 wt%, mainly Type I[6]		69 wt%	9 wt%	
	27–28 wt%, mainly Type I[7]		60 wt%	8–10 wt%	<1 wt%
	30 wt%, mainly Type I[8]		60 wt%	10 wt%	
Trabecular Bone	34.9 vol%[9]		38.1 vol%	27.0 vol%	
Tendon	75–85% dry wt, mainly Type I[8]	1–3% dry wt 2 wt%[10]		60 wt%	1–2% dry wt
Ligament	70–80% dry wt[8]	1–15% dry wt		60 wt%	1–3% dry wt
Articular Cartilage	15–20 wt%, mainly Type II[10,11]			60–80 wt%	10 wt%
	10–20 wt%, mainly Type II[8]			70–90 wt%	4–7 wt%
	10–20 wt%, mainly Type II[12]			68–85%	5–10%
Skin	70–80% dry wt[13]	1–2% dry wt			
Artery (canine)	19.6–50.7% dry wt[14]	15.6–41.1% dry wt		63.2–73.8 wt%	
Aorta (human)	19.2–29.6% dry wt[15]	27.2–44.8% dry wt			
Pulmonary Artery (human)	27.1–36.3% dry wt[15]	17.8–34.2% dry wt			

[1]Piesco and Simmelink, 2002, [2]Zaslanksy, 2008, [3]Piesco, 2002, [4]Healy, 1998, [5]Safadi et al., 2009, [6]Roverri and Palazzo, 2006, [7]Morgan et al., 2008, [8]Bartel et al., 2006, [9]Gong et al., 1964, [10]Ventre et al., 2009, [11]Parsons, 1998, [12]Mow et al., 2005, [13]Oikarinen and Knuutinen, 2002, [14]Fischer and Llaurado, 1966, [15]Hosada et al., 1984.

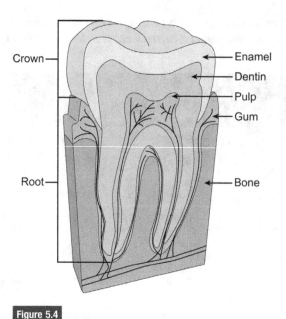

Figure 5.4

Diagram of a whole tooth, indicating the position and relative thickness of enamel and dentin on the crown.

and thicker than in bone; they are on the order of 36 nm wide, 25 nm deep, and 10 nm thick (Zaslansky, 2008). The dentinal tubules are regularly spaced (approximately 45,000 per mm^2) and have an inner diameter of around 1 μm, as shown in Figure 5.5. They radiate from the pulp space out to the dentino-enamel junction. The dentin between the tubules is termed peritubular dentin and is more mineralized than the rest of the dentin matrix.

Enamel is a brittle material due to its high mineral content. As described earlier, the elastic modulus depends on the orientation of the hydroxyapatite crystals; in the direction parallel to the mineral prisms its properties are very similar to pure hydroxyapatite. The elastic modulus of human enamel ranges from 74–88 GPa perpendicular to the surface, and the elastic modulus in the plane of the surface is quite reduced. Reported values of enamel elastic modulus vary widely between laboratories and test methods, so these data are more appropriately treated as estimates (Healy, 1998). The elastic modulus of enamel also varies between locations on the surface of the tooth; for example, chewing surface enamel is stiffer than that on the sides of the tooth. The compressive strength of enamel is 200–300 MPa, while the tensile strength is a much lower 10 MPa, as is typical for many brittle materials. The energy of fracture (also called energetic toughness) is 190 J/m^2 when notched and fractured perpendicular to the mineral prisms and tensile loaded parallel to them, and 13 J/m^2 when the fracture is parallel to the prisms (Rasmussen and Patchin, 1984). The dramatically lower energy required for fracture when the crack propagates parallel to the mineral prisms demonstrates the relative ease with which the prisms can be pulled apart.

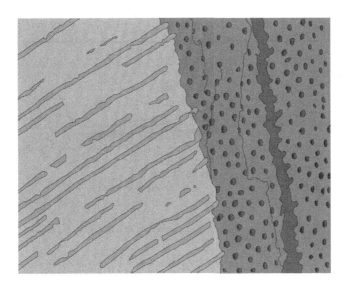

Figure 5.5

An illustration of dentinal tubules.

Dentin is more compliant than enamel, exhibiting an elastic modulus that ranges from 8–21 GPa (Healy, 1998; Zaslansky, 2008). The elastic modulus of dentin appears to be strongly influenced by location and local microstructure. The shear modulus of human dentin is 5–8 GPa and the Poisson's ratio is 0.23–0.3. The tensile strength of human dentin is 30–60 MPa, and its compressive strength is 230–350 MPa, which is similar to the compressive strength of enamel. The salient difference between enamel and dentin is not in their stiffness or strength, but in their fracture resistance or damage tolerance. The reported energy of fracture for human dentin is 550 J/m^2 when fractured parallel to the tubules and 270 J/m^2 when fractured perpendicular to them (Rasmussen and Patchin, 1984). Thus, dentin requires 1.5–200 times more energy to fracture than does enamel. It is also interesting to note that, while enamel has a very low work of fracture when the prisms are pulled apart (parallel to the tooth surface), dentin is tougher in that direction. The collagen fibers in dentin are organized perpendicular to the tubules, which increases the energetic toughness parallel to the tooth surface. Recently, fracture toughness measurements of elephant dentin in various orientations reached similar conclusions: the collagen fiber orientation dominates toughness by causing cracks to deflect or by bridging the propagating crack with unbroken ligaments (Nalla *et al.*, 2003). Koester and colleagues examined microcrack propagation in young and aged human teeth and demonstrated that many toughening mechanisms such as crack deflection, crack branching, and uncracked-ligament bridging were less effective in the aged dentin, where the tubules are filled with apatite (Koester *et al.*, 2008). Thus, the microstructure and constituents of dentin play a large role in its mechanical properties. Further discussion of dental tissues and implants can be found in Chapter 15.

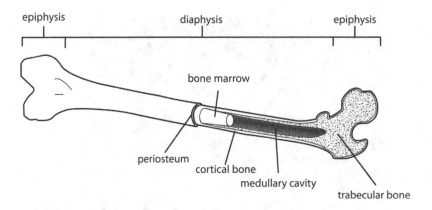

Figure 5.6

Anatomy of a long bone.

5.4.2 Cortical bone

Cortical bone tissue is the principal structural material of the skeleton, making up 80% of the skeleton's mass (Jee, 2001). Also known as compact bone, it provides the requisite strength and stiffness to support and react against muscle contraction and the loads caused by strenuous activities. It also provides protection for the organs and gives shape to the body. Bones can be classified into five types: long (femur, tibia, ulna, and radius), short (carpal), flat (skull, sternum, and scapula), irregular (vertebra and ethmoid), and sesamoid (embedded in tendon, such as the patella). The distribution of cortical bone can vary widely. For example, the ulna is 92% cortical bone, while a typical vertebra is 62% cortical bone (Jee, 2001). Long bones such as the femur or tibia have two epiphyses, wide ends that are made up of trabecular bone covered by a shell of cortical bone, and a shaft, known as the diaphysis and composed primarily of cortical bone. This geometry is depicted in Figure 5.6. Bone tissue is remarkable as a structural material in that it repairs itself by replacing old or damaged regions with new tissue through a process termed remodeling. This also allows bones to adapt to changes in activity or loading over time.

Cortical bone is a complex composite material composed chiefly of hydroxyapatite ceramic crystals bound by a collagen-based polymeric matrix. Bone tissue is approximately 60 wt% inorganic, with 8–10 wt% water content and the balance composed of organic material (Morgan *et al.*, 2008). The organic matrix phase, called the osteoid, contains 90 wt% Type I collagen, about 8 wt% non-collagenous proteins, and 2 wt% cells. It is composed of a partially oriented network of collagen fibers.

Figure 5.7 shows the structural hierarchy inside of a collagen fiber, where the collagen molecules are staggered by a quarter period and spaced by hole and pore zones. The hole zones are thought to be the first points to mineralize when new bone is being formed, with mineralization then spreading into the pore spaces between the molecules, as seen

5.4 Load-bearing tissues

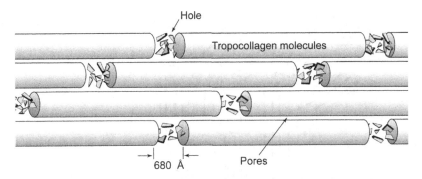

Figure 5.7

Nanoscale platelets of hydroxyapatite mineral form in the hole zones in the collagen sub-fibril in the early stages of mineralization. (After Fratzl et al., 2004.)

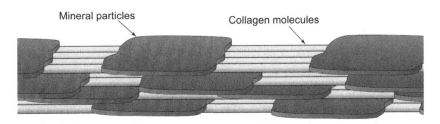

Figure 5.8

As mineralization progresses, hydroxyapatite platelets spread out between the collagen molecules into the pore space. (After Fratzl et al., 2004.)

in Figure 5.8 (Boskey, 2001; Fratzl et al., 2004). In this manner the organization of the collagenous osteoid determines the overall structure of cortical bone.

These constituents are not organized simply like engineered composites; rather, they exhibit a more complicated hierarchical structure in a basic unit termed the osteon, that may allow cortical bone to resist fracture or crack propagation in a manner unavailable to simpler types of organization (O'Brien et al., 2007; Taylor, 2007). The osteonal system is also termed the Haversian system, so named for the early work of Clopton Havers in the late 17th century (Martin et al., 1998).

As illustrated in Figure 5.9, the cylindrical osteons are comprised of individual sheets, or lamellae, of mineralized tissue. Osteons are 200–250 μm in diameter, while lamellae are 1–5 μm thick (Guo, 2001; Jee, 2001). Each osteon is made up of approximately 20–30 lamellae that wrap around the Haversian canal, which is the main conduit for blood and lymphatic vessels in the long axis of the osteon. Transverse to the Haversian canals are Volkmann's canals, which serve the same function. Ellipsoidal pores in the bone called lacunae, 5–8 μm in length, are interconnected by channels called canaliculi (approximately 0.5 μm in diameter), with 50–100 canaliculi extending from each lacuna. The lacunae contain osteocytes, the principal bone cells, and the canaliculi allow

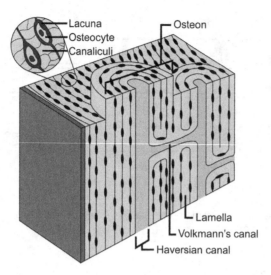

Figure 5.9

The organization of cortical bone into osteons.

transport between osteocytes. The canaliculi also allow the osteocytes to extend beyond the lacunae for the purpose of sensing the local state of deformation and communicating through the canalicular network to cells specialized to the task of structural adaptation and remodeling of the bone tissue (Park and Lakes, 2007). Osteons are separated by the cement line, which is incompletely mineralized material left at the boundary of remodeling zones. A 1 μm thick interlamellar layer separates the lamellae. This interlamellar layer has a less organized mineral phase compared to that of lamellae and may be thin enough to permit collagen fibrils to bridge from one lamella to its nearest neighbor (Martin and Burr, 1989).

As described earlier, lamellae are composed of collagen fiber bundles aligned with a dominant orientation, which in turn forces the mineral reinforcement to take on a preferred organization as well, thus making the lamellae mechanically anisotropic, meaning that their mechanical properties change according to orientation. Typically, the preferred orientation of the lamellae alternates from layer to layer, as evidenced under polarized light microscopy (Jee, 2001).

Cortical bone behaves in a linear elastic manner at moderate loads, with a clear transition at a yield point to plastic deformation. Figure 5.10 shows its uniaxial deformation behavior. Although the material properties can vary according to the type of test and the location of the bone, some representative values are given here. Human cortical bone can be tested longitudinally or transversely. From these tests, elastic modulus is approximately 17 GPa and 11.5 GPa for longitudinal and transverse testing, respectively. The shear modulus is around 3–3.5 GPa. The ultimate tensile strength is on the order of 120–130 MPa longitudinally, and just under half that value at around 51 MPa transversely. In compressive testing, the ultimate strength is higher, around 170–190 MPa

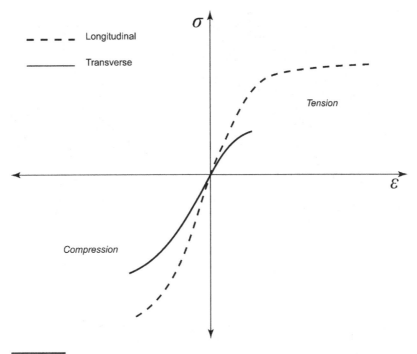

Figure 5.10

Tension and compression response of cortical bone, in the longitudinal and transverse orientations. The longitudinal orientation exhibits much more ductility in tension, and more strength in both tension and compression than the transverse orientation. (After Bartel *et al.*, 2006.)

longitudinally and 133 MPa transversely. Finally, the ultimate shear strength is approximately 68 MPa (Bartel *et al.*, 2006; Guo, 2001; Park and Lakes, 2007).

Cortical bone is mechanically anisotropic due to the orientation of the osteons and the anisotropy of the lamellae as described above. Figure 5.10 compares the resistance to deformation of cortical bone in the longitudinal and transverse directions, showing that the longitudinal direction is stiffer than the transverse direction in both tension and compression. The strength in the longitudinal direction is also greater in tension and compression, and the transverse direction tolerates very little strain in tension beyond the linear elastic region before failure (Bartel *et al.*, 2006).

5.4.3 Trabecular bone

Trabecular bone, also referred to as cancellous bone, is a bone tissue that is located principally at the ends of long bones and in the interior of vertebrae as pictured in Figure 5.11. Trabecular bone serves a different purpose from cortical bone. The pore space between the trabeculae in most trabecular bone (and particularly in the iliac

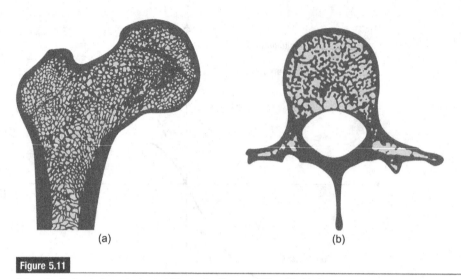

Figure 5.11

Schematic of trabecular bone structure in (a) the proximal femur and (b) vertebral body.

crest, vertebrae, and proximal femur) is filled with red bone marrow where the body's red blood cells are produced, a process called hematopoiesis. Bone tissue serves as a reservoir for calcium and phosphorous ions, with approximately 99% of the calcium in the body contained in the skeleton (Bartel *et al.*, 2006). The increased surface area of trabecular bone (nearly eight times greater than that of cortical bone) makes the removal and transport of these ions much more rapid in trabecular than cortical bone (Jee, 2001). Thus, while the majority of the body's calcium is contained in cortical bone, the most accessible fraction is in trabecular bone, which can be substantially remodeled or consumed if calcium ions are in short supply. In old age, when calcium uptake from the diet is less efficient, this can lead to excessive remodeling of trabeculae and therefore an increased risk of osteoporosis-related fractures (Bartel *et al.*, 2006). The weight percentages of water, inorganic, and organic material in trabecular bone are 27%, 38%, and 35%, respectively (Keaveny, 1998).

Trabecular bone does not exhibit a compact construction like cortical bone, but is an open three-dimensional network of interconnected rod or plate-shaped trabeculae as seen in Figure 5.12. This cellular structure is substantially less stiff than cortical bone, and is more isotropic in its overall properties although heterogeneous at the microstructure level. While trabecular bone is differentiated from cortical bone by its characteristic network structure, the properties of the actual mineralized lamellae comprising the trabeculae (and thus both the mineral and organic phases) are very similar to those of cortical lamellae (Guo, 2001). Trabeculae range in thickness from 100 to 200 μm, typically spaced 500 to 1500 μm apart, and the lamellae and osteocytes that comprise them are arranged in groups called hemi-osteons (Bartel *et al.*, 2006; Jee, 2001). Very thick trabeculae may also contain complete osteons resulting from

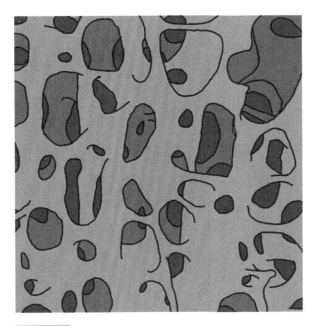

Figure 5.12

Illustration of the network structure of trabecular (cancellous) bone, showing both sheet-like (plate) and thin and beam-like (rod) morphologies.

remodeling processes identical to those that occur in cortical bone. One of the more striking aspects of trabecular bone is that the orientation of the trabeculae is coordinated (and perhaps optimized) over large distances. Trabecular bone can remodel and adapt to changes in loading conditions to maintain this organized structure, though disease states such as osteoporosis can lead to a loss of structural integrity resulting from a corrupted remodeling process.

Trabecular bone can exhibit dramatically different architectures depending on the anatomical site investigated and the health, age, and activity level of the individual. For example, vertebral trabeculae tend to be more rod-shaped, while the proximal femur contains a mixture of rod and plate-shaped trabeculae (Morgan and Keaveny, 2001). The mechanical behavior of trabecular bone is dominated by its porosity, or the volume fraction not occupied by bone tissue. The porosity of cortical bone in humans that is due to the Haversian and Volkmann's canals ranges from 5–30%, while trabecular bone porosity can range from 30% to 90%, with the higher range associated with elderly vertebrae (Samuel *et al.*, 2009).

The architecture of trabecular bone can be described as having a dominant orientation, also termed the principal material coordinate system. Trabeculae in a region tend to be aligned in this manner, though the principal material coordinate system may not correspond with any particular anatomical direction. The trabeculae in vertebrae, for

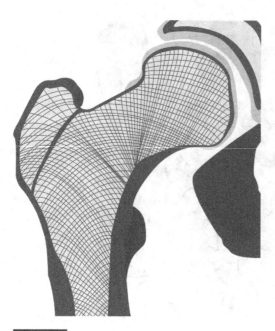

Figure 5.13

Illustration of stress trajectories in femoral trabecular bone which roughly correspond to the principal material axes in trabecular bone in Figure 11(a). (After Cowin, 2001.)

example, tend to orient along the axis of the spine, as the dominant mode of loading on vertebrae is axial compression. However, in the proximal femur the principal material coordinate system changes smoothly from point to point and is not aligned with the femoral axis. In Figure 5.13, an illustration of possible stress lines in femoral trabeculae is given. In 1866, a similar sketch was developed by Herman von Meyer based on the principal stress lines for a weight-bearing crane. A theory was then proposed linking the principal material directions of trabecular bone to the predominant "trajectory" of stress at the anatomical site. This explanation for trabecular long-range architecture remains a topic of considerable debate (Cowin, 2001).

The open network structure of trabecular bone imparts behavior characteristic of cellular solids, or materials with a porous microstructure. In these materials, the apparent macroscopic properties are related more to the deformation of the network as an assembly than to the properties of the underlying solid material. Trabeculae behave similarly to beams or plates in a frame, bending or straining to accommodate the macroscopic load state, as illustrated in Figure 5.14. The implications of the beam or plate-like behavior of trabeculae are profound. For instance, trabeculae can buckle in local compression, and individual trabeculae can be in tension when the tissue is in a global state of compression; both of these scenarios can lead to failure of the material.

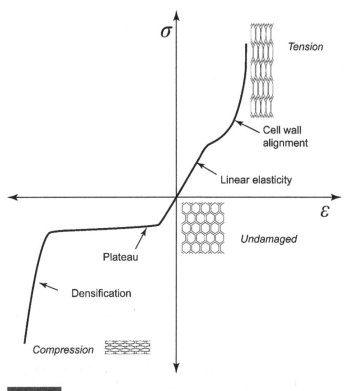

Figure 5.14

Schematic showing the typical response of an open cell foam. (After Gibson and Ashby, 1999.)

Figure 5.14 shows the deformation response typical to an open cell, foam-like material in tension and compression. In compression, the cellular structure begins to crush at a nearly constant stress after yield until at large macroscopic strains the pore space is eliminated and the cell structure compacts into a solid: a process termed densification. In tension, yield is followed by alignment and tensile loading of the cell walls, and ultimately by rupture (Gibson and Ashby, 1999).

Simplified models and experimental data demonstrate a strong dependence of the strength and stiffness of trabecular bone on its porosity and trabecular architecture, as mentioned above with the relationship schematically shown in Figure 5.15 (Martin and Burr, 1989). The porosity of trabecular bone varies not just from one anatomical site to another, but also potentially within neighboring regions in the same site. Thus, accurately predicting the behavior of whole bones is complicated by the systematic variation of trabecular bone porosity and architecture. Furthermore, mechanical properties reported for trabecular bone from a particular anatomical site (or from a non-human species) may be an order of magnitude or more different from another anatomical site or species of interest.

148 Mechanical behavior of structural tissues

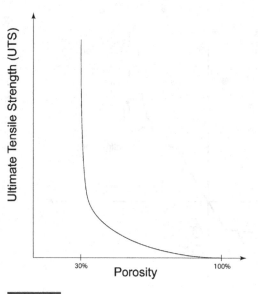

Figure 5.15

A schematic of the relationship between ultimate tensile strength and porosity of trabecular bone. (After Martin and Burr, 1989.)

The elastic modulus of human trabecular bone ranges widely, from 10 MPa to 2000 MPa, and its strength ranges from 0.1 MPa to 30 MPa (Keaveny, 1998). The macroscopic mechanical performance of human trabecular bone exhibits an approximately linear elastic region at moderate strains, followed by yielding and inelastic deformation or failure. In tension, trabecular bone is substantially less ductile; trabecular fractures occur shortly after the yield point, followed by specimen rupture (Bartel et al., 2006). The structural performance of trabecular bone is especially relevant to bone grafts used to alleviate skeletal defects.

5.4.4 Tendon and ligament

Tendon and ligament are bands of soft tissue that are made up of aligned collagen fibers. These tissues are used to transmit tensile forces between two bones (ligament) or between bone and muscle (tendon). The constituents of tendon and ligament are similar but vary in amount, as shown in Table 5.1. Tendons and ligaments have similar compositions, consisting of approximately 60 wt% water. The dry weight percentages of collagen, elastin, and proteoglycans in tendons are 75–85%, 1–3%, and 1–2%, respectively. Ligaments have a slightly lower collagen content and can have a much higher elastin content, with dry weight percentages of 70–80% collagen, 1–15% elastin, and 1–3% proteoglycans (Bartel et al., 2006). This higher elastin content allows ligaments to undergo larger deformations. Proteoglycans are complicated macromolecules with a

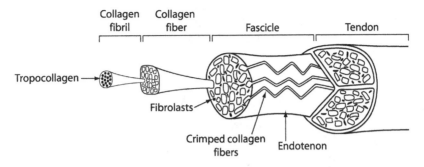

Figure 5.16

The hierarchical structure of a tendon.

heavily branched structure comprised of large brush-like arms. Cells called fibroblasts, which are embedded among the collagen fibers as depicted in Figure 5.16, maintain the tissue.

The structure of both tendon and ligament consists primarily of an assembly of collagen fibrils aligned with the length of the tissue. An important distinguishing characteristic of the structure of these tissues is a crimped or kinked form superimposed on the fibril alignment. This crimp dramatically alters the mechanical properties of the tissue. The crimped collagen fibers, 50–500 μm in diameter, are aligned and comprise a substructure of the tissue called a fascicle. The fascicles range in diameter from 0.25 mm to several millimeters, and are bound to one another by a layer of soft tissue termed the endotenon, which binds the collagen fibers and contains the nervous, lymphatic, and vascular elements of the tissue. The entire tendon is surrounded by the epitenon, which has a similar structure and function as the endotenon (Frank *et al.*, 2007; Lundon, 2003). In the toe region of the stress-strain curve, the collagen fibers uncrimp and when they are fully stretched, they can bear more load which results in the higher elastic modulus seen in the heel and linear regions.

The motive action of the muscle must be borne by the relatively small cross-section of tendon tissue, and so tendons must have a high load-carrying ability. Ligaments tie bones together, typically to constrain relative motion at joints, such as in the knee as seen in Figure 5.17. Ligaments usually take load only at extreme joint motions of flexion or extension, but in these cases the forces borne by them can be considerable. Tendon and ligament tissues perform very differently from hard tissues, undergoing large repetitive deformations and returning to their initial state without damage. Many tendons regularly slide over other structures in the body and therefore require protection from abrasion and wear. Tendons can adapt and take on cartilage-like morphology to resist direct bearing contact in these regions (Frank *et al.*, 2007).

Typical stress-strain curves for tendon and ligament are shown schematically in Figure 5.18, where the characteristic toe region is easily distinguished from the stiffer behavior at larger strains. The reported elastic moduli of tendons and ligaments show wide variation with anatomical site and deformation rate. The elastic modulus of tendons

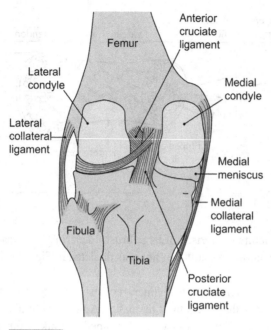

Figure 5.17

The ligaments in the knee.

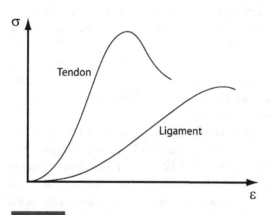

Figure 5.18

Schematic showing relative deformation response of tendon and ligament.

varies from 500–1850 MPa, while the range for ligaments is 50–541 MPa. The ultimate tensile strength for human tendon ranges from 50–125 MPa, compared to 13–46 MPa for human ligament. The ultimate strain of tendon range from 10–60%, compared to 10–120% in ligament (Bartel *et al.*, 2006; Frank *et al.*, 2007; Woo and Levine, 1998; Wren *et al.*, 2001).

The transition from compliant to stiff behavior in tendon and ligament derives primarily from the crimped form of the collagen fibers (Frank et al., 2007). This intrinsic crimp in the tissue renders it initially very compliant, but as the tissue stretches the crimps straighten out and the tissue behaves progressively more like perfectly aligned collagen fibers. This process is known as fiber recruitment. Human ligament contains substantially more elastin (1–15% by dry weight) than tendon does (1–3%). This fact, combined with the collagen alignment in ligaments being less pronounced than that of tendons, accounts for the more pronounced toe regime behavior observed in tensile tests of ligaments as compared to tendons (Ventre et al., 2009). It is important to note that tendons are structurally in parallel with muscles, which likely keeps both tissues in a steady state of tension, making the toe region irrelevant in the body under ordinary circumstances.

Tendons and ligaments are structures of discrete components, not homogeneous units. The ligaments of the knee, in a given position of flexion, can have substantial portions that are relaxed. Individual fascicles in ligaments may take precedence for certain activities (while others remain slack) and therefore take an uneven fraction of the transmitted load. Thus, prediction of the behavior of the structure from *in vitro* tensile data is problematic. Reporting values of whole ligaments and tendons avoids ambiguity of anatomical site inherent in merely reporting material values, and is recommended (Bartel et al., 2006). It is important to bear in mind that testing whole tendons or ligaments in a load frame stresses all fascicles artificially, and potentially quite differently than what would occur anatomically (Frank et al., 2007). The replacement of soft tissues including tendons and ligaments will be discussed further in Chapter 16.

5.4.5 Articular cartilage

Articular cartilage, or hyaline cartilage, is a hydrated collagenous tissue that covers the contacting surfaces of skeletal joints as portrayed in Figure 5.19. Cartilage is optimized for compressive loading. It exhibits extremely specialized characteristics, such as a much lower coefficient of friction than synthetic materials and a trapped network of charged molecules that actively recruit lubricating fluid into the tissue through osmosis. The 60–80 wt% water recruited and contained in articular cartilage is an active agent in its mechanical behavior and function, lubricating the contact surfaces for reduced friction and wear, and also providing a dynamic resistance to deformation (Parsons, 1998). The relatively small fraction of solid contained in healthy articular cartilage renders it substantially less strong or tough than most other critical structural tissues, thus motivating a detailed understanding of its behavior to predict the impact of severe or morbid conditions on its performance.

Articular cartilage has a limited capacity for self-repair. It is not vascularized, so the cells that develop and maintain cartilage, chondrocytes, receive nutrients through slow diffusion of the fluid in the tissue. Chrondrocytes are not mobile, so they cannot migrate through the tissue to repair damaged regions. The combination of relative structural

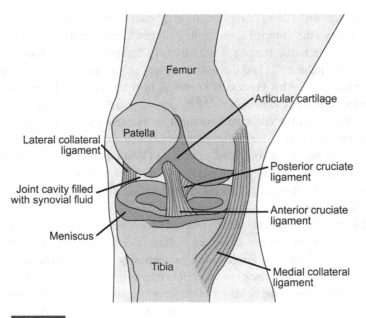

Figure 5.19

Anatomy of the knee showing the locations of articular cartilage.

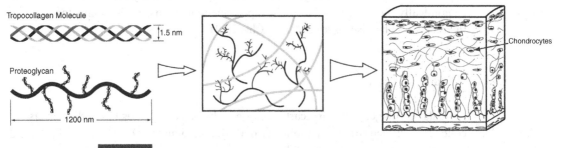

Figure 5.20

The microstructure and ultrastructure of articular cartilage, showing relative size scales and organization of the constituents. (After Mow *et al.*, 2005.)

weakness combined with poor regeneration abilities makes surgical repair of articular cartilage difficult, as wounds will not heal easily and grafts may tear out. Replacement plugs of tissue-engineered articular cartilage and synthetic hydrogels have been used recently with some success, but total removal of the joint surface and replacement with a synthetic component is currently the predominant treatment for intractable cartilage injury (Park and Lakes, 2007).

Articular cartilage consists of 60–80 wt% water, with the remainder composed of a collagen network and proteoglycans. Approximately half of the dry weight of cartilage is collagen (Parsons, 1998). Its hierarchical structure is depicted in Figure 5.20. The branched arms of proteoglycans contain the fixed electrical charges that recruit fluid by

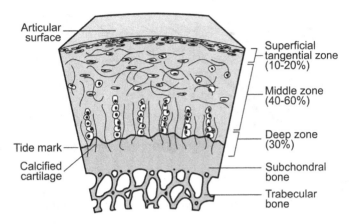

Figure 5.21

The orientation of the collagen fibers with depth in articular cartilage.

osmosis, such that the synovial fluid and its positively charged sodium ions are drawn into the tissue to cancel the negative electrical charge fixed to the proteoglycans. These molecules are large enough to entangle with the collagen superstructure of the tissue, which effectively makes the proteoglycans a constituent of the collagen network. These three essential constituents (solid, liquid, and ionic) impact the mechanical behavior of cartilage independently, and so are often treated as three distinct phases when the tissue is described mathematically.

The collagen fiber network of articular cartilage has microstructural variation with distance from the bone surface as shown in Figure 5.21. The tissue is approximately 2–4 mm thick in human joints, and can be divided into three zones with distinct fiber orientations (Mow *et al.*, 2005). The layer adjacent to the articular surface is termed the superficial tangent zone, and accounts for 10–20% of the thickness of the tissue. Here, the collagen fibers are oriented chiefly in the plane of the surface, and the collagen content is higher than in deeper layers, with a reduced proteoglycan content. Beneath the superficial tangent zone is the middle zone, which accounts for 40–60% of the thickness. The collagen fiber orientation is less organized here, so its properties are more isotropic. Below the middle zone is the deep zone, which is about 30% of the thickness of the tissue. The collagen fibers in the deep zone are oriented primarily transverse to the surface of the tissue and perpendicular to the underlying subchondral bone. The deep zone also contains the highest concentration of proteoglycan chains. Underneath the deep zone of the cartilage is a layer of partially calcified cartilage that serves as a mechanical transition to the underlying bone, and which is sometimes included as a constituent of articular cartilage.

A tensile stress-strain curve for articular cartilage exhibits the two-region response characteristic of compliant collagenous tissues. The equilibrium elastic modulus for human articular cartilage, corresponding to the elastic modulus measured under extremely slow rates of deformation, typically ranges from 1–14 MPa in tension and

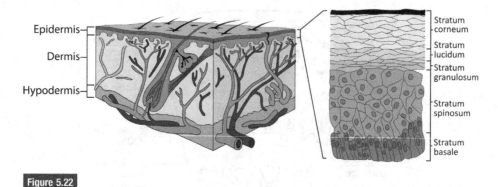

Figure 5.22

Anatomy of skin showing the cellular structure of the epidermis. A reticular collagen/elastin fiber network is located in the dermis.

0.5–0.7 MPa in compression. The equilibrium elastic modulus remains relatively constant up to approximately 15% strain. With disease, the elastic modulus can drop significantly. For example, in one study, the equilibrium tensile modulus of surface articular cartilage dropped from 7.79 MPa to 1.36 MPa when normal cartilage was compared to osteoarthritic cartilage (Mow *et al.*, 2005).

The dependence of the elastic modulus of articular cartilage on the deformation rate is due to the substantial influence of the viscous resistance to the flow of fluid on the apparent behavior of the tissue. In a test with extremely slow deformation, the fluid seeps out of the tissue with negligible resistance to flow and only the collagen network resists the applied deformation. In high-rate tests, fluid flow is more inhibited by drag of the collagen matrix and as a result the tissue exhibits greater apparent stiffness. Thus, it is difficult to use reported values or predict cartilage elastic modulus without knowing the deformation rate. High-rate loading is common in athletic activities or impacts, making the non-equilibrium properties of articular cartilage clinically significant, while low-rate loading reflects more habitual or sustained conditions. For a given constant imposed strain, the stress in articular cartilage diminishes with time from an initial response to the equilibrium value, and the elastic modulus is therefore time-dependent. The relaxation of stress under a constant strain is characteristic of a class of materials termed viscoelastic (both fluid and solid-like), which are discussed in detail in Chapter 7.

5.4.6 Skin and blood vessels (planar elastic tissues)

Skin and blood vessel walls bear a remarkable resemblance to tendon and ligament in their mechanical behavior, but are planar in their organization rather than linear. These tissues also exhibit complex, multilayered structures. Skin can be divided into two layers as shown in Figure 5.22. The outermost layer is the epidermis, which is 100–110 μm thick depending on the location in the body and contains mainly keratinocytes. These cells maintain the protein layer that protects the skin from

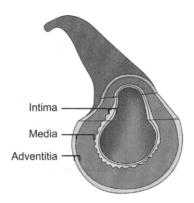

Figure 5.23

The multi-layered structure of the arterial wall.

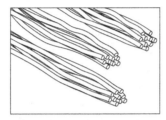
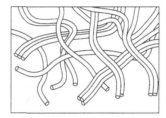

Collagen fibers   Reticular fibers

Figure 5.24

Collagen fibers are formed of aligned fibrils in linear tissues like ligament (left), but fibrils form a planar organization in skin and arterial wall (right).

environmental damage. Between the epidermis and the subcutaneous tissue is the dermis, composed of a mesh-like organization of collagen and elastic fibers and varying in thickness from 5 mm on the back of the spine to 1 mm on the eyelids (Bennett, 2009; McGrath *et al.*, 2004; Oikarinen and Knuutinen, 2002; Stamatialis, 2007).

Blood vessels have a trilaminate structure, illustrated in Figure 5.23. The interior layer, the intima, is made up of endothelial cells and does not contribute significantly to the mechanical strength of the vessel. The middle layer, the media, is made up of smooth muscle cells, collagen, and elastin in a planar reticular network as illustrated in Figure 5.24. The outer layer, the adventitia, is made up of fibroblasts in a collagen matrix. The thickness of these layers can differ from region to region in the body. The wall thickness of a blood vessel can vary from 1 μm in a capillary to 2000 μm in the aorta (Okamoto, 2008).

The primary structural attribute shared by these two tissues is their planar organization, which forces the fibers to be distributed in two dimensions as depicted in Figure 5.24, rather than in the axial structures of tendon and ligament.

Skin and arterial wall both exhibit the expected toe and heel regions in their stress-strain curves, shown schematically in Figure 5.25. Note the unusually stiff behavior of

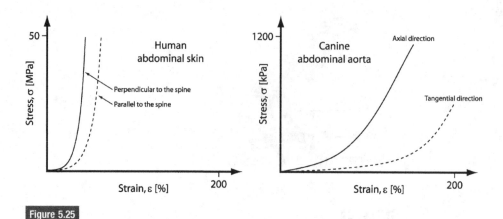

Figure 5.25

The mechanical properties of human skin (left) and canine aorta (right), demonstrating substantial anisotropic behavior within the plane of the tissue. (After Agache *et al.*, 1980; Deng and Guidon, 1998.)

arterial wall in the toe region; this tissue contains substantially more elastin than others, even exceeding the collagen content in some anatomical regions. The elastin content of arterial wall peaks in the aortic arch, where that structure acts essentially as a "secondary pump" for high-pressure blood just exiting the heart (Park and Lakes, 2007).

Both skin and arterial wall exhibit substantial anisotropy in their material planes. Human aorta has a tensile strength of 1.1 MPa in the transverse (circumferential) direction but only 0.07 MPa in the longitudinal direction (Park and Lakes, 2007). Both of these tissues are also under a steady state of pre-tension in the body, which is demonstrated in the manner that they will spring apart when cut, such as in opening angle tests for blood vessel sections. In these tests, an intact cross-section of artery is cut open and the angle at which the sample then rests is indicative of the amount of residual strain contained in the sample. The mechanical properties of skin are usually tested *in vivo* with suction, torsion, or indentation tests because the underlying substructure is so important to the mechanical properties. Reported values for elastic modulus from a torsional test are 0.42 MPa for subjects age 30 or younger and 0.85 MPa for subjects over the age of 30 (Agache *et al.*, 1980). Further discussion for replacements of these tissues can be found in Chapters 14 and 16.

5.5 Case study: creating a scaffold for tissue engineering

Tissue engineering is the use of multidisciplinary techniques to develop fully functional tissue replacements. One example in which this could have a substantial impact is in the development of vascular implants. It is common to use synthetic tubes made of expanded or woven polymers for replacements of large blood vessels. However, for replacements of small blood vessels (<5 mm diameter), only biological materials can be used, because synthetic implants are likely to become blocked by blood clots (Bhat, 2005). Thus, a

5.5 Case study: creating a scaffold for tissue engineering

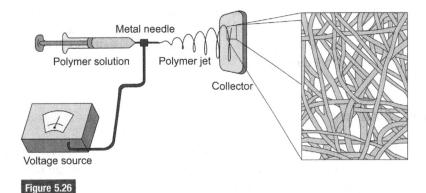

Figure 5.26

A sample setup for electrospinning, showing the jet of solution and the resulting network of fibers.

tissue-engineered vascular replacement is highly desired in order to have a controlled source. The general structure of such an implant would include a scaffold that provides geometrical form and structural support as well as a place for cells to attach and develop (Stegemann *et al.*, 2007).

There are many requirements for this scaffold. First, it must be biocompatible with any degradation products also being non-toxic even in high local concentrations. Second, the scaffold must have an open pore structure for cell nutrition and angiogenesis. It must mimic the extracellular matrix both in microstructure and mechanical properties and ideally it would be biodegradable at a rate that is comparable to the rate of neo-tissue formation (Pham *et al.*, 2007).

A simple, cost-effective method for creating such a scaffold is electrospinning. In this technique, a voltage is applied to a low-viscosity polymer solution, causing a jet of the solution to be ejected from a needle towards a collecting surface as seen in Figure 5.26. The solution jet is stretched into a thin fiber as it travels, and the fibers on the collecting surface dry and accumulate into a non-woven mat, or into a non-woven tube if wound around a mandrel. The polymer solution viscosity, solution charge density, polymer solution surface tension, flow rate, polymer molecular weight, electric field strength, distance between the tip and the collector, collector material and geometry, needle tip design and placement and ambient environment are all factors that can influence the resulting fiber morphology (Pham *et al.*, 2006).

Using this technique with collagen and elastin solutions, Boland *et al.* were able to create a trilaminate vascular replacement. The scaffold was made of two layers representing the adventitia and the media. The outer layer was composed of fibers from an 80:20 collagen-to-elastin ratio dissolved in 1,1,1,3,3,3-hexafluoro-2-propanol (HFP), while the inner layer consisted of fibers from a 30:70 collagen-to-elastin ratio dissolved in HFP. The outer scaffold layer was seeded with fibroblasts and smooth muscle cells, and then the inner layer was inserted and seeded with smooth muscle cells. Finally, endothelial cells were cultured on the inner surface to provide the intima-like layer. Later histological examination showed a trilaminate structure with complete cellular infiltration and proliferation (Boland *et al.*, 2004). Further experimentation on the mechanical properties of such a construct would of course be necessary; however, the development of a microstructure so similar to that of the natural blood vessel is very encouraging.

5.6 Summary

In this chapter, collagen, elastin, and hydroxyapatite were described as the basic building blocks of structural tissues. Indeed, the mechanical strength and resilience of most of the tissues in our body are attributable to these constituents. This chapter described hard tissues such as bone and enamel as well as softer materials including tendon and arterial wall. However different their mechanical properties, these tissues share common building blocks. In fact, elastic tissues are best thought of as comprising a family of materials exhibiting a continuous spectrum of behaviors according to the composition and alignment of the collagen, elastin, and hydroxyapatite in each. The contributions of collagen, elastin, and hydroxyapatite (strength, elasticity, and stiffness, respectively) can be traced through knowledge of their weight percentages and organizational structures in each tissue.

The organization of these elements and their relative quantities are important keys to predicting or interpreting the mechanical behavior of a structural tissue. Features such as the toe region or the stiffening at increased loads of a tendon tensile test can be easily explained once the nature of the collagen microstructure is known. Similarly, the durability of enamel is also better understood once the size and arrangement of the hydroxyapatite crystals is described.

Several values are given in this chapter as representative mechanical properties for various tissues. It is very important to understand that, unlike with materials such as metals or ceramics, the properties of natural tissues are highly dependent on factors such as type of test (tensile, compressive, torsional), test parameters (pre-load, loading rate), testing environment (temperature, hydration), and sample preparation (tissue storage method, polishing, pre-conditioning) as well as the age and health of the subject. Finally, all of these tissues have mechanical properties that are location-specific, so the mechanical properties of the anterior cruciate ligament (ACL), for example, should not be substituted for those of spinal ligaments. The values given here may be used to determine orders of magnitude, but data that are taken under conditions as close as possible to the expected use environment are of course preferred.

5.7 Problems for consideration

Checking for understanding

1. Name three types of collagen and the tissues where they can be found.

2. Why are hydroxyapatite crystals different shapes in bone and enamel? How is this reflected in the material properties of these tissues?
3. Discuss the function of arterial walls and how its anisotropic material properties reflect its functional requirements.
4. Discuss the anisotropic performance of cortical bone and how it reflects its functional requirements in tension and compression.
5. Draw a stress-strain curve for a tendon showing the arrangement of collagen at initial loading, the toe region, the heel, and the linear region.

For further exploration

6. Coral (a porous apatite mineral structure) can be used as a bone substitute. How is this done? What makes this material a candidate for implantation?
7. What are the microstructure and constituents of smooth muscle? How does this agree or disagree with the idea of structure-property relationships developed in this chapter?
8. What would be the advantages and disadvantages of having our bones be as highly mineralized as enamel?
9. Speculate on the functional advantage for the organization of collagen fibers with depth in articular cartilage.

5.8 References

Agache, P.G., Monneur, C., Leveque, J.L., and De Rigal, J. (1980). Mechanical properties and young's modulus of human skin in vivo. *Archives of Dermatology Research*, 269, 221–32.

Bartel, D.L., Davy, D.T., and Keaveny, T.M. (2006). *Orthopaedic Biomechanics*. Upper Saddle River, NJ: Pearson Prentice Hall.

Bennett, R. (2009). Anatomy and physiology of the skin. In *Facial Plastic and Reconstructive Surgery*, 3d edn., ed. I.D. Papel, J.L. Frodel, G.R. Holt, W.F. Larrabee, N.E. Nachlas, S.S. Park, J.M. Sykes, and D.M. Toriumi. New York: Thieme Medical Publishers, pp. 3–14.

Bhat, S.V. (2005). *Biomaterials*, 2d edn. Middlesex, UK: Alpha Science International Ltd.

Boland, E.D., Matthews, J.A., Pawlowski, K.J., Simpson, D.G., Wnek, G.E., and Bowlin, G.L. (2004). Electrospinning collagen and elastin: preliminary vascular tissue engineering. *Frontiers in Bioscience*, 9, 1422–32.

Boskey, A.L. (2001). Bone mineralization. In *Bone Mechanics Handbook*, ed. S.C. Cowin. Boca Raton, FL: CRC Press, pp. 5-1–5-32.

Champe, P.C., Harvey, R.A., and Ferrier, D.R. (2008). *Biochemistry*, 4th edn. Philadelphia, PA: Lippincott, Williams & Wilkins.

Cowin, S.C. (2001). The false premise in Wolff's law. In *Bone Mechanics Handbook*, ed. S.C. Cowin. Boca Raton, FL: CRC Press, pp. 30-1–30-12.

Currey, J. (1998). Cortical bone. In *Handbook of Biomaterials Properties*, ed. J. Black and G. Hastings. London: Chapman and Hall, pp. 3–14.

Deng, X., and Guidon, R. (1998). Arteries, veins and lymphatic vessels. In *Handbook of Biomaterials Properties*, ed. J. Black and G. Hastings. London: Chapman and Hall, pp. 81–104.

Fischer, G.M., and Llaurado, J.G. (1966). Collagen and elastin content in canine arteries selected from functionally different vascular beds. *Circulation Research*, 19, 394–9.

Frank, C.B., Shrive, N.G., Lo, I.K.Y., and Hart, D.A. (2007). Form and function of tendon and ligament. In *Orthopaedic Basic Science: Foundations of Clinical Practice*, 3d edn., ed. T.A. Einhorn, R.J. O'Keefe, J. A. Buckwalter. Rosemont, IL: AAOS, pp. 191–222.

Fratzl, P., Gupta, H.S., Paschalis, E.P., and Roschger, P. (2004). Structure and mechanical quality of the collagen-mineral nano-composite in bone. *Journal of Materials Chemistry*, 14(14), 2115–23.

Fung, Y.C. (1981), *Biomechanics: Mechanical Properties of Living Tissues*. New York: Springer-Verlag.

Gibson, L.A., and Ashby, M.F. (1999). *Cellular Solids: Structure and Properties*, 2d edn. Cambridge Solid State Science Series. Cambridge, UK: Cambridge University Press.

Gong, J.K., Arnold, J.S., and Cohn, S.H. (1964). Composition of trabecular and cortical bone. *Anatomical Record*, 149(3), 325–31.

Guo, X.E. (2001). Mechanical properties of cortical bone and cancellous bone tissue. In *Bone Mechanics Handbook*, ed. S.C. Cowin. Boca Raton, FL: CRC Press, pp. 10-1–10-18.

Habelitz, S., Marshall, S.J., Marshall Jr., G.W., and Balooch, M. (2001). Mechanical properties of human dental enamel on the nanometre scale. *Archives of Oral Biology*, 46, 173–83.

Healy, K.E. (1998). Dentin and enamel. In *Handbook of Biomaterials Properties*, ed. J. Black and G. Hastings. London: Chapman and Hall, pp. 24–39.

Hosada, Y., Kawano, K., Yamasawa, F., Ishii, T., Shibata, T., and Inayama, S. (1984). Age-dependent changes of collagen and elastin content in human aorta and pulmonary artery. *Angiology*, 35(10), 615–21.

Jee, W.S.S. (2001). Integrated bone tissue: anatomy and physiology. In *Bone Mechanics Handbook*, ed. S.C. Cowin. Boca Raton, FL: CRC Press, pp. 1-1–1-68.

Kadar, A. (1989). Ultrastructural properties of elastin. In *Elastin and Elastases*, Volume 1, ed. L. Robert and W. Hornebeck. Boca Raton, FL: CRC Press, pp. 21–30.

Keaveny, T.M. (1998). Cancellous bone. In *Handbook of Biomaterials Properties*, ed. J. Black and G. Hastings. London: Chapman and Hall, pp. 15–23.

Kielty, C.M., Wess, T.J., Haston, L., Ashworth, J.L., Sherratt, M.J., and Shuttleworth, C.A. (2003). Fibrillin-rich microfibrills: elastic biopolymers of the extracellular matrix. In *Mechanics of Elastic Biomolecules*, ed. W.A. Linke, H.L. Granzier, and M. Kellermayer. Dordrecht, The Netherlands: Kluwer Academic Publishers, pp. 581–96.

Koenders, M.M.J.F., Yang, L., Wismans, R.G., Van Der Werf, K.O., Reinhardt, D.P., Daamen, W., Bennink, M.L., Dijkstra, P.J., van Kuppevelt, T.H., and Feijen, J. (2009). Microscale mechanical properties of single elastic fibers: the role of fibrillin-microfibrils. *Biomaterials*, 30(13), 2425–32.

Koester, K.J., Ager III, J.W., and Ritchie, R.O. (2008) The effect of aging and crack-growth resistance and toughening mechanisms in human dentin. *Biomaterials*, 29(10), 1318–28.

Lucchinetti, E. (2001). Dense bone tissue as a molecular composite. In *Bone Mechanics Handbook*, ed. S.C. Cowin. Boca Raton, FL: CRC Press, pp. 13-1–13-5.

Lundon, K. (2003). *Orthopedic Rehabilitiation Science: Principles for Clinical Management of Nonmineralized Connective Tissue*. St. Louis, MO: Butterworth Heineman.

Martin, R.B., and Burr, D.B. (1989). *Structure, Function, and Adaptation of Compact Bone*. New York: Raven Press.

Martin, R.B., Burr, D.B., and Sharkey, N.A. (1998). *Skeletal Tissue Mechanics*. New York: Springer-Verlag.

McGrath, J.A., Eady, R.A.J., and Pope, F.M. (2004). Anatomy and organization of human skin. In *Rook's Textbook of Dermatology, Volume 1*, 7th edn., ed. T. Burns, S. Breathnach, N. Cox, and C. Griffiths. Malden, MA: Blackwell Publishing, pp. 3-1–3-76.

Morgan, E.F., Barnes, G.L., and Einhorn, T.A. (2008). The bone organ system: form and function. In *Osteoporosis, Volume 1*, 3d edn., ed. R. Marcus, D. Feldman, D.A. Nelson, and C.J. Rosen. Burlington, MA: Elsevier Academic Press, pp. 3–25.

Morgan, E.F., and Keaveny, T.M. (2001). Dependence of yield strain of human trabecular bone on anatomic site. *Journal of Biomechanics*, 34(5), 569–77.

Mow, V.C., Gu, W.Y., and Chen, F.H. (2005). Structure and function of articular cartilage and meniscus. *Basic Orthopaedic Biomechanics and Mechano-biology*, 3d edn., ed. V. C. Mow and R. Huiskes. Philadelphia, PA: Lippincott Williams & Wilkins, pp. 181–237.

Mow, V.C., and Huiskes, R. (2005). A brief history of science and orthopaedic biomechanics. In *Basic Orthopaedic Biomechanics and Mechano-biology*, 3d edn., ed. V.C. Mow and R. Huiskes. Philadelphia, PA: Lippincott Williams & Wilkins.

Nalla, R.K., Kinney, J.H., and Ritchie, R.O. (2003). Effect of orientation on the in vitro fracture toughness of dentin: the role of toughening mechanisms. *Biomaterials*, 24(22), 3955–68.

O'Brien, F.J., Taylor, D., and Lee, T.C. (2007). Bone as a composite material: The role of osteons as barriers to crack growth in compact bone. *International Journal of Fatigue*, 29(6), 1051–56.

Oikarinen, A., and Knuutinen, A. (2002). Mechanical properties of human skin: biochemical aspects. In *Bioengineering of the Skin: Skin Biomechanics, Volume 5*, ed. P. Elsner, E. Berardesca, K.-P. Wilheim, and H.I. Maibach, Dermatology: Clinical and Basic Sciences Series. Boca Raton, FL: CRC Press, pp. 3–13.

Okamoto, R.J. (2008). Blood vessel mechanics. In *Encyclopedia of Biomaterials and Biomedical Engineering*, 2d edn. *Volume 1*, ed. G.E. Wnek and G.L. Bowlin. New York: Informa Healthcare, pp. 392–402.

O'Malley, C.D. (1964). *Andreas Vesalius of Brussels, 1514–1564*. Berkeley: University of California Press.

Park, J.B., and Lakes, R.S. (2007). *Biomaterials: An Introduction*, 3d edn. New York: Plenum Press.

Parsons, J.R. (1998). *Cartilage*. In *Handbook of Biomaterials Properties*, ed. J. Black and G. Hastings. London: Chapman and Hall, pp. 40–45.

Pham, Q.P., Sharma, U., and Mikos, A.G. (2006). Electrospinning of polymeric nanofibers for tissue engineering: a review. *Tissue Engineering*, 12(5), 1197–1211.

Piesco, N.P. (2002). Histology of dentin. In *Oral Development and Histology*, 3d edn., ed. J.K. Avery, P.F. Steele, and N. Avery. New York: George Thierme Verlag, pp. 172–89.

Piesco, N.P., and Simmelink, J. (2002). Histology of enamel. In *Oral Development and Histology*, 3d edn., ed. J.K. Avery, P.F. Steele, and N. Avery. New York: George Thierme Verlag, pp. 153–71.

Rasmussen, S.T., and Patchin, R.E. (1984). Fracture properties of human enamel and dentin in an aqueous environment. *Journal of Dental Research*, 63(12), 1362–8.

Roverri, N., and Palazzo, B. (2006). Hydroxyapatite nanocrystals as bone tissue substitute. In *Tissue, cell and organ engineering*, ed. C. Kumar, Nanotechnologies for the Life Sciences Series, Volume 9. Morlenbach, Germany: Wiley- VCH Verlag GmbH & Co., pp. 283–306.

Ruppel, M.E., Miller, L.M., and Burr, D.B. (2008). The effect of the microscopic and nanoscale structure on bone fragility. *Osteoporosis International*, 19(9): pp. 1251–65.

Safadi, F.F., Barbe, M.F., Abdelmagrid, S.M., Rico, M.R., Aswad, R.A., Litvin, J., and Popoff, S.J. (2009). Bone structure, development and bone biology. In *Bone Pathology*, 2d edn., ed J.S. Khurana. New York: Humana Press, pp. 1–50.

Samuel, S.P., Baran, G.R., Wei, Y., and Davis, B.L. (2009). Biomechanics – Part II. In *Bone Pathology*, 2d edn., ed J.S. Khurana. New York: Humana Press, pp. 69–78.

Siegmund, T., Allen, M.R., and Burr, D.B. (2008). Failure of mineralized collagen fibrils: modeling the role of collagen crosslinking. *Journal of Biomechanics*, 41(7), 1427–35.

Stamatialis, D.F. (2007). Drug delivery through skin: overcoming the ultimate biological membrane. In *Membranes for Life Sciences*, ed. K.-V. Peinemann and S. Pereira-Nunez, Membrane Technology, Volume 1. Morlenbach, Germany: Wiley-VCH Verlag GmbH & Co., pp. 191–226.

Stegemann, J.P., Kaszuba, S.N., and Rowe, S.L. (2007). Review: advances in vascular tissue engineering using protein-based biomaterials. *Tissue Engineering*, 13(11), 2601–13.

Taylor, D. (2007). Fracture and repair of bone: a multiscale problem. *Journal of Materials Science*, 42(21), 8911–18.

Ventre, M., Netti, P.A., Urciuolo, F., and Ambrosio, L. (2009). Soft tissues characteristics and strategies for their replacement and regeneration. In *Strategies in Regenerative Medicine: Integrating Biology with Materials Design*, ed. M. Santin. New York: Springer, pp. 16–50.

Vincent, J.F.V. (1990) *Structural Biomaterials*, rev. edn., Princeton, NJ: Princeton University Press.

Woo, S.L.-Y., and Levine, R.E. (1998). Ligament, tendon and fascia. In *Handbook of Biomaterials Properties*, ed. J. Black and G. Hastings. London: Chapman and Hall, pp. 59–65.

Wren, T.A.L., Yerby, S.A., Beaupre, G.S., and Carter, D.R. (2001). Mechanical properties of the human Achilles tendon. *Clinical Biomechanics*, 16, 245–51.

Zaslansky, P. (2008). Dentin. In *Collagen: Structure and Mechanics*, ed. P. Fratzl. New York: Springer, pp. 421–46.

5.9 Bibliography

Park, J.B. (1979). *Biomaterials: An Introduction*. New York: Plenum Press.
(1991). *Basic Orthopaedic Biomechanics*, ed. V.C. Mow and W.C. Hayes. New York: Raven Press.

Part II **Mechanics**

6 Elasticity

Inquiry

What modifications could you make to a hip stem to minimize stress shielding of the surrounding bone while still using metals common to orthopedic implants?

An interesting property of human bones is the manner in which they efficiently build up bone density where the bones are under higher loading, and remove bone density where strength is not needed. This property can be seen in the increased bone strength of the dominant arm of tennis players as compared to the non-racquet-holding arm (Ashizawa et al., 1999). Because the dominant arm is constantly subjected to loading as a result of the impact between the ball and racquet, the bone in that arm has greater bone density than the non-dominant arm, which presumably sees only everyday loading.

This property of bones is of importance to designers of hip stems as it has been shown that patients who have hip implants with a high stiffness stem were experiencing noticeable bone loss. This phenomenon, known as **stress shielding**, was thought to be occurring because the hip implants themselves were taking up so much of the loading that the body removed from the now extraneous surrounding bone. This in turn led to implant problems, as the remaining bone was not strong enough to stabilize the hip stems, as shown in Figure 6.1. Because of this, researchers turned their attention to developing materials and implants that might provide the same strength and durability with decreased stiffness, forcing the bone to take up more of the load. By the end of this chapter, a method for determining the stresses in various configurations of materials and stem cross-sections, in order to reduce stress shielding, will be developed.

6.1 Overview

All biomaterials experience forces or loads during their lifetimes, whether they are natural tissues or synthetic materials as parts of medical implants. The resulting deformations are particularly important in implant design, as a knowledge of the material characteristics combined with an understanding of the loads and deformations is necessary to prevent device failure through yield or fracture. These deformations can fall into many categories including elastic, plastic, or time-dependent deformation. For this chapter, the focus will be on elastic, or totally reversible, deformation and its role in

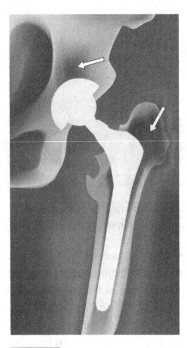

Figure 6.1

Illustrated example of an x-ray image showing bone resorption.

medical device design. In medical device applications, the loading situations are also three-dimensional, as opposed to simple uniaxial configurations that are often replicated in basic mechanical tests. This chapter will describe one-dimensional (uniaxial), two-dimensional (planar), and three-dimensional (body) analysis techniques.

Important mechanical properties used in describing linear elastic, isotropic deformation are elastic modulus, E, and Poisson's ratio, ν. Other parameters, such as shear modulus G and bulk modulus K, can be calculated from the values of E and ν. These mechanical properties can be determined using standardized tests and basic testing equipment. A simple tensile test and measurements of the sample geometry can be used to determine these properties after the test has been completed. Other common metrics that are derived from a tensile test include Ultimate Tensile Strength (UTS), Yield Strength (S_y), Failure Strength (S_f), and strain to failure (ε_f). These parameters and their usages will be described in further detail later in the chapter.

Because mechanical properties can be sensitive to different test parameters, it is necessary to use a standardized test if the results are to be compared between laboratories. These testing standards, commonly from international standards development organizations, give detailed test protocols that reduce the variability in test results between different laboratories and test runs. The challenges of achieving repeatable results between laboratories can be compounded when testing tissues, as there are often large inter-sample inhomogeneities and frequently, no accepted testing standards. Materials that are easily characterized are known as **standard engineering materials**, and include metals, ceramics, and even some polymers. Soft polymers and most natural tissues fall outside this definition. Due to these issues, it is particularly important to report all aspects of test methodology and sample preparation when summarizing mechanical testing results. For example, large differences can be seen when the same test is run on fresh tissue versus frozen and then thawed tissue (Nazarian *et al.*, 2009; Venkatasubramanian *et al.*, 2010). In this chapter, conventional test methods and mechanical property definitions for standard engineering materials will be discussed.

A review of bending stresses and beam theory will also be given in this chapter. Beam theory is useful for modeling the loading situation of certain types of medical implants *in vivo*, such as in the stem of a hip implant. Finally, the elastic properties of composite materials will be discussed, as the use and analysis of composite materials become increasingly relevant in medical device development.

6.2 Learning objectives

This chapter provides a framework for characterizing elastic behavior in natural and synthetic biomaterials. By the end of this chapter, students should be able to:

1. define stress, strain, elastic modulus, Poisson's ratio, yield strength, ultimate tensile strength, failure strength, strain to failure
2. derive the stress tensor from the traction vector
3. find principal stresses using two methods
4. solve simple 3D problems using Hooke's Law
5. create and use simple stress-strain curves for various materials in tension and compression
6. give the number of independent constants for some specific examples of isotropy/anisotropy
7. compare true and engineering stress and strain
8. use the beam equation to model medical device applications
9. find the upper and lower limits for E in composite materials

6.3 Stress and strain

6.3.1 Definition of strain

When a component is loaded, there is normally some deformation in response to this loading. This deformation is captured in the concept of **strain**, a non-dimensional quantity that describes the change in a component's physical configuration during loading. There are two types of strain: normal (or extensional) strain and shear strain. Normal

Elasticity

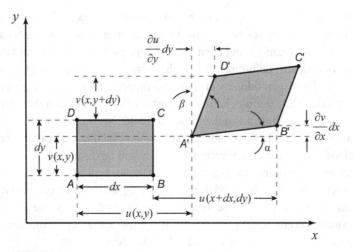

Figure 6.2

Two-dimensional strain definitions.

strain is defined as the extension or contraction in the direction that the load is applied, normalized by the unit length of fibers in that direction. **Shear strain** is defined as the change in the angle between two originally orthogonal directions. Shear strain is measured in radians. For the purposes of this chapter, it is assumed that all strains are small.

Figure 6.2 illustrates the parameters used in defining the two types of strain. The object on the left is an undeformed rectangle with sides of length dx and dy, while the object on the right is the same rectangle after deformation. Reference Point A has been moved by an amount $u(x, y)$ in the x-direction and $v(x, y)$ in the y-direction to Point A'. Similarly, Reference Point B has been moved by an amount $u(x + dx, dy)$ and $v(x + dx, dy)$ to Point B'. According to small deformation theory,

$$u(x + dx, dy) \cong u(x, y) + \frac{\partial u}{\partial x} dx \tag{6.1}$$

From the definition given above, it can be seen that strain in the x-direction, termed ε_x, is defined as:

$$\varepsilon_x = \frac{A'B' - AB}{AB} \tag{6.2}$$

Using simple geometry, the following is found:

$$A'B' = \sqrt{\left(dx + \frac{\partial u}{\partial x}dx\right)^2 + \left(\frac{\partial v}{\partial x}dx\right)^2} = \sqrt{1 + 2\frac{\partial u}{\partial x} + \left(\frac{\partial u}{\partial x}\right)^2 + \left(\frac{\partial v}{\partial x}\right)^2} \, dx \tag{6.3}$$

When higher order terms are dropped and small displacements are taken into account, this equation reduces to

$$A'B' = \left(1 + \frac{\partial u}{\partial x}\right) dx. \tag{6.4}$$

Since $AB = dx$, the strain equation becomes

$$\varepsilon_x = \frac{\left(1 + \frac{\partial u}{\partial x}\right) dx - dx}{dx} = \frac{\partial u}{\partial x}. \tag{6.5}$$

Similarly,

$$\varepsilon_y = \frac{\partial v}{\partial y}. \tag{6.6}$$

To develop the shear strain equation, begin with the definition above. Shear strain, termed γ_{xy}, is defined:

$$\gamma_{xy} = \frac{\pi}{2} - \angle D'A'B' = \alpha + \beta. \tag{6.7}$$

For small deformations, assume $\alpha = \tan \alpha$ and $\beta = \tan \beta$.

Thus

$$\gamma_{xy} = \frac{\frac{\partial v}{\partial x} dx}{dx + \frac{\partial u}{\partial x} dx} + \frac{\frac{\partial u}{\partial y} dy}{dy + \frac{\partial v}{\partial y} dy} = \frac{\partial u}{\partial y} + \frac{\partial v}{\partial x} \tag{6.8}$$

after dropping higher order terms. It can also be shown that $\gamma_{xy} = \gamma_{yx}$.

These analyses can be extended to the y-z and x-z planes such that the complete equations for strain are defined as follows:

$$\text{Normal strain: } \varepsilon_x = \frac{\partial u}{\partial x}, \varepsilon_y = \frac{\partial v}{\partial y}, \varepsilon_z = \frac{\partial w}{\partial z} \tag{6.9}$$

$$\text{Shear strain: } \gamma_{xy} = \frac{\partial u}{\partial y} + \frac{\partial v}{\partial x}, \gamma_{yz} = \frac{\partial v}{\partial z} + \frac{\partial w}{\partial y}, \gamma_{zx} = \frac{\partial w}{\partial x} + \frac{\partial u}{\partial z} \tag{6.10}$$

For a simple example, uniaxial displacement is often described. In Figure 6.3, a bar is placed under a uniaxial load. In this case, strain is defined as:

$$\varepsilon = \frac{l_0 - l_f}{l_0} = \frac{\Delta l}{l_0} \tag{6.11}$$

Normal strain is unitless, and conventionally, positive values for strain are used to describe extension, while negative values for strain describe compression.

6.3.2 Definition of stress

Stress is the transmission of force through deformable materials. Given the object in Figure 6.4, the force vector ΔP can be resolved into three coordinate directions. The

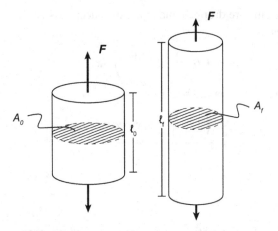

Figure 6.3

A bar under uniaxial tension, before (left) and after (right) loading.

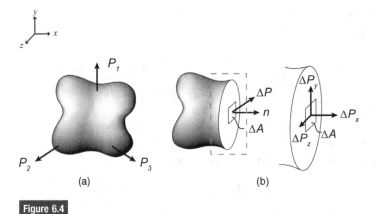

Figure 6.4

(a) An object under loading, showing (b) the definition of the traction vector.

traction vector is then defined as follows:

$$\tau_{xx} = \lim_{\Delta A \to 0} \frac{\Delta P_x}{\Delta A}, \tau_{xy} = \lim_{\Delta A \to 0} \frac{\Delta P_y}{\Delta A}, \tau_{xz} = \lim_{\Delta A \to 0} \frac{\Delta P_z}{\Delta A}. \quad (6.12)$$

In this notation, the first subscript of the traction vector refers to the plane on which the force is acting, and the second subscript refers to the direction of the force. By convention, $\tau_{xx} = \sigma_x$, the normal stress, while τ_{xy} and τ_{xz} are shear stresses. It is important to note here that the orientation of the surface ΔA will have an effect on the magnitude and direction of the stresses, even if the load ΔP remains the same. Stress is measured in Pascals [Pa] or pounds per square inch [psi]. Similarly to the sign conventions for strain, positive values for stress indicate tensile stress, while negative values indicate compressive stress. In a simple uniaxial case shown in Figure 6.3, stress is defined simply as

$$\sigma = \frac{F}{A_0}. \tag{6.13}$$

6.3.3 Stress tensor

Instead of using the object in Figure 6.4, imagine that there is an infinitesimal cube inside of an object under loading, whose sides are parallel with the three coordinate planes as illustrated in Figure 6.5. In this configuration, there are three normal stresses (σ_x, σ_y, and σ_z) and six shear stresses ($\tau_{xy} = \tau_{yx}$, $\tau_{yz} = \tau_{zy}$, and $\tau_{zx} = \tau_{xz}$). The stress state of this object demonstrates observer invariance – that is, it does not change when viewed by different observers. However, the components of stress, the normal and shear stresses that act on the orthogonal sides of the infinitesimal element, can be different in different configurations. If this infinitesimal cube is rotated, the stresses in the new configuration will be related to the stresses in the original configuration, but they will generally not be of the same magnitude. This change in stresses made by changing from one set of coordinate axes to another is known as a **stress transformation**.

It is common to group the values for stress together in the **stress tensor**, defined as σ_{ij}, or:

$$\underline{\underline{\sigma}} = \begin{bmatrix} \sigma_{11} & \sigma_{12} & \sigma_{13} \\ \sigma_{21} & \sigma_{22} & \sigma_{23} \\ \sigma_{31} & \sigma_{32} & \sigma_{33} \end{bmatrix} \tag{6.14}$$

This is sometimes written as

$$\underline{\underline{\sigma}} = \begin{bmatrix} \sigma_x & \tau_{xy} & \tau_{xz} \\ \tau_{yx} & \sigma_y & \tau_{yz} \\ \tau_{zx} & \tau_{zy} & \sigma_z \end{bmatrix}. \tag{6.15}$$

It can be shown that this tensor must be symmetric, or that $\tau_{ij} = \tau_{ji}$. This is accomplished by looking at each plane in turn. First, take the x-y plane, as shown in Figure 6.6. Sum the moments around Point A, remembering that static equilibrium requires this sum to be zero.

$$\sum M_A = \tau_{xy}h + \sigma_y\frac{h}{2} + \sigma_x\frac{h}{2} - \sigma_y\frac{h}{2} - \sigma_x\frac{h}{2} - \tau_{yx}h = 0 \tag{6.16}$$

and thus

$$\tau_{xy} = \tau_{yx}. \tag{6.17}$$

This can be repeated in the x-z and y-z planes to prove $\tau_{zx} = \tau_{xz}$ and $\tau_{yz} = \tau_{zy}$.

When an object is under stress, this stress state can be broken into **dilatational** and **deviatoric** components. The dilatational component is responsible for volume change,

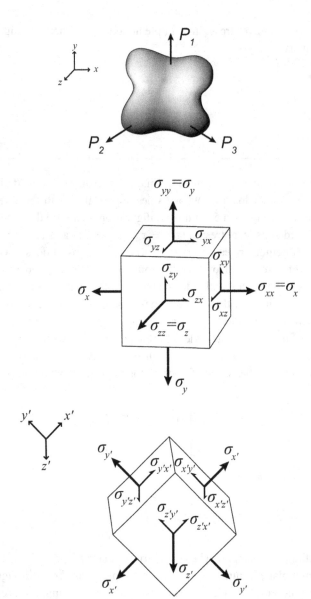

Figure 6.5

An infinitesimal cube inside an object under loading (top), before (middle), and after (bottom) rotation.

and is sometimes referred to as the hydrostatic component, while the deviatoric component is responsible for shape change, or distortion. The dilatational stress tensor, $p\delta_{ij}$, is written as:

$$\begin{bmatrix} p & 0 & 0 \\ 0 & p & 0 \\ 0 & 0 & p \end{bmatrix} \qquad (6.18)$$

6.3 Stress and strain

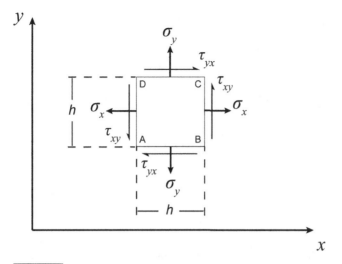

Figure 6.6

Summation of moments for the x-y plane of an infinitesimal element.

where

$$p = \frac{\sigma_{11} + \sigma_{22} + \sigma_{33}}{3} \tag{6.19}$$

and the deviatoric stress tensor, s_{ij}, is defined as the difference between the stress tensor and the dilatational stress tensor as follows:

$$\begin{bmatrix} s_{11} & s_{12} & s_{13} \\ s_{21} & s_{22} & s_{23} \\ s_{31} & s_{32} & s_{33} \end{bmatrix} = \begin{bmatrix} \sigma_{11} & \sigma_{12} & \sigma_{13} \\ \sigma_{21} & \sigma_{22} & \sigma_{23} \\ \sigma_{31} & \sigma_{32} & \sigma_{33} \end{bmatrix} - \begin{bmatrix} p & 0 & 0 \\ 0 & p & 0 \\ 0 & 0 & p \end{bmatrix}. \tag{6.20}$$

Sometimes it is useful to look at the stresses in several different sets of coordinate axes. It is common to apply coordinate transformations in order to find the set of coordinate axes that have the highest stress values (either normal or shear) for yield or failure predictions. It can also be useful to know in which coordinate systems normal or shear stresses are minimized. An example of this can be seen in the analysis of a uniaxial loading situation. Imagine an infinitesimal element in plane stress as given in Figure 6.7(a). If it is examined in a different coordinate system rotated by θ as shown in Figure 6.7(b), the stresses σ_x, σ_y, and τ_{xy} are transformed to σ'_x, σ'_y, and τ'_{xy}, respectively. It is important to remember that these quantities are merely a new (and equivalent) representation of the initial stress state. To find the values of the stresses in

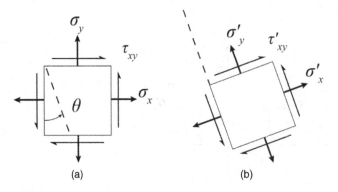

Figure 6.7

(a) An infinitesimal element, (b) undergoing rotation by an angle θ.

the new coordinate system, use simple geometry to find:

$$\sigma'_x = \frac{\sigma_x + \sigma_y}{2} + \frac{\sigma_x - \sigma_y}{2}\cos 2\theta + \tau_{xy}\sin 2\theta \qquad (6.21)$$

$$\sigma'_y = \frac{\sigma_x + \sigma_y}{2} - \frac{\sigma_x - \sigma_y}{2}\cos 2\theta - \tau_{xy}\sin 2\theta \qquad (6.22)$$

$$\tau'_{xy} = -\frac{\sigma_x - \sigma_y}{2}\sin 2\theta + \tau_{xy}\cos 2\theta. \qquad (6.23)$$

These equations give the variation of normal and shear stress in the material as a function of the angle of rotation between the new coordinate system and the initial coordinate system. It is not necessary to memorize these derivations, as a convenient graphical method also exists for determining the stresses in a system after a change of coordinate axes. **Mohr's circle**, introduced by Otto Mohr in the 1880s, can be used in any loading situation, although each plane rotation must be addressed separately. Recall the infinitesimal element illustrated in Figure 6.7, and its coordinate transformation by an amount θ. To show this in Mohr's circle, plot the normal and shear stresses on the x and y faces as shown in Figure 6.8. Use these two points to draw a circle. Rotation of an angle 2θ in Mohr's circle space represents a rotation of θ in actual space. For example, a rotation of 30° in actual space to reach maximum shear stress is represented by a rotation of 60° in Mohr's circle space. Again, geometry can be used to find the values for the normal and shear stresses after a coordinate transformation.

Example 6.1 Forces in an intramedullary rod

A tibial fracture occurs at 80° to the longitudinal axis of a tibia and is repaired using an intramedullary rod as shown in Figure 6.9. When the person is standing on both feet, assuming her body weight, BW, is split evenly between her right/left tibias, what are the normal and shear forces in the intramedullary rod in the plane of the fracture? Also assume for this example that the rod has a diameter, d, and that it is carrying all of the load in the tibia.

6.3 Stress and strain

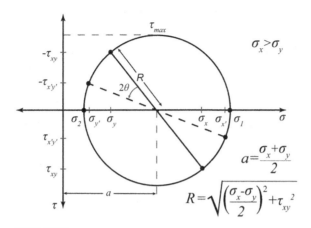

Figure 6.8

Mohr's circle representation of a two-dimensional problem.

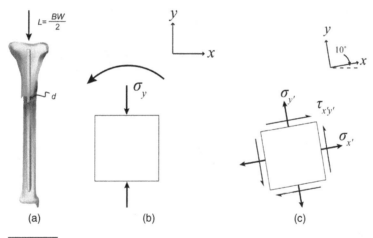

Figure 6.9

(a) A bone with fracture and intramedullary rod, (b) an infinitesimal element in this loading scheme, and (c) the infinitesimal element rotated to match the plane of the crack.

Solution

$$\sigma_y = \frac{-\frac{BW}{2}}{\frac{1}{4}\pi d^2} = -\frac{2BW}{\pi d^2}, \sigma_x = 0, \tau_{xy} = 0$$

$$\sigma_{x,80°} = \frac{\sigma_y}{2} + \frac{-\sigma_y}{2}\cos 2(80°) = -\frac{BW}{\pi d^2}(1 - \cos(160°))$$

$$\sigma_{y,80°} = \frac{\sigma_y}{2} + \frac{\sigma_y}{2}\cos 2(80°) = -\frac{BW}{\pi d^2}(1 + \cos(160°))$$

$$\tau_{xy,80°} = \frac{\sigma_y}{2}\sin 2(80°) = -\frac{BW}{\pi d^2}\sin(160°).$$

178 Elasticity

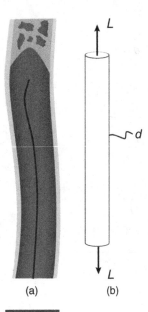

Figure 6.10

(a) A microguidewire in an artery, and (b) the same guidewire under tensile loading during retraction.

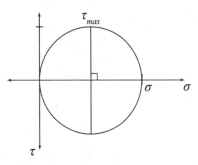

Figure 6.11

Mohr's circle representation of the stress state described in Example 6.2.

Example 6.2 **Using Mohr's circle to determine maximum shear stress**

A microguidewire used in catheterization is made from a ductile metal (illustrated schematically in Figure 6.10). Knowing that if it fails, it will fail in shear, use Mohr's circle to draw the coordinate system that maximizes shear stress in order to find the weakest plane and the value for maximum shear stress.

Solution From Mohr's circle (given in Figure 6.11),

$$\tau_{max} = \frac{\sigma}{2} = \frac{2L}{\pi d^2} \text{ and occurs at } 45°.$$

6.3 Stress and strain

As described earlier, finding the highest stresses is necessary when exploring failure or yield situations. Another way to represent this is to rotate the stress tensor such that all of the shear stresses are eliminated and the stress tensor can be written as

$$\underline{\underline{\sigma}} = \begin{bmatrix} \sigma_1 & 0 & 0 \\ 0 & \sigma_2 & 0 \\ 0 & 0 & \sigma_3 \end{bmatrix} \quad (6.24)$$

with σ_1, σ_2, and σ_3 referred to as the **principal stresses** corresponding to **principal directions** p_1, p_2, and p_3. One method for finding the principal stresses is to take the derivative of Equations (6.21) and (6.22) above with respect to θ and set equal to zero. The results will give the principal stresses in two dimensions:

$$\sigma_1, \sigma_2 = \frac{\sigma_x + \sigma_y}{2} \pm \sqrt{\left(\frac{\sigma_x - \sigma_y}{2}\right)^2 + \tau_{xy}^2}. \quad (6.25)$$

In the coordinate transformation that results in the principal stresses, the shear stresses are found to be zero. Conversely, if the shear stresses in a representation are zero, then the normal stresses are the principal stresses. It may also be useful to know the direction and value for maximum shear stress. Using a similar technique, the maximum shear stress is found to be

$$\tau_{max} = \left| \frac{\sigma_x - \sigma_x}{2} \right| \quad (6.26)$$

and the rotation necessary to achieve this is $\theta = 45°$. Mohr's circle can also be used to find the principal stresses. An example is provided here.

Example 6.3 **Principal stresses in an artificial spinal disk**

An infinitesimal element in an artificial spinal disk is loaded as shown in Figure 6.12 with $\sigma_x = -2.2$ MPa, $\sigma_y = -1.1$ MPa, $\sigma_z = -0.58$ MPa, $\tau_{xy} = -0.57$ MPa, $\tau_{yz} = -0.33$ MPa, and $\tau_{zx} = -0.79$ MPa. Use Mohr's circle to find the principal stresses in this situation.

Solution From Mohr's circle as shown in Figure 6.13, the principal stresses are $\sigma_1 = -0.25$ MPa, $\sigma_2 = -0.86$ MPa and $\sigma_3 = -2.77$ MPa.

The eigenvalues of the stress tensor are the principal stresses and the eigenvectors are the principal directions. Given the stress tensor in Equation (6.15), the principal stresses can be found by solving

$$\det[\sigma - \lambda I] = 0. \quad (6.27)$$

The principal directions are then easily found by solving

$$[\sigma - \lambda I][p] = 0. \quad (6.28)$$

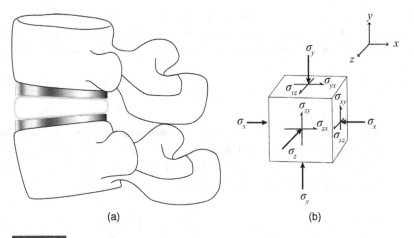

Figure 6.12

(a) An artificial intervertebral disk, and (b) the stress tensor associated with its loading.

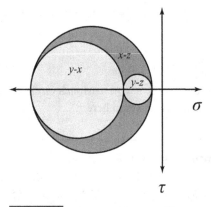

Figure 6.13

Mohr's circle representation of the stress state described in Example 6.3.

Example 6.4 Using eigenvalues and eigenvectors

Given the same loading scenario as in Example 6.3, use eigenvalues and eigenvectors to find the principal stresses and principal directions. Check that the eigenvectors are orthogonal and unit vectors.

Solution

$$\det \begin{bmatrix} -2.2 - \lambda & -0.57 & -0.79 \\ -0.57 & -1.1 - \lambda & -0.33 \\ -0.79 & -0.33 & -0.58 - \lambda \end{bmatrix} = 0$$

$$(-2.2 - \lambda)[(-1.1 - \lambda)(-0.58 - \lambda) - 0.33^2]$$
$$+ 0.57[-0.57(-0.58 - \lambda) + 0.33(-0.79)]$$
$$- 0.79[-0.57(-0.33) + 0.79(-1.1 - \lambda)] = 0$$

$$\lambda_1 = -0.25 \text{ MPa}$$
$$\lambda_2 = -0.86 \text{ MPa (Principal stresses)}$$
$$\lambda_3 = -2.77 \text{ MPa}$$

$$\begin{bmatrix} 2.2 - \lambda_i & 0.57 & 0.79 \\ 0.57 & 1.1 - \lambda_i & 0.33 \\ 0.79 & 0.33 & 0.58 - \lambda_i \end{bmatrix} \begin{bmatrix} p_{i,1} \\ p_{i,2} \\ p_{i,3} \end{bmatrix} = 0$$

$$p_1 = \begin{bmatrix} -0.86 \\ -0.36 \\ -0.36 \end{bmatrix}, \quad p_2 = \begin{bmatrix} 0.39 \\ -0.92 \\ 0.01 \end{bmatrix}, \quad p_3 = \begin{bmatrix} 0.34 \\ 0.13 \\ -0.93 \end{bmatrix} \text{ (Principal directions)}$$

Check orthogonality by verifying that $p_1 \cdot p_2 = p_2 \cdot p_3 = p_3 \cdot p_1 = 0$.
Check unit length by verifying that $|p_1| = |p_2| = |p_3| = 1$.

6.3.4 Constitutive behavior

During uniaxial elastic deformation of an isotropic material, the relationship between stress and strain is linear and is known as **Hooke's Law** and is given as:

$$\sigma = E\varepsilon \quad (6.29)$$

E, also known as Young's modulus or elastic modulus, is the mechanical property that describes the ratio between stress and strain during elastic deformation. Its units are [Pa] or [psi]. The range of values for E is quite large, from 1000 GPa for covalently bonded solids like diamond to $\approx$1 MPa for soft tissues like cartilage. The theoretical value for E was derived in Chapter 1. For most standard engineering materials, E is a material constant, meaning that all materials with the same chemical makeup and microstructure will have the same value for elastic modulus. In some cases, such as metals, even different microstructures will not affect the elastic modulus. However, for polymers and soft tissues, the measured value for E is highly dependent on test conditions such as strain rate as discussed in Chapters 4 and 5. Thus, when using reported values for E for these materials, it is important to note the conditions under which the tests were run and where possible, ensure that these conditions match the conditions for use of the material.

The Poisson's ratio, ν, defines the amount that a material contracts in the transverse direction as a response to a normal strain. For example, if an isotropic material is

Table 6.1 Development of 3D Hooke's Law

	Strain response to:		
	σ_x	σ_y	σ_z
ε_x	$= \dfrac{1}{E}\sigma_x$	$= -\nu\varepsilon_y = -\dfrac{\nu}{E}\sigma_y$	$= -\nu\varepsilon_z = -\dfrac{\nu}{E}\sigma_z$
ε_y	$= -\nu\varepsilon_x = -\dfrac{\nu}{E}\sigma_x$	$= \dfrac{1}{E}\sigma_y$	$= -\nu\varepsilon_z = -\dfrac{\nu}{E}\sigma_z$
ε_z	$= -\nu\varepsilon_x = -\dfrac{\nu}{E}\sigma_x$	$= -\nu\varepsilon_y = -\dfrac{\nu}{E}\sigma_y$	$= \dfrac{1}{E}\sigma_z$

stretched in the z direction, it will (with few exceptions) contract in the x and y directions. For this material, it is defined as:

$$\nu = -\frac{\varepsilon_x}{\varepsilon_y} = -\frac{\varepsilon_x}{\varepsilon_z}. \tag{6.30}$$

The Poisson's ratio is typically between 0 and 0.5, with the value for most metals around 0.3. Post-yield, ν is close to 0.5, and deformation occurs with no change in volume. The shear modulus, G, is the ratio between the shear stress and shear strain, analogous to the elastic modulus.

$$\tau = G\gamma. \tag{6.31}$$

The shear modulus can be written as a function of E and ν as follows:

$$G = \frac{E}{2(1+\nu)}. \tag{6.32}$$

6.3.5 Multiaxial loading

Uniaxial deformation is not a realistic model for many medical device applications. For this reason, it is important to consider how equations such as Hooke's Law can be applied to a multiaxial loading situation. The principle of **linear superposition**, which states that for a linear system, the overall response to two or more stimuli is equal to the sum of responses to those stimuli individually, will be used in this proof. In this case, this means that if the strain responses to applied loads in the x-, y-, and z-directions are analyzed individually, and then these responses are summed, the result is the strain responses for an object under multiple applied loads.

Figure 6.14 shows a block under a summation of applied loads (or stresses). Table 6.1 can be used to develop three-dimensional Hooke's Law. The strain responses are derived either from one-dimensional Hooke's Law (Equation 6.30) or the definition for Poisson's ratio (Equation 6.31).

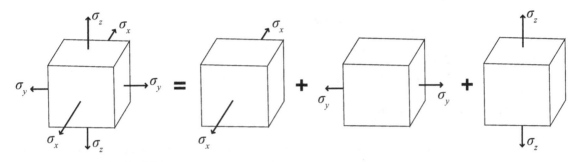

Figure 6.14

The summation of applied stresses leading to the development of three-dimensional Hooke's Law.

To derive three-dimensional Hooke's Law, sum the strain contributions from each stress:

$$\varepsilon_x = \frac{1}{E}[\sigma_x - \nu(\sigma_y + \sigma_z)]$$
$$\varepsilon_y = \frac{1}{E}[\sigma_y - \nu(\sigma_x + \sigma_z)] \quad (6.33)$$
$$\varepsilon_z = \frac{1}{E}[\sigma_z - \nu(\sigma_x + \sigma_y)]$$

Example 6.5 **Constrained loading in a tibial plateau**

Imagine a design for a tibial plateau that can be modeled for simplicity as shown in Figure 6.15. It is made of UHMWPE in a constraining frame of CoCr. Given a load of 30 kN which is evenly spread across the tibial plateau, what are the stresses and strains that develop in this implant? Use $E = 1$ GPa and $\nu = 0.4$ for UHMWPE and assume that the CoCr acts as a rigid constraint around the polymer.

Solution

$$\varepsilon_x = 0 = \frac{1}{E}[\sigma_x - \nu(\sigma_y + \sigma_z)]$$
$$\varepsilon_y = 0 = \frac{1}{E}[\sigma_y - \nu(\sigma_x + \sigma_z)]$$
$$\varepsilon_z = \frac{1}{E}[\sigma_z - \nu(\sigma_x + \sigma_y)]$$

Solve to get $\sigma_x = \sigma_y = 16.7$ MPa, $\varepsilon_z = 0.012$.

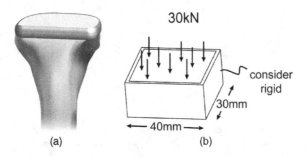

Figure 6.15

(a) A tibial plateau, which can be modeled by fully constrained loading as shown in (b).

6.3.6 Isotropy/anisotropy

The equations in three-dimensional Hooke's Law assume that the material is homogeneous and **isotropic**, meaning that the deformation in response to load is invariant with respect to direction. If a material is anisotropic, the general form of Hooke's Law is given as follows:

$$\begin{Bmatrix} \varepsilon_x \\ \varepsilon_y \\ \varepsilon_z \\ \gamma_{yz} \\ \gamma_{zx} \\ \gamma_{xy} \end{Bmatrix} = \begin{bmatrix} S_{11} & S_{12} & S_{13} & S_{14} & S_{15} & S_{16} \\ S_{21} & S_{22} & S_{23} & S_{24} & S_{25} & S_{26} \\ S_{31} & S_{32} & S_{33} & S_{34} & S_{35} & S_{36} \\ S_{41} & S_{42} & S_{43} & S_{44} & S_{45} & S_{46} \\ S_{51} & S_{52} & S_{53} & S_{54} & S_{55} & S_{56} \\ S_{61} & S_{62} & S_{63} & S_{64} & S_{65} & S_{66} \end{bmatrix} \begin{Bmatrix} \sigma_x \\ \sigma_y \\ \sigma_z \\ \tau_{yz} \\ \tau_{zx} \\ \tau_{xy} \end{Bmatrix} \quad (6.34)$$

$S_{ij} = S_{ji}$ in the compliance matrix $\underline{\underline{S}}$. For the anisotropic case, there are 21 independent constants needed to fully define the interactions between stress and strain. Many standard engineering materials are considered to be isotropic. As stated earlier, the isotropic case can be defined by two independent constants, E and ν. This special case is shown in the matrix form as:

$$\begin{Bmatrix} \varepsilon_x \\ \varepsilon_y \\ \varepsilon_z \\ \gamma_{yz} \\ \gamma_{zx} \\ \gamma_{xy} \end{Bmatrix} = \begin{bmatrix} \frac{1}{E} & -\frac{\nu}{E} & -\frac{\nu}{E} & 0 & 0 & 0 \\ -\frac{\nu}{E} & \frac{1}{E} & -\frac{\nu}{E} & 0 & 0 & 0 \\ -\frac{\nu}{E} & -\frac{\nu}{E} & \frac{1}{E} & 0 & 0 & 0 \\ 0 & 0 & 0 & \frac{1}{G} & 0 & 0 \\ 0 & 0 & 0 & 0 & \frac{1}{G} & 0 \\ 0 & 0 & 0 & 0 & 0 & \frac{1}{G} \end{bmatrix} \begin{Bmatrix} \sigma_x \\ \sigma_y \\ \sigma_z \\ \tau_{yz} \\ \tau_{zx} \\ \tau_{xy} \end{Bmatrix}. \quad (6.35)$$

For simplicity, the shear modulus, G, is used instead of its representation using E and ν.

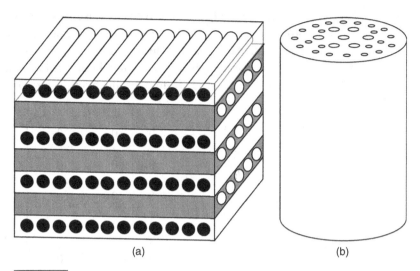

Figure 6.16

Special cases of symmetry: (a) orthotropy and (b) transverse isotropy.

Two other special cases are of interest when considering biomaterials. The first is **orthotropy**. An orthotropic material possesses symmetry about three orthogonal planes, such as a composite material with fibers of different strengths laid 90° to each other, as illustrated in Figure 6.16(a). In this case, there will be three elastic moduli, E_x, E_y, and E_z, each associated with one plane of symmetry. Although this example uses x, y, and z for the planes of symmetry, the planes do not need to be tied to the coordinate axis system. There will also be three shear moduli, G_{xy}, G_{yz}, and G_{zx} and three Poisson's ratios, v_{xy}, v_{yz}, and v_{zx}. It is important to remember that $v_{xy} = v_{yx}$ due to symmetry, which is how the total number of independent constants for an orthotropic material is reduced to nine. The matrix form of Hooke's Law for an orthotropic material is given below:

$$\begin{Bmatrix} \varepsilon_x \\ \varepsilon_y \\ \varepsilon_z \\ \gamma_{yz} \\ \gamma_{zx} \\ \gamma_{xy} \end{Bmatrix} = \begin{bmatrix} \frac{1}{E_x} & -\frac{v_{yx}}{E_y} & -\frac{v_{zx}}{E_z} & 0 & 0 & 0 \\ -\frac{v_{xy}}{E_x} & \frac{1}{E_y} & -\frac{v_{zy}}{E_z} & 0 & 0 & 0 \\ -\frac{v_{xz}}{E_x} & -\frac{v_{yz}}{E_y} & \frac{1}{E_z} & 0 & 0 & 0 \\ 0 & 0 & 0 & \frac{1}{G_{yz}} & 0 & 0 \\ 0 & 0 & 0 & 0 & \frac{1}{G_{zx}} & 0 \\ 0 & 0 & 0 & 0 & 0 & \frac{1}{G_{xy}} \end{bmatrix} \begin{Bmatrix} \sigma_x \\ \sigma_y \\ \sigma_z \\ \tau_{yz} \\ \tau_{zx} \\ \tau_{xy} \end{Bmatrix}. \quad (6.36)$$

The final special case to be considered here is **transverse isotropy**. In transverse isotropy, the mechanical properties are the same in a single plane (for example, the x-y plane) and different in the z direction. An example of this is shown in Figure 6.16(b). Transversely isotropic materials have five independent constants: in this example, they are E_x and ν_{xy} for the x-y plane, and E_z, ν_{xz}, and G_{zx} for the z direction. The matrix form is shown below:

$$\begin{Bmatrix} \varepsilon_x \\ \varepsilon_y \\ \varepsilon_z \\ \gamma_{yz} \\ \gamma_{zx} \\ \gamma_{xy} \end{Bmatrix} = \begin{bmatrix} \frac{1}{E_x} & -\frac{\nu_{yx}}{E_x} & -\frac{\nu_{zx}}{E_z} & 0 & 0 & 0 \\ -\frac{\nu_{xy}}{E_x} & \frac{1}{E_x} & -\frac{\nu_{zx}}{E_z} & 0 & 0 & 0 \\ -\frac{\nu_{xz}}{E_x} & -\frac{\nu_{xz}}{E_x} & \frac{1}{E_z} & 0 & 0 & 0 \\ 0 & 0 & 0 & \frac{1}{G_{zx}} & 0 & 0 \\ 0 & 0 & 0 & 0 & \frac{1}{G_{zx}} & 0 \\ 0 & 0 & 0 & 0 & 0 & \frac{2(1+\nu_{xy})}{E_x} \end{bmatrix} \begin{Bmatrix} \sigma_x \\ \sigma_y \\ \sigma_z \\ \tau_{yz} \\ \tau_{zx} \\ \tau_{xy} \end{Bmatrix} \quad (6.37)$$

Example 6.6 **Finding strains in cortical bone**

Cortical bone can be thought of as a transversely isotropic material. The compliance matrix for dry human femur is as follows (Yoon and Katz, 1976), with values in GPa:

$$\underline{\underline{S}} = \begin{bmatrix} 0.053 & -0.017 & -0.010 & 0 & 0 & 0 \\ -0.017 & 0.053 & -0.010 & 0 & 0 & 0 \\ -0.010 & -0.010 & 0.037 & 0 & 0 & 0 \\ 0 & 0 & 0 & 0.115 & 0 & 0 \\ 0 & 0 & 0 & 0 & 0.115 & 0 \\ 0 & 0 & 0 & 0 & 0 & 0.139 \end{bmatrix}.$$

If the femur is loaded such that it experiences 5 MPa in compressive stress along the z-axis, what are the resulting strains?

Solution

$$\begin{Bmatrix} \varepsilon_x \\ \varepsilon_y \\ \varepsilon_z \\ \gamma_{yz} \\ \gamma_{zx} \\ \gamma_{xy} \end{Bmatrix} = \begin{bmatrix} 0.053 & -0.017 & -0.010 & 0 & 0 & 0 \\ -0.017 & 0.053 & -0.010 & 0 & 0 & 0 \\ -0.010 & -0.010 & 0.037 & 0 & 0 & 0 \\ 0 & 0 & 0 & 0.115 & 0 & 0 \\ 0 & 0 & 0 & 0 & 0.115 & 0 \\ 0 & 0 & 0 & 0 & 0 & 0.139 \end{bmatrix} \begin{Bmatrix} 0 \\ 0 \\ -0.005 \\ 0 \\ 0 \\ 0 \end{Bmatrix} = \begin{Bmatrix} 0.00005 \\ 0.00005 \\ -0.000185 \\ 0 \\ 0 \\ 0 \end{Bmatrix}$$

The bone will experience slight expansion in the x-y plane, and an even smaller compression in the z-direction.

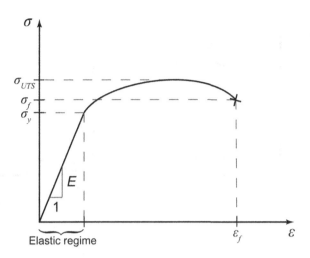

Figure 6.17

A traditional stress-strain curve for a ductile material.

6.3.7 Stress-strain curves

Basic material properties, such as elastic modulus and yield strength, can be measured in a uniaxial test. These are scalar quantities and are intrinsic to the material. The uniaxial test does not replace the need for multiaxial testing of a material, which has more relevance to the design of a device. The results of uniaxial tensile tests are traditionally plotted with stress on the ordinate and strain on the abscissa. Strain may be plotted as its dimensionless value, or as a percentage, where $\%\varepsilon$ is 100ε. A traditional stress-strain curve is given in Figure 6.17.

In this plot, several important parameters are labeled. The slope of the initial, linear portion of the stress-strain curve is E, the elastic modulus of the material. The linear portion can be described by Hooke's Law. The point at which the linear portion ends marks the yield strength, or σ_y. The yield strength marks the point at which plastic (or non-recoverable) deformation begins to occur. The maximum stress value on the plot is also known as the ultimate tensile strength, or UTS. Finally, failure occurs and σ_f and ε_f are given as the **failure strength** and **strain to failure**, respectively. Sometimes ε_f is referred to as εab – elongation at break.

Although this schematic stress-strain curve shows a pronounced change in slope at the point of yield, the change in slope is rarely so drastic in actual materials testing. In these cases, E and σ_y can be defined in other ways. A common method involves drawing a line that extends from $0.2\%\varepsilon$ and is parallel to the initial portion of the stress-strain curve. E is then defined as the slope of this line, while σ_y is the point where this line crosses the stress-strain curve. If the stress-strain curve does not exhibit a noticeable linear portion, it is traditional to take a tangent at a specific point, such as the origin, and call the slope of this line the **tangent modulus**, E_t. In this case, the strain value at the point of

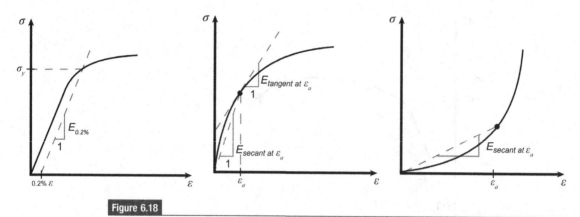

Figure 6.18

Alternative methods for calculating the elastic modulus.

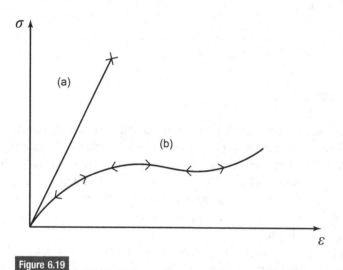

Figure 6.19

Stress-strain curves for (a) brittle and (b) elastomeric materials.

tangency should be noted along with the results. Another method for defining the elastic modulus is to use the **secant modulus**, which is the slope of a line connecting the origin to a point on the stress-strain curve. These methods are illustrated in Figure 6.18(a) and Figure 6.18(b). In natural biomaterials, it can be much more difficult to define these parameters than in standard engineering materials. An example of this is shown in Figure 6.18(c). This curve, which schematically shows the stress-strain response for tendon, demonstrates how the elastic modulus changes as the collagen fibers become aligned. In this case, it would be most appropriate to use the secant modulus, and to select that secant modulus to correspond to a stress or strain of interest.

Materials may exhibit stress-strain curves that are drastically different from Figure 6.17. In Figure 6.19, sample stress-strain curves for brittle and rubbery materials

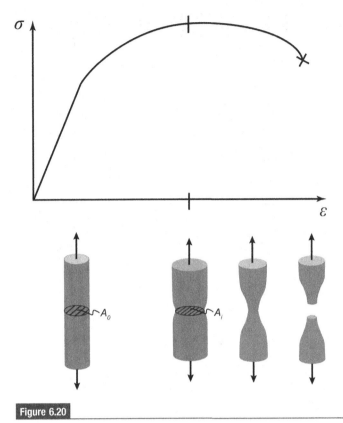

Figure 6.20

The evolution of necking during a uniaxial tensile test.

are given. The arrow markings indicate that if the material is unloaded before failure, it will return along the loading curve.

Certain ductile materials will exhibit **necking** behavior during tensile testing, which is shown in Figure 6.20. Inhomogeneities in the material (as mentioned in Chapter 2) can cause local fluctuations in stress and strain. These fluctuations in combination with strain-hardening can lead to a reduction in sample cross-sectional area that abruptly increases the local stress. This effect also leads to non-uniform deformation across the length of the sample. A typical stress-strain curve for a material undergoing necking will show stiffening in the increased slope of the stress-strain curve before fracture.

Example 6.7 Finding the elastic modulus from a stress-strain curve

Given the following points from a tensile test of stainless steel, create a stress-strain curve and use it to find the 0.2% offset modulus.

Elasticity

Stress (MPa)	Strain
0	0
38	0.0002
115	0.0006
160	0.002
180	0.004
190	0.006
210	0.01
225	0.015
245	0.02
270	0.03
290	0.05
320	0.1
400	0.2

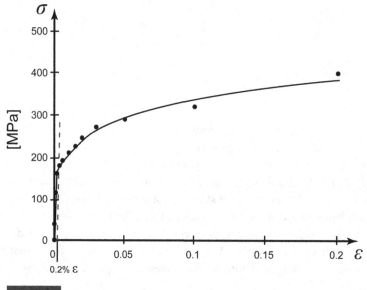

Figure 6.21

A stress-strain curve plotted using data given in Example 6.7.

Solution From the plot given in Figure 6.21 and using 0.2% offset, the elastic modulus is 190 GPa.

True stress and strain

Stress and strain as defined in Equations (6.13) and (6.11) are also known as "engineering" stress and strain. Because the length and cross-sectional area of a sample change during a tensile test, these are not measures of the true stress and strain at a given moment. They do, however, provide good estimates assuming that the changes to the sample in length and cross-sectional area are very small, so that the initial values can be used for stress and strain calculations without loss of accuracy. When there are large strains (greater than 1%), it becomes useful to use **true stress**, $\bar{\sigma}$, and **true strain**, $\bar{\varepsilon}$. However, for a material that exhibits necking behavior or experiences substantial strain before failure, the change in the cross-sectional area is significant, and a large difference arises between engineering and true stress and strain. True stress and strain use instantaneous values for area and length change as follows:

$$\bar{\sigma} = \frac{F}{A_i} \tag{6.38}$$

$$\bar{\varepsilon} = \int \frac{dl}{l} = \ln \frac{l_i}{l_0}. \tag{6.39}$$

True stress can be related to engineering stress by substituting $F = \sigma A_0$. This gives:

$$\bar{\sigma} = \sigma \frac{A_0}{A_i}. \tag{6.40}$$

Further, if constant volume can be assumed during plastic deformation (recall that post-yield materials are incompressible as demonstrated with $\nu = 0.5$), such that

$$A_0 l_0 = A_i l_i \tag{6.41}$$

then this leads to

$$\frac{A_0}{A_i} = \frac{l_i}{l_0} = \frac{l_0 + dl}{l_0} = 1 + \varepsilon \tag{6.42}$$

Substituting this result into Equations (6.39) and (6.40) gives the following relationships between true and engineering stress and strain.

$$\bar{\sigma} = \sigma (1 + \varepsilon) \tag{6.43}$$

$$\bar{\varepsilon} = \ln (1 + \varepsilon) \tag{6.44}$$

A comparison of an engineering stress-strain curve and a true stress-strain curve is given in Figure 6.22.

Stress-strain curves in natural tissues

In an ideal material, at small strains, the tensile and compressive engineering stress-strain curves are very similar (although with opposite signs) as shown in Figure 6.23.

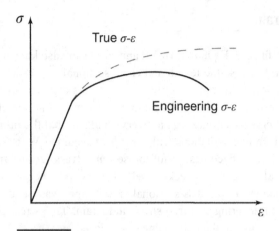

Figure 6.22

True stress and strain as compared to engineering stress and strain.

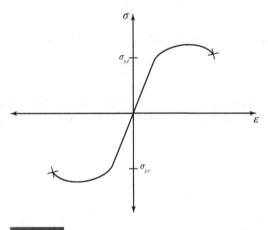

Figure 6.23

Idealized stress-strain curves in tension and compression.

However, this is not always the case for natural or synthetic biomaterials. These materials may exhibit different behavior in compression than in tension, leading to a curve similar to that in Figure 6.24. In this schematic, the tension-compression stress-strain curve shows that $\sigma_{y,\text{tensile}} < \sigma_{y,\text{compressive}}$. This curve, which represents the stress-strain curve for compact bone, demonstrates how the microstructure of bone allows it to carry higher compressive than tensile stresses. For this reason, it is important to verify if the values used in calculations for medical devices were taken from tensile or compressive tests, and match the test type to the dominant type of deformation or loading the material will see during use.

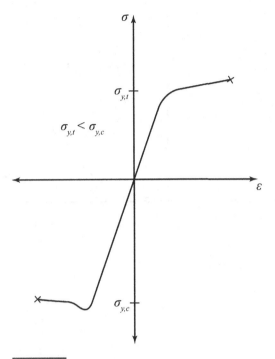

Figure 6.24

Stress-strain curves for cortical bone demonstrating differences in tensile and compressive loading.

Recall the stress-strain curves for natural tissues from Chapter 5, presented again in Figure 6.25. Note the change in slope with increasing strain, which is typical of biomaterials whose microstructure evolves during tension to provide increased stiffness. An example of this is the organization of crimped collagen bundles aligned with the direction of loading in tendon. As these bundles are pulled and extended, the full strength of the collagen is manifested in an increased elastic modulus. Also interesting is the difference between the circumferential and longitudinal stress-strain behavior of the human artery, which is necessary due to the very different requirements of the tissue in those two directions. Further discussion of the structure-property relationships of biological tissues can be found in Chapter 5.

6.4 Bending stresses and beam theory

6.4.1 Basics of beam theory

Development of beam equation

An intramedullary rod, discussed earlier in this chapter, is an example of a device that can be modeled using beam theory, which develops the response of a beam to a bending

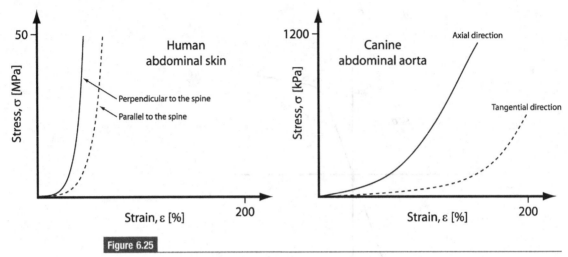

Figure 6.25

Stress-strain curves for skin (left) and artery (right).

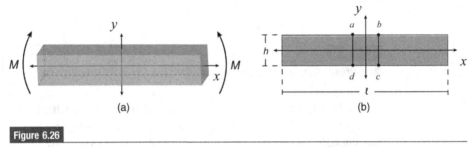

Figure 6.26

(a) A beam under bending moment with (b) marked segments that are perpendicular to the axis.

moment. These long, slender rods, made of stainless steel, titanium alloys, or nickel-titanium alloys, are used to stabilize long bones such as the femur or tibia after fracture. The derivation for beam theory (developed by Leonhard Euler and Daniel Bernoulli in the mid-1700s) is outlined here. Consider a horizontal prismatic beam under moment M_z, as shown in Figure 6.26(a). Bending moment has units of [N-m] or [lb-in]. It should be clear from the picture of this beam that the stresses which develop due to the bending moment will be tensile on the convex side of the beam and compressive on the concave side of the beam. The plane at which the stresses induced by M_z are zero is known as the **neutral axis**.

Assume that the beam has a cross-section with a vertical axis of symmetry and consider an element of the beam defined by the lines $\overline{ad}$ and $\overline{bc}$ as shown in Figure 6.26(b). When the beam is subjected to a bending moment, these lines become $\overline{a'd'}$ and $\overline{b'c'}$, as seen in Figure 6.27. These lines remain straight and perpendicular to the beam axis. This result, that plane sections remain plane during bending, is fundamental to the following

6.4 Bending stresses and beam theory

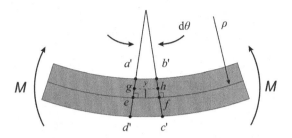

Figure 6.27

The same beam illustrating that plane sections remain plane during bending.

derivation of the beam equation. Generally it is also assumed that the length of the beam is much greater than the dimensions of the cross-section (usually $l/h > 20$).

In the pure bending scenario with no axial or shear forces applied, the beam axis will deform into an arc with radius of curvature ρ as shown in Figure 6.27. The segment described by the line $\overline{ef}$ or the angle $d\theta$ will deform to the length $ds = \rho \, d\theta$. Thus

$$\frac{d\theta}{ds} = \frac{1}{\rho} = \kappa \tag{6.45}$$

where κ is the curvature of the axis. The desired result is the variation of ε_x with y. The segment described by the line $\overline{gh}$ can be similarly defined as $dr = (\rho - y)d\theta$ The difference between these two, defined as $d\widehat{u}$, is:

$$d\widehat{u} = \rho d\theta - (p - y)d\theta = y d\theta. \tag{6.46}$$

Divide both sides by ds to bring the right-hand side to a similar form as Equation (6.45). Furthermore, if we assume that the angles involved are so small that the projections of $d\widehat{u}$ and ds onto the x-axis will have the same value, they can be replaced in Equation (6.46) by du and dx, achieving

$$\frac{du}{dx} = y \frac{d\theta}{ds} \tag{6.47}$$

$$\varepsilon_x = \kappa y. \tag{6.48}$$

This is the variation of strain with y. Incorporating Hooke's Law gives

$$\sigma_x = E\varepsilon_x = E\kappa y. \tag{6.49}$$

In order to reach the goal of finding the stress due to the applied bending moment, it is assumed that the beam is in equilibrium and that the sum of moments is zero.

$$-M_z + \int_A (E\kappa y dA)(y) = 0 \tag{6.50}$$

$$M_z = E\kappa \int_A y^2 dA. \tag{6.51}$$

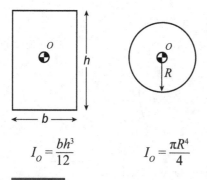

Figure 6.28

Area moments of inertia for rectangular and circular cross-sections.

The integral is the **area moment of inertia**, which depends only on the cross-sectional geometry. This is normally referred to as I with a subscript referring to a specific axis. This axis should be the neutral axis of the cross-section. Equation (6.51) can be re-written as

$$\kappa = \frac{M_z}{E I_z}. \tag{6.52}$$

Substitute Equation (6.49) into Equation (6.52) and rearrange to get the final result:

$$\sigma_x = \frac{M_z y}{I_z}. \tag{6.53}$$

Often, the maximum stress, σ_{max}, is the value of interest when analyzing a beam under a bending moment because this value provides the worst-case scenario for the internal stresses in the beam. As described earlier in the chapter, the maximum stress might also be useful when examining failure scenarios. To find this value, simply find the maximum value of y (sometimes this maximum value is denoted c, as below) and the equation for bending stress becomes

$$\sigma_{max} = \frac{Mc}{I}. \tag{6.54}$$

Equations for I of selected shapes are given in Appendix A. However, two of the most common are given here in Figure 6.28.

Example 6.8 Designing a hip stem cross-section

A designer is trying to decide between a rectangular cross-section and a circular cross-section for a hip stem. She would like to model them using beam theory in order

6.4 Bending stresses and beam theory

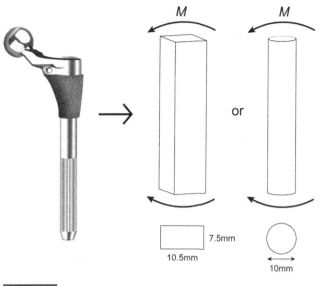

Figure 6.29

Circular and rectangular cross-sections for the hip stem described in Example 6.8.

to determine if the maximum stresses are different under a 50 N-m bending moment. In order to keep the cross-sectional area the same, she decides to use the cross-sections shown in Figure 6.29. Assuming they are both made of Ti-4Al-6V with $E = 114$ GPa, what are the maximum stresses in each hip stem? Are there other things to consider in selecting a cross-section beyond simply reducing stress?

Solution   Using the dimensions given in the problem, $I_{rect} = 7.18$ m^4 and $I_{circle} = 4.91$ m^4. Substituting these values along with the bending moment and maximum distance from the neutral axis into Equation (6.54), the result is $\sigma_{max,rect} = 365$ MPa and $\sigma_{max,circle} = 509$ MPa. The maximum stress is larger in the circular cross-section. Other things to consider in selecting a cross-sectional geometry include ease of manufacture, reduced stress-concentrations, and stabilization inside the femur.

6.4.2 Composite beam

Finding the neutral axis of a composite beam

It may become necessary to find the stresses that develop in a composite beam under applied bending moment, as given in Figure 6.30. This could be important in a device

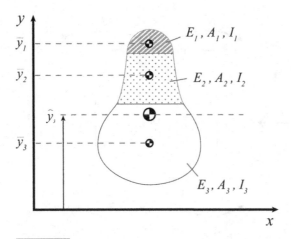

Figure 6.30

A generic composite beam, used to demonstrate the method for determining the location of the neutral axis.

made up of layered materials. In this beam, it is assumed that all n sections are firmly bonded and that the cross-section is still symmetric in the z-plane.

If the cross-section is placed in an arbitrary coordinate system as shown, use $\hat{y}$ to refer to the distance to the neutral axis of the overall beam, $\bar{y}_i$ to represent the distance to the centroid of the individual sections, and t to represent general vertical distance from the point of interest to the neutral axis. To find the neutral axis, use the following formula:

$$\hat{y} = \frac{\sum_{i=1}^{n} E_i \bar{y}_i A_i}{E_i A_i}. \tag{6.55}$$

This formula gives the location of the neutral axis with respect to the arbitrary coordinate system. The next step is to find the moment area of inertia for each section with respect to the neutral axis. Using $\bar{I}_j$ to represent the moment area of inertia for the section around its own centroid, the following equation, known as the Parallel Axis Theorem, gives the moment area of inertia around the neutral axis, $\hat{I}_j$:

$$\hat{I}_j = \bar{I}_j + (\bar{y}_j - \hat{y})^2 A_j. \tag{6.56}$$

Finally, the stress induced by the bending moment within a particular section of the composite beam is given as

$$\sigma_{i,\text{bending}} = \frac{M t E_i}{\sum_{j=1}^{n} E_j \hat{I}_j}. \tag{6.57}$$

6.4 Bending stresses and beam theory

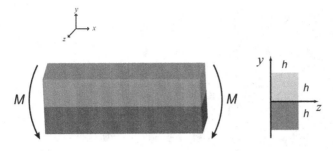

Figure 6.31

A composite beam for a custom implant described in Example 6.9. The upper half is UHMWPE while the lower half is Ti-6Al-4V.

Example 6.9 **Bending of a composite beam**

Consider a composite beam which consists of an UHMWPE and a Ti-6Al-4V beam perfectly bonded together for a custom implant that requires a polymer surface on one side and a metal surface on the other. Both beams have length and width $h = 20.0$ mm and are bonded as shown in Figure 6.31.

$E_{UHMWPE} = 1$ GPa and $E_{Ti-6Al-4V} = 114$ GPa. What is the maximum stress in the UHMWPE beam under bending moment $M = 100.0$ N-m? What is the maximum stress in the Ti-6Al-4V beam?

Solution First, use Equation (6.55) to find the neutral axis.

$$\hat{y} = \frac{E_1 \bar{y}_1 A_1 + E_2 \bar{y}_2 A_2}{E_1 A_1 + E_2 A_2} = \frac{E_1 \left(\frac{h}{2}\right) h^2 + E_2 \left(-\frac{h}{2}\right) h^2}{E_1 h^2 + E_2 h^2} = -9.83 \text{ mm}$$

Substitute this value into Equation (6.57).

$$\sigma_{UHMWPE,\max} = \frac{M(h - \hat{y}) E_1}{E_1 \left[\frac{h(h^3)}{12} + \left(\frac{h}{2} - (\hat{y})\right)^2 h^2\right] + E_2 \left[\frac{h(h^3)}{12} + \left(-\frac{h}{2} - (\hat{y})\right)^2 h^2\right]}$$
$$= -1.76 \text{ MPa}$$

$$\sigma_{Ti-6Al-4V,\max} = \frac{M(-h - \hat{y}) E_2}{E_1 \left[\frac{h(h^3)}{12} + \left(\frac{h}{2} - (\hat{y})\right)^2 h^2\right] + E_2 \left[\frac{h(h^3)}{12} + \left(-\frac{h}{2} - (\hat{y})\right)^2 h^2\right]}$$
$$= -68.5 \text{ MPa}.$$

If the composite beam is under purely axial loading, the formula for calculating stress in a composite beam is

$$\sigma_{i,axial} = \frac{F E_i}{\sum_{j=1}^{n} E_j A_j}. \tag{6.58}$$

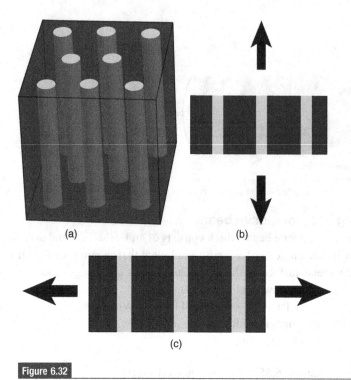

Figure 6.32

(a) A composite undergoing (b) longitudinal or (c) transverse loading.

6.5 Composites

6.5.1 Finding upper and lower limit of E

Composite materials, as described in Chapter 1, are highly desired because their mechanical properties can be tailored to fit their application. In medical devices, they are found in dental fillings and are increasingly considered for orthopedic implants. When composites are composed of unidirectional fibers, as shown in Figure 6.32(a), they exhibit orthotropic behavior. Of particular interest in these cases are the upper and lower bounds for elastic modulus.

If it is assumed that fibers are stiffer than the matrix, then intuitively it is clear that the upper bound for the elastic modulus of the composite as a whole will be in the direction parallel to the fiber axis and that the lower bound will be found perpendicular to this direction. Also assume that the fibers are perfectly bonded such that there is no delamination and that the matrix is an isotropic material. For these two derivations, the total cross-sectional area will be referred to as A, while the total cross-sectional area of the fibers is A_f and the total cross-sectional area of the matrix is A_m. Thus

$$A = A_f + A_m. \tag{6.59}$$

6.5 Composites

When a tensile force is applied in the direction parallel to the fiber axis as shown in Figure 6.32(b), the assumption of perfect bonding gives

$$\varepsilon = \varepsilon_f = \varepsilon_m \tag{6.60}$$

and the applied force will be a sum of the force in the fibers and the matrix

$$F = F_f + F_m. \tag{6.61}$$

Substituting in Equation (6.13) gives

$$\sigma A = \sigma_f A_f + \sigma_m A_m \tag{6.62}$$

$$E_{\text{upper}} \varepsilon A = E_f \varepsilon_f A_f + E_m \varepsilon_m A_m. \tag{6.63}$$

Using Equation (6.60) and volume fraction definitions $V_f = \dfrac{A_f}{A}$ and $V_m = \dfrac{A_m}{A}$, divide both sides of Equation (6.63) by A and find

$$E_{\text{upper}} = E_f V_f + E_m V_m. \tag{6.64}$$

This result shows that for uniaxial fibers, the upper bound for elastic modulus can be determined through a simple rule of mixtures.

To find the lower bound, imagine transverse loading as shown in Figure 6.32(c). In this case, the stress in the fibers and matrix must be equal to the total stress.

$$\sigma = \sigma_f = \sigma_m. \tag{6.65}$$

The total length in this direction can be written as the sum of the total lengths of the fibers and matrix:

$$l = l_f + l_m. \tag{6.66}$$

Furthermore, the total change in length in this direction can be written as the sum of the changes in the fibers and in the matrix:

$$\Delta l = \Delta l_f + \Delta l_m. \tag{6.67}$$

As described earlier, the definitions for strain can be written

$$\varepsilon = \frac{\Delta l}{l} \quad \varepsilon_f = \frac{\Delta l_f}{l_f} \quad \varepsilon_m = \frac{\Delta l_m}{l_m} \tag{6.68}$$

Combining Equations (6.67) and (6.68) gives

$$\varepsilon = \frac{\varepsilon_f l_f + \varepsilon_m l_m}{l}. \tag{6.69}$$

Using Equations (6.13) and (6.69) and volume fraction definitions $V_f = \dfrac{l_f}{l}$ and $V_m = \dfrac{l_m}{l}$, the following result is found:

$$\frac{1}{E_{\text{lower}}} = \frac{V_f}{E_f} + \frac{V_m}{E_m} \quad \text{or} \quad E_{\text{lower}} = \frac{E_f E_m}{E_f V_m + E_m V_f}. \tag{6.70}$$

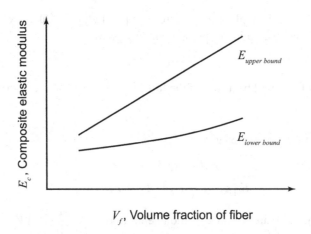

Figure 6.33

The difference between upper and lower bound elastic modulus for increasing fiber volume fraction in a fiber-reinforced composite.

These upper and lower bound predictions provide a range for estimates of E in uniaxial fiber composites. Figure 6.33 shows the range between upper and lower bound moduli for increasing fiber volume fractions in a representative fiber-reinforced composite. Short fibers, such as those found in UHMWPE or PEEK for implants, will not necessarily achieve maximal values for E. For particulate reinforcement, the elastic modulus estimate will follow the rule of mixtures. If working with more complex composites, it may be useful to consult a text specifically on that subject for more detailed derivations of elastic modulus estimates.

Example 6.10 Developing optimal thicknesses in layered devices

An engineer is designing a hydroxyapatite-coated polymer sleeve to assist with bone ingrowth in a total joint replacement. The cross-section of the proposed device is shown in Figure 6.34. Although the inner and outer diameters of the device are fixed at 6 mm and 12 mm, respectively, the thickness of the two layers needs to be optimized such that the overall elastic modulus of the device matches that of bone. Using these parameters, find the optimal thickness for each layer, using the following values: $E_\text{polymer} = 10$ GPa, $E_{HA} = 27$ GPa, $E_\text{bone} = 17$ GPa.

Solution Assume that the overall elastic modulus follows the rule of mixtures, that is

$$E_\text{device} = \frac{E_\text{polymer} A_\text{polymer} + E_{HA} A_{HA}}{A_\text{polymer} + A_{HA}}.$$

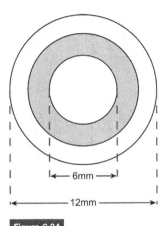

Figure 6.34

The cross-section of a hydroxyapatite-coated polymer sleeve as described in Example 6.10.

The area of the polymer cross-section and the HA cross-section both depend on the desired value, the thickness of the polymer layer, t.

$$A_{\text{polymer}} = \pi \lfloor (r_{\text{polymer}} + t)^2 - r_{\text{polymer}}^2 \rfloor$$

$$A_{HA} = \pi [r_{HA}^2 - (r_{\text{polymer}} + t)^2].$$

Plug these into the equation for E of the device and set equal to 17 GPa. Solve for t to find $t = 2$ mm. The optimal thickness of the polymer layer is 2 mm, leaving 1 mm for the HA layer.

6.6 Case study: modifying material and cross-section to reduce bone absorption

When considering orthopedic device design, a primary consideration is to ensure that the bone is carrying enough load that it does not suffer from stress shielding. As discussed in the inquiry for this chapter, stress shielding leads to bone resorption and eventual need for revision surgery. In this case study, two factors in the design of a hip stem will be considered: material choice and hip stem diameter. The material choices are common for orthopedic implants: stainless steel, a CoCr alloy and a titanium alloy. The outer diameter of the bone is assumed to be 2.5 cm, and the inner diameter of the bone is 1.0 cm. The hip stem will have an outer diameter of 1.1 cm (d_1) or 1.5 cm (d_2). Using the geometry and material properties given in Figure 6.35, the stresses in the bone with each stem can be calculated by evaluating the stresses for an axial load of 2851 N (4 × Body Weight for a 160 lb person) and a bending moment of 30 N-m separately and

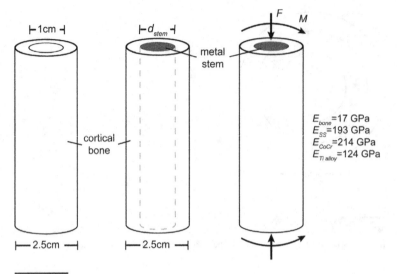

Figure 6.35

A schematic representation of a hip stem fixed into a femur.

then summing them.

$$\sigma_{\text{bone,axial}} = \frac{E_{\text{bone}} F}{E_{\text{bone}} A_{\text{bone}} + E_{\text{stem}} A_{\text{stem}}}$$

$$\sigma_{\text{bone,bending}} = \frac{E_{\text{bone}} M \left(\dfrac{d_{\text{bone}}}{2}\right)}{E_{\text{bone}} I_{\text{bone}} + E_{\text{stem}} I_{\text{stem}}}$$

Stress due to axial load [MPa]		
bone alone	6.91	
	d_1	d_2
bone with SS	1.93	1.23
bone with CoCr	1.79	1.12
bone with Ti	2.62	1.78

Stress due to bending load [MPa]		
bone alone	20.07	
	d_1	d_2
bone with SS	14.09	8.35
bone with CoCr	13.63	7.82
bone with Ti	15.82	10.77

Total stress [MPa]		
bone alone	26.98	
	d_1	d_2
bone with SS	16.02	9.58
bone with CoCr	15.43	8.94
bone with Ti	18.44	12.55

From these calculations, it is obvious that the worst-case scenario from a stress-shielding point of view is the CoCr stem with the 1.5 cm diameter. The best-case scenario is the titanium alloy stem with the 1.1 cm diameter, which allows the bone to experience a stress closest to what it would experience without any stem present. A set of calculations as quick and simple as in this analysis could help to set design parameters or guide research directions. Using the techniques described in this chapter, a possible solution to a well-known medical device issue was found.

6.7 Summary

The basic Hooke's Law describes the linear relationship between stress and strain during uniaxial testing. This assumes that the material is isotropic and homogeneous. A more complex version can be used to describe deformations during multiaxial loading using the compliance matrix, $\underline{S}$. The compliance matrix describes all possible interactions between stress and strain. Fortunately, most synthetic biomaterials and even many natural tissues are not fully anisotropic, allowing the compliance matrix to collapse to a more manageable number of independent constants that can be found using mechanical testing techniques. These constants are then used for comparing and evaluating materials. Elastic modulus, yield strength, and Poisson's ratio are some of the most commonly compared mechanical properties. Elastic modulus can be used as a measure of stiffness and is a material constant for standard engineering materials. Yield strength gives information about the transition between elastic (recoverable) deformation and plastic (non-recoverable) deformation. Poisson's ratio defines the deformations that occur in directions transverse to the loaded direction.

True stress and strain are used to give a more realistic picture of deformation behavior when the sample cross-section changes dramatically during testing. True stress and strain can be related to engineering stress and strain. It is more common to use engineering stress and strain particularly if small deformations are expected.

Beam theory provides a valuable model for certain loading situations *in vivo*. The derivation of the beam equation is provided along with examples in which it can be used to determine the maximum stress in a component. A method for determining the elastic modulus of composite materials is given for simple, uniaxial fibers in an isotropic matrix. For all of these subjects, there are also many reference books that can provide more detailed information, listed in the reference section below.

The concepts described here are fundamental to any analysis of the mechanical and material aspects of a medical device. Knowing the stresses and strains that are acting on a component is a crucial part of determining whether or not that component will fail. Understanding how to do basic mechanical analysis is a necessary skill in any device designer's repertoire.

6.8 Problems for consideration

Checking for understanding

1. Prove that $-1 < \nu < 1/2$.
2. A company wants to test the change in dimensions of its coronary guidewire under tensile loading. It has decided a reduction in diameter of 0.05% is acceptable, but anything greater will not do. The guidewire is made of 316L stainless steel with $E = 193$ GPa and $\nu = 0.35$ and has a diameter of 0.36 mm. What is the maximum axial load this guidewire can sustain while staying within the acceptable range?
3. Imagine a large tooth filling as shown below. It can be modeled as a rigid block with an elastic filling as illustrated in Figure 6.36. Given a load σ_z, what are the strains that will develop in the x-, y-, and z-directions, assuming that the filling is isotropic? Leave your answer in terms of σ_z and ν.
4. An intragastric balloon made of silicone with $E = 90$ MPa and nu $= 0.48$ is inflated to 0.4 MPa. The balloon is initially 1 mm thick with a radius of 6 cm. What is the inflated diameter?

For further exploration

5. From a study of excised human Achilles tendons by Wren and co-workers, the Achilles tendon fails at approximately 5000 N and has a cross-sectional area of

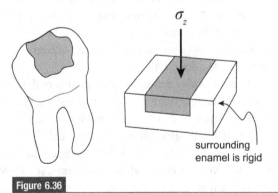

Figure 6.36

The filling in a tooth modeled as a solid with a rigid boundary.

127 mm² (Wren *et al.*, 2001). Give an example of one polymer and cross-sectional design that might be an appropriate replacement, and explain your choice.
6. How might the failure conditions for a balloon catheter be determined if the yield strength, wall thickness, and radius are known?

6.9 References

Ashizawa, N., Nonaka, K., Michikami, S., Mizuki, T., Amagai, H., Tokuyama, K., and Suzuki, M. (1999). Tomographical description of tennis-loaded radius: reciprocal relation between bone size and volumetric BMD. *Journal of Applied Physiology*, 86: 1347–51.

Nazarian, A., Hermannsson, B.J., Muller, J., Zurakowski, D., and Snyder, B.D. (2009). Effects of tissue preservation on murine bone mechanical properties. *Journal of Biomechanics*, 42(1): 82–6.

Venkatsubramanian, R.T., Wolkers, W.F., Shenoi, M.M., Barocas, V.H., Lafontaine, D., Soule, C.L., Iaizzo, P.A., and Bischof, J.C. (2010). Freeze-thaw induced biomechanical changes in arteries: role of collagen matrix and smooth muscle cells. *Annals of Biomedical Engineering*, 38(3): 694–706.

Wren, T.A.L., Yerby, S.A., Beaupre, G.S., and Carter, D.R. (2001). Mechanical properties of the human Achilles tendon. *Clinical Biomechanics*, 16: 245–51.

Yoon, H.S., and Katz, J.L. (1976). Ultrasonic wave propagation in human cortical bone, II: Measurements of elastic properties and microhardness. *Journal of Biomechanics*, 9: 459–64.

6.10 Bibliography

Courtney, T.H. (2000). *Mechanical Behavior of Materials*. Boston, MA: McGraw-Hill.

Dowling, N.E. (2007). *Mechanical Behavior of Materials, Engineering Methods for Deformation, Fracture and Fatigue*, 3d edn. Upper Saddle River, NJ: Pearson Education.

Gibson, L.J., and Ashby, M.F. (1997). *Cellular Solids, Structure and Properties*, 2d edn. Cambridge, UK: Cambridge University Press.

Popov, E.P. (1990). *Engineering Mechanics of Solids*. Englewood Cliffs, NJ: Prentice-Hall.

Sadd, M.H. (2005). *Elasticity, Theory, Applications and Numerics*. Burlington, MA: Elsevier.

7 Viscoelasticity

Jevan Furmanski and Lisa A. Pruitt

Inquiry

How are the material properties of cartilage and polymer bearing materials affected by the rate and duration of applied loading? How then does one design with biomaterials, accounting for time-dependent material properties?

The majority of important biomaterials, polymers, and tissues, for example, express some of the most complicated mechanical behaviors. When subjected to a constant load these materials will relax and continue to deform potentially indefinitely in a process called creep, making their use in structures problematic. Further, their response to high-frequency loading may differ dramatically from that observed under quasi-static conditions, directly impacting their performance in service. For instance, imagine the dynamic loading experienced by natural cartilage or a total knee replacement during downhill skiing, and how this rapid and severe loading could differ from the carefully controlled quasi-static testing typically performed in a laboratory. Comprehensive and effective modeling techniques are needed to properly account for time-dependent material properties in viscoelastic materials.

7.1 Overview

Viscoelastic materials, displaying both solid-like and fluid-like characteristics, are common in biomedical applications, from polymeric structures and coatings, to load-bearing connective tissues. The behavior of these materials cannot be described using only linear elasticity or plasticity theory, as is typical for many engineering materials. Rather, they are best described as having time-dependent material properties, such as an effective elastic modulus that is a function of the duration of applied loading. This time-dependence is usually associated with a characteristic time scale, called the **relaxation time**, and for time scales much greater or less than the relaxation time the material may not appear viscoelastic at all. Mathematical models of varying complexity can capture this behavior well, many times with only two or three material constants. Energy dissipation from the fluid-like part of the response can be separated from solid-like energy storage using a complex modulus, where they are represented by an imaginary and real part (respectively) of a complex material property (discussed below). In some cases, the material behavior is not simple enough to be represented by linear superposable functions, and nonlinear methods are especially suitable, for example when the unloading response does not match the loading response of the material.

7.2 Learning objectives

This chapter provides a broad overview of the fundamentals of viscoelasticity as applied to biomaterials. At the completion of this chapter students will be able to:

1. describe the physical meaning of viscoelasticity
2. define creep and stress relaxation
3. explain time-dependent material parameters such as glassy, transition, and rubbery plateaus as well as the Deborah number
4. calculate time-dependent creep and stress relaxation using Maxwell, Kelvin, and Standard Linear Solid models
5. define generalized linear viscoelasticity and how it is implemented in creep and stress relaxation
6. discuss the meaning of frequency domain analysis and the use of storage and loss modulus as material parameters
7. illustrate time-temperature superposition and shift factors
8. calculate the shift factor for an amorphous solid using the WLF (Williams, Landel, and Ferry) equation
9. describe the limitations of linear viscoelasticity and the attributes of using nonlinear viscoelasticity

7.3 Introduction to viscoelasticity

Engineering materials exhibit certain characteristics that are functionally important for design engineers: elastic modulus for structural rigidity; as well as yield, ultimate, and fatigue strength for long-term performance. For many materials, these properties remain unchanged over long durations of time, which makes the process of predicting performance from tabulated laboratory measurements straightforward. However, if these properties vary continuously over the life of a device, there is no single value of the property that can be referenced for design. This is the case with **viscoelastic** solids, or materials that exhibit both solid-like and fluid-like characteristics. Furthermore, when material properties are time-dependent they are more sensitive to the particular testing methods employed, which can complicate comparisons between independently measured properties. Thus, when viewed merely in terms of traditional engineering issues, time-dependent material properties make the design process more challenging.

Viscoelasticity is a physical phenomenon representative of the molecular and microstructural nature of a material. In some cases, the fluid-like nature of the deformation comes from a coherent fluid phase within the material that flows during deformation, such as in hydrated gels, cartilage, or other tissues. In other cases it reflects the inelastic flow of a single-phase solid, such as in polymers or macromolecules. In many instances it can be described quite accurately using simple mathematical models that are informed by the deformation mechanisms that constitute the flow in the material. These models make predictions of the real-world behavior of such time-dependent materials relatively straightforward (Ward, 1983).

Viscoelastic behavior profoundly affects the performance of materials used in load-bearing applications. There are two structural relaxation effects that are typically of concern: creep and stress relaxation. **Creep** is the continuous accumulation of strain under a sustained applied load, and is a concern for loaded structures that are required to maintain their geometry without extensive deformation, such as a suture. **Stress relaxation** occurs when a structure has a fixed configuration resulting in an initial imposed stress (such as a press-fit component), where the initial stress relaxes over time and may result in the assembly coming apart. Either of these two phenomena can result in an adverse outcome in an implant system and therefore must be anticipated and analyzed in viscoelastic materials. Given the predominance of viscoelastic materials used or treated in biomedical applications, viscoelasticity is central to biomaterials science.

7.3.1 Time-dependent material properties

One of the simplest descriptions of viscoelastic solids regards the apparent difference in the material behavior as measured over very long and very short time periods or above and below its viscoelastic (glass) transition temperature. As seen in Figure 7.1, the instantaneous elastic modulus of a polymer typically is several orders of magnitude higher than its apparent modulus at long time scales (or its behavior at elevated temperature). The short-term behavior is relatively unchanged over a wide time scale, and this is termed the **glassy plateau**. Similarly, if the material behaves like a solid, the long-term properties are also constant over a wide time scale, and this is the **rubbery plateau**. The constant value of the modulus in either of these regimes results in an apparently ordinary elastic material response for moderate changes in time scale, and viscoelastic effects can often be neglected. However, between these two extremes of behavior lies a transition region, in which the material is strongly affected by changes to temperature, the loading duration, or frequency, and thus the material behavior is highly **time-dependent**. This regime of behavior is termed the **viscoelastic transition**.

The characteristic times associated with each of the three regimes of behavior (glassy, rubbery, and transition) for a viscoelastic material are critical for determining its performance. The mean time of the viscoelastic transition is often designated the **relaxation time**, which is a rough delineation between the glassy and rubbery plateaus. Similarly, the characteristic temperature associated with the viscoelastic transition is known as the glass transition temperature (discussed below). The material behavior for time scales

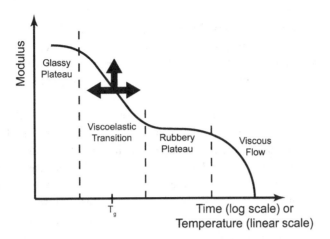

Figure 7.1

Elastic modulus as a function of time (log scale) or temperature (linear scale) for a viscoelastic material showing regions of glassy plateau, viscoelastic transition, rubbery plateau, and viscous flow. T_g is the viscoelastic transition temperature and is often denoted as the glass transition temperature.

that are an order of magnitude or two shorter than the relaxation time or concomitantly below the glass transition temperature gives the impression of being glassy, while behavior for time scales that are correspondingly longer than the relaxation times (or above glass transition temperature) appears rubbery. Although this distinction may not appear sufficiently quantitative for engineering analysis, it is a preliminary determinant of the expected behavior of a material and is useful for initial consideration of performance in a given application.

A valuable quantity in determining the overall behavior of a viscoelastic solid in a particular application is the **Deborah number**, D_e, which is simply a non-dimensional parameter that is defined as the normalized value of relaxation time (t) relative to the time duration τ of a given experiment or loading event

$$D_e = t/\tau. \qquad (7.1)$$

Thus, for $D_e \ll 1$ the behavior of the material is glassy or stiff, and for $D_e \gg 1$ the behavior is rubbery or fluid-like. The key to this distinction is the relative time scale of the relaxation and the loading event. For some materials, the relaxation is extremely fast (such as in water), and only very rapid loading events can elicit a solid-like response. For many solids, the transition is very slow (such as in ice or rock), and the material may behave like a fluid only over very long (geological) time scales. Materials that experience a relaxation over time scales of interest in practical applications (physiological time scales for biomaterials) are regarded as viscoelastic, and the particular value of D_e and thus the overall qualitative material response will depend on the specifics of a given mode of loading as well as material characteristics.

Example 7.1 **Deborah number and relaxation time of Silly Putty™**

Silly Putty™ is a viscoelastic polymer that bounces when dropped with a near ideal elastic response, but is easily plastically deformed, and when left for an hour or more flows into a puddle like a fluid. This is an ideal example of the applicability of the Deborah number, D_e.

The relaxation time for Silly Putty is approximately 1 second. The duration of an impact event, such as being dropped on a hard surface, is around 1 ms. Thus, for an impact, $D_e \sim 0.001$, and the expected behavior is that of a stiff solid. However, after resting for an hour, $D_e \sim 3600$, and the expected behavior in this case is that of a fluid (Silly Putty does not have a rubbery plateau, but rather behaves as a viscous fluid at long times). For intermediate times, such as when pressing a form of Silly Putty flat, the apparent stiffness of the material depends on how hard or rapidly one presses on it. This behavior is shown in Figure 7.2.

A relaxation time of 1 second places Silly Putty directly in the center of the time scales of interest for typical human experience, which makes it interesting as a toy, and problematic as a model material for technical or structural applications. The relaxation time provides a basic but critical measure of the suitability of a given material for a particular application.

7.3.2 Polymers as viscoelastic materials

The above example of the deformation of Silly Putty™ motivates the complex nature of employing viscoelastic materials in load-bearing applications. Viscoelastic polymers behave as a solid for shorter time scales, and flow like a fluid at elongated time scales, requiring a greater range of analysis for a component that must remain structurally viable for long periods of service (years or decades for many medical devices). Polymer viscoelastic behavior for a given backbone chemistry is also dependent upon molecular variables such as molecular weight, crosslink density, entanglement density, and crystallinity (Chapter 4). Modeling techniques have been developed to predict the behavior of polymers over wide ranges of temperature and frequency conditions, and to reduce the experimental complexity involved in engineering with polymers. These techniques focus on the temperature and rate sensitivity of polymer mechanical behavior, though other processes such as environmental degradation also affect the observed deformation response and temperature sensitivity. In some applications, the rate sensitivity is extremely important to quantify correctly, such as in gaskets between dynamically loaded parts, and in others the creep response is vital to the overall assessment of performance, as in total joint replacements where creep deformation can be mistaken for problematic bearing wear (Sychterz et al., 1999). Finally, in grafts that reinforce or replace tissues, the viscoelastic response is important to match to the greatest extent possible to maintain compatibility over all time scales of interest.

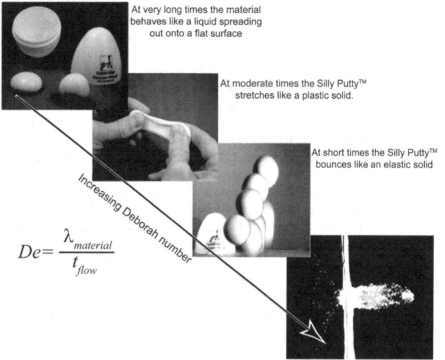

Figure 7.2

Silly Putty shows both elastic and viscous properties depending on the length scale of deformation. The image on the far left shows viscous flow (small Deborah number) while the image on the right shows the result of high-impact (elastic) loading associated with a high value of the Deborah number. Image courtesy of the Cambridge Polymer Group (Cambridge, MA) and Dr. James Bale at MIT (Cambridge, MA).

7.3.3 Tissues as viscoelastic solids

Most tissues at their essential microstructural size scale are composed of an entangled polymer network containing collagen and other long-chain molecules (see Chapter 5). The extent of entanglement, interpenetration of multiple networks, and the crosslinking between these molecules affects their character, especially when considering viscous effects. Structural arrangements of the molecules, such as crimping in tendons, may result in a nonlinear instantaneous elastic response (Chapter 5), which is then compounded by time-dependent effects stemming from the molecular interactions of the constituents. To complicate matters further, the extent of crosslinking in the collagen network can be affected by the environment, as in the case of enzymatic crosslinks in the organic phase of bone tissue being dependent on the local sugar concentration (Saito *et al.*, 2006; Saito and Marumo, 2010). Thus, both the instantaneous and time-dependent mechanical properties of tissues can be complex and highly dependent on the dynamic character of the constituent molecules and their interactions.

Most tissues are hydrated to maintain their function, and in some cases the mechanical behavior of the fluid component directly impacts the deformation response of the whole tissue. This is a special case of viscoelastic behavior, where part of the response is actually from a separate material phase behaving as a true viscous medium. For example, in articular cartilage, the tissue swells with fluid when unloaded and puts the molecular microstructure into a steady state of tension. When mechanical loads are applied to the tissue, the tissue is deformed, and the fluid is forced out through pore space in the tissue (see Example 7.2). Thus, the temporarily trapped fluid bears pressure and contributes to the mechanical response, and the friction resulting from the flow of the fluid over the internal pore surfaces also dissipates energy and restricts the flow. In this manner, both the intrinsic viscoelastic character of the tissue and the viscosity of the fluid phase contribute to the dynamic response of the tissue. Fluid constituent behavior is also important for the function of other tissues such as the annulus fibrosis and nucleus pulposus in the intervertebral disk (Riches *et al.*, 2002; Silva *et al.*, 2005; Williams *et al.*, 2007). There are dedicated models to describing the multi-phase behavior of hydrated tissues, such as the poroelastic model used in soft tissues (Mow *et al.*, 1980), which can be extended to include more complicated viscoelastic character (Wilson *et al.*, 2005), or even an additional phase accounting for osmotic effects (Lai *et al.*, 1991).

Example 7.2 Viscoelastic behavior of articular cartilage

Articular cartilage, as described above, provides cushioning for the end of long bones in synovial joints and facilitates complex lubrication schemes that enable healthy joints to function with minimal friction. Articular cartilage comprises a functionally graded collagen fiber orientation that is contained within a hydrated gel (Figure 7.3). This hydrated tissue responds much like a viscoelastic solid wherein the strain response under load depends very much on the rate of loading. Figure 7.3 shows the response of an idealized form of articular cartilage in which fluid exudes into the joint space as a function of loading state (Mow *et al.*, 1980). This behavior is often modeled with poroelasticity theory; however, a first approximation of the strain response of cartilage under compressive load can be described by using a simple viscoelastic model with a spring and dashpot in parallel (Kelvin model).

7.4 Linear viscoelastic networks

Linear viscoelasticity is used widely to model deformation due to its inherent ability to treat the individual components of a spectral response as additive through the property

7.4 Linear viscoelastic networks

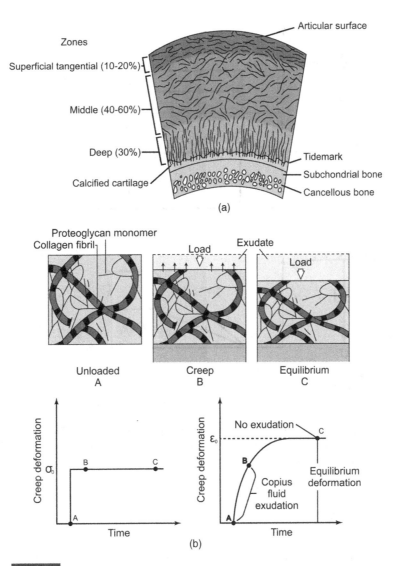

Figure 7.3

Articular cartilage (a) comprises collagen fibers distrusted in an orientation that varies with thickness of the tissue and (b) exhibits time-dependent strain under deformation owing to the fluid constituent of the tissue.

of linear superposition. The overall strain response, for example, can be represented by a series summation of individual linear component functions:

$$\varepsilon_{(t)} = f_1(A\sigma_1 + B\sigma_2, t) + f_2\ldots = Af_1(\sigma_1, t) + Bf_1(\sigma_2, t) + f_2\ldots \quad (7.2)$$

where each function in the series represents a separate viscoelastic relaxation function corresponding to a distinct deformation process in the material. In practice, Equation (7.2) is the solution to an ordinary differential equation, and each component in this expression takes the form of a simple mathematical expression. In linear viscoelasticity,

216 Viscoelasticity

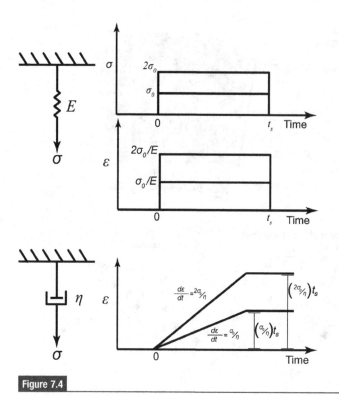

Figure 7.4

Schematic showing spring and dashpot, and their individual responses to the application of an applied stress.

mathematically linear individual model elements are assembled into a network model that describes the overall behavior of the solid. The most basic first-order viscoelastic elements are introduced first and then higher order networks composed of these elements are evaluated.

7.4.1 Linear viscoelastic elements

Linear viscoelastic models are composed of a network of linear elastic and linear viscous elements (Figure 7.4). The linear elastic element represents the classical linear elastic solid (for nominal stress, though it is often instead written in terms of shear deformation):

$$\sigma = E\varepsilon \tag{7.3}$$

while the linear viscous response corresponds to Newtonian fluid flow (with μ representing the viscosity)

$$\sigma = \mu \frac{d\varepsilon}{dt}. \tag{7.4}$$

Equation (7.3) returns a fixed deformation for a constant applied stress, while Equation (7.4) will flow indefinitely at a set strain rate for a constant applied stress. Put another

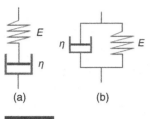

Figure 7.5

Illustration of the two basic models used in viscoelasticity: (a) Maxwell with the spring and dashpot in series, and (b) Kelvin with the spring and dashpot in parallel.

way, at long times under load, the damper effectively has no stiffness. This represents the essential difference between solid-like and fluid-like behavior.

These fundamental elements are typically represented by a spring and a damper (also called a dashpot), respectively. It is clear from Equations (7.3) and (7.4) that linearity is satisfied, and thus Equation (7.2), so linear networks can be assembled from them. Also, note that viscoelastic response composed of some combination of these completely partitions the elastic from the viscous component (i.e., the elements are either purely solid or fluid, but not both). It is uncommon in viscoelasticity to be concerned with a particular element of the response; spring-damper combinations are more common in viscoelasticity for use as basic building blocks of higher order networks. These are discussed next.

7.4.2 First-order viscoelastic material models

There are two methods of combining basic linear elastic and viscous elements – by imposing either equal stress or equal strain across each element. This corresponds to a series or parallel network relationship (respectively), shown in Figure 7.5.

The series network is denoted the **Maxwell model**, while the parallel network is the **Kelvin model**. Intuitive insight into their behavior can be gained by inspecting the two models. For the Maxwell model, each element experiences the entire applied stress at all times. Thus, the damper element strains indefinitely under load, and the Maxwell model behaves like a fluid at long times. Alternatively, the Kelvin model distributes the stress between the elements, with the fraction of the applied stress in each corresponding to their relative stiffness. At long times, when the damper has no effective stiffness, the Kelvin model degenerates to the linear elastic solid of the spring element, and thus ultimately behaves like a solid.

The response of the Maxwell and Kelvin models is now developed in the time domain, armed with the knowledge that the Maxwell model is fluid-like and the Kelvin model is solid-like. First the constant stress (creep) response for the Maxwell model is derived. For strain compatibility, the total strain is given as

$$\varepsilon_M = \varepsilon_1 + \varepsilon_2 \tag{7.5}$$

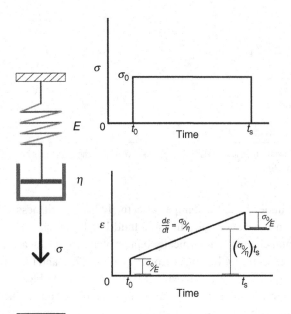

Figure 7.6

Creep (time-dependent strain) response of the Maxwell model subjected to a constant stress.

and the two elements experience the same stress at all times. Integrating Equation (7.4) and invoking linear superposition, one can write

$$\varepsilon_{M,Cr}(t) = \left(\frac{1}{E} + t\frac{1}{\mu}\right)\sigma. \tag{7.6}$$

The strain response as a function of time of the Maxwell model is then easily separated into the elastic and viscous components in creep (Figure 7.6).

For an applied constant strain boundary condition, the viscous component of the Maxwell model relaxes and illustrates the phenomenon of **stress relaxation**.

Taking Equations (7.3) and (7.4) and equating them (recall the equal stress condition),

$$E\varepsilon_1 = \mu\frac{d\varepsilon_2}{dt}. \tag{7.7}$$

Considering the total strain is constant, the time derivative vanishes, and one obtains

$$\frac{d\varepsilon_{M,SR}}{dt} = 0 = \frac{d\varepsilon_1}{dt} + \frac{d\varepsilon_2}{dt}. \tag{7.8}$$

Substituting Equation (7.7) into (7.8), the differential equation for the strain in the spring element is found

$$\frac{d\varepsilon_1}{dt} = -\frac{E}{\mu}\varepsilon_1 \tag{7.9}$$

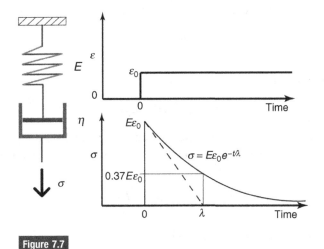

Figure 7.7

Schematic of the stress relaxation behavior (effective stiffness) for the Maxwell model subjected to a constant displacement.

which is satisfied for an exponential function

$$\varepsilon_{1,SR}(t) = C_1 e^{-t\frac{E}{\mu}}. \tag{7.10}$$

The relaxing stress in the spring element (and thus across the entire model) is simply the $E\varepsilon_1$, and the time-dependent modulus of the Maxwell element is this stress divided by the imposed strain

$$E_{M,SR}(t) = \frac{E\varepsilon_{1,SR}}{\varepsilon_{M,SR}} = E e^{-t/\tau} \tag{7.11}$$

where τ is a time constant (the relaxation time referred to earlier, and represents the time for a 63% drop in effective modulus); in this case $\tau = \mu/E$, and the coefficient E is the instantaneous ($t = 0$) elastic modulus of the Maxwell model, which corresponds to the stiffness of the spring E (at $t = 0^+$, the damper remains undeformed and has "infinite" stiffness). Mathematically, the effective elastic modulus of the Maxwell model decays exponentially from the spring stiffness E exponentially down to zero at times much longer than the time constant τ (Figure 7.7).

The behavior of the Maxwell model fits with the fluid-like description of viscoelasticity from the discussion of the Deborah number, where the Maxwell model is a fluid for $t \gg \tau$ (or $D_e \gg 1$) but an elastic solid for $t \ll \tau$ ($D_e \ll 1$).

The Kelvin model assumes equal strain in its two parallel elements; its response can be easily derived by equating their strain rates. Taking the derivative of Equation (7.3) yields

$$\frac{d\sigma}{dt} = E\frac{d\varepsilon}{dt} \tag{7.12}$$

and one can combine this with Equation (7.4) to obtain

$$\frac{1}{E}\frac{d\sigma_1}{dt} = \frac{1}{\mu}\sigma_2. \quad (7.13)$$

For equilibrium the stresses in the elements are added:

$$\sigma_K = \sigma_1 + \sigma_2. \quad (7.14)$$

For an imposed constant stress in creep, Equations (7.13)–(7.14) give the distribution of the total stress between the two elements as a function of time. First, rewrite Equation (7.14) as

$$\sigma_K = \sigma_1 + \frac{\mu}{E}\frac{d\sigma_2}{dt} \quad (7.15)$$

and solve for the stress in the spring element

$$\sigma_1 = \sigma_K - C_2 e^{(-t/\tau)} \quad (7.16)$$

as well as the stress in the damper element

$$\sigma_2 = C_2 e^{(-t/\tau)}. \quad (7.17)$$

Observing that the damper has "infinite" stiffness at $t = 0$ for a rapidly applied load, we know that $C_2 = \sigma_{K,Cr}$. Since all of the stress is borne by the damper at $t = 0$, the Kelvin model has theoretically infinite instantaneous elastic modulus, which is not realistic but can be corrected with the addition of another elastic spring to create a standard linear solid model.

For a rapidly applied and held load, we see from combining Equations (7.3) and (7.15) that

$$\sigma_{K,Cr} e^{-t/\tau} = \mu \frac{d\varepsilon_K}{dt} \quad (7.18)$$

which can be integrated to give the creep strain for the Kelvin model

$$\varepsilon_{K,Cr}(t) = \frac{\sigma}{E}\left(1 - e^{-t/\tau}\right). \quad (7.19)$$

The creep (time-dependent) strain response for the Kelvin model is shown in Figure 7.8.

The Kelvin model relaxes from a highly stiff response, until the strain rate decreases to zero at long times, and the material reaches a final strain dictated solely by the linear elastic element. Hence, the Kelvin model is not useful at short times for analyzing material deformation due to a rapidly applied constant load (creep) but nevertheless still illustrates the essential behavior of a viscoelastic solid. For similar reasons, it is not realistic to use the Kelvin model to illustrate a rapidly applied strain, as this model would exhibit infinite stiffness (viscosity) at high rates of deformation. Hence the Kelvin model is incapable of predicting stress relaxation behavior.

7.4 Linear viscoelastic networks

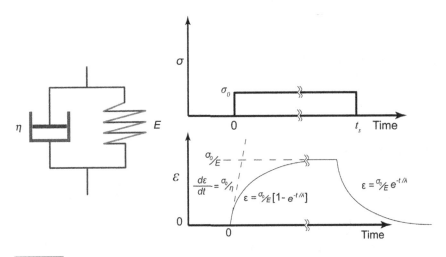

Figure 7.8
Creep strain as a function of time and in response to an applied constant stress for the Kelvin model.

Example 7.3 **Stress relaxation in a polymeric implant**

Consider a suture anchor that undergoes stress relaxation behavior that is captured by the Maxwell model where the spring element has a modulus of 4 GPa and the dashpot element has a viscosity of 10 GPa·s. For this viscoelastic material, what is the stress relaxation time? If an instantaneous, constant strain of 0.5% is applied to the polymer at $t = 0$ and is followed by an additional strain of 3% at $t = 30$ seconds, what is the stress in the polymer after 50 seconds? What are the implications for stress relaxation in such an implant?

Solution The characteristic time as defined by the Maxwell model above is given as:

$$\tau = \eta/G = 10\,\text{GPa} \cdot \text{s}/4\,\text{GPa} = 2.5\,\text{s}.$$

The form of the stress relaxation for a Maxwell material is given as:

$$\sigma = \sigma_o e^{\frac{-t}{\tau}}$$

Because the system is described by linear viscoelasticity, the stresses at each increment of time may be superposed. $\varepsilon_1 = 0.5\%$ (applied at $t = 0$), $\varepsilon_2 = 3.0\%$ (applied at $t = 30$ s), and hence the stress is given as

$$\sigma_{t=50} = \sigma_1 e^{\frac{-t_1}{\tau}} + \sigma_2 e^{\frac{-t_2}{\tau}} = 4\,\text{GPa} \cdot 0.005 e^{\frac{-30}{2.5}} + 4\,\text{GPa} \cdot 0.03 e^{\frac{-20}{2.5}}$$
$$= 0.041\,\text{Pa} + 40,225\,\text{Pa} = 40.3\,\text{kPa}.$$

Stress relaxation can result in loosening of the implant and may result in insufficient stress transfer to tissue for wound healing.

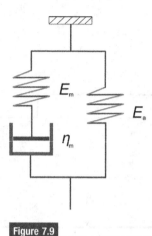

Figure 7.9

Schematic illustration of the Standard Linear Solid model.

7.4.3 The Standard Linear Solid Model

The most useful concept motivated by the Kelvin model is the sharing of stress over time between the two elements that both experience the imposed strain equally. This is physically interesting, as relaxation processes in a material could shift the stress from one constituent or phase to another over time. The limitation of the Kelvin model comes from the unrealistic behavior of the damper at high rates of deformation, which can be alleviated by increasing the complexity of the model slightly while retaining the concept of variable stiffness elements in parallel. This is done by adding an additional spring as shown in Figure 7.9.

The **Standard Linear Solid** (SLS) can be thought of as an improved Kelvin model solid, as it is the simplest solid model that can exhibit both realistic creep and stress relaxation behavior. The stress in the parallel network is distributed between the spring E (which can be thought of as a Maxwell model with very high viscosity), and a complete Maxwell model that exhibits a time-dependent effective modulus, as discussed above. Thus, the overall modulus of the SLS is time-dependent, and the distribution of stress between the two arms of the network also varies with time. Yet, it exhibits only one relaxation transition, as there is only one dissipative element in the model, though the relaxation time is slightly different for creep and stress relaxation scenarios.

The creep response of the SLS is considered first. The strain across each arm of the network corresponds to the overall strain in the network, as in the Kelvin model, and is written as

$$\varepsilon_{SLS} = \varepsilon_a = \varepsilon_m. \tag{7.20}$$

Taking the derivative, and substituting the behavior of each element, yields

$$\frac{d\varepsilon_{SLS}}{dt} = \frac{1}{E_a}\frac{d\sigma_a}{dt} = \frac{1}{E_m}\frac{d\sigma_m}{dt} + \frac{1}{\mu}\sigma_m \tag{7.21}$$

and recalling the equilibrium requirement, one has the function determining the partitioning of the stress into the Maxwell arm of the SLS:

$$\frac{1}{E_a}\frac{d\sigma_{SLS}}{dt} = \left(\frac{1}{E_m} + \frac{1}{E_a}\right)\frac{d\sigma_m}{dt} + \frac{1}{\mu}\sigma_m. \qquad (7.22)$$

In creep the applied stress is constant (the left term of Equation (7.22) is zero), yielding the differential equation for creep stress in the Maxwell arm of the SLS

$$\tau_{SLS,Cr}\frac{d\sigma_m}{dt} + \sigma_m = 0 \qquad (7.23)$$

where the $\tau_{SLS,Cr}$ is the creep time constant for the SLS

$$\tau_{SLS,Cr} = \mu\left(\frac{1}{E_m} + \frac{1}{E_a}\right) \qquad (7.24)$$

and the time-varying stress in the Maxwell element during creep of the SLS is

$$\sigma_{m,Cr} = C_{Cr}e^{-t/\tau_{SLS,Cr}}. \qquad (7.25)$$

To solve for the initial value C_{Cr}, note that as $t \to 0$ the damper has not yet deformed, and the strain in both springs is identical to the total strain. Thus, at $t \to 0$,

$$\frac{\sigma_a}{E_a} = \frac{\sigma_m}{E_m} \qquad (7.26)$$

and thus one substitutes into the equilibrium condition and finds

$$C_{Cr} = \sigma_{SLS}\frac{E_m}{E_a + E_m}. \qquad (7.27)$$

Combining Equations (7.27) and (7.25) yields the stress in the Maxwell arm of the SLS during creep,

$$\sigma_m = \sigma_{SLS}\frac{E_m}{E_a + E_m}e^{-t/\tau_{SLS,Cr}}. \qquad (7.28)$$

The overall strain response of the SLS in creep is then easily derived from the spring arm of the network to give

$$\varepsilon_{SLS,Cr}(t) = \frac{\sigma_{SLS}}{E_a}\left(1 - \frac{E_m}{E_a + E_m}\right)e^{-t/\tau_{SLS,Cr}} \qquad (7.29)$$

and thus the effective creep compliance of the SLS is given as

$$J_{SLS}(t) = \frac{\varepsilon_{SLS,Cr}}{\sigma_{SLS}} = \frac{1}{E_a}\left(1 - \frac{E_m}{E_a + E_m}\right)e^{-t/\tau_{SLS,Cr}}. \qquad (7.30)$$

Inspecting the creep compliance of the SLS, it is seen that at short times ($t \ll \tau$) its effective modulus is $E_{SLS} = E_a + E_m$, while at long times ($t \gg \tau$) the stiffness of the network relaxes to E_a. This relaxation from a stiff solid to a more compliant solid is the quintessential viscoelastic response. Recalling the original motivation for viscoelasticity from the transition in polymers from stiff glassy behavior to compliant rubbery behavior, the SLS embodies this observation especially well when $E_m \gg E_a$. In

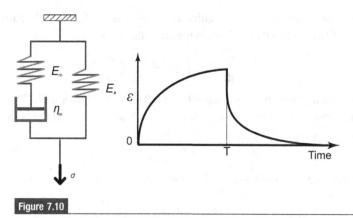

Figure 7.10

Illustration of the SLS and its concomitant strain response (creep) under constant load conditions.

this case, the response of the SLS relaxes from the instantaneous elastic modulus $E_0 \sim E_m$ down to the long-term elastic modulus $E_\infty \sim E_a$.

The stress relaxation response of the SLS can be found easily, by first recognizing that the stress experienced by the spring arm of the network is constant, since the strain in the model is held constant. Therefore, the dynamics in the stress relaxation of the SLS are solely due to the Maxwell model arm of the network. This results in a different relaxation time for stress relaxation of the SLS than for creep,

$$\tau_{SLS,SR} = \frac{\mu}{E_m} \qquad (7.31)$$

and the overall stress relaxation response of the SLS is

$$\sigma_{SLS,SR}(t) = \varepsilon_0 \left[E_a + E_m e^{-t/\tau_{SLS,SR}} \right]. \qquad (7.32)$$

Another form of the standard linear solid (also known as the Kelvin-type SLS) is composed of the Kelvin model in series with another spring. A schematic illustration of the Kelvin-type SLS and its concomitant strain response (creep) under constant load conditions is shown in Figure 7.10.

Example 7.4 Stiffness evolution in a viscoelastic biomaterial

A common technique in fatigue research is to track the change in elastic modulus of a material or structure subjected to dynamic loading and infer from this the accumulation of damage in the material. This is clear-cut in the case of a true linear-elastic material like a ceramic, where changes to the effective modulus come from internal cracks that have potentially disastrous implications for a flaw-intolerant material. However, in the case of viscoelastic materials, the apparent modulus is subject to the duration of the test. For the case of linear viscoelasticity, where the relaxation spectrum of the material is unchanged with deformation, the overall material stiffness will reduce over time,

7.4 Linear viscoelastic networks

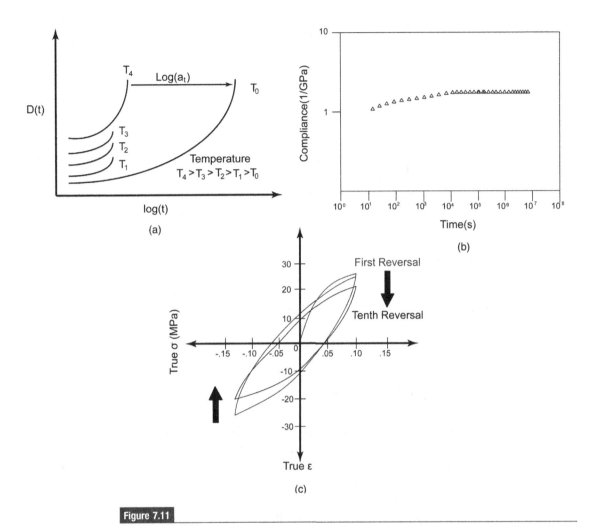

Figure 7.11

Illustration of (a) the master curve correction factors for UHMWPE, (b) creep compliance over time for UHMWPE at 37°C loaded at 1 MPa, and (c) cyclic softening in UHMWPE subjected to fully reversed cyclic strain levels of 0.12, which is comparable to physiological strains for a joint replacement. After Deng *et al.* (1998) and Krzypow and Rimnac (2000).

which is not a damage process. Beyond creep effects, changes to the dynamic response can also be caused by deformation and reorganization of the material as a nonlinear viscoelastic effect, which is also not a damage mechanism.

UHMWPE used in total joint reconstruction is a viscoelastic material. Long-term creep tests on this material have shown an order of magnitude increase in the compliance (loss of stiffness) over the time frame of one month (Deng *et al.*, 1998). Similarly, this polymer is known to cyclically soften owing to its viscoelastic behavior (Krzypow and Rimnac, 2000). This creep and cyclic strain softening behavior is shown in Figure 7.11. Thus, inferring damage in biomaterials from changes in the dynamic

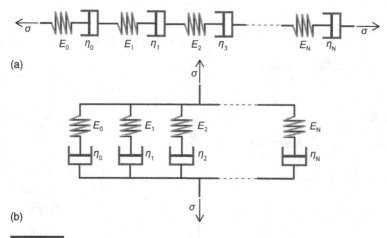

Figure 7.12

Generalized forms of (a) the Maxwell model and (b) the Kelvin model.

response is confounded with the underlying changes to the deformation stemming from other time-dependent physical phenomena, and must be undertaken with care.

7.4.4 Generalized linear viscoelasticity

The Standard Linear Solid has only one transition time, which is an idealization of the more complicated relaxation behavior of polymers, where five or more discernable transitions are common. Additional terms can be added in a series representation of the relaxation to reflect other transitions with distinct transition times, leading to a generalized Kelvin or generalized Maxwell model (Figure 7.12).

Specifically, Kelvin models are chained in series to produce models of arbitrary complexity for creep deformation, and Maxwell models are combined in parallel for stress relaxation analysis. Using a weight factor for the magnitude of the relaxation for each transition, we obtain the series representations for relaxation behavior

Generalized Maxwell model stress relaxation:

$$\sigma(t) = \varepsilon \sum_n g_n e^{-t/\tau_n}. \tag{7.33}$$

Generalized Kelvin model creep compliance:

$$J_{GenK}(t) = J_0 + \sigma \left[h_n \left(1 - e^{-t/\tau_n} \right) \right] \tag{7.34}$$

where g_n and h_n are weight coefficients that specify the magnitude of relaxation associated with each transition.

It is often tempting to increase the number of parameters in a viscoelastic model, as this will always lead to a better apparent fit to experimental data. Increasing the complexity of the model comes at the expense of a clear picture of the relative contributions of each individual element, as the transition times begin to run together and the weight terms become less certain. The strength of the generalized linear models is that they transparently represent the distinct characteristic times corresponding to molecular relaxations; overly complicated models lose their interpretability while simultaneously increasing their ambiguity. If the material response of interest cannot be represented by a few terms, then it may be that the response is nonlinear and no number of elements will satisfactorily capture the phenomena of interest. It should also be borne in mind that each term of the series contains two free parameters; meaning that increasing the complexity of the model necessitates additional experimental effort to probe the full range of relaxation dynamics.

7.5 Frequency domain analysis

The duration of a load event can be thought of roughly as the inverse of a characteristic test frequency. In fact, most loading events except for sinusoidal waveforms produced in the laboratory contain a spectrum of individual frequency components, though they will still generally have an overall dominant frequency component. Understanding the material behavior over a range of loading frequencies is therefore analogous to the treatment above for relaxation time scales, but provides additional opportunity for mathematical analysis utilizing the spectral mechanical response. Identifying the spectrum of relaxation by sinusoidally deforming the material at a single frequency at a time over the full spectrum of interest is an equivalent but technically advantageous method for quantifying the relaxation spectrum. This approach is termed **Dynamic Mechanical Analysis (DMA)**, and uses testing frames designed to accurately force sinusoidal deformation over a wide range of frequencies.

7.5.1 Storage and loss modulus

Linear systems driven at steady state by a sinusoidal input must return a sinusoidal output of the same frequency, modulated by a change in magnitude and a time delay, or **phase shift**. Specifically, for an imposed sinusoidal strain input

$$\varepsilon(t) = \varepsilon_0 \sin(\omega t) \tag{7.35}$$

the resulting stress response (after any transients have decayed) can be written as

$$\sigma(t) = \sigma_0 \sin(\omega t + \delta). \tag{7.36}$$

This can be expanded to a component in phase with the strain and one orthogonal to it

$$\sigma(t) = \sigma_0 \left[\sin(\omega t) \cos(\delta) + \cos(\omega t) \sin(\delta) \right]. \tag{7.37}$$

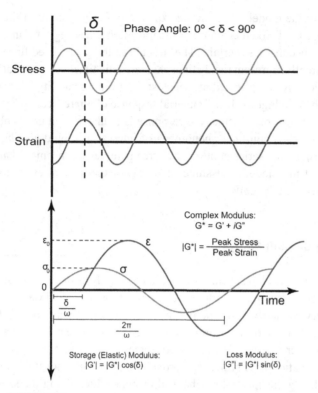

Figure 7.13

Schematic showing the in-phase and out-of-phase constituents resulting in a complex modulus in a viscoelastic material.

The stress-strain response can then be written

$$\sigma(t) = \varepsilon_0 \left[G_1 \sin(\omega t) + G_2 \cos(\omega t) \right] \quad (7.38)$$

with the in-phase modulus, G_1, defined as the **storage modulus**, the out of phase modulus, G_2, defined as the **loss modulus**, and G^* defined as the **complex modulus**

$$G_1 = \frac{\sigma_0}{\varepsilon_0} \cos(\delta), \quad (7.39)$$

$$G_2 = \frac{\sigma_0}{\varepsilon_0} \sin(\delta), \quad (7.40)$$

and

$$G^* = G_1 + i G_2 = \frac{\sigma_0}{\varepsilon_0} \left[\cos(\delta) + i \sin(\delta) \right]. \quad (7.41)$$

The decomposed in-phase and out-of-phase components of the modulus can be simply represented by a complex modulus, since the components of the response are orthogonal (Figure 7.13).

7.5 Frequency domain analysis

The in-phase component of the response is referred to as the energy storage component, while the out-of-phase part is the viscous loss component, as can be shown by calculating the peak and dissipated strain energy during the loading cycle. Taking the integral of the stress-strain response, one obtains the total change in strain energy during the load cycle, which is the lost energy due to viscous dissipation

$$\Delta E = \oint \sigma\, d\varepsilon = \omega \varepsilon_0^2 \int_0^{2\pi/\omega} \left[G_1 \sin(\omega t) \cos(\omega t) + G_2 \cos^2(\omega t) \right] dt \quad (7.42)$$

which can be solved to find the dissipated energy from a single load cycle,

$$\Delta E = \pi G_2 \varepsilon_0^2 \quad (7.43)$$

demonstrating that the dissipated energy in a complete load cycle is determined solely by G_2.

The stored energy component is maximum at the peak strain condition and occurs at $t = \pi/(2\omega)$. The integral up to this point for the second term drops out, and so the peak stored energy is given as

$$E_{\text{peak}} = \omega \varepsilon_0^2 \int_0^{\pi/2\omega} \left[G_1 \sin(\omega t) \cos(\omega t) \right] dt. \quad (7.44)$$

Hence, the peak energy is yielded solely as a function of G_1 and is written as

$$E_{\text{peak}} = \frac{1}{2} G_1 \omega \varepsilon_0^2 \quad (7.45)$$

where G_1 is referred to as the storage modulus, and G_2 as the loss modulus, with a clear physical interpretation from the stored and dissipated cyclic strain energy. Returning to the concept of relaxation spectra, G_1 and G_2 have a single value for a given frequency, but in general vary over the spectrum of applied frequencies. The storage modulus follows the overall response we have seen for viscoelastic transitions, while the loss modulus is approximately zero outside the viscoelastic transition region, and peaks at each characteristic relaxation frequency (Figure 7.14).

The viscous dissipation is often represented by G_2 normalized by G_1, which corresponds to the ratio of the dissipated and stored energy over the relaxation spectrum.

$$\tan(\delta) = \frac{G_2}{G_1} = \frac{\Delta E}{2\pi E_{\text{peak}}} \quad (7.46)$$

Tan (δ), which is a function of the test frequency ω, is often reported as the relative strength of the viscous component of the spectral response, and its peak values are used for engineering calculations.

A typical plot obtained under dynamic loading conditions of a viscoelastic material provides the storage modulus, loss modulus, and tan(δ) over a range of temperatures or frequencies. The dynamic mechanical properties for polyethylene terephthalate (PET)

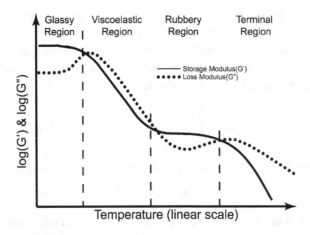

Figure 7.14

Illustration of the in-phase and out-of-phase components of the complex modulus, known as the storage and loss modulus, over the viscoelastic transition.

measured at a frequency of 1 Hertz are shown in Figure 7.15. PET is commonly used as a suture material employed in biomedical implants.

7.5.2 Complex response of Maxwell, Kelvin, and Standard Linear Solid models

One of the advantages of representing the linear response with complex numbers is that springs and dampers can be treated identically as complex springs, making the solution of the equilibrium network response straightforward. First, let us examine the complex response of spring and damper elements. It is clear that a linear spring alone dissipates no energy, and has no associated phase lag, and the complex modulus is simply the storage (elastic) modulus. Likewise, a damper under forced sinusoidal strain is given as

$$\varepsilon(t) = \varepsilon_0 \mu \omega \cos(\omega t). \tag{7.47}$$

The complex modulus of the Kelvin model under imposed sinusoidal strain is found by adding the stress response of the spring and damper, as they both experience the imposed strain:

$$G^* = E + i\mu\omega. \tag{7.48}$$

This immediately yields the interpretation that at low frequency ($\omega = 0$) the modulus of the Kelvin model is purely elastic, and at very high frequency the damping component dominates the elastic part and is unbounded. Unbounded stiffness at high frequency is

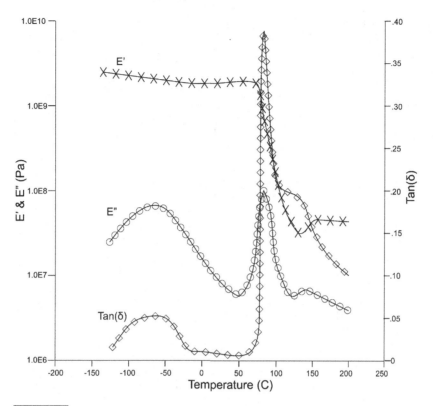

Figure 7.15

Dynamic scan of PET showing the storage modulus, loss modulus, and tan(δ) over a range of temperatures. (After Tobolsky and Mark, 1971.)

physically unrealistic (as discussed above), and this highlights the limitations of the Kelvin model.

An imposed sinusoidal strain in the Maxwell model is distributed between the spring and damper elements; the overall stress response is found by combining the spring and damper output in series:

$$G_M^* = \left[\frac{1}{E} + \frac{1}{i\mu\omega}\right]^{-1} = \frac{iE\mu\omega}{E + i\mu\omega}. \tag{7.49}$$

Multiplying through by the complex conjugate of the denominator, one obtains

$$G_M^* = \frac{iE\mu\omega}{E + i\mu\omega}\frac{E - i\mu\omega}{E - i\mu\omega} = \frac{E\mu^2\omega^2 + iE^2\mu\omega}{E^2 + \mu^2\omega^2}. \tag{7.50}$$

The real part of Equation (7.50) is the storage modulus of the Maxwell model,

$$G_{1,M} = \frac{E\mu^2\omega^2}{E^2 + \mu^2\omega^2} \tag{7.51}$$

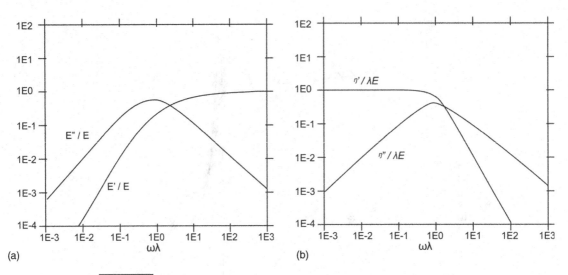

Figure 7.16

Illustration showing (a) the storage and loss modulus as well as (b) the in-phase and out-of-phase constituents of viscosity for a Maxwell model subjected to dynamic loading.

and the imaginary part is the loss modulus

$$G_{2,M} \frac{E^2 \mu \omega}{E^2 + \mu^2 \omega^2}. \tag{7.52}$$

At low frequency the storage modulus vanishes, as expected, since the Maxwell model behaves like a fluid. At high frequency, the storage modulus approaches the elastic modulus of the spring element, as found before. The loss factor $\tan(\delta)$ for the Maxwell model is simply

$$\tan(\delta)_M = \frac{E}{\mu \omega}. \tag{7.53}$$

Plotting G_2 as a function of frequency, we see that the dissipated energy peaks at the transition frequency of the Maxwell model, E/μ, which can be verified by taking the derivative of G_2. An illustration showing the storage and loss modulus as well as the in-phase and out-of-phase constituents of viscosity for a Maxwell model subjected to dynamic loading is provided in Figure 7.16.

The stress response of the SLS is easily found by adding an additional spring to the real part of the Maxwell model, thus

$$G^*_{SLS} = \left[E_1 + \frac{E \mu \omega}{E^2 + \mu^2 \omega^2} \right] + i \left[\frac{E^2 \mu \omega}{E^2 + \mu^2 \omega^2} \right]. \tag{7.54}$$

From the complex modulus, one sees that the dissipated energy is unchanged from the Maxwell model (peaking at the transition frequency), and the storage modulus is always finite and bounded, as required for a realistic material model. Again, the Standard

Linear Solid model demonstrates the essential behavior expected of a simple viscoelastic material.

7.6 Time-temperature equivalence

Relaxation transitions in polymers are often associated mechanistically with a particular molecular dynamic motion, such as the slipping of molecular backbones and the reptation and disentanglement of intertwined molecules (Chapter 4). Each of these motions has a characteristic time scale associated with it, which is in turn related to the likelihood of such an event occurring within a given time frame through the random thermal motions of the molecules. As the temperature of the material rises, the rate at which the thermal processes occur increases, effectively reducing the characteristic transition times or increasing the transition frequencies for the material. What this means is that a given observed viscoelastic phenomenon is the product of both the loading frequency and temperature of the material, and is not unique, but can be reproduced for a range of frequency and temperature combinations. This is important mechanistically for interpreting the meaning of viscoelastic data, and is commonly exploited experimentally to probe the relaxation spectrum by varying temperature and frequency together. Such information is also of great value in predicting behavior at body temperature over extended periods of time, as is the case with medical devices employing polymeric components.

7.6.1 Time-temperature superposition and shift factors

As discussed above, the viscoelastic transition is centered around the transition time (Figure 7.1); however, this transition time is itself a function of the temperature. The alteration to the time dependence is represented by the **shift factor a_T**, which accounts for the time-temperature equivalence in viscoelasticity. The shift factor can be exploited to chain together multiple experiments into one master curve representing the complete material relaxation spectrum. This is useful when the spectrum contains transitions that are outside the testable frequencies of a given test instrument, or if a test is limited to a maximum duration. For instance, if a creep experiment is repeated at several temperatures but for a limited duration, the result from each can be chained together to form one effective response over several more orders of magnitude of time (Figure 7.16). An illustration of the time-temperature superposition behavior for an elastomeric polymer is shown in Figure 7.17.

In principle, the individual responses can be combined by observation through translating in time without any predefined shift factor required for finding a continuous master curve. From this process, the effective shift factor for each test condition can be found. However, for many materials the shift factor is in close agreement with

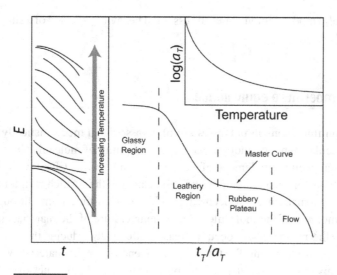

Figure 7.17

Time-temperature superposition behavior of an elastomeric polymer (adapted from Tobolsky and Mark, 1971).

analytical predictions, and this empirical approach can be replaced by quantitative methods.

7.6.2 The Williams, Landel, and Ferry (WLF) equation

Williams, Landel, and Ferry examined the shift factors from master curve constructions for a variety of non-crystalline polymers, and found that one empirical equation predicted the shift factor accurately for nearly all cases (Williams *et al.*, 1955). Namely, the shift factor a_T was found as a function of the test temperature relative to a reference temperature, and was found to be accurate to within $+/-$ 50 K of the reference temperature. While the reference temperature was arbitrary originally, the **WLF equation** is typically employed with respect to the **glass transition temperature** T_g, as this is a physically meaningful and objective quantity for polymers and generally is situated centrally in the temperature range of interest. The WLF equation for the temperature-dependent shift factor, relative to the glass transition temperature, is given as

$$\log(a_T) = \frac{C_1^g (T - T_g)}{C_2^g + (T - T_g)} \tag{7.55}$$

where C_1^g and C_2^g are fit constants (denoted with a "g" referring to the glass transition reference temperature) that have been shown to be nearly invariant for amorphous and melted polymers,

$$C_1^g = 17.4, \quad C_2^g = 51.6 \, \text{K}. \tag{7.56}$$

The observed generality of the WLF coefficients has been derived from analytical treatments of the mechanics associated with changes in the free volume between chains in amorphous or molten polymers near a glass transition temperature (Williams *et al.*, 1955), providing a physical interpretation of their origin.

It should be borne in mind that materials that do not behave like a polymer solution (i.e., semi-crystalline polymers or composites) will not follow idealized WLF behavior, and the shift factors should be verified by or found empirically from the master curve. Attempts have been made to identify models for the prediction of shift factors for semi-crystalline polymers, such as for UHMWPE where creep at body temperature over 20 years was sought from room-temperature creep experiments conducted at much shorter times (Deng *et al.*, 1998). Such models must be carefully constructed on a case-by-case basis, but can still be useful for predicting viscoelastic behavior far outside the measureable spectrum.

7.7 Nonlinear viscoelasticity

Many viscoelastic materials do not exhibit true linear characteristics, owing to complicated viscous phenomena or dramatically different loading and unloading responses. Additionally, some nonlinear models (like power-law viscoelasticity) are mathematically advantageous for analysis and are remarkably accurate. The nuances of nonlinear viscoelasticity are manifold and for the most part are beyond the scope of this text; the interested reader can find useful details in a number of sources (Ferry, 1980; Ward, 1983). Thus, nonlinear viscoelasticity is not as clean as its linear counterpart, but a basic treatment is useful for practical applications involving sophisticated biomaterials such as tissues, especially if complicated loading schemes or cyclic loading responses are required.

7.7.1 Common nonlinear models

One of the most common empirical correlation schemes is the power law. In viscoelasticity, the elastic modulus (or compliance) is often related through a time power law,

$$E(t) = \left(\frac{t}{\tau}\right)^n. \tag{7.57}$$

This relationship can be used for a nonlinear model of viscous deformation near a crack tip (Chapter 9), which has proven effective due to its simplicity and convenience for carrying out fracture mechanics calculations.

A common nonlinear damping element employs a stress power law governing the strain rate

$$\frac{d\varepsilon}{dt} = \left(\frac{\sigma}{\sigma'}\right)^n \tag{7.58}$$

making it potentially much more sensitive to the applied stress than the linear damping element. The phenomenon associated with this element is termed power-law viscoplasticity. This model of viscous behavior has been effectively employed in nonlinear network models of polymer behavior, primarily for use in finite element code, alongside nonlinear (hyperelastic) spring elements (Bergstrom and Boyce, 1998; Bergstrom et al., 2004).

There have been other nonlinear creep functions proposed to match experimental data. For instance, in UHMWPE, good agreement was found for a power-law correlation of the creep compliance to the logarithmic time (Deng et al., 1998)

$$J(t) = A \left[\log(t)\right]^n. \tag{7.59}$$

This modification to the power law reflects an experimentally observed deceleration of the strain rate with increasing deformation. It is important to consider such a decelerating strain rate in the case when predictions of creep are to be extrapolated far beyond the tested regime, as in the aforementioned case where Deng and co-workers found good agreement at long times with this model in conventional UHMWPE.

7.7.2 Nonlinear relaxation

For a linear material, the dynamics of the unloading response must mirror those of the loading response. An unusual characteristic of nonlinear relaxation is that the relaxation time is often dependent on the duration of the applied load, such that a longer creep experiment will require longer to recover. One approach that is effective in such a situation is to relate the fractional recovery,

$$\varepsilon_r = \frac{\varepsilon_{rec}}{\varepsilon_{Cr}} \tag{7.60}$$

to the reduced time

$$t_r = \frac{t_{rec}}{t_{Cr}} \tag{7.61}$$

resulting in a master curve for the relaxation transition for experiments of varying duration, which would not coincide without the application of Equations (7.60)–(7.61).

Figure 7.18 shows such a master curve obtained for UHMWPE, demonstrating that a creep specimen must be allowed to relax approximately 70 times the duration of the experiment if it is to be considered at equilibrium (Zapas and Crissman, 1984). This is useful for planning intermittent experiments if unloading is required, but the deformation from each loading segment must be measured individually. Similarly, if a

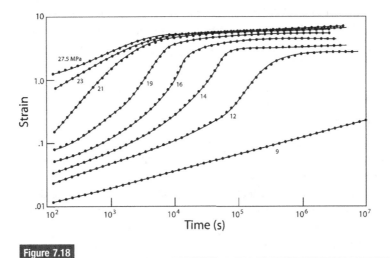

Figure 7.18

Master curve for UHMWPE showing creep behavior over a range of applied stresses (after Zapas and Crissman, 1984).

material exhibits nonlinear recovery characteristics, the initial relaxation response could easily be misinterpreted as the dominant relaxation mechanism.

7.8 Case study: creep behavior of UHMWPE used in total joint replacements

Wear of UHMWPE bearings in total joint arthroplasty is linked to periprosthetic osteolysis; yet, assessing the amount of wear occurring at the articulating surface of prosthesis while *in vivo* is challenging. X-ray methods are often employed by surgeons to determine the amount of femoral head penetration into an acetabular cup with the presumption that displacement is owed to the material loss (wear) of the softer (UHMWPE) component. However, UHMWPE is a viscoelastic material and displacement of the metal head into the polymer cup can be the result of creep deformation in addition to material loss associated with wear. The amount of true wear in a clinical setting is often not known until a device is retrieved and the retrieved bearing surface is topographically mapped for all modes of surface damage.

Recently, Fisher and coworkers compared wear rates in a number of crosslinked UHMWPE resins for both *in vitro* hip simulator and *in vivo* clinical wear studies (Galvin et al., 2006). In this study, a surface measuring system was used to assess material loss (wear) and measurements of the UHMWPE were taken every million cycles The findings from this study showed a decreasing wear rate with increased crosslink level. All materials showed initial creep deformation in the first million cycles owing to the viscoelastic nature of UHMWPE. Surface topography revealed that the materials with highest levels of crosslinking had the lowest surface roughness, and it was speculated by the authors that the wear resistance of these materials was improved by the enhanced lubrication schemes available at the articulating surface. Viscoelastic behavior and creep deformation of UHMWPE is not without benefits; many surgeons feel that the ability of the UHMWPE to flow under load is a useful property that enables the polymer to alter its conformity to accommodate unique geometric requirements and specific patient anatomy.

7.9 Summary

Designing implants with viscoelastic materials can be confidently undertaken when the relevant component loading and material relaxation time scales are understood, and a suitable predictive mathematical model is matched to the deformation phenomena of interest. Likewise, tissues can be modeled and understood using simplified techniques, often requiring only two or three material constants to accurately capture their observed behavior. Models should always be verified to predict deformation accurately before being applied, as the material may behave in a nonlinear manner, or there could be additional viscoelastic transitions not considered in the model development.

7.10 Problems for consideration

Checking for understanding

1. A total knee replacement experiences a spectrum of loading events, some rapid and some nearly steady state (but for a limited duration). Estimate the duration of the possible loading events, and compute the Deborah number for each, assuming a transition time for the polymer bearing of one day. Are any of these expected to occur in a fluid-like regime? If so, what is your recommendation for how this should be accounted for in these devices?
2. A polymer screw is used as an anchor for a tendon replacement, and is heavily loaded. It is found that the polymer creeps and slips out of the hole bored for it after an unacceptably short time. Experiments show that the Kelvin model describes the behavior of the anchor well over the time scale of interest, with $E_m = 500$ MPa and $\eta_m = 20E6$ MPa-s. What is the transition time of this material, and is it of interest? The borehole is 10 mm wide, and the interference fit is approximately 1 mm at time $t = 0$. How long will it take for the screw to relax until the interference effectively reduces to zero when the interference pressure relaxes to 10% of its original value and the anchor slips out? Recommend some strategies to correct for this relaxation failure mechanism.
3. An experiment is performed at 5 Hz at room temperature in a linear viscoelastic amorphous polymer. What is the equivalent frequency of the experiment if performed at body temperature? Is this difference enough to change the

expected behavior of an implant that experiences a wide spectrum of loading frequencies?

For further exploration

4. Using complex methods, derive the creep response of the Four Element model, which consists of a Maxwell model in series with a Kelvin model. Compare and contrast its expected behavior (storage and dissipative) with that of the SLS.
5. Consider the nonlinear relaxation behavior of UHMWPE shown in Figure 7.16. Laboratory A uses a testing procedure where a specimen is loaded sinusoidally at 5 Hz for 10,000 cycles for an experiment and is allowed to rest for an equal duration between experiments. Laboratory B performs a similar experiment except the duration is not fixed, but decreases as the material deforms, and there are no rest periods. Assuming a transition time of 1 day, and also that the total experimental time does not exceed 2 days, are these two experiments expected to yield similar results? If a steady-state condition from a single experiment in Laboratory A is desired, how long must the wait period be? Should wait periods be considered for predicting actual material response in an implant (recall Problem 7.1)?
6. A high-order linear model is calibrated to an artificial ligament material, designed to match the loading behavior of the natural ligament it is replacing. However, the relaxation behavior of the material is nonlinear, while the natural ligament relaxation was approximately linear. What are the concerns here for performance? Do you recommend matching the loading response precisely, or approximately matching the loading and unloading responses with error in both?

7.11 References

Bergstrom, J.S., and Boyce, M.C. (1998). Constitutive modeling of the large strain time-dependent behavior of elastomers. *Journal of the Mechanics and Physics of Solids* 46(5): 931–54.

Bergstrom, J.S., Rimnac, C.M., and Kurtz, S.M. (2004). An augmented hybrid constitutive model for simulation of unloading and cyclic loading behavior of conventional and highly crosslinked UHMWPE. *Biomaterials* 25(11): 2171–78.

Deng, M., Latour, R.A., Ogale, A.A., and Shalaby, S.W. (1998). Study of creep behavior of ultra-high-molecular-weight polyethylene systems. *Journal of Biomedical Materials Research*, 40(2), 214–23.

Ferry, J.D. (1980). *Viscoelastic Properties of Polymers*. New York: Wiley.

Galvin, A.L., Ingham, E., Stone, M.H., and Fisher, J. (2006). Penetration, creep and wear of highly crosslinked UHMWPE in a hip joint simulator. *Journal of Bone and Joint Surgery*, (British), 88B, 236.

Krzypow, D.J., and Rimnac, C.M. (2000). Cyclic steady state stress-strain behavior of UHMW polyethylene. *Biomaterials*, 21, 2081–87.

Lai, W.M., Hou, J.S., and Mow, V.C. (1991). A triphasic theory for the swelling and deformation behaviors of articular-cartilage. *Journal of Biomechanical Engineering-Transactions of the ASME*, 113(3), 245–58.

Mow, V.C., Kuei, S.C., Lai, W.M., and Armstrong, C.G. (1980). Biphasic creep and stress-telaxation of articular-cartilage in compression-theory and experiments. *Journal of Biomechanical Engineering-Transactions of the ASME*, 102(1), 73–84.

Riches, P.E., Dhillon, N., Lotz, J., Woods, A.W., and McNally, D.S. (2002). The internal mechanics of the intervertebral disc under cyclic loading. *Journal of Biomechanics*, 35(9), 1263–71.

Saito, M., Fujii, K., Mori, K., and Marumo, Y. (2006). Role of collagen enzymatic and glycation induced cross-links as a determinant of bone quality in spontaneously diabetic WBN/Kob rats. *Osteoporosis International*, 17(10), 1514–23.

Saito, M., and Marumo, Y. (2010). Collagen cross-links as a determinant of bone quality: a possible explanation for bone fragility in aging, osteoporosis, and diabetes mellitus. *Osteoporosis International* 21(2), 195–214.

Silva, P., Crozier, S. Veidt, M., and Pearcy, M.J. (2005). An experimental and finite element poroelastic creep response analysis of an intervertebral hydrogel disc model in axial compression. *Journal of Materials Science-Materials in Medicine*, 16(7), 663–69.

Sychterz, C.J., Yang, A., and Engh, C.A. (1999). Analysis of temporal wear patterns of porous-coated acetabular components: Distinguishing between true wear and so-called bedding-in. *Journal of Bone and Joint Surgery (American)*, 81A(6), 821–30.

Tobolsky, A.V., and Mark, H.F. (1971). *Polymer Science and Materials*. New York: Wiley.

Ward, I.M. (1983). *The Mechanical Properties of Solid Polymers*. New York: Wiley.

Williams, J.R., Natarajan, R.N., and Andersson, G.B.J. (2007). Inclusion of regional poroelastic material properties better predicts biomechanical behavior of lumbar discs subjected to dynamic loading. *Journal of Biomechanics*, 40(9), 1981–87.

Williams, M.L., Landel, R.F., and Ferry, J.D. (1955). The temperature dependence of relaxation mechanisms in amorphous polymers and other glass-forming liquids. *Journal of the American Chemical Society*, 77, 3701–06.

Wilson, W., van Donkelaar, C.C., van Rienumberrgen, B., and Huiskes, R. (2005). A fibril-reinforced poroviscoelastic swelling model for articular cartilage. *Journal of Biomechanics*, 38(6), 1195–1204.

Zapas, L.J., and Crissman, J.M. (1984). Creep and recovery behavior of ultrahigh molecular-weight polyethylene in the region of small uniaxial deformations. *Polymer*, 25(1), 57–62.

8 Failure theories

Inquiry

How would you safely design a tibial insert of a total knee replacement that is known to experience a complex loading state with a normal stress component that is on the order of the uniaxial strength for this material?

The inquiry posed above represents a realistic design challenge that one might face in the field of orthopedics. Many of the **tibial** components used in total knee arthroplasty utilize ultra-high molecular weight polyethylene with a uniaxial **yield stress** on the order of 20 MPa; yet, the contact pressures for many of the clinical designs exceed this value. In order to assess the likelihood for failure owing to yield or plastic deformation, it is important to calculate the **effective stress** that provides a scalar representation of the multiaxial stress state acting on the implant. It is the effective stress that must be compared to the uniaxial yield strength as an assessment for the **factor of safety** against failure. Furthermore, localized plastic damage due to the presence of a notch or stress concentration can serve as a nucleation site for cracks if the component undergoes cyclic loading conditions. All of these factors must be considered when designing the implant.

8.1 Overview

The process of material failure depends upon the stress state of the system as well as whether its microstructure renders it ductile, brittle, or semi-brittle. In general, ductile materials *yield before fracture* while brittle materials *fracture before yield*. A semi-brittle system offers a small amount of plastic or permanent deformation prior to fracture. In the broad spectrum of materials behavior, metals are generally considered strong, tough, and ductile; ceramics are known to be strong in compression but weak in tension, and are notoriously brittle; and polymers are usually compliant, resilient, and highly sensitive to strain rate. Composites and tissues are typically anisotropic and are highly dependent upon the distribution of constituents. The most commonly employed mechanical test for material characterization is the uniaxial tensile test, which provides several important material properties including elastic modulus, yield strength, ultimate tensile strength, fracture stress, energetic toughness, and ductility (as shown in Figure 8.1).

242 Failure theories

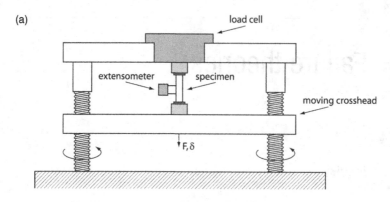

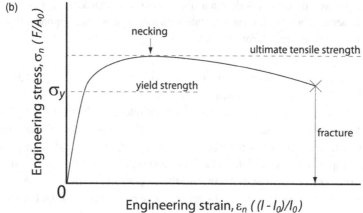

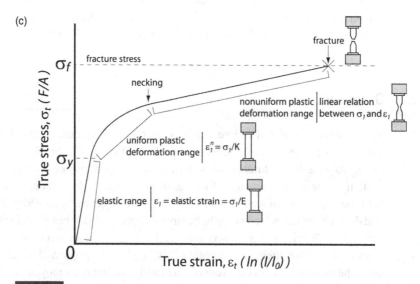

Figure 8.1

Schematic illustration showing (a) the typical tensile testing configuration used to measure material properties, (b) the engineering stress-strain behavior, and (c) the true stress-strain behavior.

8.1 Overview

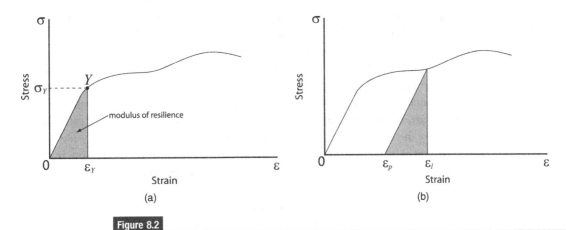

Figure 8.2

(a) Illustration of stress-strain behavior for a material that shows the modulus of resilience. This parameter provides a measure of the material's ability to store elastic energy. This energy is fully recoverable until the stress reaches the value of the yield stress (denoted by Y on the plot). Once permanent deformation, or plastic strain (ε_p), occurs in the system, some portion of the energy is lost when the material is unloaded as shown in (b).

The **yield strength** of a material is defined as the stress at which plastic (permanent) deformation begins (as discussed in the previous chapter). Similarly, the **modulus of resilience** for a material is defined as the energy that is stored in a material until the onset of yielding (Figure 8.2(a)). Once yielding has occurred in a material, only elastic strain can be recovered upon unloading; plastic strain results in permanent deformation (Figure 8.2(b)) that is not recovered upon removal of the stress.

As previously discussed, the tensile yield strength provides the stress level at which permanent deformation will occur in an isotropic material subjected to a one-dimensional (axial) tensile stress. This material property serves as an important design parameter, as it represents the upper limit of stress that can be applied without incurring plastic deformation to the component. A challenge in medical devices, however, is that most components experience complex states of stress (as described in the previous chapter) and the yield condition depends upon the complete stress state of the material. For this reason, a universal criterion is needed that addresses the requirements for yielding under multiaxial loading conditions.

Ductile materials generally deform through shear in response to generalized states of stress. Many alloy systems have a large number of slip systems available, and dislocation movement (slip) is initiated on planes of critically resolved shear stress (Chapter 2). Consequently the yield criteria developed for ductile metals are based on localized maximum shear stress (planes of maximum shear stress) or distortional energy (planes of maximum octahedral stress). Henri Tresca developed the first criterion for yield in 1864 – this theory utilizes the maximum shear stress as the predictor of plastic deformation in metals. The other well-known criterion was established in 1913 by Richard von Mises, who utilized the distortional energy as a basis for yield in ductile materials. The Tresca and von Mises yield criteria are commonly employed to this day and are examined in detail below.

Brittle systems, in contrast to ductile materials, are weak in tension and their failure modes utilize normal stresses rather than shear stresses. In brittle materials, the generalized failure criterion is based on the normal stress, or principal stresses, reaching the fracture stress or ultimate strength of the material. Rankine initially developed the normal stress failure theory in 1850 for brittle solids. He stated that failure occurs when the magnitude of the major principal stress reaches that which causes fracture in the uniaxial tension test. The normal stress criteria can be utilized as a first estimate of failure using the tensile or compressive strength of the material, but this premise can be severely limited by the flaw distributions within the material and, for this reason, fracture mechanics is more widely used in the design of components employed in safety-critical applications.

This chapter provides an overview of the failure theories employed for the primary material behaviors (ductile, brittle, and semi-brittle) and classifications (metals, ceramics, polymers, composites, and tissues). Additionally, this chapter addresses methodologies employed for design against plastic (permanent) deformation for multiaxial loading conditions and in components where local stress concentrations (notches) are expected.

8.2 Learning objectives

This chapter provides a broad overview of failure criteria utilized in engineering design. The failure criterion for each theory is presented and the benefits and limitations of these philosophies are discussed. At the completion of this chapter the student should be able to:

1. describe the difference between a ductile and brittle material
2. discuss the historical origins of yield theories
3. illustrate various stress states and their effect on yielding
4. employ the maximum shear (Tresca) criterion to predict yield in a structure
5. utilize the distortional energy (von Mises) criterion to calculate the effective stress and to predict yield in a structure
6. explain the basis for modified versions of the von Mises yield criterion
7. use the maximum normal stress criterion to predict fracture in a brittle solid
8. determine the factor of safety against yield or fracture
9. elucidate the role of notches or stress concentrations
10. describe the micromechanisms of failure in metals, ceramics, and polymers
11. perform a stress analysis and assess the likelihood for failure in an implant

8.3 Yield surfaces

A **yield surface** is the surface within the space of stresses that defines the boundary between elastic and plastic behavior for a material. The state of stress that is contained

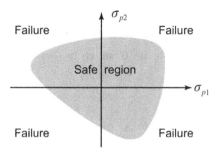

Figure 8.3

Hypothetical yield surface for a planar (two-dimensional) stress state.

inside this surface represents elastic behavior of the material. The boundary represents stress states at which the material has reached its yield strength and is behaving in a plastic manner. The yield surface may change shape and size as the plastic deformation evolves, but stress states that lie outside the yield surface are mathematically non-permissible. For a generalized state of stress, there will be a number of yield points that together form the yield surface for the material; Figure 8.3 illustrates a hypothetical yield surface for a planar (two-dimensional) stress state.

The yield surface is generally represented in principal stress space or space defined by the stress invariants. The stress invariants used to describe the yield surface are given as:

$$I_1 = \sigma_1 + \sigma_2 + \sigma_3$$
$$J_2 = \frac{1}{6}[(\sigma_1 - \sigma_2)^2 + (\sigma_2 - \sigma_3)^2 + (\sigma_3 - \sigma_1)^2] \quad (8.1)$$
$$J_3 = \det(s) = s_1 s_2 s_3$$

where σ is the Cauchy stress (true stress), $\sigma_1, \sigma_2, \sigma_3$ are the principal values of σ, and s is the deviatoric part of the stress whose principal values are s_1, s_2, s_3. These stress invariants are utilized in the fundamental failure theories discussed below – Tresca employs the use of the maximum shear stress, von Mises makes use of the deviatoric part of the stress (J_2 invariant), and the normal stress criteria utilize the principal normal stresses.

8.4 Maximum shear stress (Tresca yield criterion)

The **maximum shear stress criterion** is founded upon the early work of Tresca, who recognized that metals plastically deform primarily through shear processes. These shear processes utilize dislocation slip systems to accommodate plastic deformation once the elastic range of deformation is exhausted. This criterion is premised on the notion that yielding (slip) occurs when the maximum shear stress reaches the yield stress determined

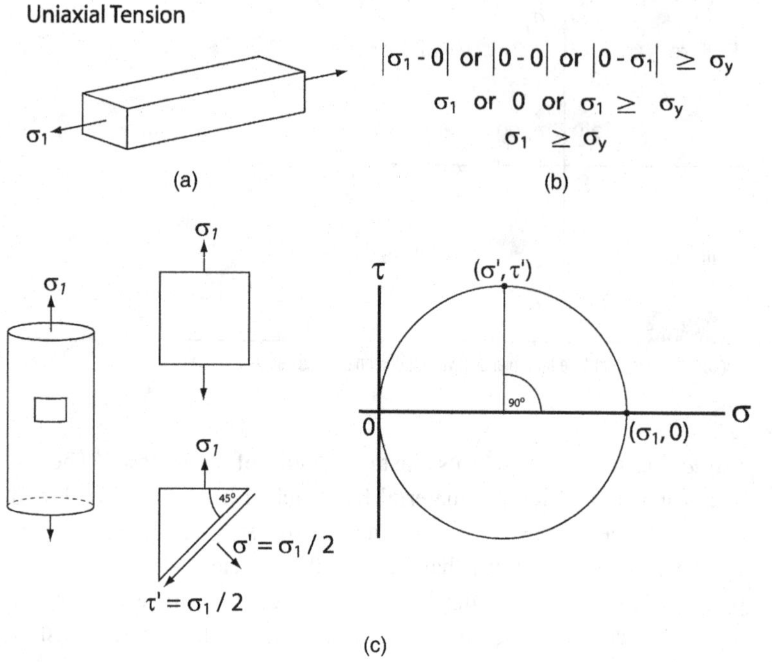

Failure theories

Uniaxial Tension

(a)

$|\sigma_1 - 0|$ or $|0 - 0|$ or $|0 - \sigma_1| \geq \sigma_y$
σ_1 or 0 or $\sigma_1 \geq \sigma_y$
$\sigma_1 \geq \sigma_y$

(b)

$\sigma' = \sigma_1/2$
$\tau' = \sigma_1/2$

(c)

Figure 8.4

Illustration shows the Mohr circle for uniaxial loading and the relationship between maximum shear stress and the principal (normal) stress. Yielding occurs when the principal stress is greater than or equal to the yield strength measured in the tensile test. The figure shows (a) uniaxial tension, (b) resulting equations, and (c) relationship depicted using Mohr's circle.

from the uniaxial tensile test. The relationship between the principal stresses and planes of maximum shear stress are readily visualized using the Mohr circle (as developed in Chapter 6). Figure 8.4 schematically illustrates the relationship between the principal stresses, the planes of maximum shear stress, and the uniaxial yield stress measured from a tensile test. The planes of maximum shear stress are oriented at $\theta = 45°$ to the principal stresses (recall that the Mohr circle plots shear as an angular function of normal stress using 2θ and that maximum planes of shear stress are oriented at $90°$ to the principal stress direction).

The Mohr circle plot for uniaxial loading shown in Figure 8.4 indicates that the maximum shear stress occurs at the radius that is equal to one-half of the principal stress difference. For uniaxial loading, the maximum shear stress occurs at $\sigma_y/2$, and thus we can write:

$$\tau_{max} = \frac{|\sigma_1 - 0|}{2} = \frac{\sigma_y}{2}$$
$$\text{for}$$
$$\sigma_1 = \sigma_y, \sigma_2 = \sigma_3 = 0$$

(8.2)

8.4 Maximum shear stress (Tresca yield criterion)

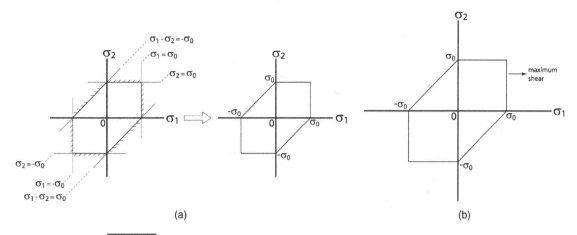

Figure 8.5

Illustration of the Tresca yield criterion developed for a plane stress state. The Tresca equations (a) are employed to generate the hexagonal yield surface (b). This surface represents the boundary between elastic behavior (within the hexagon) and the plastic deformation (the hexagonal boundary) for a ductile material that fails through maximum shear (c). Note that the Tresca yield surface assumes isotropic behavior.

A state of plane stress is often used to develop the well-known two-dimensional Tresca yield surface (hexagon) that is schematically illustrated in Figure 8.5. For this biaxial stress state (where $\sigma_3 = 0$, and $\sigma_o = \sigma_y =$ uniaxial flow or yield strength), the general equations for maximum shear stress using the principal stress differences defined by the Tresca yield criteria are plotted in $\sigma_1 - \sigma_2$ space to generate the boundary of the yield surface. For cases where the stresses are less than the maximum shear yield criteria, the material will be contained within the envelope and will be safe from plastic deformation. The boundary of the yield surface (hexagon) is defined by the equations below:

$$\begin{aligned} |\sigma_1| &= \sigma_y \\ |\sigma_2| &= \sigma_y \\ |\sigma_1 - \sigma_2| &= \sigma_y \\ \sigma_y &= \sigma_o \end{aligned} \qquad (8.3)$$

In three-dimensional space, the yield surface is a hexagonal prism with a hexagon projected down the hydrostatic axis where $\sigma_1 = \sigma_2 = \sigma_3$ as shown in Figure 8.6. Note that this indicates that there is no change in material yield response with the application of a **hydrostatic stress**. The general three-dimensional Tresca yield criterion is founded upon the notion that the plane of greatest shear stress dictates the maximum overall shear stress, and is given as:

$$\tau_{\max} = \tau_f = \frac{\sigma_y}{2} = MAX \left\{ \frac{|\sigma_1 - \sigma_2|}{2}, \frac{|\sigma_2 - \sigma_3|}{2}, \frac{|\sigma_1 - \sigma_3|}{2} \right\}. \qquad (8.4)$$

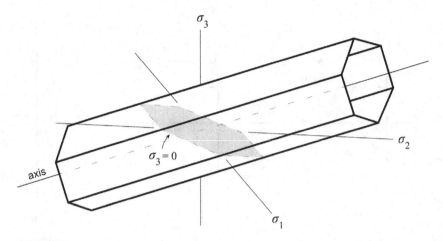

Figure 8.6

Illustration of the three-dimensional Tresca yield envelope which takes the form of a hexagonal prism that is projected down the hydrostatic stress axis.

Example 8.1 **Determining the Tresca stress in a spinal implant**

Determine the Tresca stress for the infinitesimal element in an artificial spinal disk that is loaded as shown in Example 6.3 (Figure 6.12). If the uniaxial yield strength of the implant material is 8 MPa, is the device safe from yielding?

Solution

$$\sigma_x = -2.2 \text{ MPa}, \sigma_y = -1.1 \text{ MPa}, \sigma_z = -0.58 \text{ MPa}$$
$$\tau_{xy} = -0.57 \text{ MPa}, \tau_{xz} = -0.79 \text{ MPa}, \tau_{yz} = -0.33 \text{ MPa}$$

Mohr's circle was used in Example 6.3 to find the principal stresses:

$$\sigma_1 = -0.25 \text{ MPa}, \sigma_2 = -0.86 \text{ MPa}, \sigma_3 = -2.77 \text{ MPa}$$

Using the expression for the Tresca yield criterion (Equation 8.4), the effective Tresca stress is:

$$\tau_{max} = \frac{\sigma_y}{2} = MAX \left\{ \frac{|\sigma_1 - \sigma_2|}{2}, \frac{|\sigma_2 - \sigma_3|}{2}, \frac{|\sigma_1 - \sigma_3|}{2} \right\}$$
$$= MAX \left\{ \frac{|-0.25 - (-0.86)|}{2}, \frac{|-0.86 - (-2.77)|}{2}, \frac{|-0.25 - |(-2.77)|}{2} \right\}$$
$$= MAX \{0.305 \text{ MPa}, 0.95 \text{ MPa}, 1.26 \text{ MPa}\} = 1.26 \text{ MPa}$$
$$\sigma_{Tresca} = 2\tau_{max} = 2(1.26) = 2.52 \text{ MPa}$$

Yielding occurs when the Tresca stress reaches the uniaxial yield strength. If the Tresca stress is less than the yield strength, then the material is safe from yield:

$$\sigma_{Tresca} = \sigma_{yield} \text{ (Yields)}$$
$$\sigma_{Tresca} = 2.52 \text{ MPa} < \sigma_{yield} = 8 \text{ MPa (No Yielding)}$$

The factor of safety (FS) against yielding is defined as the ratio of the yield stress normalized by the effective (Tresca) stress:

$$FS = \frac{\sigma_{yield}}{\sigma_{Tresca}} = \frac{8 \text{ MPa}}{2.52 \text{ MPa}} = 3.17.$$

The factor of safety for this material utilized in the spine application is 3.

8.5 Maximum distortional energy (von Mises yield criterion)

The maximum **distortional energy** criterion is founded upon the early work of Huber, von Mises, and Hencky, who utilized strain energy methods as a basis for yield in isotropic, ductile materials. This criterion is based on the view that yielding occurs when the maximum distortional energy associated with the combined stress state reaches the uniaxial yield strength. The symmetric stress tensor comprises both normal stress components and shear stress components and can be decomposed into dilatational and distortional components (Figure 8.7). The dilatational portion of stress is responsible for volume change and is controlled by the normal stresses (imagine a cube that becomes a larger cube under the action of hydrostatic stresses). The distortional portion of stress results in shape change but no volume change and is controlled by the shear stresses (imagine a deck of cards that is sheared).

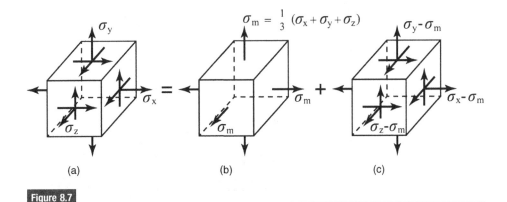

Figure 8.7

Illustration depicting the decomposition of (a) state of stress into (b) dilatational stresses and (c) distortional or deviatoric stresses.

The distortional strain energy is obtained using the deviatoric portion of the normal and shear stresses in the system. The basic form of strain energy is given as:

$$U = \int \sigma \, d\varepsilon. \tag{8.5}$$

For elastic behavior, this can be re-written as:

$$U = E \int \varepsilon \, d\varepsilon = \frac{E\varepsilon^2}{2} = \frac{\sigma \varepsilon}{2} \tag{8.6}$$

or in terms of the principal stresses:

$$U = \frac{1}{2}(\sigma_1 \varepsilon_1 + \sigma_2 \varepsilon_2 + \sigma_3 \varepsilon_3). \tag{8.7}$$

If the generalized Hooke's Law is utilized in terms of the principal stresses and strains, we get the basic form of the strain energy:

$$\varepsilon_1 = \frac{1}{E}(\sigma_1 - \nu(\sigma_2 + \sigma_3)); \; \varepsilon_2 = \frac{1}{E}(\sigma_2 - \nu(\sigma_1 + \sigma_3));$$

$$\varepsilon_3 = \frac{1}{E}(\sigma_3 - \nu(\sigma_1 + \sigma_2))$$

$$\rightarrow U = \frac{1}{2E}\left(\sigma_1^2 + \sigma_2^2 + \sigma_3^2 - 2\nu(\sigma_1 \sigma_2 + \sigma_1 \sigma_3 + \sigma_2 \sigma_3)\right). \tag{8.8}$$

For ductile metals, the role of hydrostatic stresses on the yield behavior is assumed negligible (as was the case with Tresca) and the principal stresses can be written in terms of the pressure to obtain the hydrostatic portion of the strain energy:

$$\sigma_1 = \sigma_2 = \sigma_3 = -p$$

$$U_p = \frac{1}{2E}(3p^2 - 6\nu p^2). \tag{8.9}$$

If the hydrostatic pressure is taken as:

$$p = -\frac{(\sigma_1 + \sigma_2 + \sigma_3)}{3} \tag{8.10}$$

then we get:

$$U_p = \frac{1 - 2\nu}{6E}(\sigma_1 + \sigma_2 + \sigma_3)^2. \tag{8.11}$$

We then write the distortional energy as:

$$U_d = U - U_p = \frac{1}{2E}\left(\sigma_1^2 + \sigma_2^2 + \sigma_3^2 - 2\nu(\sigma_1 \sigma_2 + \sigma_1 \sigma_3 + \sigma_2 \sigma_3)\right)$$

$$- \frac{1 - 2\nu}{6E}(\sigma_1 + \sigma_2 + \sigma_3)^2$$

$$= \frac{1 + \nu}{3E}\left(\sigma_1^2 + \sigma_2^2 + \sigma_3^2 - \sigma_1 \sigma_2 - \sigma_1 \sigma_3 - \sigma_2 \sigma_3\right). \tag{8.12}$$

8.5 Maximum distortional energy (von Mises yield criterion)

The yield stress measured in uniaxial tension is used to establish the failure criteria:

$$\sigma_1 = \sigma_y$$
$$\sigma_2 = \sigma_3 = 0 \quad (8.13)$$
$$U_d = \frac{1+\nu}{3E}\sigma_y^2.$$

Hence, the general failure criterion developed using the distortional energy is given as:

$$\frac{1+\nu}{3E}\left(\sigma_1^2 + \sigma_2^2 + \sigma_3^2 - \sigma_1\sigma_2 - \sigma_1\sigma_3 - \sigma_2\sigma_3\right) = \frac{1+\nu}{3E}\sigma_y^2$$
$$\rightarrow \sigma_y^2 = \sigma_1^2 + \sigma_2^2 + \sigma_3^2 - \sigma_1\sigma_2 - \sigma_1\sigma_3 - \sigma_2\sigma_3 \quad (8.14)$$
$$\sigma_y = \sqrt{\sigma_1^2 + \sigma_2^2 + \sigma_3^2 - \sigma_1\sigma_2 - \sigma_1\sigma_3 - \sigma_2\sigma_3}.$$

This failure stress is often denoted as the von Mises effective stress and can be written also as:

$$\sigma_{\mathit{eff}} = \frac{1}{\sqrt{2}}\sqrt{(\sigma_1 - \sigma_2)^2 + (\sigma_2 - \sigma_3)^2 + (\sigma_3 - \sigma_1)^2}. \quad (8.15)$$

The von Mises yield criterion is sometimes termed the octahedral shear yield criterion as it can also be derived using the planes of maximum **octahedral shear stress**. This method is similar in premise to the maximum shear stress theory but instead makes use of the octahedral shear plane (the plane that is oriented at 54.7° to the principal stress and which is concomitant with the orientation of a close-packed plane (111) in an FCC crystal structure (Chapter 2)). Again, the assumption presumes that dislocation slip is the mechanism for plastic deformation in the system, and that the material behavior is isotropic. Figure 8.8 shows the orientation of the octahedral shear plane with respect to the principal stress and its location in Mohr circle space.

The shear stress on the octahedral shear plane is given as:

$$\tau_{oct} = \frac{1}{3}\sqrt{(\sigma_1 - \sigma_2)^2 + (\sigma_2 - \sigma_3)^2 + (\sigma_3 - \sigma_1)^2}. \quad (8.16)$$

Substituting the uniaxial yield strength for the principal stress, we can find the critical value of octahedral shear stress resulting in plastic deformation:

$$\tau_{oct,f} = \frac{1}{3}\sqrt{2\sigma_y^2} \quad (8.17)$$

and the same failure criterion is developed using the maximum distortional energy condition:

$$\frac{1}{3}\sqrt{2\sigma_y^2} = \frac{1}{3}\sqrt{(\sigma_1 - \sigma_2)^2 + (\sigma_2 - \sigma_3)^2 + (\sigma_3 - \sigma_1)^2}$$
$$\rightarrow \sigma_{\mathit{eff}} = \frac{1}{\sqrt{2}}\sqrt{(\sigma_1 - \sigma_2)^2 + (\sigma_2 - \sigma_3)^2 + (\sigma_3 - \sigma_1)^2}. \quad (8.18)$$

252 Failure theories

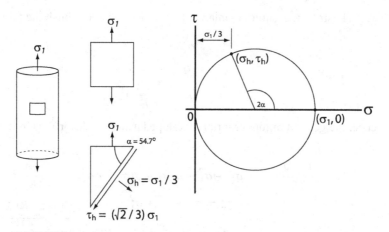

Figure 8.8

Schematic showing the octahedral shear plane plotted in Mohr space.

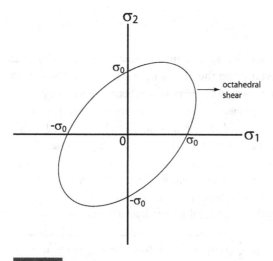

Figure 8.9

Schematic of the von Mises ellipse yield surface.

Another advantage of the von Mises yield criterion is that it can also be written in general stress component terms:

$$\sigma_{eff} = \frac{1}{\sqrt{2}}\sqrt{(\sigma_x - \sigma_y)^2 + (\sigma_y - \sigma_z)^2 + (\sigma_z - \sigma_x)^2 + 6\left(\tau_{xy}^2 + \tau_{yz}^2 + \tau_{zx}^2\right)}. \quad (8.19)$$

A state of plane stress is often used to develop the well-known two-dimensional von Mises yield surface (ellipse) that is schematically illustrated in Figure 8.9. In this case, the yield surface is plotted in $\sigma_1 - \sigma_2$ space using the general equations for the

8.5 Maximum distortional energy (von Mises yield criterion)

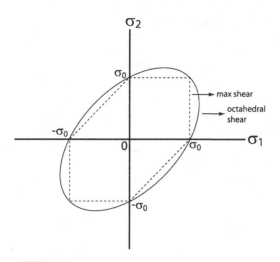

Figure 8.10

Illustration shows the overlay of the von Mises (octahedral shear or maximum distortional energy) and Tresca (maximum shear) yield surfaces.

von Mises (distortional energy) yield criterion:

$$\sigma_y^2 = \sigma_1^2 - \sigma_1\sigma_2 + \sigma_2^2 \qquad (8.20)$$

Note that Equation (8.20) is an equation for an ellipse, and is used to generate the yield envelope for the plane stress case as is shown in Figure 8.10. Any stress state contained within the ellipse is considered safe while any stress state on the boundary has yielded and plastically deformed. The proximity to the yield boundary provides a measure of safety factor and can be used in design as a parameter for preventing yield.

It is readily seen by comparing the Tresca and von Mises yield surfaces (Figure 8.10) that the von Mises ellipse circumscribes the Tresca hexagon for an isotropic solid with the same uniaxial yield strength. In this regard, the maximum shear stress criterion is more conservative than the criterion using the distortional energy; however, the von Mises yield criterion is known to better match experimental data for many alloy systems and is commonly employed in design for the calculation of an effective stress for a component subjected to multiaxial loading.

In three dimensions, the von Mises yield space is a cylinder that is centered upon the hydrostatic stress axis (Figure 8.11). Like the maximum shear stress theory, the distortional energy criterion does not depend upon hydrostatic stress. The yield envelope becomes an ellipse where the cylinder intersects the $\sigma_1 - \sigma_2$ plane, as shown above.

Example 8.2 **Determining the von Mises effective stress in a tibial plateau**
Determine the von Mises effective stress for the tibial plateau presented in Example 6.5. This implant is made of UHMWPE in a constraining frame of CoCr and loaded

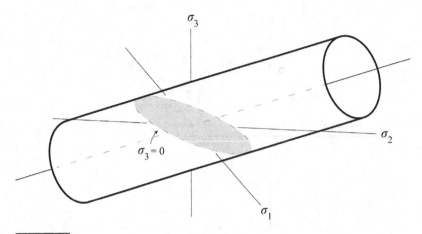

Figure 8.11

Schematic illustration of the three-dimensional von Mises yield envelope that takes the form of a cylinder oriented along the axis of hydrostatic stress.

with a force of 30 kN that is evenly spread across the tibial plateau. UHMWPE has a uniaxial yield strength of 22 MPa. Is the device safe from yielding?

Solution Applying the 3D Hooke's Law and geometric constraints provides the stresses and strains:

$$\sigma_x = \sigma_y = 16.7 \, \text{MPa}$$
$$\sigma_z = 25 \, \text{MPa}$$
$$\tau_{xz} = \tau_{xy} = \tau_{yz} = 0 \, \text{MPa}$$
$$\varepsilon_x = \varepsilon_y = 0$$
$$\varepsilon_z = 0.012$$

The effective von Mises stress is found by employing Equation (8.19):

$$\begin{aligned}
\sigma_{\mathit{eff}} &= \frac{1}{\sqrt{2}} \sqrt{(\sigma_x - \sigma_y)^2 + (\sigma_y - \sigma_z)^2 + (\sigma_z - \sigma_x)^2 + 6\left(\tau_{xy}^2 + \tau_{yz}^2 + \tau_{zx}^2\right)} \\
&= \frac{1}{\sqrt{2}} \sqrt{(16.7 - 16.7)^2 + (16.7 - 25)^2 + (25 - 16.7)^2} \\
&= \frac{1}{\sqrt{2}} \sqrt{2(8.3)^2} = 8.3 \, \text{MPa}
\end{aligned}$$

The material is safe from yielding if the effective (von Mises) stress is less than the uniaxial strength:

$$\sigma_{eff} < \sigma_y$$
$$8.3 \text{ MPa} < 22 \text{ MPa}$$

The factor of safety (FS) against yielding is found by normalizing the yield strength with the effective stress:

$$\text{FS} = \frac{\sigma_y}{\sigma_{eff}} = \frac{22 \text{ MPa}}{8.3 \text{ MPa}} = 2.65.$$

8.6 Predicting yield in multiaxial loading conditions

The primary benefit of developing a general yield criterion, either through maximum shear stress or maximum octahedral stress, is that a scalar representation of the three-dimensional stress state can be used in combination with the uniaxial yield strength to predict the likelihood of failure. The **factor of safety** (FS) against plastic deformation can be calculated from the ratio of the uniaxial yield strength to the effective von Mises stress or by normalizing the uniaxial yield stress with the maximum shear stress for the component:

$$\begin{aligned}
\text{FS}_{\text{Tresca}} &= \frac{\sigma_y}{\sigma_s} \\
\sigma_s &= MAX\left\{|\sigma_1 - \sigma_2|, |\sigma_2 - \sigma_3|, |\sigma_3 - \sigma_1|\right\} \\
\text{FS}_{VM} &= \frac{\sigma_y}{\sigma_{eff}}. \\
\sigma_{eff} &= \frac{1}{\sqrt{2}}\sqrt{(\sigma_1 - \sigma_2)^2 + (\sigma_2 - \sigma_3)^2 + (\sigma_3 - \sigma_1)^2}
\end{aligned} \qquad (8.21)$$

In general, the factor of safety should be such that the effective stress (or maximum shear stress) is at least a factor of 2–3 lower than the uniaxial yield strength. For safety-critical applications, however, this is often insufficient, and factors of safety are often extended and additional analysis pertaining to fracture, fatigue, and wear are employed for complete structural analysis.

Multiaxial loading has an interesting result on the yield behavior of ductile materials. It is worth exploring the effect of common loading scenarios using both the Tresca and von Mises yield theories. An isotropic material (with anequal compressive and tensile yield strength) subjected to any one-dimensional normal stress will plastically deform when this stress reaches the uniaxial yield strength for the material, as shown in Figure 8.12. Uniaxial loading is not a typical stress state found in structural medical devices; an exception is a suture.

256 Failure theories

Uniaxial Tension

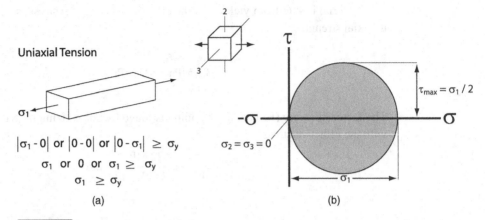

$|\sigma_1 - 0|$ or $|0 - 0|$ or $|0 - \sigma_1| \geq \sigma_y$

σ_1 or 0 or $\sigma_1 \geq \sigma_y$

$\sigma_1 \geq \sigma_y$

(a)

$\sigma_2 = \sigma_3 = 0$

$\tau_{max} = \sigma_1 / 2$

(b)

Figure 8.12

Schematic illustration of (a) uniaxial stress state in an isotropic ductile material and (b) the Tresca yield criterion for this stress state.

Equi-Biaxial Tension

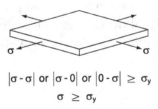

$|\sigma - \sigma|$ or $|\sigma - 0|$ or $|0 - \sigma| \geq \sigma_y$

$\sigma \geq \sigma_y$

Figure 8.13

Equi-biaxial stress has the same yield criteria as that for uniaxial stress.

Biaxial stress states can be found in thin-walled pressure vessels and may be relevant in angioplasty balloons, wound healing, artificial skin, intra-aortic balloon pumps, and intra-gastric balloons. In the case of an equi-biaxial stress state (Figure 8.13), the effective von Mises or Tresca stress at which yield will occur is equal to the yield stress calculated for uniaxial loading. If instead we consider a thin-walled cylindrical pressure vessel, where the axial stress is twice that of the hoop stress, then we find that the von Mises effective stress needed for yield is 16% greater than that for uniaxial loading (Figure 8.14).

Components may also experience torsional loading or pure shear loading and this can strongly affect the yield behavior of a given material. Figure 8.15 illustrates the stress state for pure shear and shows the conditions for yield using both the Tresca and von Mises yield criteria. In the former, the shear stress needed for plastic deformation is one-half of the uniaxial yield strength, and in the latter, the required shear

8.6 Predicting yield in multiaxial loading conditions

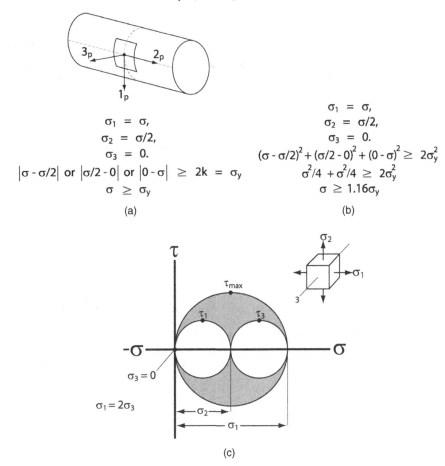

Figure 8.14

Thin-walled cylindrical pressure vessel with (a) equations for the pressure vessel using the Tresca criterion, (b) equations for the pressure vessel using the von Mises criterion, and (c) the Mohr's circle depicting the stress state. Note that the von Mises effective stress needed for yield is 16% greater than that for uniaxial loading.

stress for permanent strain is approximately 58% of the uniaxial yield strength for the material.

Another interesting stress state to examine with the von Mises and Tresca yield criteria is three-dimensional hydrostatic stress, or pressure. Recall that both yield criteria are founded upon the notion that shear is the mechanism for plastic deformation and that the application of a hydrostatic stress has no effect on yield behavior. Indeed, both predict that a hydrostatic stress state does not incur any plastic deformation (Figure 8.16). This suggests that the mechanism of plastic deformation associated with hydrostatic loading is not shear but a shear cavitational fracture process. In such cases

Pure Shear

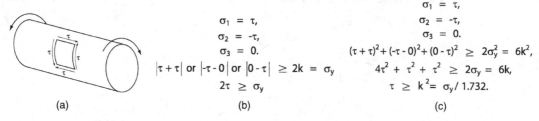

(a)

$\sigma_1 = \tau,$
$\sigma_2 = -\tau,$
$\sigma_3 = 0.$
$|\tau + \tau|$ or $|-\tau - 0|$ or $|0 - \tau| \geq 2k = \sigma_y$
$2\tau \geq \sigma_y$

(b)

$\sigma_1 = \tau,$
$\sigma_2 = -\tau,$
$\sigma_3 = 0.$
$(\tau + \tau)^2 + (-\tau - 0)^2 + (0 - \tau)^2 \geq 2\sigma_y^2 = 6k^2,$
$4\tau^2 + \tau^2 + \tau^2 \geq 2\sigma_y = 6k,$
$\tau \geq k^2 = \sigma_y / 1.732.$

(c)

Figure 8.15

Illustration of (a) the stress state for pure shear (torsion loading of tube), (b) the stresses captured using the Tresca criterion, and (c) the stresses captured using the von Mises criterion. The Tresca yield criterion indicates that the shear stress needed for plastic deformation is one-half of the uniaxial yield strength, and von Mises yield criterion indicates that the required shear stress for permanent strain is approximately 58% of the uniaxial yield strength for the material.

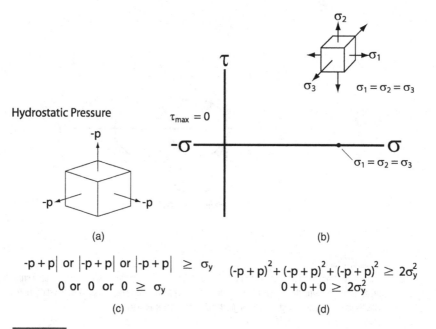

$-p + p|$ or $|-p + p|$ or $|-p + p| \geq \sigma_y$
0 or 0 or 0 $\geq \sigma_y$

(c)

$(-p + p)^2 + (-p + p)^2 + (-p + p)^2 \geq 2\sigma_y^2$
$0 + 0 + 0 \geq 2\sigma_y^2$

(d)

Figure 8.16

Illustration of (a) hydrostatic pressure or triaxial stress state showing (b) the Mohr's circle, (c) Tresca yield criterion, and (d) von Mises yield criterion. But criteria predict no yield (both the Tresca stress and the von Mises effective stress are equal to zero).

the yield criteria no longer serve as the best assessment for failure. Instead, the critical normal stresses become better predictors of failure. Furthermore, it should be noted that the presence of a defect in the material subjected to triaxial tensile stresses warrants the use of fracture mechanics, rather than yield criteria, for the safe design of the component.

Example 8.3 Multiaxial loading in the femoral stem

(i) Consider a cemented metallic femoral stem that is isotropic and is subjected to a multiaxial stress state as shown in Figure 8.17(a). These multiaxial stresses arise from several sources. From the joint contact force on the head, there is a net tensile bending stress (48 MPa) on the lateral side of the stem, a shear force of 16 MPa due to the load transmitted to the bone through the bone cement, and a compressive force in the transverse direction due to press fitting the implant in the bone (-32 MPa).

Assume that the alloy used in this device is made of an isotropic alloy that yields through a ductile shear mechanism. The orthopedic alloy has a tensile yield strength of 900 MPa.

Calculate the effective (von Mises) stress for this implant – how does this compare to the magnitude of the individual stress components? Determine whether this component is safely designed against yielding (calculate the factor of safety). Comment on your findings.

Solution (i) Using the schematic illustration of the stresses in Figure 8.17(a), one can identify the components of the stress matrix:

$$\sigma_{xx} = -32 \text{ MPa}, \; \sigma_{yy} = 48 \text{ MPa}, \; \sigma_{zz} = -32 \text{ MPa}, \; \tau_{xy} = 16 \text{ MPa}, \; \tau_{yz} = 16 \text{ MPa},$$
$$\tau_{zx} = 0 \text{ MPa}$$

Using Equation (8.19), we can determine the effective von Mises stress for this loading state:

$$\sigma_{\text{eff}} = \frac{1}{\sqrt{2}} \sqrt{(\sigma_x - \sigma_y)^2 + (\sigma_y - \sigma_z)^2 + (\sigma_z - \sigma_x)^2 + 6\left(\tau_{xy}^2 + \tau_{yz}^2 + \tau_{zx}^2\right)}$$
$$= \frac{1}{\sqrt{2}} \sqrt{(-32 - 48)^2 + (48 - (-32))^2 + (-32 - (-32))^2 + 6(16^2 + 16^2 + 0^2)}$$
$$= 89 \text{ MPa}.$$

The effective stress is well above all the constituent stresses; however, the effective scalar stress for the system is 89 MPa. The yield strength for this alloy is 900 MPa. The factor of safety against yielding is:

$$\text{FS}_{VM} = \frac{\sigma_y}{\sigma_{\text{eff}}} = \frac{900}{89} \cong 10.$$

Though a factor of safety of 10 against yielding is computed, it should be noted that this analysis assumes there is no inherent stress concentration and neglects the effects of any stresses that may be cyclical in nature.

(ii) Assume now that the stress state can be reduced to two dimensions, as shown in Figure 8.17(b), by neglecting the stresses acting in the z-direction.

Determine the principal stresses and maximum shear stress for this 2D stress state. Calculate the effective stress for this implant.

How is the factor of safety affected by reducing this problem to a 2D stress state?

The von Mises effective stress is a scalar representation of the stress tensor and the effect of coordinate rotation has no effect on the effective stress state. Show that this is true using a rotation angle of 30° to the x'-y' coordinate axes.

Assess what is lost in reducing the problem to 2D for this case.

Solution (ii) Using the schematic illustration of the stresses in Figure 8.17(b), which neglects stresses in the z-axis, we can identify the components of the stress matrix:

$$\sigma_{xx} = -32 \text{ MPa}, \sigma_{yy} = 48 \text{ MPa}, \sigma_{zz} = 0 \text{ MPa}, \tau_{xy} = 16 \text{ MPa}, \tau_{yz} = \tau_{zx} = 0 \text{ MPa}$$

The principal stresses are found using the basic equations for the Mohr circle:

$$\sigma_1, \sigma_2 = \left(\frac{\sigma_x + \sigma_y}{2}\right) \pm \sqrt{\left(\frac{\sigma_x - \sigma_y}{2}\right)^2 + \tau_{xy}^2}$$

$$= \left(\frac{-32 + 48}{2}\right) \pm \sqrt{\left(\frac{-32 - 48}{2}\right)^2 + 16^2}$$

$$= 8 \pm 43 \text{ MPa}$$

$$\sigma_1 = 51 \text{ MPa}$$

$$\sigma_2 = -35 \text{ MPa}$$

The maximum shear stresses are found using the Tresca yield criteria:

$$\tau_{max} = MAX \left\{ \frac{|\sigma_1 - \sigma_2|}{2}, \frac{|\sigma_2 - \sigma_3|}{2}, \frac{|\sigma_1 - \sigma_3|}{2} \right\}$$

$$= MAX \left\{ \frac{|51 - (-35)|}{2}, \frac{|-32|}{2}, \frac{|51|}{2} \right\}$$

$$\tau_{max} = 43 \text{ MPa}.$$

The effective stress can be determined using the von Mises yield criteria (using either the principal stresses or all stress components) and is given as:

The effective stress using the 2D analysis is 75 MPa, which is lower than the 89 MPa calculated using the 3D analysis. The former results in a factor of safety of 12 while the latter predicts a factor of safety of approximately 10. Thus, the 3D analysis is more conservative.

The effective stress is a scalar quantity and it is independent of coordinate frame. Hence, if we calculate the effective stress for the x'-y' axes (30° rotation) we find the same value of stress:

$$\sigma_{x'}, \sigma_{y'} = \left(\frac{\sigma_x + \sigma_y}{2}\right) \pm \left(\frac{\sigma_x - \sigma_y}{2} \cos 2\theta + \tau_{xy} \sin 2\theta\right)$$

$$= \left(\frac{-32 + 48}{2}\right) \pm \left(\left(\frac{-32 - 48}{2}\right) \cos 60° + 16 \sin 60°\right)$$

$$\sigma_{x'} = 1.86 \text{ MPa}$$

$$\sigma_{y'} = 14.14 \text{ MPa}$$

$$\tau_{x'y'} = \left(\frac{\sigma_x + \sigma_y}{2}\right) \pm \left(\frac{\sigma_x - \sigma_y}{2}\sin 2\theta + \tau_{xy}\cos 2\theta\right)$$

$$= \left(\frac{-32 + 48}{2}\right) \pm \left(\left(\frac{-32 - 48}{2}\right)\sin 60° + 16\cos 60°\right)$$

$$\tau_{x'y'} = 42.6 \text{ MPa}$$

$$\sigma_{\mathit{eff}} = \frac{1}{\sqrt{2}}\sqrt{(\sigma_x - \sigma_y)^2 + (\sigma_y - \sigma_z)^2 + (\sigma_z - \sigma_x)^2 + 6\left(\tau_{xy}^2 + \tau_{yz}^2 + \tau_{zx}^2\right)}$$

$$\sigma_{\mathit{eff}} = \frac{1}{\sqrt{2}}\sqrt{(1.86 - 14.14)^2 + (14.14)^2 + (1.86)^2 + 6(42.6)^2}$$

$$\sigma_{\mathit{eff}} = 75 \text{ MPa}.$$

This example demonstrates how the von Mises yield criteria can provide for us a scalar quantity that is invariant to the coordinate axes and captures the complete loading state of an isotropic material system. This example also shows us that we can simplify a 3D problem into a 2D problem but at the expense of the safety factor. Also, this analysis only predicts safety from yielding failure under static loading conditions and offers no insight into the effects of cyclic loading on fatigue or fracture of the component. A full structural analysis necessitates consideration of both static and dynamic loads – though service stresses must be calculated and compared to the yield strength, the safety of the device under cyclic loading or in the presence of a stress concentration must also be assessed.

8.7 Modified yield criteria

The yield theories developed above are widely utilized in design calculations for the determination of effective stresses and in predicting factors of safety against permanent deformation. However, the Tresca and von Mises yield criteria are relevant primarily for isotropic, ductile materials that deform through shear processes and possess equivalent yield strengths in both tension and compression. These assumptions hold for many metals or alloys, though even in these materials anisotropy can develop under certain processing conditions, such as rolling or drawing. Moreover, many structural materials and nearly all tissues are anisotropic in nature. To accommodate such directional dependence, modified yield criteria were developed for anisotropic yield behavior (Hill, 1950). One form of this yield equation utilizes material symmetry; for an orthotropic system that has symmetry about three orthogonal planes one can write:

$$H(\sigma_x - \sigma_y)^2 + F(\sigma_y - \sigma_z)^2 + G(\sigma_z - \sigma_x)^2 + 2N\tau_{xy}^2 + 2L\tau_{yz}^2 + 2M\tau_{zx}^2 = 1 \quad (8.22)$$

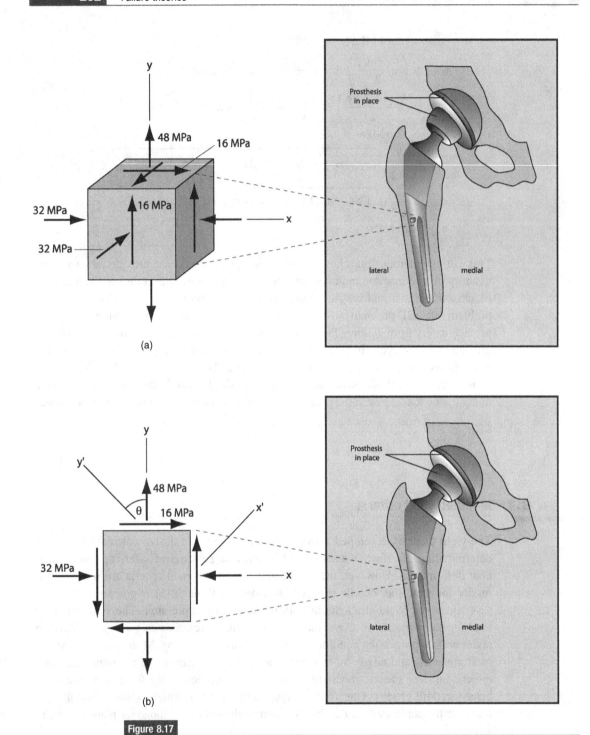

Figure 8.17

Schematic of a metallic femoral stem subjected to (a) a 3-dimensional loading state and (b) a simplification to a 2-dimensional stress state that neglects stresses in the z-axis and a potential for coordinate transformation is indicated through x' and y'.

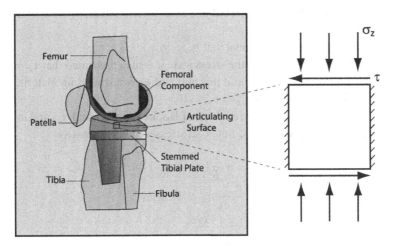

Figure 8.18

Schematic of a polymeric tibial insert subjected to a combined normal stress and shear stress.

where X, Y, and Z refer to the planes of symmetry and the other terms, H, F, G, N, L, M, are material constants that are determined empirically (Hill, 1950).

Another challenge in developing yield criteria for biomaterials is that many of these material systems have yield strengths that widely differ between compression and tension. This is true in many biological tissues, brittle systems, and polymers. It is believed that for these materials there is a dependence on hydrostatic stress, and the von Mises yield criterion can be modified accordingly to accommodate this behavior:

$$(\sigma_1 - \sigma_2)^2 + (\sigma_2 - \sigma_3)^2 + (\sigma_3 - \sigma_1)^2 + 2(|\sigma_{y,c}| - \sigma_{y,t})(\sigma_1 + \sigma_2 + \sigma_3)$$
$$= 2|\sigma_{y,c}|\sigma_{y,t} \tag{8.23}$$

where $\sigma_{y,t}$ is the yield stress in tension and $\sigma_{y,c}$ is the yield stress in compression (Dowling, 2007). Many polymers used in biomaterial applications have ratios of $|\sigma_{y,c}|/\sigma_{y,t}$ ranging from 1.0–1.5. Example 8.4 makes use of a modified von Mises yield criteria, incorporating the differences in yield strength in tension and compression for a polymeric component used in an orthopedic device.

Example 8.4 Use of a modified yield criteria for a polymeric implant

Consider a polymeric tibial insert that is subjected to a simplified stress state involving a compressive normal stress of 20 MPa and an applied shear stress of 2 MPa as shown in Figure 8.18. The compressive yield strength for this polymer is 30 MPa and the tensile yield strength is 20 MPa. The Poisson ratio is 0.46 and its elastic modulus is 1 GPa.

Calculate the effective (von Mises) stress for this implant. What is the factor of safety against yield? How will yield by affected by the unevenness of the yield strengths?

Solution The implant is directly subjected to a *compressive* stress, $\sigma_{zz} = 20$ MPa, and a shear stress, $\tau_{xy} = 2$ MPa.

From symmetry it is known that $\sigma_{xx} = \sigma_{yy}$.

Due to the 2D nature of the stress state it is also presumed that $\tau_{yz} = \tau_{zx} = 0$.

Also owing to the constraint in x and y (due to surrounding material) the strains in these directions are zero, $\varepsilon_{xx} = \varepsilon_{yy} = 0$.

First we use the generalized form of Hooke's Law to find σ_{xx} and σ_{yy} (which are equal due to symmetry) and note that stresses are compressive.

$$\varepsilon_{xx} = \frac{\sigma_{xx}}{E} - \frac{\nu}{E}(\sigma_{yy} + \sigma_{zz}) = \frac{\sigma_{xx}}{E}(1-\nu) - \frac{\nu}{E}\sigma_{zz} = 0$$

$$\Rightarrow \sigma_{xx} = \frac{\nu}{(1-\nu)}\sigma_{zz} = 17\,\text{MPa}$$

$$\Rightarrow \sigma_{xx} = \sigma_{yy} = 17\,\text{MPa}$$

Using Equation (8.19), the effective von Mises stress for this compressive loading state is:

$$\sigma_{\text{eff}} = \frac{1}{\sqrt{2}}\sqrt{(\sigma_x - \sigma_y)^2 + (\sigma_y - \sigma_z)^2 + (\sigma_z - \sigma_x)^2 + 6\left(\tau_{xy}^2 + \tau_{yz}^2 + \tau_{zx}^2\right)}$$

$$= \frac{1}{\sqrt{2}}\sqrt{(17-17)^2 + (17-20)^2 + (20-17)^2 + 6(0^2 + 2^2 + 0^2)}$$

$$= 4.58\,\text{MPa}.$$

The effective stress is well below the yield strength for this material. If we employ the compressive yield strength, we can determine the factor of safety:

$$\text{FS}_{VM} = \frac{\sigma_y}{\sigma_{\text{eff}}} = \frac{30}{4.58} \cong 6.55.$$

If one wants to account for the asymmetry of the yield strengths, then a modified von Mises yield criterion should be used. For this material, we have $|\sigma_{y,c}|/\sigma_{y,t} = 30/20 = 1.5$, and thus $|\sigma_{y,c}| = 1.5\,\sigma_{y,t}$. Hence, we expect that the hydrostatic component of stress becomes important for this material. If we use the ratio of the yield stresses in Equation (8.23), we get a general form of the modified von Mises criteria that incorporates the effect of hydrostatic stresses:

$$(\sigma_1 - \sigma_2)^2 + (\sigma_2 - \sigma_3)^2 + (\sigma_3 - \sigma_1)^2 + 2(|\sigma_{y,c}| - \sigma_{y,t})(\sigma_1 + \sigma_2 + \sigma_3)$$
$$= 2\,|\sigma_{y,c}|\,\sigma_{y,t}$$

$$\Rightarrow (\sigma_1 - \sigma_2)^2 + (\sigma_2 - \sigma_3)^2 + (\sigma_3 - \sigma_1)^2 + 2(|1.5\sigma_{y,t}| - \sigma_{y,t})(\sigma_1 + \sigma_2 + \sigma_3)$$
$$= 2|1.5\sigma_{y,t}|\sigma_{y,t}$$

$$\Rightarrow (\sigma_1 - \sigma_2)^2 + (\sigma_2 - \sigma_3)^2 + (\sigma_3 - \sigma_1)^2 + (\sigma_1 + \sigma_2 + \sigma_3) = 3\sigma_{y,t}^2$$

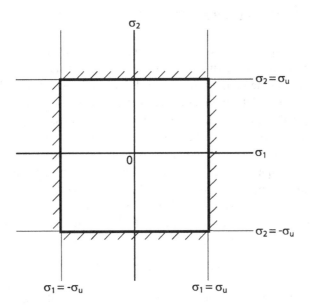

Figure 8.19

The failure envelope for the normal stress criterion (plotted in $\sigma_1 - \sigma_2$ space) is a square that intersects the conditions of principal stresses equal to the ultimate strength, σ_u.

8.8 Maximum normal stress failure theory

The failure theories above work well for ductile materials; however, when a material is brittle in nature it is highly sensitive to tensile stresses. The normal stress criterion is founded upon the concept that failure occurs when one of the principal stresses reaches the ultimate strength for the material. The criterion states that failure occurs when:

$$\sigma_f = \sigma_{ult} = MAX\ \{\sigma_1,\ \sigma_2,\ \sigma_3\}. \tag{8.24}$$

The failure stress is found by measuring the ultimate tensile strength or fracture stress from a uniaxial tensile test (Figure 8.1). The basic form of the failure criterion, like that of Tresca and von Mises, assumes isotropic behavior, and equivalent strengths in tension and compression. In two dimensions, plotted in $\sigma_1 - \sigma_2$ space, the failure envelope is a square (in three dimensions the failure envelope is a cube) that intersects the axes when principal stresses equal to the ultimate strength or fracture strength of the material, as is shown in Figure 8.19.

The criterion is employed as a first approximation for failure in brittle solids. However, it is limited in that it does not account for any stress concentrations or defects that can nucleate conditions for fracture at much lower stress levels. In such cases, fracture mechanics must be employed to safely design the component.

8.9 Notches and stress concentrations

The presence of notches and stress concentrations can cause localization of stresses and concomitant strains that can contribute to local plastic deformation and can also lead to crack nucleation in the component. It is well known from elasticity theory that the presence of a hole in a plate provides a threefold increase in local stresses at the edge of the hole oriented orthogonally to the applied normal stress. Stress concentrations can occur at screw threads, geometric reductions of cross-sectional areas, rivets, holes, and notches. These stress concentrations serve as considerable design challenges for engineers as such design features are often necessary for proper anatomical fit, fracture fixation, or for locking mechanisms of medical devices. A geometric or **stress concentration factor**, K_t, is used to relate the local stress at the discontinuity, σ_{loc}, to that of the stress that is applied far-field, σ^∞, and is defined as follows (Schigley and Mischke, 1989):

$$K_t = \frac{\sigma_{loc}}{\sigma^\infty}. \tag{8.25}$$

It should be noted that K_t is not a material property but rather a scalar entity that provides the magnitude of stress near a structural discontinuity. K_t is highly dependent upon geometry and is often determined experimentally. Many of these empirical values are found in table format in reference handbooks such as *Stress Concentration Factors* (Peterson, 1974). One geometry that can be solved numerically is the case of an embedded ellipse in an infinite plate (Inglis, 1913):

$$K_t = 1 + \frac{2b}{a} \tag{8.26}$$

where b is the half-width (half of the long axis) of the ellipse and a is the half-height (half of the short axis). For the case of an embedded hole in the plate where $a = b$, the stress concentration factor is 3. Also, the stress concentration factor K_t increases as the value of a decreases. For an atomically sharp crack, a tends toward zero and K_t becomes infinitely large *in the absence of plasticity*. Thus, for sharp notches in brittle solids that *fracture before yield*, it is important to perform a fracture mechanics analysis using a stress intensity factor which accounts for the applied stress and geometry of the flaw (see Chapter 9). In ductile materials that *yield before fracture*, the sharp notches blunt in the presence of plastic flow (yielding) and dissipate the effect of the stress concentration and the local stresses ahead of the notch. Hence, the stress distribution around the flaw is strongly influenced by whether the material behaves in a brittle, ductile, or semi-brittle fashion (discussed below).

Example 8.5 Determining the coupled effect of geometry and notch on stresses

Consider the two different hip implant geometries presented in Example 6.8. If the design of the rectangular cross-section is further altered such that it has an embedded hole (elliptical notch with $a = b$ and that of the rectangular cross section has an elliptical flaw with $b = 2a$), then how are the stresses of the implant affected locally for each case?

Solution In Example 6.8, the stresses were solved for these designs using beam theory:

$$\sigma_{rect} = 365 \text{ MPa}$$
$$\sigma_{circ} = 509 \text{ MPa}$$

The stress concentration factor for the notches can be found using Equation (8.26):

$$K_{t,rect} = 1 + \frac{2(2a)}{a} = 5$$
$$K_{t,circ} = 1 + \frac{2(a)}{a} = 3$$

and the local stresses can be found using Equation (8.25):

$$K_t = \frac{\sigma_{loc}}{\sigma^\infty}$$
$$\Rightarrow \sigma_{loc} = \sigma^\infty K_t$$
$$\sigma_{loc,circ} = \sigma_{circ}^\infty K_{t,circ} = 509 \cdot 3 = 1527 \text{ MPa}$$
$$\sigma_{loc,rect} = \sigma_{rect}^\infty K_{t,rect} = 365 \cdot 5 = 1825 \text{ MPa}$$

This finding indicates that the local stresses are controlled by the presence of a notch and that the overall stress in the implant is strongly affected by the geometric design. Consider that the implant is machined from a forged, heat-treated Ti6Al4V alloy with a yield strength of 1034 MPa. In this case both implant designs would be susceptible to local plastic flow. Under cyclic conditions this would very likely result in a low cycle fatigue failure.

Some materials are more sensitive than others to the presence of a notch and this becomes most evident under cyclic loading conditions. For such materials, a notch sensitivity factor must be incorporated into the total life fatigue analysis. Usually an experimentally determined notch sensitivity factor is employed into the equations for stress-life or strain-life calculations as the effect of the notch lowers the number of loading cycles to failure for a given stress state. The fatigue stress concentration factor, K_f, is defined as:

$$K_f = \frac{\sigma_{end,unnotched}}{\sigma_{end,notched}} \quad (8.27)$$

where $\sigma_{end,unnotched}$ is the endurance limit for an unnotched bar and $\sigma_{end,notched}$ is the endurance limit for a notched bar.

The notch sensitivity, q, for a material is defined as:

$$q = \frac{K_f - 1}{K_t - 1} \quad (8.28)$$

where q ranges between $0 - 1$. If $q = 0$ then there is no sensitivity to notches and $K_f = 1$. On the other end of the spectrum, if $q = 1$, then $K_f = K_t$ and there is full notch sensitivity in the material. In order to find the fatigue sensitivity to notches, one must first find K_t for the geometry of the notch. The value of q is then found specifically for the material in question (many steels have a notch sensitivity factor on the order of 0.4–0.9 depending upon its composition and severity (radius) of the notch (Sines and Waisman, 1959)). The fatigue stress concentration factor is then found by reorganizing the above equation:

$$K_f = 1 + q(K_t - 1). \tag{8.29}$$

A familiar form of the fatigue stress concentration factors is the Peterson equation that was empirically developed for steels (Peterson, 1959):

$$K_f \approx 1 + \frac{(K_t - 1)}{1 + \frac{A_n}{\rho}} \tag{8.30}$$

where ρ is the notch root radius and A_n is a constant that depends on the strength and ductility of the ferrous alloy (A_n is on the order of 0.025 mm for high-strength steels and an order of magnitude higher for lower strength steels).

Another interesting design challenge with notches that undergo loading and unloading is the role of the residual stresses that develop at the notch tip upon unloading (Figure 8.20). This is of considerable interest for the case of cyclic compression loading where a zone of residual tension that forms ahead of a sharp notch may be sufficient to initiate the growth of a crack from the notch-tip. In fact, fatigue cracks have been shown to nucleate under cyclic compression loading of tibial inserts comprised of ultra-high molecular weight polyethylene (see the case study below).

It has been established for some time that cyclic compression loading to a plate containing a sharp notch results in the inception of crack growth ahead of the notch (Gerber and Fuchs, 1968; Hubbard, 1969). In the absence of any other stress state these cracks become self-saturating; however, if these flaws are nucleated in a component that is subjected to complex loading states, then the crack that has initiated in cyclic compression may grow into a region of tensile strength sufficient to cause failure in the structure. Suresh and co-workers provided great insight into the mechanisms of crack nucleation from a sharp notch under the conditions of cyclic compression loading for metals, ceramics, polymers, and composites (Suresh, 1998). Suresh used finite element methods to show that a region of reversed plastic flow developed at the notch tip upon unloading from far-field compression (Holm et al., 1986). This zone of residual tensile stress was later quantitatively measured in situ for a number of amorphous polymer systems using interferometry and photoelasticity methods (Pruitt and Suresh, 1993). The size of the residual stress zone at the notch tip is typically one-fourth the size of a monotonic plastic zone for an elastic-perfectly plastic material (Suresh, 1998); and thus

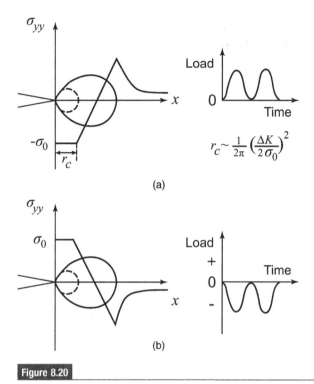

Figure 8.20

Illustration showing (a) region of residual compression that develops ahead of a sharp notch upon unloading from far-field cyclic tension and (b) zone of residual tension that is created upon unloading from far-field cyclic compression. (After Suresh, 1998.)

cracks nucleated under far-field compression are limited in length due to the size of the reversed plasticity zone.

These findings indicate that a crack can be nucleated due to local stress and strain conditions in the presence of a notch. This has a considerable effect on the structural lifetime of components that are subjected to variable amplitude loading in the presence of a notch. Often failure times for such loading conditions are short and the component failure is typically characterized as "low-cycle fatigue." Note that while the failure mode is labeled as a fatigue process, the micromechanisms of fracture are predominantly controlled through local plasticity mechanisms that may include microvoid coalescence as the process for fracture (Figure 8.21).

8.10 Failure mechanisms in structural biomaterials

The mechanical performance of a system depends strongly on whether its material microstructure renders it ductile, brittle, or semi-brittle. As stated above, **ductile** materials *yield before fracture* while **brittle** materials *fracture before yield*. A **semi-brittle**

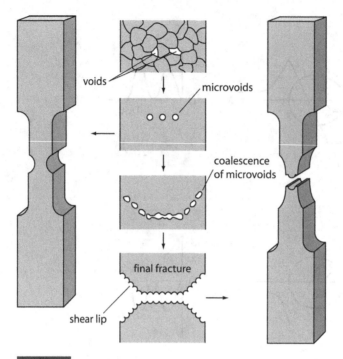

Figure 8.21

Microvoid coalescence in a ductile metal.

system offers a small amount of plastic or permanent deformation prior to fracture. Normally we think of metals as ductile, ceramics as brittle, and polymers or tissues ranging between the two boundaries (depending on their structures). Metals that have limited slip systems may in fact behave as a semi-brittle material and may fail through a cleavage mechanism rather than a slip mechanism (discussed below). Ductile systems generally utilize shear deformation in response to a stress on the system and accordingly the yield criterion for plasticity is based on the octahedral stress (von Mises) or maximum shear stress (Tresca). Brittle solids fail through a decohesion mechanism and are most sensitive to normal stresses (normal stress failure criteria). Also of interest is the orientation of the fracture surface owing to different failure modes. Fracture occurs orthogonal to the loading axis in a cylinder of a brittle solid subjected to tensile loading, while the fracture axis is oriented at 45° relative to the torsion axis. In both of these cases, the material fails on planes of maximum principal stress. On the other hand, a cylinder of a ductile material subjected to a tensile load will fail at 45° to the loading axis (where maximum shear stresses occur) and will fail on an orthogonal axis (where maximum shear stresses occur) when subjected to torsional loading. Hence, fractography (imaging of a failed surface) can reveal insight into the mechanism of failure and the nature of the stresses. Figure 8.22 shows how the fracture surface of a cavitating ductile material can vary for different modes of loading.

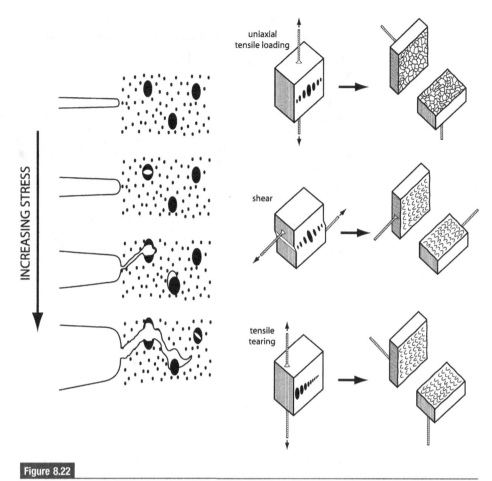

Figure 8.22

Schematic of fracture surface of a cavitating ductile material can vary for different modes of loading.

8.10.1 Failure mechanisms in metals

Most alloys utilized in biomedical device designs are sufficiently ductile to meet the yield before fracture condition. Alloys comprising close-packed or closest-packed crystalline structures facilitate the movement of dislocations on slip systems. Recall from Chapter 2 that the microstructure and defects present in a metal dictate the yielding or failure process (Figure 8.23).

Substitional atoms, interstitial atoms, and precipitates can pin or interact with dislocations in such a way that more stress is needed for yielding (dislocation movement). Edge dislocations and screw dislocations (Chapter 2) may interact with each other in such a way that the system strain hardens once yielding has initiated. For an FCC metal, the primary slip systems comprise the {111} family of planes and the <110> family of directions. As dislocations pile up at the grain boundaries (or other dislocation locking

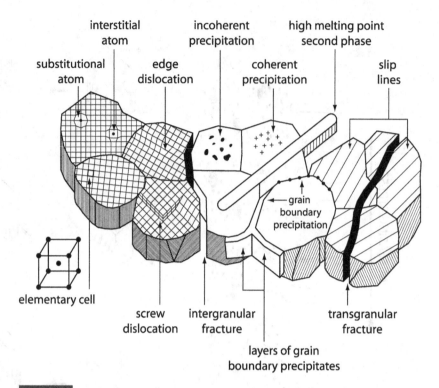

Figure 8.23

Illustration of defect types and sources of microstructural features associated with damage process found in metal systems. (After Suresh, 1998.)

mechanisms), the stress required for continued deformation is increased. This strengthening behavior is a measure of strain hardening for the alloy system. Grain boundaries make it difficult to maintain a resolved critical shear stress and can also increase the yield strength of a metal (recall that the Hall-Petch relationship states that the yield strength is inversely proportional to the square root of the grain size). If the alloy has a limited number of slip systems, then the material may behave in a semi-brittle manner and fail through decohesion of cleavage planes (Figure 8.24).

Similarly, in the presence of a notch or small crack, the plastic zone size can be on the length scale of the grain size and this can lead to crack growth that follows crystallographic planes of slip or cleavage, or with sufficient triaxial stresses may coalesce voids ahead of the stress concentration (Figure 8.25).

8.10.2 Failure mechanisms in ceramics

Most structural ceramics employed in medical devices are brittle systems that *fracture before yield*. The **fracture strength** of ceramics are dictated by the coupled effects of

8.10 Failure mechanisms in structural biomaterials

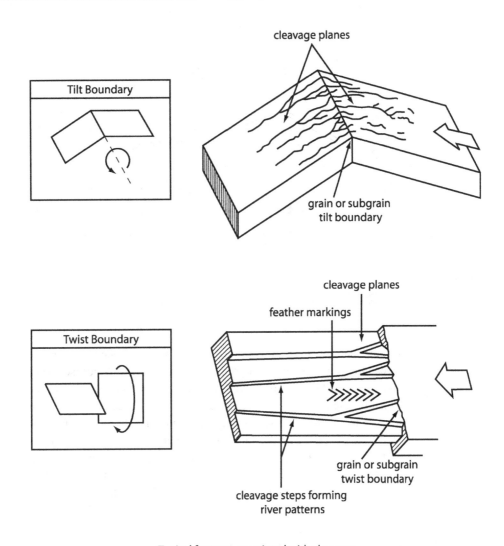

Typical features associated with cleavage.

Figure 8.24

Illustration of cleavage mechanisms activated in a semi-brittle solid.

normal stresses and flaws, and for this reason ceramics are significantly stronger in compression than in tension. The presence of a stress concentration in a ceramic system can have a profound effect on its structural integrity. Unlike ductile materials, these systems cannot rely on plasticity mechanisms to dissipate their stress fields ahead of the notch or other stress concentration. The presence of sharp cracks poses concern for fracture as these materials are known to have very low fracture toughness. Medical ceramics such as alumina, zirconia, pyrolytic carbon, or ceramic glasses (Chapter 3) are utilized in many medical implants that cannot tolerate fractures, such as orthopedic bearing

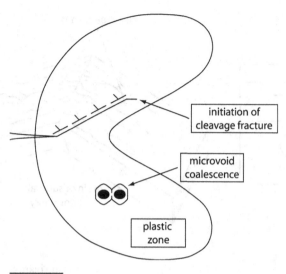

Figure 8.25

Illustration of competition between microvoid coalescence in a plastic zone ahead of a notch. (After Suresh, 1998.)

surfaces and heart valves. For this reason, designs using ceramics generally use fracture mechanics to ensure an appropriate structural analysis. Because plasticity mechanisms are not available to toughen ceramics, these materials utilize damage processes that can consume energy and reduce the effective stress intensity of the system. These mechanisms may include microcracking, phase transformations, and crack deflection (Suresh, 1998). However, the greatest improvements in improving fracture resistance in ceramic systems utilize damage mechanisms in the wake of crack tip including oxide wedging, interlocking grains, and crack bridging (Figure 8.26). The damage processes utilized in ceramics to enhance fracture toughness are further addressed in the following chapter.

8.10.3 Failure mechanisms in polymers

Polymers may offer a large amount of ductility or very little, depending on their molecular structure. The mechanical behavior of macromolecular structures depends on many variables such as chain chemistry and configuration; chain length and molecular weight distribution; degree of branching, entanglement, or crosslinking; degree of crystallinity; strain rate; temperature (relative to glass transition temperature); and local loading conditions (Chapter 4). The yield behavior of polymer systems may also be affected by hydrostatic stress or pressure and is best described using a modified yield criterion that accounts for both shear and pressure effects, as discussed above.

Many amorphous and some semicrystalline polymers undergo a cavitational process such as **crazing** when subjected to hydrostatic stress states. Crazing involves formation of voids followed by the fibrillation of the bulk polymer between the voids in a linear

8.10 Failure mechanisms in structural biomaterials

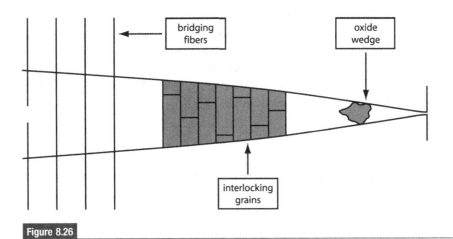

Figure 8.26

Illustration of damage mechanisms in the wake of a crack in a brittle solid.

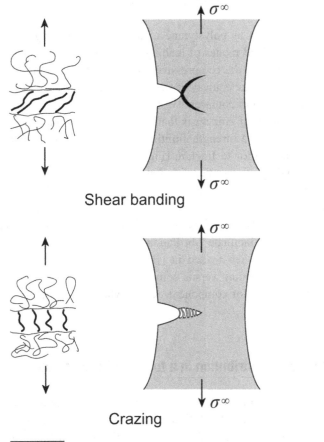

Figure 8.27

Illustration of shear banding and crazing in a semibrittle polymer.

localized zone of deformation oriented normal to the maximum principal stress. Crazing is fundamentally different than shear banding, which is oriented on planes of maximum shear stress much like shear deformation in metals. These mechanisms are illustrated in Figure 8.27. Crazing requires tensile stresses and a (hydrostatic) stress state that facilitates formation of voids between the fibrils. The stress criteria for crazing utilize the first stress invariant (Kausch, 1978):

$$\sigma_c = A(T) + \frac{B(T)}{\sigma_1 + \sigma_2} \tag{8.31}$$

where $A(T)$ and $B(T)$ are material constants that are temperature dependent. The criterion for shear banding is more similar to that for a ductile solid and includes a term for pressure effects:

$$\tau_{oct} = \tau_o - \mu \sigma_m \tag{8.32}$$

where τ_o is the shear yield strength, μ is the pressure sensitivity coefficient, and σ_m is the mean stress.

Glassy polymers such as polystyrene undergo shear banding in compression and crazing in tension. Mixed modes of loading can result in both modes of deformation being activated; however, the compressive portion of variable amplitude loading may be more damaging than the equivalent stress amplitude applied in pure tension, as the craze fibrils cannot support compressive stresses.

An interesting aspect of crazing is that a singular craze ahead of a notch or stress concentration serves as a strength limiting defect and limits the amount of strain that can be sustained prior to fracture (Figure 8.28(a)). However, if that same polymer is modified with particles that facilitate crazing (such as rubber), then a region of multiple crazing within a process zone is activated ahead of the notch or stress concentration and this process dissipates energy and results in a toughened system capable of sustaining plastic deformation (Figure 8.28(b)). Because crazing requires a hydrostatic stress to facilitate fibrillation, it is a micromechanism of failure that can be activated only when loaded in tension. Hence, many polymers behave differently under tensile loading versus compression loading and this phenomenon has important consequences for component design when the stresses are expected to be variable.

8.11 Case study: stress distribution in a total joint replacement

Complex stress distributions develop in ultra-high molecular weight polyethylene (UHMWPE) owing to the **articulation** with the metallic components of a total joint replacement. It has been estimated that the maximum principal stress ranges from 10 MPa of tension to 40 MPa of compression as the contact area sweeps across the surface during flexion for a total **condylar** type tibial component (Bartel *et al.*, 1986).

8.11 Case study: stress distribution in a total joint replacement

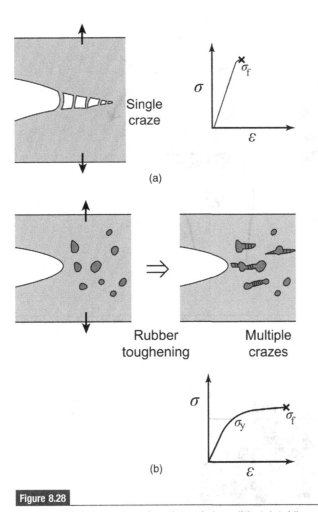

Figure 8.28

Illustration of (a) single craze in a polymer that results in very little strain to failure and (b) the use of rubber particles to nucleate multiple crazes and to improve toughness of the material.

Figure 8.29 shows how the contact stresses between the tibial insert and femoral component can develop with normal articulation (**extension-flexion**) of the joint. The kinematics of the joint result in cyclic stresses that are highly compressive in nature. In fact, the average contact pressure for many of these devices exceeds the uniaxial yield strength of UHMWPE (Figure 8.30).

For **pitting** and **delamination** to occur on articulating surfaces of knee joint components, fatigue cracks nucleate under conditions of cyclic compression loading. Prior to 1994, the propagation of fatigue cracks under cyclic tensile loading in UHMPWE had been studied extensively, but little was known about fatigue behavior under cyclic compressive loading in the presence of a notch or stress concentration. Using notched specimens and stress ranges representative of clinical loading in total

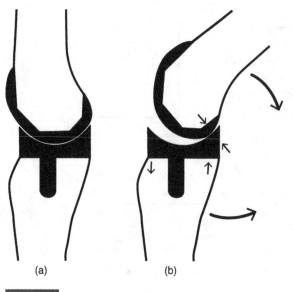

Figure 8.29

Contact loading between the tibial insert and femoral component change as the joint articulates between (a) extension and (b) flexion.

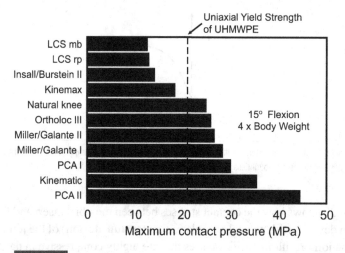

Figure 8.30

Maximum contact pressures experienced in clinical total knee replacement designs. Note most of these designs exceed the uniaxial yield strength of UHMWPE.

knee replacements, Pruitt *et al.* (1995) showed that fatigue cracks could initiate under the application of fully compressive and compressive-tensile loading of UHMWPE. Figure 8.31(a) shows a sharp crack that has nucleated in the plane of the notch orthogonal to the applied cyclic compressive stress. Additionally, this study investigated the

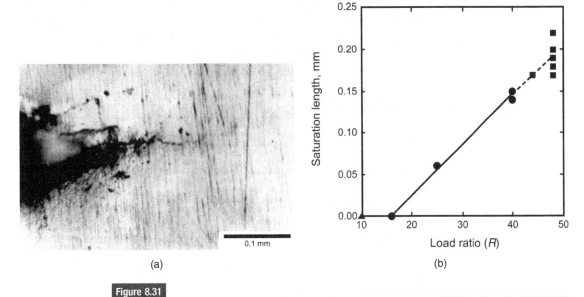

Figure 8.31

(a) shows the sharp fatigue crack that has nucleated ahead of a notch in UHMWPE and (b) shows the effect of load ratio on the saturation crack length. (After Pruitt et al., 1995.)

role of clinically relevant **stress ratios** for fully compressive loading and their effect on the saturation length of cracks. Larger peak compressive stresses resulted in longer crack lengths that were on the order of 0.2 mm (Figure 8.31(b)). This study provides insight into how cracks nucleate and how delamination (fatigue wear) may occur when the contact surfaces of a total knee replacement are exposed to fully or predominantly compressive stresses.

8.12 Summary

Medical devices are subjected to multiaxial stresses that are complex in nature. For the structural integrity of the implant it is imperative to ensure that the component will not inadvertently yield or fracture under quasi-static loading. Failure theories are utilized to provide a criterion for yield or fracture in the material. The two most commonly employed failure theories for ductile materials are the Tresca (maximum shear stress) and von Mises (maximum octahedral stress or distortional energy) yield criteria. Both of these theories presume plastic deformation through a shear process, *yielding before fracture*, and yielding failure when the effective stress in the system reaches the uniaxial yield strength. In contrast, brittle materials *break before yield* and are most sensitive to normal stresses. The normal stress criterion employs the ultimate strength or fracture stress obtained from a tensile test as the critical stress for failure. The normal stress criterion does not account for the presence of stress concentrations or defects in the material. Notches and other stress concentrations serve as sources of stress localization and nucleation sites for cracks. At a minimum, the safe design of biomaterial components requires a thorough structural analysis that provides a quantitative measure of effective stress and assessment of local stresses ahead of stress concentrations.

8.13 Problems for consideration

Checking for understanding

1. Describe the different failure theories philosophies utilized in the design of medical implants.
2. Based on the various failure design philosophies, describe the process by which you would safely design a total mandibular joint (TMJ) replacement composed of articulating CoCr alloy constituents. Would your design methodology change if instead of CoCr you were utilizing a fluoroploymer as a bearing surface in the joint design? What, if anything, is different between these two situations?
3. Explain the sources and significances of statistical scatter in yield data. How might this affect the use of Tresca or von Mises yield criteria for medical device design purposes?
4. A polymeric material made of nylon is chosen for a balloon angioplasty procedure. The balloon is spherical and thin-walled. The diameter of the balloon is 1.2 cm and it must withstand an internal pressure of 0.7 MPa. The polymer has yield strength of 26 MPa, an elastic modulus of 1 GPa, and a Poisson's ratio of 0.4.
5. What are the stresses and strains that develop in the balloon?
6. What must the thickness of the balloon be if we are to design against yielding of the polymer with a factor of safety of two?

For further exploration

Consider an artificial disk replacement for use in the spine that is composed of two titanium alloy end plates and a polymeric insert (Figure 8.32). Stress analysis of the polymer insert reveals that the material is subjected to a compressive load of 30 MPa in the axial direction and 26 MPa of shear along the interface with the metal backing. The elastic modulus of the polymer is 2 GPa, the shear modulus is 0.5 GPa, and the uniaxial yield stress is 22 MPa.

For the polymeric insert:

7. Evaluate and quantify the stresses and strains that develop as a result of the multiaxial loading. State all assumptions.
8. Calculate the maximum shear stress and the von Mises effective stress. Assess the factor of safety against yielding.
9. Determine the maximum compressive stress that can be sustained without yielding.
10. Discuss the consequences of the presence of a notch or stress concentration in the implant design.

Figure 8.32

Schematic illustration of a spinal implant.

8.14 References

Bartel, D.L., Bicknell, V.L., and Wright, T.M. (1986). The effect of conformity, thickness and material on stresses in ultra high molecular weight polyethylene components for total joint replacements. *Journal of Bone and Joint Surgery*, 68, 1041–51.

Dowling, N.E. (2007). *Mechanical Behavior of Materials*. Upper Saddle River, NJ: Pearson Education.

Gerber, T.L., and Fuchs, H.O. (1968). Analysis of non-propagating cracks in notched parts with compressive mean stress. *Journal of Materials*, 3, 359–74.

Hill, R. (1950). *The Mathematical Theory of Plasticity*. Oxford, UK: Clarendon Press.

Holm, D.K., Blom, A.F., and Suresh, S. (1986). Growth of cracks under far-field compressive loads: Numerical and experimental results. *Engineering Fracture Mechanics*, 23, 1097–1106.

Hubbard, R.P. (1969). Crack growth under cyclic compression. *Journal of Basic Engineering*, 91, 625–31.

Inglis, C.E. (1913). Stresses in a plate due to the presence of cracks and sharp corners. *Transactions of the Institute of Naval Architects*, 55, 219–41.

Kausch, H.H. (1978). *Polymer Fracture*. Berlin: Springer-Verlag.

Peterson, R.E. (1959). *Notch Sensitivity*. In *Metal Fatigue*, ed. G. Sines and J.L. Waisman. New York: McGraw-Hill, pp. 293–306.

Peterson, R.E. (1974). *Stress Concentration Factors*. New York: Wiley.

Pruitt, L., Koo, K., Rimnac, C.M., Suresh, S., and Wright, T.M. (1995). Cyclic compressive loading results in fatigue cracks in ultra high molecular weight polyethylene. *Journal of Bone and Joint Surgery*, 13, 143–46.

Pruitt, L., and Suresh, S. (1993). Cyclic stress fields for fatigue cracks in amorphous solids: Experimental measurements and their implications. *Philosophical Magazine A*, 67, 1219–45.

Schigley, J.E., and Mischke, C.R. (1989). *Mechanical Engineering Design*. New York: McGraw-Hill.

Sines, G., and Waisman, J.L. (1959). *Metal Fatigue*. New York: McGraw-Hill.

Suresh, S. (1998). *Fatigue of Materials*. Cambridge, UK: Cambridge University Press.

9 Fracture mechanics

Jevan Furmanski

> How might implants catastrophically fail when experiencing stresses far below the yield or ultimate tensile strength? How could this outcome be affected by material behavior, manufacturing, and quality concerns?

Biomedical implants are often complicated systems with multiple interacting components made of different materials. Thus, designing them successfully to function in the body and not fail even in predictable ways can be difficult. Basic structural analysis can quantify the expected potential of overcoming the strength of the implant under load, but such analyses presume the implant materials to be essentially unflawed or to remain flaw-free prior to global failure. In practice, medical devices fail in more complicated modes than exceeding the strength, and one of the most catastrophic of these modes is fracture. Medical device fractures occur when cracks or sharp flaws cause a local extreme stress that causes material breakdown, even under relatively mild loading conditions. Flaws or cracks can either exist inherently in the material or be caused by manufacturing, which can be mitigated through inspection and quality control. Cracks can also form during loading near sharp notches designed into the component, even if the component is flaw-free. The key is to consider fracture a catastrophic outcome for medical devices, one that could endanger the lives of patients if incompletely understood. The field of fracture mechanics has been developed to prevent such failures in a wide range of materials through analysis, and is very relevant to the design of implanted medical devices.

9.1 Overview

The origin of modern fracture mechanics is mainly traced to the contributions and insights of two well-known scientists: A.A. Griffith after the First World War, and G.R. Irwin after the Second. In 1921, Griffith published his results on the wide discrepancy observed between the actual strength of brittle materials and the theoretically obtainable strength (as discussed in Chapter 3). His work showed that reducing the size of glass fibers increased their strength directly according to predictions, nearly achieving the theoretical strength of the material (as described in Chapter 3). The central role of sharp flaws in the strength of materials was established along with the energetic description for the driving force for fracture. In 1948, Irwin evaluated crack tip conditions, and ultimately developed the more engineering-oriented approach of studying the extreme stresses at crack tips, and relating them to simple geometry-independent parameters. The field of fracture mechanics has remained active since that time, investigating more complicated materials and part configurations, but the underlying principles developed by Griffith and Irwin have endured.

Fracture is a failure mode distinct from the concept of strength that is commonly applied in engineering design. While strength can be thought of as a value of homogenous stress corresponding to material failure, fracture is governed by localized processes that occur near the site of flaws. These extreme but local stresses then strongly impact the overall failure potential of a component. For example, introducing a microscopic flaw into a component that is stressed far below its failure strength could result nevertheless in catastrophic failure. While designing for strength might involve increasing the size of a component to reduce the stresses during service (also called the **strength of materials** approach), fracture processes occur locally near a crack or other flaw, and so they are less sensitive to component size or the global severity of stress. It is critical therefore to distinguish the material resistance to fracture, termed the fracture toughness, as fundamentally different from mechanical strength, and to develop specialized techniques for understanding and mitigating the risk of fracture in engineered components. **Fracture mechanics** is the collection of techniques used to study and predict the local behavior of a material in the presence of a flaw or crack, and the overall risk of catastrophic failure resulting from local fracture phenomena.

Fracture has historically been associated with brittle materials, but few truly brittle materials are used in biomedical devices, where polymers and metals are instead dominant. Nevertheless, there are numerous applications today where fracture mechanics can be applied to improve implant function and patient outcomes. For instance, fracture of pyrolytic carbon heart valves or ceramic total hip replacement femoral heads occur clinically and certainly can be controlled using elementary linear elastic fracture mechanics, but ductile or viscoelastic materials require a more advanced treatment. Hydrogels that are candidates for cartilage replacement exhibit limited fracture toughness and cannot yet withstand the extreme loading that occurs in the knee for a sufficient lifetime of service. Ultra-high molecular weight polyethylene bearing components in total hip replacements may fracture, and the reported failures in these devices are likely tied to a reduced fracture toughness in modern crosslinked formulations of the material. Even thin membranes such as contact lenses may tear during ordinary use – a process essentially governed by the material resistance to fracture. Virtually any implantable device may tear or fracture in service, and fracture as a failure mode must not be overlooked in the design and quality control of biomedical implants.

9.2 Learning objectives

This chapter provides the motivation and development of fracture mechanics, including linear elastic fracture mechanics (LEFM), elastic-plastic fracture mechanics (EPFM), and viscoelastic fracture mechanics (VEFM). The key learning objectives of the chapter are that at its completion the student will be able to:

1. give examples of fracture in biomedical applications
2. describe the crack tip stresses in LEFM
3. define the stress intensity factor and explain how it is used to predict fracture
4. describe and diagram the three primary loading modes in fracture
5. verify that fracture stress is inversely proportional to the square root of flaw size
6. define the energy release rate, G, and its relationship to the stress intensity factor, K
7. depict the physical basis of growth resistance curves
8. use G and K as failure criteria, derive and use the critical flaw size
9. describe limitations of LEFM, when it is invalid, and the meaning of K dominance
10. calculate the plastic zone size and how it affects fracture prediction
11. differentiate between K_c and K_{IC}
12. define CTOD and describe its use in defining a fracture criterion
13. explain cohesive zone models and their use in various materials
14. describe how the J-integral is quantified, and when it is invalid
15. delineate the meaning of J-integral as the energy release rate and fracture criterion
16. identify the meaning of J-dominance and compare it to K dominance
17. illustrate crack tip stress state in elastic-plastic fracture mechanics (HRR fields)
18. draw a J_R curve, also derive it, describe where it is applicable, and how it is used
19. portray when VEFM is necessary and how crack initiation time in this theory compares to other fracture mechanics theories
20. explain how the fracture behavior of a material is linked to its extrinsic and intrinsic damage mechanisms

9.3 Linear elastic fracture mechanics (LEFM)

It has long been understood that the theoretical strength of materials is in practice much greater than the apparent strength, and the discrepancy is generally attributed to the inherent presence of flaws in the microstructure of the material that were not accounted for in the hypothetical "perfect" solid. In Chapter 2 the effect of dislocation defects in crystals in reducing the flow stress was introduced, and in Chapter 3 the theoretical strength for ceramics was shown to be much higher than the actual tensile strength due to the effect of small inherent cracks. It was the deficiency in the theoretical appreciation of the strength of brittle materials that motivated A. A. Griffith to investigate the apparent strength of glass fibers, teasing out the role of inherent flaws in the material on its failure properties.

The theoretical strength of an atomically bonded solid was approximated in Chapter 3 as $\sigma_f = E/\pi$, which is several orders of magnitude greater than the measured strength in bulk engineering ceramics. Using conservation of energy arguments and the crack tip

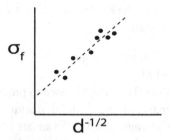

Figure 9.1

Schematic of experimental data from Griffith's glass fiber experiment, showing the expected dependence of observed fiber strength on the diameter.

stress analysis of Inglis (Inglis, 1913), Griffith derived that the actual tensile strength of a brittle material containing crack-like flaws is (Griffith, 1921)

$$\sigma_f = \frac{2}{\pi} \left(\frac{E}{\gamma a} \right)^{1/2} \tag{9.1}$$

where the stress to cause fracture is now also dependent on the crack size, a, and the surface energy, γ. The assumptions underlying this result are that all the energy required to generate new crack surface is accounted for by the surface energy of an increment new crack area, and that fracture occurs when the rate of potential energy supplied to the body exceeds that dissipated through crack advance. The remarkable result of Griffith's work is its agreement with experimental testing of drawn glass fibers of progressively smaller diameters (Figure 9.1) (Griffith, 1921).

As can be seen in Figure 9.1, making the diameter of the glass fibers larger rapidly decreases the strength for a bulk specimen, and making the diameter smaller increases the apparent strength up to the theoretical limit. The simple way to understand this result is to note that a fiber cannot harbor a flaw of size greater than its diameter, and thus the thinner fibers will be stronger in a predictable inverse square root dependence on the fiber diameter. Griffith's work became the progenitor both of energetic methods in fracture (Section 9.3.4) and the later engineering stress-based approaches developed and popularized by Irwin and others (Section 9.3.1). The work of G. I. Irwin resulted in a tractable analytical approach for understanding the localized and extreme nature of the deformation in materials near a sharp flaw, providing a direct link from traditional concepts of material strength to small-scale material failure in the mechanics of fracture (Irwin, 1948).

Finally, the effectiveness of reducing the size of brittle material elements to improve their strength is highly relevant to the study of biomaterials. It is this phenomenon that has led to the proliferation of high-strength fiber composites, and by analogy, is the basis of the material properties of many structural tissues in the body. Bone is an excellent example of a composite employing nanoscale ceramic reinforcement in a polymer matrix, and also exhibits other crack-stopping strategies to increase its fracture toughness. In a more extreme case, mussel shells are 95% microscopic calcium carbonate platelets, but the 5% of organic material and the material microstructure boosts the toughness from a value of ~2 J/m² for pure calcium carbonate to 1500 J/m² (Atkins and Mai, 1985). The latter is also a good example of how energy dissipation beyond surface energy contributions is important in the fracture of many biomaterials (Sections 9.4–9.6).

9.3.1 Cracks as extreme stress concentrators

Fracture mechanics is concerned with the locally extreme stresses that occur near the tip of a crack or flaw under stress, which is typically much more severe than the stress experienced by material remote from the flaw. As discussed in Chapter 8, geometric discontinuities (notches, holes, etc.) in components can result in local concentrations of stress up to five times more severe than in the bulk material. As the discontinuity becomes more pronounced, the stress concentration factor increases, and the local stress (exacerbated by the stress concentrator) may exceed the strength of the material under loading. The concept of severe stress caused by an extreme stress concentrator under load is a useful place to start to understand the fundamentally localized nature of fracture phenomena.

The first quantification of the local stress state near a crack tip is due to Inglis (Inglis, 1913), who took the mathematical formulation for the stress concentration factor resulting from an elliptical hole in a plate, and took its aspect ratio to the limit to approximate a two-dimensional crack. Consider a plate with a uniform global tensile stress, σ_∞, applied perpendicular to the long axis of an elliptical through-hole of length $2a$ and width $2b$. At the tip of the ellipse (denoted as point A), the tensile stress in the plate is maximal,

$$\sigma_A = \sigma_\infty \left(1 + 2\sqrt{\frac{a}{\rho}}\right) \tag{9.2}$$

where the radius of curvature, ρ, at Point A is

$$\rho = \frac{b^2}{a}. \tag{9.3}$$

Thus, if $b \ll a$, then $\rho \ll a$, and Equation (9.2) at this limit gives the elastic stress at the tip of a sharp elliptical through-crack in a plate under remote loading

$$\sigma_A = 2\sigma_\infty \sqrt{\frac{a}{\rho}}. \tag{9.4}$$

The first consequence of Equation (9.4) requires that an atomically sharp crack experience near-infinite stress at the crack tip for a finite applied stress, which is of course impossible for real materials to bear, and thus it is clear how the global performance of materials and structures can be dependent on microscopic sharp cracks. The existence of a stress singularity near a crack tip is the foundation of LEFM, and any material breakdown near the crack tip due to the extreme stress condition there is commonly ignored if it is limited in its extent. In Section 9.6, corrections to the local conditions near the crack tip for limited material plasticity are explored.

9.3.2 The stress intensity factor, K

The peak stress near a sharp crack in an elastic material is singular; however, the stress concentration factor does not give us insight on what the distribution of the extreme stress is near a flaw. LEFM is only valid if the local material breakdown at the crack tip occurs in a relatively small region compared with the zone of validity for the LEFM equations, and so it is essential to know how the singularity stresses extend beyond the crack tip.

From the mathematical theory of elasticity it can be shown that the stress at the tip of the crack in linear elastic isotropic materials decays as a strong function of distance from the crack, r, (in polar coordinates),

$$\sigma_{ij} = A \frac{1}{\sqrt{r}} f_{ij} + \sum_{m=0} A_m r^{1/2} g_{ij(\theta)}^{(m)} \qquad (9.5)$$

where f_{ij} and $g_{ij}(m)$ are dimensionless functions giving the angular distribution (out of the crack plane) of the stress, and A and A_m are coefficients amplifying the magnitude of stress resulting from each term. The leading term in Equation (9.5) becomes singular as r approaches zero and the higher order terms are vanishingly small in comparison. Thus, in LEFM, it is typically just the first term of Equation (9.5) that we use to describe the distribution of stresses in the singularity region we predicted in Equation (9.4) to exist at the tip of a sharp crack,

$$\sigma_{ij} = \frac{K}{\sqrt{2\pi r}} f_{ij(\theta)}. \qquad (9.6)$$

The coefficient K (Equation 9.6) specifies the relative strength of the stress singularity outside the material breakdown region; we call this parameter the **stress intensity factor**. Thus, the stress intensity factor compares the relative behavior of two cracks in a material, which is at the heart of LEFM. K is therefore a *single parameter* that describes completely the severity of the stress in the singularity region near a crack tip, which is of primary interest in fracture mechanics. The strength of this approach is that all sharp cracks express the same singular stress behavior, regardless of the mode of application of the stress or crack geometry, and thus all cracks in linear elastic isotropic media can be compared through K. The nature of the stress singularity is also invariant as a crack grows under load, a concept termed **self-similarity** (also known as the concept of similitude). Self-similarity is a precondition for the applicability of most fracture mechanics analyses, and is typically only violated with dramatic changes in crack length or geometry.

A model of how stresses are distributed in material adjacent to sharp cracks under remote loading is illustrated in Figure 9.2. This illustration shows the three regions that

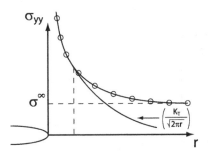

Figure 9.2

The stress becomes singular near the crack tip, according to the inverse root dependence expected for the use of K. Far from the crack tip, the tensile stress approaches the far-field value σ_∞. The region where K accurately describes the stress is called the singularity dominated zone. Very near the crack tip the stress exceeds what the material can withstand, and there the material deforms or breaks down.

have been identified so far: a singular region described by Equation (9.6), a region of material breakdown adjacent to the crack tip, and a far field where the stress approaches the global value.

Inspecting Equation (9.4) one sees that the stress at a sharp crack tip ($\rho \ll a$) is proportional to the global stress and the square root of crack length, a. It can be shown that K follows exactly the same dependence on global stress and crack length, i.e.,

$$K = Y\sigma_\infty\sqrt{\pi a} \qquad (9.7)$$

which it must, since it completely characterizes the magnitude of the stress singularity near the crack tip. Equation (9.7) contains a geometric coefficient Y, which is unity when the body surrounding the crack is infinite, and is greater for finite component geometries. For instance, $Y = 1.12$ for a small planar through-crack in the edge of a plate.

The stress intensity factor formula in Equation (9.7) is often compiled in tabular form for experimental specimens using an alternative expression,

$$K = F(a/W)\frac{P}{B\sqrt{W}} \qquad (9.8)$$

where P is the applied load, B is the specimen thickness, and W is the effective specimen length. The remaining material ahead of the crack tip, $W - a$, is called the ligament and is also a useful quantity in fracture and crack growth experiments. The function $F(a/W)$ is tabulated for each specimen geometry, and is analogous to the geometric factor Y in Equation (9.7). The polynomial $F(a/W)$ for the compact tension (CT) specimen of Figure 9.3 is shown in Equation (9.9).

$$F\left(\frac{a}{W}\right) = \frac{2 + (a/W)}{(1 - (a/W))^{3/2}}[0.866 + 4.64(a/W) - 13.32(a/W)^2 \\ + 14.72(a/W)^3 - 5.60(a/W)^4] \qquad (9.9)$$

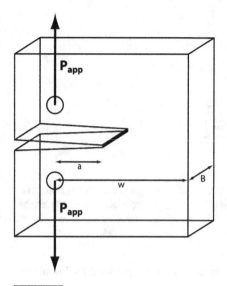

Figure 9.3

Compact tension (CT) specimen geometry with crack length *a* and specimen length *W*, as measured from the load line. The specimen thickness (out of the page) is denoted *B*.

Example 9.1 Estimating fracture toughness from fractography of a tibial implant

Consider the tibial plateau presented in Examples 6.5 and 8.2. This implant is made of UHMWPE in a constraining frame of CoCr and loaded with a force of 30 kN that is evenly spread across the tibial plateau. UHMWPE has a yield strength of 22 MPa and an elastic modulus of 1 GPa. A fractured component reveals an embedded penny-shaped flaw that served as the initiation site for fast fracture. The diameter of this flaw is measured to be 2mm using electron microscopy. Estimate the fracture toughness for this material using the fractography (fracture surface image) and knowledge of stresses on the system. State all assumptions.

Solution Applying the 3D Hooke's Law and geometric constraints provides the stresses and strains (Example 6.5):

$$\sigma_x = \sigma_y = 16.7 \text{ MPa}; \sigma_z = 25 \text{ MPa } \tau_{x_z} = \tau_{x_y} = \tau_{yz} = 0 \text{ MPa}$$
$$\varepsilon_x = \varepsilon_y = 0; \varepsilon_z = 0.012$$

The effective von Mises stress was found in Example 8.2: $\sigma_{eff} = 8.3$ MPa.

The fractography reveals that the radius of the flaw is 25 mm. The maximum far field stress is 25 MPa.

The form of the stress intensity factor for an embedded penny-shaped flaw is estimated as (Anderson, 2004):

$$K_I = \sigma^\infty \sqrt{\pi a}$$

solving for K_I provides an estimate of the fracture toughness:

$$K_I = \sigma^\infty \sqrt{\pi a} = 25\,\text{MPa}\sqrt{\pi}\sqrt{25 \times 10^{-6}(\text{m})} = 0.221\,\text{MPa}\sqrt{\text{m}}.$$

Assumptions: It is assumed that linear elastic fracture mechanics is valid and that any yielding in the material is small scale in comparison to the region of K-dominance. The maximum stress applied in the quasi-static condition is the axial stress; however, the effective stress is representative of the multiaxial loading conditions. It is assumed that the flaw advances through mode I fracture mechanisms. Also the geometric factor is neglected for simplicity and hence the fracture estimate is low in comparison to published values.

9.3.3 Loading modes and mixed-mode fracture

Crack growth is most critical under the application of a tensile stress; however, cracks may grow under shear stress or combinations of these components. Shear stresses are common at interfaces, such as in the bond layer between implants and bone. Medical devices often experience complicated stress states, for instance through the mechanical interaction of components, contact stresses in total joint replacements, or multiaxial loads caused by structural service in the skeleton. Therefore, the driving force for crack growth under generalized states of stress is important for fracture control in load-bearing medical devices.

Because the loading in the body is complex and medical devices are often subjected to multiaxial stress states, it is useful to describe the behavior of cracks under complicated global stress states. In considering general states of loading we first need to differentiate the global loading into three principal and independent parts. These three independent loading modes are shown in Figure 9.4.

Mode I is the crack-opening mode, where the applied global stress is perpendicular to both the crack plane and crack front. Mode II is the shearing mode, with a global stress resulting in shear displacement parallel to the crack plane and perpendicular to the crack front. Mode III is the twisting (or tearing) mode, where the displacement is a shear parallel to both the crack face and crack front. The three loading modes for cracks are distinguished not by the quality of the singularity at the crack tip (which is common to all cracks in LEFM) but instead by the magnitudes and distributions of the components of stress. Thus, through superposition, the stresses can be determined individually and then combined to give the result due to an arbitrary loading condition. In this manner,

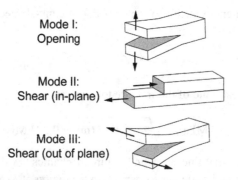

Figure 9.4

Principal crack loading Modes I, II, and III.

Equation (9.6) can be rewritten to reflect the individual loading modes (here shown for Mode I):

$$\sigma_{ij}^{(I)} = \frac{K_I}{\sqrt{2\pi r}} f_{ij(\theta)}^{(I)} \tag{9.10}$$

where the $f_{ij}^{(I)}$ function is distinct for each loading mode describing the mode-dependent stress distribution with the angle θ from the crack plane. The f_{ij} functions are found elsewhere (Anderson, 2004; Suresh, 1998). Note that in the crack plane ($\theta = 0$) $f_{xx}^{(I)} = f_{yy}^{(I)} = 1$ and so

$$\sigma_{ij}^{(I)} = \frac{K_I}{\sqrt{2\pi r}} \text{ for } \theta = 0. \tag{9.11}$$

The form of the stress intensity factor (Equation 9.7) can also be rewritten for individual loading modes:

$$K_{(I)} = Y\sigma_{\infty}^{(I)}\sqrt{\pi a} \tag{9.12}$$

Note that Equation (9.12) only admits the global stress in the loading mode under consideration (in this case, Mode I); other modes satisfy exactly the same relation but admitting only the corresponding portion of the global stress. When a global condition contains more than one loading mode, it is called **mixed-mode** loading, which can occur because of multiple applied remote stresses, or, for example (appealing to Mohr's circle), when a crack is propagating in a plane skewed to the applied stress.

It is critical to note that multiple loading states in the same mode can be superimposed due to the inherent linear superposition properties of linear elastic stress and strain fields, but multiple distinct loading modes cannot be superimposed to compute an overall K. That is:

$$K_I^{Tot} = K_I^A + K_I^B + \cdots \tag{9.13}$$

but

$$K^{Tot} \neq K_I^A + K_{III}^B + \cdots \qquad (9.14)$$

Thus, multiple conditions with the same loading mode can be superimposed to yield an overall stress intensity factor, but mixed-mode loading of cracks must be handled separately. We will see in the next section, however, that the energetic contributions of each individual mode are in fact additive, even though the stress intensity factors are not.

Example 9.2 **Determining the critical flaw size in component loaded in shear**

Consider the artificial spinal disk from Example 6.3. Consider that the insert of the implant is made from a polymeric material whose mode II fracture toughness is 0.1 MPa√m. What is the largest size of an embedded, penny-shaped flaw that can be tolerated without fracturing the component?

The stress intensity for the embedded penny-shaped flaw (mode II) can be assumed to take the simplified form (neglecting the geometric factor):

$$K_{II} = \tau_{\max}^\infty \sqrt{a}$$

Solution The stresses on the implant as shown in Figure 5.11 are:

$$\sigma_x = -2.2\,\text{MPa},\ \sigma_y = -1.1\,\text{MPa},\ \sigma_z = -0.58\,\text{MPa}$$
$$\tau_{xy} = -0.57\,\text{MPa},\ \tau_{xz} = -0.79\,\text{MPa},\ \tau_{yz} = -0.33\,\text{MPa}$$

The Tresca yield criterion (Equation 8.4) was employed in Example 8.1 to determine the effective stress on the system and the maximum shear stress:

$$\sigma_{\text{Tresca}} = 2.52\,\text{MPa} \Rightarrow \tau_{\max,\text{Tresca}} = \frac{\sigma_{\text{Tresca}}}{2} = 1.26\,\text{MPa}.$$

The critical flaw size is found by rearranging the form of the stress intensity:

$$a_C = \left(\frac{K_{IIC}}{\tau_{\max}^\infty}\right)^2 = \left(\frac{0.1\,\text{MPa}\sqrt{\text{m}}}{1.26\,\text{MPa}}\right)^2 = 0.00629\,\text{m} = 6.29\,\text{mm}.$$

9.3.4 Energetic methods in fracture mechanics

There are significant practical and philosophical limitations to the treatment of cracks as extreme stress concentrators. Extreme stresses resulting in local material breakdown at crack tips are predicted even for relatively low stresses, which could appear to result in material instability in the presence of any crack. However, the energy dissipated through the creation of new crack surface is not unbounded, which makes it an attractive alternative measure of localized crack phenomena. Such an observation served as the motivation for Griffith to describe fracture mechanics through an energetic equilibrium approach (Griffith, 1921).

The Griffith energy balance holds that for a crack to grow and remain in equilibrium, the following must be true:

$$\frac{dE_P}{dA} + \frac{dW_S}{dA} = \frac{dE_{Tot}}{dA} = 0 \qquad (9.15)$$

where the increment in work required to create new crack surface area dW_S then must be balanced by a change in potential energy, dE_P, which is supplied by internal and external work done on the body. Following the nomenclature developed by Irwin (Irwin, 1956) $\frac{dE_P}{dA}$ denotes the change in potential energy, the **energy release rate**, G:

$$G = \frac{dE_P}{dA} \qquad (9.16)$$

which is a metric of the available internal energy that can be supplied to produce a new crack surface. One then rewrites Equation (9.15) as:

$$G = \frac{dW_S}{dA}. \qquad (9.17)$$

Thus, crack equilibrium holds that any extension balances the energy release rate against energy consumed in producing new crack surface (in the purely elastic case, only surface energy). It is seen in Section 9.5.5 that if the material has a constant dissipated energy for any amount of crack growth, the equilibrium of Equation (9.17) is unstable and so yields the condition for unbounded growth and thus catastrophic fracture. It is noteworthy that the energy release rate in Equation (9.16) takes the form of the negative of the derivative of a potential, which is the physical definition of a force. Thus, we say that the energy release rate **G** is a crack growth **driving force**.

9.3.5 Crack growth and resistance

The energetic treatment above revealed that a certain value of crack driving force must be overcome to extend a crack under equilibrium conditions. The equilibrium condition in Equation (9.17) can be rewritten as

$$G = R \qquad (9.18)$$

where the material resistance to crack growth is denoted as R. However, this only describes the condition at incipient crack growth without determining whether the process is stable or unstable. If we take another derivative of the equation with respect to crack length, we can evaluate the stability of crack equilibrium. If the crack

9.3 Linear elastic fracture mechanics (LEFM)

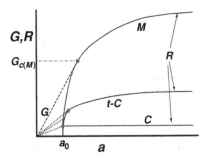

Figure 9.5

R curves (R) for a typical metal (M), ceramic (C), and toughening ceramic (t-C), with a linear G shown for increasing applied loads. When the G curve is tangent to the R curve, fracture occurs (star marker) and that value of G is the fracture toughness G_c. The ceramic displays constant crack resistance, and no stable crack extension prior to fracture. The toughening ceramic shows both higher fracture toughness than the untoughened ceramic and some stable crack propagation. The metal shows a much higher fracture toughness and extent of stable crack propagation than either ceramic.

growth resistance increases at a greater rate than the driving force, we have a stable equilibrium

$$\frac{dG}{da} \leq \frac{dR}{da} \qquad (9.19)$$

and otherwise the driving force exceeds the growth resistance and crack growth is unbounded.

The energetic description of crack growth resistance and stability provides another view of how fracture differs from material strength concepts. It has been shown that increasing the size of components may not substantially affect the risk of fracture in the same way that it might prevent yielding. Examining Equation (9.15), one sees that the crack driving force comes from the stored internal energy in the body that comes from the global loading. If the stored energy in the body is sufficient to start a crack propagating, so long as it continues to exceed the crack growth resistance the crack will continue to propagate, driven solely by the stored potential energy in the body. Put another way, the energy dissipated by crack growth comes from stored energy or boundary work that has already been provided prior to the onset of propagation, and the remaining potential energy in the system may exceed that required for catastrophic fracture. In this way, large bodies with small cracks can supply all the energy to sustain crack growth to failure, in a way that is not the least bit improved by their extent, even if they are stressed far below the ultimate strength.

Materials that demonstrate an increasing resistance to crack extension after the onset of crack growth may demonstrate a finite amount of stable crack propagation prior to catastrophic fracture. This phenomenon is captured with crack growth resistance curves (also known as R curves). Figure 9.5 shows the R curve for a material with a constant crack growth resistance, and another that demonstrates increasing resistance.

It can be seen in Figure 9.5 that a material with a constant crack growth resistance immediately becomes unstable when $G = R$, if G is an increasing function of a. For another material with the same load applied but an increasing R curve, we see that crack growth will arrest after a stable extension of the crack. After some increasing of the load, the instability condition is met and the crack is driven to unbounded growth. The flat R curve reflects the perfectly brittle fracture assumed to occur by Griffith, where the energy consumed is derived from the surface energy of the material, and is therefore a material constant. Increasing R curves may reflect either intrinsic or extrinsic toughening mechanisms (e.g., plastic hardening, crack closing, crack tip bridging). The instability condition shown in Figure 9.5 corresponds to a critical value of the energy release rate, denoted G_c. It is apparent that G_c may or may not be preceded by an amount of stable crack growth. However, the value of G_c may vary for a given increasing R curve, depending on the functional form of $G(a)$, which is dependent on the geometry and boundary conditions of the loaded body. Thus, the R curve is a material property, but G_c is in general not, though it should be noted that the R curve shape is sensitive to stress triaxiality (i.e., plane stress vs. plane strain). Other practical scenarios resulting in varying crack tip stress triaxiality include short (initiated) cracks and the differences between test specimen configurations.

9.3.6 Superposition of *G* and relationship to *K*

The energy release rate describes an energetic condition due to the presence of a crack, and such a quantity is inherently additive across multiple or mixed-mode loading conditions. Individual modes of K are not additive, which makes the energetic description of fracture seem attractive for analyzing cracks under complicated load conditions. However, G and K can be directly linked, and thus the energetic contribution from each mode of K in fact can be directly summed and then used to evaluate fracture risk. Consider the energy release rate for an unbounded center-cracked plate,

$$G = \frac{\pi \sigma^2 a}{E'} \tag{9.20}$$

where $E' = E$ for plane stress and $E' = E/(1 - \nu^2)$ for plane strain. For an infinite plate one can combine Equation (9.20) with Equation (9.8) ($Y = 1$),

$$G = \frac{K_I^2}{E'}. \tag{9.21}$$

Equation (9.21) holds in general for other body configurations loaded in Mode I (Irwin, 1948), and can be shown to be analogously true for other modes as well. Thus, according to the additive property of G we can write

$$G = \frac{K_I^2}{E'} + \frac{K_{II}^2}{E'} + \frac{K_{III}^2}{2\mu} \tag{9.22}$$

9.3 Linear elastic fracture mechanics (LEFM)

which provides a clear link between independent modes of K and an overall metric of crack driving force under multiaxial states of stress.

9.3.7 K_c and G_c as intrinsic failure criteria

When LEFM is applicable, both K and G completely characterize the local conditions near a crack tip, and therefore both are sufficient indicators of the critical condition that results in unstable crack growth. The critical value of K is called the **fracture toughness**, K_{Ic}, which is a material property and is not dependent on the global geometry due to its localized nature. However, K is dependent on the stress triaxiality at the crack tip and is sensitive to the conditions of plane stress or plane strain. This variation of K_c as a function of constraint is discussed in Section 9.6.2 as a modified method in LEFM. It was noted above that the measured value of G_c can vary somewhat according to the test method, but for relatively limited crack tip plasticity this should be negligible.

The critical value of K_I is typically much greater than that for K_{II} or K_{III}, in that fracture under Modes II and III is uncommon, and under mixed-mode loading crack growth tends to be dominated by Mode I behavior. Thus, mixed or non-tensile modes of loading may be important in specific applications or for incipient crack growth, but it is Mode I that is typically the most critical and that therefore is used to describe the fracture failure condition.

A critical sized flaw can be defined relative to the fracture toughness, and the critical crack length resulting in fracture for an applied global stress σ^∞ is obtained by rewriting Equation (9.10), as

$$a_C = \frac{1}{\pi}\left(\frac{K_{IC}}{Y\sigma_\infty}\right)^2. \tag{9.23}$$

This value is of utmost importance in medical devices that may contain flaws, but for which failure cannot be tolerated, such as heart valves comprising pyrolytic carbon or acetabular cups made with zirconia or alumina. In this case components can be non-destructively inspected and those with initial cracks near this size (with an appropriate margin of safety) are then discarded. Alternatively, small flaws may grow in fatigue until they reach a_c; and so a_c provides a failure condition for determining the fatigue life of flawed components. This approach for fatigue life estimation is called the defect tolerant method and is detailed in the next chapter.

Example 9.3 **The effect of geometry of the critical flaw size**
Consider the rectangular and circular hip implant designs presented in Example 6.8. If both of these designs utilize a titanium alloy with a fracture toughness of 25 MPa√m,

then what is the critical flaw size in each of these cases? Assume that the flaw will behave like an edge-notched crack whose stress intensity factor is defined as:

$$K_I = 1.12\sigma^\infty \sqrt{\pi a} \Rightarrow \Delta K_I = 1.12\Delta\sigma^\infty \sqrt{\pi a}$$

The stresses in both designs were calculated in Example 5.8 using beam theory:

$$\sigma_{rect} = 365\,\text{MPa}$$
$$\sigma_{circ} = 509\,\text{MPa}$$

Solution The critical flaw size can be determined by rearranging the expression for the stress intensity above:

$$K_I = 1.12\sigma^\infty \sqrt{\pi a} \Rightarrow \frac{K_{IC}}{1.12\sigma^\infty \sqrt{\pi}} = \sqrt{a_C} \Rightarrow$$

$$a_C = \frac{K_{IC}^2}{(1.12\sigma^\infty)^2 \pi}$$

$$a_{C,circ} = \frac{(25)^2}{(1.12(509))^2 \pi} = 0.000612\,\text{m} = 612\,\mu\text{m}$$

$$a_{C,rect} = \frac{(509)^2}{(365)^2} \cdot a_{C,circ} = 2.44 a_{C,circ} = 1492.5\,\mu\text{m}$$

An interesting finding is that the ratio of maximum stresses is:

$$\frac{\sigma_{rect}}{\sigma_{circ}} = \frac{365\,\text{MPa}}{509\,\text{MPa}} = 0.717$$

while the ratio of the critical flaw size is inversely proportional to the ratio of the square of the stresses:

$$\frac{a_{C,rect}}{a_{C,circ}} = \frac{(509)^2}{(365)^2} = 1.944$$

The stress in the rectangular cross-section is about 72% of the stress in the circular cross-section; for the same materials with the same type of embedded flaw, the critical flaw size in the rectangular section is nearly double that of the circular cross-section.

9.3.8 Summary and limitations of LEFM

Linear elastic fracture mechanics provides a mathematically compelling explanation for how sharp flaws in bodies can produce locally extreme stresses under load, and result in catastrophic failure of components loaded far below their material strength. Crack growth may be stable or unstable, depending on whether the material toughness increases with crack length faster than the crack driving force, or not (respectively). The parameters K and G describe the conditions near the crack tip where the local effects dominate the global behavior of the body. An essential assumption in this analysis was that the local breakdown of material at the crack tip does not substantially affect the local stress field and that the energy required to extend a crack in an elastic material is dominated by work required to sever bonds near the crack tip (and not due to other modes of energy dissipation). Both slight and extensive alterations to the LEFM approach to fracture mechanics have been developed, depending on the extent to which a material diverges from this idealized behavior. These fall under the classifications of modified methods in LEFM (Section 9.4), elastic-plastic fracture mechanics (Section 9.5), or time-dependent fracture mechanics (Section 9.6).

9.4 Modified methods in LEFM

The basic formulation of LEFM is an important building block for the understanding of crack behavior in engineering solids, but is only an accurate description of situations where there is no inelastic contribution to the overall energy required for crack propagation. Fortunately, LEFM works very well in many situations where inelasticity is present but limited. With some modest modifications to the theory, accurate analyses can be conducted for many problems of interest in biomedical devices.

9.4.1 Small-scale plasticity

In LEFM it is assumed that the extent of material breakdown near the crack tip was insignificant compared to the region dominated by the stress singularity (Figure 9.2). All materials will deform inelastically in this region, and in materials that can deform plastically, as do metals, this inelastic region may in fact engulf the singularity zone and nullify the validity of LEFM. In situations where the inelastic region (called the **plastic zone** hereafter) is limited in extent, LEFM is still valid, and we denote this condition **small-scale plasticity** (also called small-scale yielding).

In small-scale plasticity, even though the plastic zone is small, its existence nevertheless alters the distribution of stress in the singularity. The plastic zone is depicted in Figure 9.6. If the plastic zone can only support stress at the yield strength in plane stress, σ_Y (or $3\sigma_Y$ in plane strain due to a reduced deviatoric stress), we can estimate its size, r_y, by appealing to Equation (9.10),

$$r_{y-P\sigma} = \frac{1}{2\pi}\left(\frac{K_I}{\sigma_Y}\right)^2 \text{ for plane stress} \quad (9.24)$$

$$r_{y-P\varepsilon} = \frac{1}{6\pi}\left(\frac{K_I}{\sigma_Y}\right)^2 \text{ for plane strain.} \quad (9.25)$$

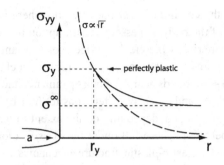

Figure 9.6

The first plastic zone size approximation, also called the Irwin Plastic Zone Size r_y, assuming a perfectly plastic material yielding near the crack tip.

This estimate is in error, however, as Equation (9.10) assumed a purely elastic material, and so the presumed stress within r_y that is in excess of the yield strength in Equation (9.8) must be borne by adjacent material. This results in additional material outside of r_y yielding, and so a correction to the plastic zone can be estimated; assuming a perfectly plastic (non-hardening) material,

$$r_{p-P\sigma} = \frac{1}{\pi}\left(\frac{K_I}{\sigma_Y}\right)^2 \text{ for plane stress} \qquad (9.26)$$

$$r_{p-P\varepsilon} = \frac{1}{\pi}\left(\frac{K_I}{\sigma_Y}\right)^2 \text{ for plane strain.} \qquad (9.27)$$

The shift of stress from r_y to cause yielding out to r_p is depicted in Figure 9.7.

It can be seen in Figure 9.7 that the singularity stress solution of Equation (9.10) has shifted away from the true crack tip by a distance of r_y but is otherwise unchanged past r_p. Thus, the correction of the plastic zone size results in an effective increase in crack length,

$$a_{\text{eff}} = a + r_y. \qquad (9.28)$$

It is noted that a_{eff} should be used in place of a in analyses when r_y is a substantial fraction of a_{eff} (i.e., non-negligible). It should be noted also that recalculating K based on the effective crack length then produces a slightly increased value for r_y, and so iteration to a convergent value is required.

9.4.2 K_c as a function of plastic constraint

Small-scale plasticity is valid when the plastic zone is small with respect to the singularity zone size, and it must also be small with respect to the dimensions of the body or it

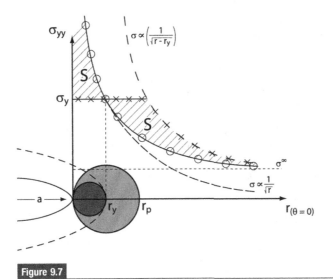

Figure 9.7

To achieve equilibrium, the total stress S truncated in Figure 9.6 above the yield strength must be borne by the remaining material ahead of the plastic zone. Thus S is pushed forward, where more material in turn yields and the Irwin Corrected Plastic Zone Size r_p is obtained. Also shown is the new "effective" crack tip centered now at r_y from the true crack tip, so called because the LEFM singular stress distribution takes its origin at that location ($\sigma \sim 1/(r - r_y)^{1/2}$).

further alters the conditions at the crack tip. For instance, as the thickness of a cracked plate approaches the plastic zone size and therefore plane stress conditions, the plastic zone is three times larger than for a thick plate in plane strain, and the stress at the crack tip is biaxial. In thick specimens, the free surfaces are effectively in a state of plane stress, and so the plastic zone size varies through the thickness as is shown in Figure 9.8.

The relaxation of stress triaxiality from the plane strain condition commonly results in a suppression of crack growth phenomena and an apparent improvement in fracture toughness, K_c. Figure 9.9 shows how K_c drops from a peak value in plane stress to an asymptote in plane strain, where the contributions of the plane stress zones on the surface of the body are negligible.

The asymptotic value of K_c in plane strain is denoted as K_{Ic} and is also known as the **plane strain fracture toughness**. K_{Ic} is the intrinsic measure of the material fracture toughness due to its invariance for a variety of body configurations, and the fact that it is a conservative limit for the range of potential values of K_c. This, however, should not be interpreted as challenging the validity of K under plane stress conditions. Rather, the mechanisms locally governing the critical fracture condition typically depend directly on stress triaxiality (as mentioned for the R curve above), and therefore so must the fracture toughness K_c.

In order to obtain a credible measure of K_{Ic}, the dimensions of a test specimen (e.g., Figure 9.3) must be substantially greater in every extent than the plastic zone size. ASTM E1820 requires that the specimen thickness (B), crack length (a), and ligament

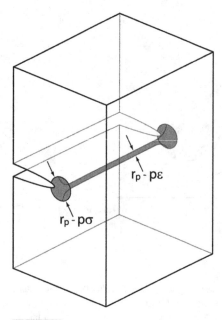

Figure 9.8

In a thick specimen, the interior of the body is in plane strain, but the exterior surfaces are effectively in plane stress. The plastic zone size varies according to Equations (9.26)–(9.27).

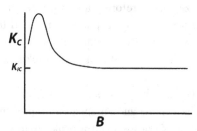

Figure 9.9

Variations in K_c with specimen thickness B. The value of K_c plateaus at large specimen thicknesses (where the effects of the plane stress regions seen in Figure 9.8 are negligible), and this is denoted K_{IC}.

ahead of the crack tip (b) all exceed approximately 25 times the corrected plane strain plastic zone size:

$$[B, a, W] \geq 2.5 \left(\frac{K_I}{\sigma_Y}\right)^2. \tag{9.29}$$

If this condition is met in a fracture test, then a legitimate measurement of K_{Ic} can be made. It is worth noting that this condition can be unreasonable to meet for tough materials, even if they meet the other conditions for LEFM validity (Example 9.1).

Example 9.4 **Specimen configuration requirements for four biomaterials**
Given the specimen configuration requirements in Equation (9.29), determine the minimum specimen dimension for a fracture toughness test for each of the following materials:

a. High-strength Ti alloy: $\sigma_Y = 1060$ MPa, $K_{IC} = 73$ MPa$\sqrt{m}$
b. High-toughness CoCr alloy: $\sigma_Y = 470$ MPa, $K_{IC} = 208$ MPa$\sqrt{m}$
c. Highly crosslinked UHMWPE: $\sigma_Y = 22$ MPa, $K_{IC} = 2.5$ MPa$\sqrt{m}$
d. Stabilized zirconia: $\sigma_Y = 1000$ MPa, $K_{IC} = 7$ MPa$\sqrt{m}$

Appealing to Equation (9.29), the minimum specimen dimension for each is (a) 9.9 mm, (b) 490 mm, (c) 32.2 mm, and (d) 0.122 mm. The titanium specimen width is reasonable. However, the necessary specimen width for the CoCr alloy is excessive, as material billets in this size may be difficult to obtain, and force necessary to conduct the test would be extremely high. For such a material, elastic-plastic fracture mechanics methodologies may be required, as most components would likely become fully plastic prior to fracture. The dimensional requirements for the ceramic clearly are automatically satisfied as it is a highly brittle material. The required thickness for the UHMWPE is somewhat higher than is typically employed in many fatigue or fracture tests, and in this case LEFM is also perhaps not the best choice for critical analysis of crack phenomena.

9.5 Elastic-plastic fracture mechanics (EPFM)

For materials capable of extensive deformation prior to failure, the inelastic zone may completely engulf the singularity zone derived in LEFM, and in such situations even modifications to the theory are no longer reliable. The mathematics of LEFM lend themselves to closed-form solutions, but this is no longer the case when nonlinearities in the material behavior are present. Nevertheless, fracture behavior comprising both an elastic and inelastic component is still analytically tractable in many cases of engineering interest, and the literature on the subject is evolving, particularly in the field of natural materials and tissues.

In this section it will be shown that plastic blunting of the crack tip does not contradict the concept of cracks as extreme concentrators, and that the resulting stress singularity with moderate levels of plasticity can be predicted with realistic assumptions of material behavior. The elastic-plastic generalization of the crack driving force (or energy release rate) is also developed and reconciled with the elastic case. While many analyses in EPFM must be approximate or numerical, the concepts of fracture mechanics motivated by LEFM become stronger through generalization to the inelastic case, and therefore the foundation is laid for other developments in fracture that account for other material nonlinearities (for example, viscous time-dependence of properties in polymers). Due to the inelastic nature of many biomaterials, EPFM is therefore of great interest in the development of biomedical devices.

9.5.1 Crack tip plasticity and the crack opening displacement, δ

In Section 9.4 it was shown that, even in the realm of small-scale plasticity, the size of the plastic zone affected the distribution of stresses ahead of the crack tip and also impacted the measured fracture toughness in a test. Modified LEFM methods still inherit the assumption that the crack tip is sharp (e.g., Equation 9.4), an assumption that was not addressed when admitting crack tip plasticity. In modified LEFM we are concerned with an effective crack tip, which is then assumed to remain "sharp." The actual physical crack tip is a distance behind the effective crack tip equal to the plastic zone size r_y (Figure 9.7).

As shown in Figure 9.7, it is observed that the physical crack tip may indeed be quite blunted, while the effective crack tip is embedded in the center of the plastic zone. Since the crack is assumed to be closed when unloaded, the displacement of the crack faces at the physical crack tip can be computed from LEFM for the effective crack tip. The displacement of the crack faces at the physical crack tip is called the crack opening displacement (COD) or δ, and has been widely used as a characterizing parameter analogous to G and K for conditions with substantial local plasticity. The functional form of the crack face vertical displacement with distance r behind the effective crack tip (Figure 9.7) is given as

$$u_y = \frac{\kappa+1}{2\mu} K_I \sqrt{\frac{r}{2\pi}} \qquad (9.30)$$

where κ is the bulk (volumetric) modulus and μ is the shear modulus. The crack opening displacement is twice the value of u_y (Equation 9.25), a distance r_y behind the effective crack tip; so, substituting Equation (9.20) (for plane stress) yields

$$\delta = 2u_y = \frac{4}{\pi} \frac{K_I^2}{\sigma_Y E}. \qquad (9.31)$$

The crack opening displacement can also be directly related to G (Equation 9.21):

$$\delta = \alpha \frac{G}{\sigma_Y} \qquad (9.32)$$

where according to Equation 9.31 $\alpha = 4/\pi$, yet according to an energy balance it may be preferable to use $\alpha = 1$ for plane stress and $\alpha = 1/2$ for plane strain in Equation (9.32) (Kanninen and Poplar, 1985). Thus, δ corresponds to K and G, and therefore can serve as a single-parameter determinant of a critical crack opening displacement, δ_{crit}, corresponding precisely with K_c and G_c in LEFM.

COD methods were developed and popularized by Wells (Wells, 1963) as an alternative approach to modified LEFM. It was postulated that δ would continue to be a

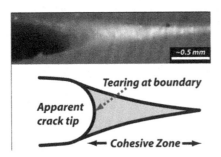

Figure 9.10

A wedge-shaped cohesive zone for a propagating crack in UHMWPE, where the finite crack tip radius is evident. Note that the zone is over 2 mm long, which violates small-scale plasticity assumptions (see Example 9.1). Material failure occurs at the apparent crack tip, where new crack surface is created.

valid parameter even after extensive plasticity violated even modified LEFM analyses. In fact, it was hoped that δ would remain valid up to and beyond the point where the entire body was deforming plastically. In situations where large-scale plastic collapse of a body is potentially competing with fully plastic crack propagation, the two may be considered simultaneously, and this is the basis of two-parameter failure assessment diagrams, such as the $R - \delta$ curve (Kanninen and Poplar, 1985). The δ concept is indeed valid even under large-scale plasticity, but mainly through a specialized treatment of the theory. Nevertheless, δ remains popular and effective as it is simple to experimentally quantify and meaningful for a wide range of conditions of interest to engineers.

9.5.2 Strip-yield crack tip cohesive zones

The discussion of plastic zones so far has been limited either to a small hypothetical material breakdown region or to a rounded zone extending beyond the crack tip. A model of crack tip inelasticity confined to a wedge-shaped zone was proposed separately by Dugdale (for plane stress steel sheets) and Barenblatt (for material breakdown in brittle materials) (Barenblatt, 1962; Dugdale, 1960). The presumption of a wedge-shaped plastic zone crack tip zone matches experimental observations in many cases (Figure 9.10), and also allows detailed insights into how the behavior of the failing material near a crack tip affects the fracture process. Such models are also used extensively to describe time-dependent crack propagation in polymers (Section 9.8).

Dugdale studied the Mode I strip-yield zones that formed in slits in thin steel sheets in plane stress under conditions ranging from small-scale yielding to global yielding (Dugdale, 1960). The novel idea posed by Dugdale held that the plastic zone ahead of the apparent crack tip can be modeled as a traction pushing the adjacent boundaries of the plastic zone together with a stress equivalent to the yield strength of the material. This phenomenon is illustrated in Figure 9.11.

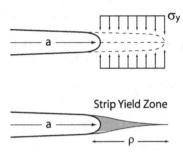

Figure 9.11

Strip-yield zone diagram, showing a constant cohesive stress (the yield strength) distributed ahead of the true crack tip incident on a virtual crack surface. The resulting plastic zone ahead of the crack tip (of length ρ) takes the cusp-shaped form of a free surface crack with an applied (negative) pressure.

Dugdale then postulated that the stress at the effective crack tip (at the tip of the wedge) due to the crack closing traction nullifies the stress singularity predicted from LEFM. Put another way, for an initial crack under load, the plastic zone would grow until the singular stress (and therefore the crack driving force) was extinguished. The process zone size required for this condition to be met can be calculated by linear superposition of the known K solution for a constant pressure applied to the crack tip balancing the K contribution from the global stress, and it then can be shown the length of the Dugdale plastic zone is

$$r_d = 2a \sin^2\left(\frac{\pi}{4}\frac{\sigma}{\sigma_Y}\right) \tag{9.33}$$

which for small-scale yielding reduces to:

$$r_d \approx \frac{\pi}{8}\left(\frac{K_I}{\sigma_Y}\right)^2. \tag{9.34}$$

The Dugdale plastic zone size is similar in form to the Irwin corrected plane stress plastic zone size, $r_{p-P\sigma}$ (Equation 9.26). As with the modified LEFM methods, an effective K can be computed for the effective crack tip at the end of the plastic zone (Burdekin and Stone, 1966) and is written as

$$K_{\textit{eff}} = \sigma_Y\sqrt{\pi a}\left\{\frac{8}{\pi^2}\ln\left[\sec\left(\frac{\pi\sigma}{2\sigma_Y}\right)\right]\right\}^{1/2}. \tag{9.35}$$

Dugdale experimentally found that Equation (9.33) accurately predicted the plastic zone length in mild steel sheets for applied global stress from initial plasticity to just below global yield, in both central and edge cracked sheets. The applicability of this approach from small- to large-scale plasticity matches the expectations for δ, and we will see in the next section that the Dugdale strip-yield model provides a critical link between δ and energetic methods within elastic-plastic fracture mechanics.

9.5 Elastic-plastic fracture mechanics (EPFM)

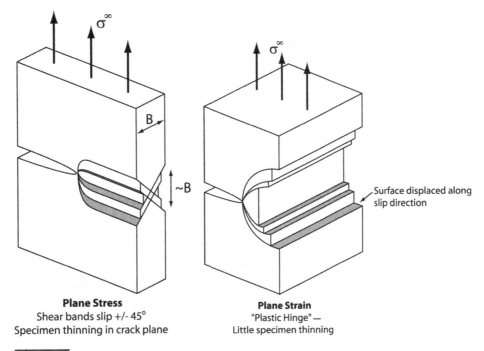

Figure 9.12

Plastic constraint in plane stress (left) and plane strain (right). The deformation in plane stress is in the principal shear stress direction within the sheet ($+/- 45°$ to the crack plane), and is constrained therefore to a region approximately the thickness of the sheet. The plane strain shear deformation occurs in the ligament direction and is independent of specimen thickness. After (Hahn and Rosenfield, 1965).

It should be noted that the strip-yield shape of the plastic zone in Dugdale's experiment is an outcome of the limited extent of plastic deformation permitted in a thin sheet and the limited plastic hardening exhibited by the mild steel he studied (Hahn and Rosenfield, 1965). Strip-yield zones, however, are observed in a variety of materials (Figure 9.10 above), and the Dugdale analysis nevertheless provides substantial insight into crack behavior under moderate and severe crack tip plasticity, regardless of the limitations of its origins. Additionally, the constraint of the plastic zone into a strip-yield wedge under plane stress conditions provides additional insight into how the near-tip stresses can depend on specimen thickness through alteration of the local deformation mode. Figure 9.12 illustrates the plastic constraint behavior in plane stress (left) and plane strain (right) condition.

Barenblatt independently studied strip-yield models (Barenblatt, 1962), chiefly to understand how brittle materials with limited local ductility accommodate the stress singularity at the crack tip. It was his notion that there is some cohesive stress (such as the atomic bond strength) that holds the material together adjacent to the crack, and this cohesive stress exerts a closing pressure on the effective crack faces in the same way that

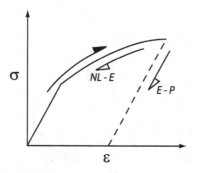

Figure 9.13

Deformation plasticity (P) and the equivalent nonlinear (NL) elasticity (E) response during the loading portion of the curve, also showing the difference in their unloading responses. Deformation plasticity EPFM ignores the unloading behavior.

the flow stress did in Dugdale's model. The advantage of Barenblatt's approach is that it does not strictly assume any particular mechanism of material cohesion, which leaves it open to a wide range of deformation phenomena in the cohesive zone. Strip-yield models are often used to analyze the fracture behavior of polymers, where the behavior in the cohesive zone may be nonlinear and time-dependent.

9.5.3 Nonlinear material behavior and deformation plasticity

Linear elastic fracture mechanics is based fundamentally on stress and strain field solutions appealing to the theory of linear elasticity, which when modified to account for limited amounts of inelastic deformation provides a powerful toolset for the analysis of many engineering problems. However, a number of structures of components may be loaded globally in a relatively extreme manner, resulting in conditions that violate LEFM. Recall, Example 9.1 shows that the plastic zone in high toughness materials with a low yield strength may also be so large as to violate the small-scale yielding assumption and thus even modified LEFM methods are insufficient. Elastic-plastic fracture mechanics admits inelastic behavior both within the singularity zone and in the body, providing a means to overcome the limitations inherent in LEFM for describing inelastic deformation states.

A prominent fracture researcher by the name of Rice first considered the general case of nonlinear elasticity as a special application for analyzing plasticity in fracture mechanics (Rice and Rosengren, 1968). The insight, shown in Figure 9.13, is that a nonlinear elastic material under increasing load is indistinguishable from an elastic-plastic material with the same stress-strain behavior, so long as the load is not relaxed. The theory of plasticity that determines a single stress-strain relationship for a material is called **deformation plasticity**.

In many cases this simplification of elastic-plastic behavior is a good approximation to the actual three-dimensional deformation in a component. With deformation plasticity

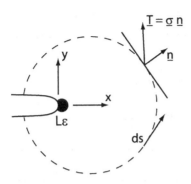

Figure 9.14

Integral contour path for the *J*-integral around the crack tip.

as the beginning point, one can revisit the scheme developed in LEFM and produce both analytical singular stress methods (utilizing **HRR** singularities) and energetic methods (through the ***J*-integral**) for fracture in elastic-plastic materials.

9.5.4 The *J*-integral, the nonlinear energy release rate

Rice developed a contour integral that, when taken around a crack tip, was path independent and therefore a unique indicator of the crack tip conditions (Rice and Rosengren, 1968). The *J* contour integral, also called the *J-integral* or just *J*, is given as

$$J = \oint_\Gamma \left(w dy - T_i \frac{\partial u_i}{\partial x} ds \right). \tag{9.36}$$

The *J* contour integral is illustrated in Figure 9.14 on a closed curve Γ with arc length ds, where w is the strain energy density

$$w = \int_0^{\varepsilon_{ij}} \sigma_{ij} d\varepsilon_{ij}, \tag{9.37}$$

T_i are the components of the traction vector with the stress incident on a plane with unit normal vector *n* that is normal to Γ, i.e.,

$$T_i = \sigma_{ij} n_j \tag{9.38}$$

and u_i are the corresponding components of the displacement.

Rice showed that *J* is path independent, meaning it has a single value for all closed paths in the body. In a critically important interpretation, Rice also found that *J* is the energy release rate (i.e., a nonlinear *G*) when Γ contains the crack tip, which is demonstrated next.

If one considers a planar body, bounded by the curve Ω, with area A', the total potential energy in the body is the strain energy and the boundary work, where the boundary traction is defined on a portion of the boundary, Γ',

$$E_{PE} = \int_{A'} w\, dA - \int_{\Gamma'} T_i u_i\, ds. \tag{9.39}$$

Then consider a change in the potential energy due to an increment of crack growth,

$$\frac{dE_{PE}}{da} = \int_{A'} \frac{dw}{da} dA - \int_{\Omega} T_i \frac{du_i}{da} ds \tag{9.40}$$

where we can integrate over the whole boundary contour. Now, considering crack growth in the $+x$ direction (moving the origin with the crack tip), one has $\partial x / \partial a = -1$, such that

$$\frac{d}{da} = \frac{\partial x}{\partial a} \frac{\partial}{\partial x} - \frac{\partial}{\partial a} - \frac{\partial}{\partial x}. \tag{9.41}$$

Continuing on, one can rewrite Equation (9.40) as

$$\frac{dE_{PE}}{da} = \int_{A'} \left(\frac{\partial w}{\partial a} - \frac{\partial w}{\partial x} \right) dA - \int_{\Omega} T_i \left(\frac{\partial u_i}{\partial a} - \frac{\partial u_i}{\partial x} \right) ds. \tag{9.42}$$

It can be shown that

$$\int_{A'} \left(\frac{\partial w}{\partial a} \right) dA = \int_{\Omega} T_i \left(\frac{\partial u_i}{\partial a} \right) ds \tag{9.43}$$

and giving

$$\frac{dE_{PE}}{da} = -\int_{A'} \frac{\partial w}{\partial x} dA + \int_{\Omega} T_i \frac{\partial u_i}{\partial x} ds. \tag{9.44}$$

Employing the divergence theorem gives

$$-\frac{dE_{PE}}{da} = \int_{\Omega} \left(w n_x - T_i \frac{\partial u_i}{\partial x} \right) ds \tag{9.45}$$

and since $n_x ds = dy$, this yields

$$-\frac{dE_{PE}}{da} = \int_{\Omega} \left(w\, dy - T_i \frac{\partial u_i}{\partial x} \right) ds. \tag{9.46}$$

Recalling Equation (9.32), it is thus shown that the J-integral is the nonlinear energy release rate,

$$-\frac{dE_{PE}}{da} = J. \tag{9.47}$$

It is worth noting in passing that the J contour integral is relatively straightforward to compute numerically in finite element analyses, provided some technical details are satisfied (the value must be ensured to not be substantially contour- or mesh-dependent, for instance, and this can be less straightforward in three dimensions). Many commercial modeling packages can compute the *J*-integral at blunt or sharp crack tips for a multitude of material behaviors.

The result in Equation (9.47) then implies that for small-scale plasticity one can equate $J = G$ and $J = K^2/E'$. The real utility of *J*, however, is in its applicability beyond small-scale yielding conditions. If *J* is computed for the Dugdale strip-yield case (see Figure 9.11) integrating around the wedge, for $\delta \ll r_d$, one obtains

$$J = \int_0^\delta \sigma_{yy(\delta)} d\delta. \tag{9.48}$$

For $\sigma_{yy} = \sigma_Y$ (yield strength), there exists a simple relation between *J* and δ for the Dugdale zone case

$$J = m\sigma_Y \delta \tag{9.49}$$

with $m = 1$ for plane stress (recall Equation 9.32) which is valid beyond the limits of LEFM, as one might expect. Equations (9.44) and (9.46) fully connect the *J-integral* and EPFM with small-scale plasticity LEFM, showing *J* to be a generally applicable and practical parameter. Next, the stress fields in EPFM are examined and it is shown that *J* uniquely characterizes the stress singularity under plastic conditions.

9.5.5 Elastic-plastic crack tip singularity – the HRR field

The crack tip stresses in a hardening elastic-plastic material were developed concurrently by three researchers working in the fracture mechanics field: Hutchinson (Hutchinson, 1968) and Rice and Rosengren (Rice and Rosengren, 1968). These researchers employed deformation plasticity to develop the near-tip stress field solutions and this formulation is named the **HRR field** after their work. Considering a power-law hardening material, according to the Ramberg-Osgood equation for elastic-plastic deformation,

$$\frac{\varepsilon}{\varepsilon_0} = \frac{\sigma}{\sigma_0} + \alpha \left(\frac{\sigma}{\sigma_0}\right)^n \tag{9.50}$$

where the strain is relative to a reference strain ε_0, which occurs at a chosen reference stress σ_0 (typically σ_Y); α is a fit coefficient and **n** is the plastic strain-hardening exponent. Note that this equation is not valid upon any decrease of an applied stress. Linear elastic strains are negligible compared to plastic strains in the stress singularity zone, so power-law plasticity can be considered near the crack tip,

$$\frac{\varepsilon_p}{\varepsilon_0} = \alpha \left(\frac{\sigma}{\sigma_0}\right)^n. \tag{9.51}$$

Hutchinson showed that *J* could only be path independent in such a material if the strain energy density *w* (Equations 9.33–9.34) was singular near the crack tip with a $1/r$

dependence (Hutchinson, 1968). Armed with this finding, he showed that the stresses and strains near the crack tip in a power-law deformation plasticity material are

$$\sigma_{ij} = C_1 \left(\frac{J}{r}\right)^{1/(n+1)} \tilde{\sigma}_{ij(n,\theta)} \qquad (9.52)$$

$$\varepsilon_{ij} = C_2 \left(\frac{J}{r}\right)^{1/(n+1)} \tilde{\varepsilon}_{ij(n,\theta)} \qquad (9.53)$$

where $\tilde{\sigma}_{ij(n,\theta)}$ and $\tilde{\varepsilon}_{ij(n,\theta)}$ are dimensionless angular distribution functions.

The coefficients in Equations (9.33)–(9.34) are given by

$$C_1 = \sigma_0 \left(\frac{E}{\alpha \sigma_0 I_n}\right)^{1/(n+1)} \qquad (9.54)$$

$$C_2 = \frac{\alpha \sigma_0}{E} \left(\frac{E}{\alpha \sigma_0 I_n}\right)^{1/(n+1)} \qquad (9.55)$$

and I_n is a scalar that depends on n, varying from approximately $4 < I_n < 6$ for plane strain and $3 < I_n < 4$ in plane stress, for $2 < n < 13$.

The strength of the HRR stress singularity (Equation 9.52) is dependent on n, varying from $1/r^{1/2}$ for $n = 1$ (i.e., linear plasticity) to r^0 (non-singular) for perfectly plastic materials. Thus, since $n > 1$, there exists a zone of HRR singularity that is applicable in the plastic zone that exists in small-scale yielding (recall that Irwin assumed r^0 behavior in the plastic zone, which matches the assumption of Dugdale of perfect plasticity). We can see how the regions of J and K dominance nest in small-scale plasticity (Figure 9.15), where there still exists a region near the crack tip of extreme stress and material breakdown, where deformation plasticity is not applicable.

For more extensive plasticity, LEFM no longer applies, as there is no region described uniquely by K. However, EPFM is still valid for moderate yielding, and the HRR field solution, J, and δ characterize the crack tip (Figure 9.13). It is possible that for even more extensive plasticity there is no J-dominated region, such that there is no simple intrinsic determinant of the crack tip state.

A more general form of the relationship between J and δ, based on HRR conditions, was developed by Shih (Shih, 1981). He established a form similar to Equation (9.49):

$$\delta = d_n \frac{J}{\sigma_0} \qquad (9.56)$$

where d_n is a scalar function of n, which (for $\alpha = 1$) in plane stress varies monotonically from $1 > d_n > 0.1$ for $0 < (1/n) < 0.5$ (respectively), and for plane strain $0.8 > d_n > 0.1$ for $0 < (1/n) < 0.5$ (Shih, 1981). For other values of α, multiply d_n by $\alpha^{1/n}$. The HRR fields for plane stress and strain conditions are shown in Figure 9.16. Thus, Equation (9.56) agrees with the small and strip-yield plasticity relationships between J and δ (i.e.,

9.5 Elastic-plastic fracture mechanics (EPFM)

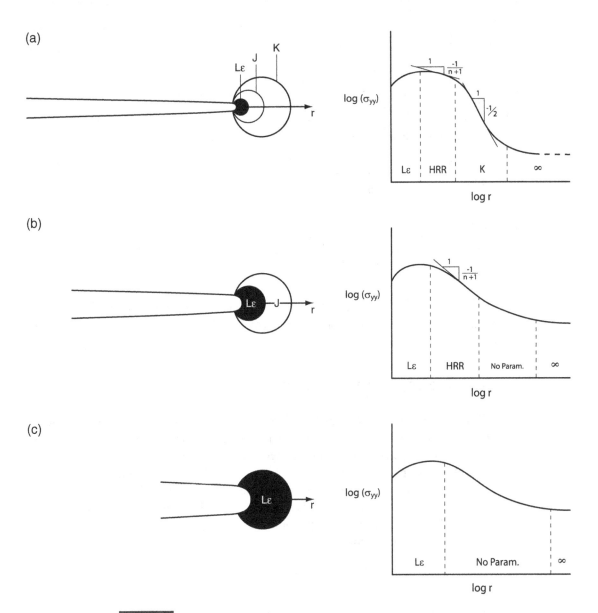

Figure 9.15

Increasingly severe plastic deformation and regions of applicability of various analyses. (a) In small-scale plasticity, there is an LEFM zone outside the plastic zone, an HRR zone where J is applicable, and a large strain ($L\varepsilon$) zone where the stresses are lower due to material break down. (b) At greater plastic deformations there is no LEFM zone (no K dominated region), but there is still an HRR singularity and an $L\varepsilon$ domain. (c) Plasticity is so advanced that there is no fracture mechanics parameter than describes the crack tip state, and the $L\varepsilon$ domain is quite pronounced. After Anderson (2004).

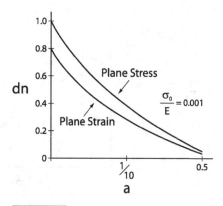

Figure 9.16

Parameter *dn* for increasing values of $(1/n)$, where n is the plastic hardening exponent. For other values of σ_0/E the values can differ, and these can be found in Anderson's text (Anderson, 2004).

Equations 9.31 and 9.49), but shows that it continues to apply to plastically deforming and hardening bodies far beyond the limits of LEFM.

9.5.6 Approximation of a true elastic-plastic *J*

There are a number of methods for computing J for both linear elastic solids and fully plastic solids. For true elastic-plastic materials, though, neither of these approaches fully characterizes the crack tip state or the energy in the body, which in general experiences both elastic and plastic strains (Figure 9.13). This problem was addressed by the Electric Power Research Institute (EPRI) at General Electric, which published an elastic-plastic fracture handbook designed to accommodate true EPFM conditions (Kumar *et al.*, 1981). Their approach was to approximate the *J*-integral as the sum of the modified elastic and fully plastic contributions:

$$J_{Tot} = J_{el(a_{eff})} + J_{pl(a,n)}. \tag{9.57}$$

It can be seen in Figure 9.17 that this approach provides a good approximation to the actual *J*-integral in EPFM. It is clear from this diagram that the elastic *J* dominates in LEFM (low loading), but that the plastic contribution becomes dominant as the loading approaches general yielding and the fully plastic limit.

Recall that the a_{eff} in modified LEFM was based on perfect plasticity (Equation 9.24), and that a smaller zone is expected with hardening. Thus, for the modified elastic J (J_{el}), one utilizes an effective flaw size of the form

$$a_{eff} = a + \lambda r_y \tag{9.58}$$

9.5 Elastic-plastic fracture mechanics (EPFM)

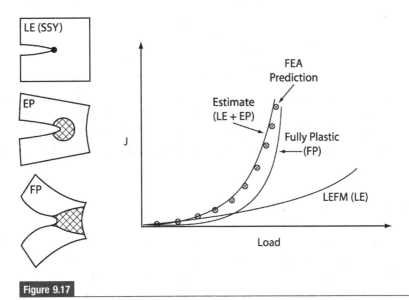

Figure 9.17

EPRI estimated linear elastic (LE) + fully plastic (FP) J-integral, compared to FEA predictions (Kumar et al., 1981). This analysis works extremely well considering the arbitrary nature of the assumption. (After Anderson, 2004.)

where

$$\lambda = \frac{1}{1 + (p/p_0)^2} \tag{9.59}$$

for an applied load P and the limit load for general plasticity P_o. The plastic zone radius is

$$r_y = \frac{1}{\beta\pi}\left(\frac{n-1}{n+1}\right)\left(\frac{K}{\sigma_Y}\right)^2 \tag{9.60}$$

with $\beta = 2$ for plane stress and $\beta = 6$ for plane strain, as seen for the Irwin correction.

The EPRI estimation procedure and tabulated solutions are slightly different than the formalism we presented above, where the tables for the plastic contribution to the driving force are referenced to specific test specimen configurations. Much the same approach is taken for the tabulation of geometric parameters for the computation of K (e.g., Equation 9.8). The tabulated form of J_{pl} is given as

$$J_{pl} = \alpha\varepsilon_0\sigma_0 b h_{1_{(a/W,n)}}\left(\frac{p}{p_0}\right)^{n+1} \tag{9.61}$$

where a and W are shown for a compact tension specimen in Figure 9.3, and $b = W - a$. The parameter h_1 is computed and tabulated for a range of a/W and n values for each given configuration. The limit load P_0 is also given for each configuration; again, for the CT specimen

$$p_{0-CT} = 1.455\eta_{CT} B b \sigma_0 \tag{9.62}$$

where

$$\eta_{CT} = \sqrt{\left(\frac{2a}{b}\right)^2 + \frac{4a}{b} + 2} - \left(\frac{2a}{b} + 1\right) \quad (9.63)$$

Using handbooks of tabulated results for computing J_{pl}, one can find the total EPFM J for many engineering problems of interest. To predict fracture failure under elastic-plastic conditions we will still need to appeal to both crack growth resistance curves (J_R curves), fracture instability, and plastic collapse at the limit load, p_o.

9.5.7 J_R crack growth resistance curves

The resistance to crack propagation in EPFM can be extended directly from the G-R curve concept to J through the interpretation of J as the energy release rate. Materials that harden in plasticity show a rising crack growth resistance curve, meaning that they will typically undergo some stable crack propagation prior to instability and fracture. The J_R curve is considered a material property, though it is somewhat sensitive to test specimen configuration.

In a clear departure from what we saw in $G - R$ curves for linear elastic materials, the first part of the J_R curve corresponds to a plastic blunting of the crack tip and not actual material failure resulting in new crack surface. The blunting region of the J_R curve is shown as Regime I in Figure 9.18. Blunting gives way to actual crack tip material failure, corresponding to the initiation of crack growth. This phenomenon is of critical importance, and so crack initiation occupies Regime II in Figure 9.18. The J-integral at crack initiation (in Mode I) is denoted J_{Ic}, and is often utilized as the EPFM measure of fracture toughness. Regime III then corresponds to the rising portion of the J_R curve, where stable propagation occurs until an instability condition is reached. If fracture instability does not occur the J_R curve often saturates, reaching a constant value for steady-state crack growth J_{ss} (Regime IV). While fracture instabilities in EPFM are of interest, we will see in the next section that limitations to the theory make crack initiation criteria a more robust choice for use in design, and the generation of J_R curves is often undertaken just to determine a value of J_{Ic}.

Fracture instability in EPFM occurs when the energy release rate increases more rapidly than the material crack growth resistance (just as in LEFM). Thus, crack growth is stable if

$$\frac{dJ}{da} \leq \frac{dJ_R}{da} \quad (9.64)$$

and otherwise crack growth is unstable.

It should be noted that the tearing instability may be predicted to occur at advanced stages of plasticity. For materials with a high toughness and low strength, this instability may not occur before plastic collapse at the limit load, resulting in a competition of

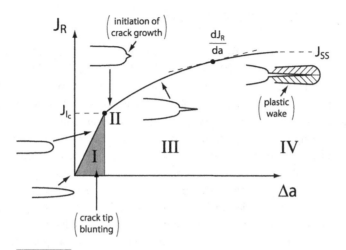

Figure 9.18

J_R curve showing the four regimes of behavior: (I) crack tip blunting without crack growth, (II) crack initiation at J_{Ic}, (III) crack growth with increasing resistance, and (IV) steady-state crack growth at J_{ss}. Also shown is the derivative dJ_R/da, which determines the point of instability and fracture.

terminal mechanisms that can depend on the specifics of how the load is applied. Two-parameter failure assessment diagrams can be constructed for EPFM, extending the R-6 approach mentioned above but using pseudo-elastic modified K_c methods (Kanninen and Poplar, 1985; Kumar et al., 1981).

9.5.8 J_c as a crack initiation based failure criterion

The development of the J-integral requires that the stresses only increase due to the governing assumptions inherited from deformation plasticity. However, when the crack extends beyond the blunted initial crack, the material adjacent to the new crack surface unloads elastically (leaving what is termed the plastic wake), violating the simplification of negligible unloading inherent in deformation plasticity. Thus, the use of J to characterize propagating cracks is not in general rigorously supported by the analysis. Typically, conditions are considered to only involve small deviations to local behavior in the plastic zone due to elastic unloading during crack growth, and hence J is applied to crack growth problems (growth under cyclic stress, however, is another matter). The ambiguity in the applicability of J for growing cracks, particularly under varying stresses or cyclic loading, motivates the preference for the J_{Ic} as a crack initiation and failure criterion. Put another way, J_{Ic} is the greatest rigorously applicable (and therefore intrinsic) value of J for design and fracture control purposes. In many engineering applications, particularly when the extent of stable crack growth is limited, designing against crack initiation may be preferred in any case as a conservative fracture control approach.

9.5.9 Summary of elastic-plastic fracture mechanics

The development of EPFM, from the COD, to cohesive zone modeling, to the development of the *J*-integral and HRR singularity fields, was conducted to further refine the successful but ultimately incomplete analysis available with modified LEFM methods. The picture that emerges is a more general and stronger picture of what is occurring near a crack in a body where potentially large amounts of inelastic deformation are occurring in the neighborhood of the crack. EPFM therefore provides a more nuanced and insightful foundation for understanding crack phenomena in materials, even if it still contains limitations inherent to its governing assumptions. Such a foundation is strongly recommended for practitioners in the field of biomaterials, where LEFM is unlikely to provide much insight into the behavior of many of the nonlinear materials of biomedical interest. The framework of EPFM is directly applicable to many of these materials. However, an extension of it to include time-dependent material properties and viscous effects provides a more satisfying and accurate framework for natural and synthetic biomaterials. A combination of elastic-plastic and time-dependent fracture mechanics comprise the analytic toolset for the biomaterials design engineer designing seeking to prevent fracture in nonlinear materials.

9.6 Time-dependent fracture mechanics (TDFM)

As discussed above, elastic-plastic fracture mechanics (EPFM) and linear-elastic fracture mechanics (LEFM) both presume time-independent material behavior, which is typically sufficient for engineering design applications. With the exception of metals and ceramics, many biomaterials (e.g., tissues, polymers, and gels) exhibit viscous or fluid-like characteristics in their deformation, limiting the applicability of the foregoing treatment of fracture mechanics for many biomedical applications. The literature on *time-dependent fracture mechanics* (TDFM) is extensive, and much of it is concerned with creep of metals at high temperatures and therefore is outside the scope of this book. However, many of the analyses of EPFM can be adapted to a time-dependent framework with good results.

Viscous solids differ from other inelastic solids that they continually flow under a constant stress, a process that may or may not saturate after accumulating a finite amount of creep strain. Viscous dissipation at a crack tip implies that the material resistance to deformation diminishes with time, and at the long time limit the effective yield strength could be vanishingly small. Time-dependent vanishing yield strengths substantially alter the picture of the crack tip processes from one of static equilibrium and load-bearing capacity in LEFM and EPFM to one dominated by local relaxation and viscous flow in viscoelastic fracture mechanics (VEFM). Fortunately, simple descriptions of material viscosity result in analytically tractable solutions for fracture mechanics analyses. There have been a number of significant contributions to VEFM (Schapery, 1975; Schapery,

1975; Williams, 1984; Wnuk, 1971); however, only the model due to Williams (Williams, 1984) is discussed for brevity.

9.6.1 Viscoelastic fracture mechanics (VEFM)

If one considers a viscoelastic body as responding in a pseudo-elastic manner, but with a time-dependent compliance, we can adapt the formalism of LEFM and EPFM to produce measures of the nonlinear energy release rate J. Consider material with a power-law time-dependent compliance,

$$C_{(t)} = \frac{1}{E_0}\left(\frac{t}{\tau_0}\right)^d \qquad (9.65)$$

for time $t > 0$, an instantaneous ($t \to 0$) elastic modulus E_0, and fit parameters τ_0 and d. For a constant applied global stress we then can write

$$J_{(t)} = J_0 \left(\frac{t}{\tau_0}\right)^d \qquad (9.66)$$

where the J-integral inherits the time-dependence as the compliance, increasing from the approximately linear-elastic value $J_0 = J_{el}$. Equation (9.49) implies that the energy release rate is continually increasing after the first instant of the load, as strain is continually accumulating due to local flow at the crack tip. Assuming the material to flow in a strip-yield zone manner, we can then predict the length of the cohesive zone

$$r_{d(t)} = \frac{\pi}{8}\left(\frac{K_t}{p_c}\right)^2 \qquad (9.67)$$

for a constant (with time and position) flow stress p_c transmitting the applied stress across the cohesive zone and a pseudo-elastic $K_{(t)}$, i.e., analogous to the Dugdale solution.

If crack initiation occurs at a critical value of the energy release rate, J_c, one can rewrite Equation (9.66) to find the time under a constant load to initiate crack propagation

$$t_i = \tau_0 \left(\frac{J_C}{J_0}\right)^{1/d}. \qquad (9.68)$$

The **crack initiation time**, t_i, is given in Equation (9.68) as a function the compliance (through τ_0 and d), the fracture toughness J_c, and the applied crack driving force J_0. While the applied driving force J_0 is composed of extrinsic factors (body geometry, loading severity, etc.), since $J_0 = K_0^2/E_0'$ we also see that J_0 depends on the material stiffness.

Crack initiation time can serve as a failure criterion in design of components made from viscous solids, particularly if the parts have notches or stress concentrations. For a desired minimum t_i as a designed life criterion, there are a variety of strategies available to the engineer. Increasing t_i can be accomplished by employing materials with increased J_c, or reduced time dependence (d) can be employed to slow the accumulation of J.

Additionally, the driving force J_0 can be reduced by altering the component geometry to reduce local or global stresses, or by increasing the elastic modulus. Thus, by using t_i as a fracture failure criterion, we are not restricted to only metrics of fracture toughness for design. Rather, Equation (9.51) implies that for a given load and geometry, t_i still is not uniquely defined by or even necessarily monotonically related to J_c.

9.7 Intrinsic and extrinsic fracture processes

The process of crack growth in a material is a competing mechanism between the work in the system providing the crack driving force (the stress intensity factor or energy release rate) and the materials resistance to extension of the crack and the creation of new crack surface. Any deformation process of a material that enhances its resistance to crack advance is considered a toughening mechanism. There are two basic types of toughening mechanisms (fracture processes) utilized in structural materials: **intrinsic toughening** and **extrinsic toughening**. The micromechansims of damage associated with both intrinsic and extrinsic processes are illustrated in Figure 9.19.

Intrinsic mechanisms are inherent to a material and are directly a function of the driving force. Under monotonic or quasi-static loading conditions, the intrinsic microstructural processes control the stress intensity at the crack tip and are generally active regardless of the length of the flaw in the material system. Intrinsic mechanisms are what one would think of as "true" material properties, and involve fundamental phenomena that control the plastic deformation, nonlinear elasticity, or damage ahead of the crack tip, and are controlled by the microstructure (heat treating, grain size, precipitate distribution, etc.), as the extent of local deformation that a material can tolerate is increased as the intrinsic toughness is enhanced. Intrinsic dissipation mechanisms include plasticity processes, shear banding, phase transformation, microcracking, and crazing (Pearson and Pruitt, 1999; Ritchie, 1988; Suresh, 1998).

Extrinsic mechanisms, on the other hand, alter the effective driving force experienced at the crack tip, and may involve induced processes that are activated in the crack wake or in the plastic zone (Figure 9.19). Extrinsic toughening involves the development of an inelastic region around the crack wake, a crack surface contacting mechanism such as crack wedging with oxide formation, or crack bridging (Ritchie, 1999). For this reason, the extrinsic mechanisms depend on the history of crack growth and crack length, and provide an increasing R-curve for monotonic (quasi-static) loading. The extrinsic toughening mechanisms shield the crack tip from the global stress and associated intrinsic driving force and can be brought about through three basic types of processes: (i) *crack deflection* in which the crack tip is forced out of the critical propagation plane as it advances and the effective crack tip stress intensity is reduced; (ii) *contact shielding* in which the mechanisms work behind the advancing crack and typically produce traction or closure in the crack wake through particle or fiber bridging; and (iii) *zone shielding* in which growth mitigating processes such as shear banding at second-phase additions,

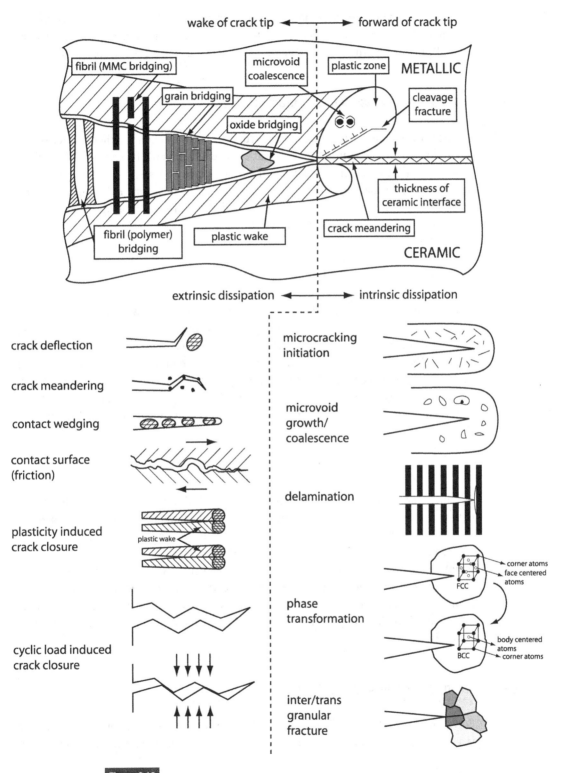

Figure 9.19

Illustration of micromechanisms associated with both intrinsic and extrinsic processes. (After Ritchie, 1999.)

phase transformation, void formation around particles, multiple crazes around reinforcing phases, or microcracking are activated within the crack tip process zone (Evans, 1986; Pearson and Pruitt, 1999; Ritchie, 1999; Ritchie *et al.*, 2000; Suresh, 1998).

9.8 Fracture mechanisms in structural materials

The fracture resistance of a material system depends heavily upon whether its microstructure renders it ductile, brittle, or semi-brittle. In general, ductile materials are capable of plasticity that facilitate the dissipation of near-tip crack stresses, while brittle materials have no plasticity and rely on damage mechanisms to mitigate crack driving force. Natural tissues are known to be damage tolerant and offer unique combinations of materials that enable energy dissipation in the system.

9.8.1 Fracture mechanisms in metals

Most metals and alloys are inherently tough and tend to have high values of K_{IC} and J_{IC}. Metal systems make use of plasticity mechanisms such as dislocation slip, shear, and microvoid coalescence (as described in the previous chapter) to dissipate crack tip stresses. The primary concern with metals utilized in biomaterial applications is their sensitivity to corrosion and the fact that many of these metal systems have a fracture toughness associated with stress corrosion cracking (K_{ISCC}) that is substantially lower than their fracture toughness in the presence of a corrosive environment (Chapter 2).

9.8.2 Fracture mechanisms in ceramics

Ceramics are brittle materials with diminutive ability for plastic deformation (Chapter 3) and consequently exhibit a low fracture toughness that is dominated by the surface energy of newly formed fracture surface. Such materials have given way in recent years to toughening ceramics in engineering applications, which employ additional phases to elicit extrinsic toughening mechanisms. As a result, the toughness of a ceramic may be improved tenfold via microstructural engineering of the material (Ritchie, 1999). Such micromechanisms are used in structural ceramics and include phase transformation toughening, microcracking, crack bridging, and crack deflection methods.

9.8.3 Fracture mechanisms in polymers

Polymers can exhibit both ductile and brittle fracture processes, and for this reason there is a wide range of fracture toughness (K_{IC}, G_{IC}, or J_{IC}) observed in these materials. Much of the disparity is attributed to the vast differences in molecular variables such as molecular weight, crystallinity, crosslink density, and entanglement density (as discussed in Chapter 4). Polymers can exhibit ductile processes such as chain slip, shear banding, and fibrillation or semi-brittle processes such as crazing, as discussed in the previous chapter focusing on failure mechanisms. Like ceramic materials, the toughness of multiphase polymers can be improved tenfold via microstructural engineering (Pearson and Pruitt, 1999). Polymers can be substantially toughened with the addition of a second phase, such as rubber particulate, that serve to nucleate shear processes and fibrillation of the matrix as well as provide a bridging mechanism between crazes (Chapters 4 and 8).

9.8.4 Fracture mechanisms in tissues

Structural tissues do not have a wide variety of constituents available as building blocks (Chapter 5). This has resulted in a multitude of strategies for combining these constituents in hierarchical structures that are highly tolerant of damage. Heterogeneous mixtures allow for the use of brittle phase materials (hydroxyapatite in bone) in submicroscopic crystals that approach the theoretical maximum strength that cannot be attained in macroscopic flawed crystals. Further, the rubbery matrix material that binds the crystals together is generally very ductile and can absorb an enormous amount of strain energy before decohering and allowing a flaw to propagate. This preferential localization of high-energy deformation increases the effective work to fracture, and hence the fracture toughness, in a manner analogous to the second-phase craze-toughening strategy employed in polymers (Chapter 8).

Cortical bone is an example of a hierarchical material, composed of concentrically nested bone lamella in osteons separated internally by thin layers of collagenous matrix, and the osteons separated from each other by relatively soft cement lines. The separation of structural layers by a soft interspersed phase is thought to serve as a crack-arresting strategy. A key concept for bone is that it need not have an exceptionally high resistance to fracture, but rather need only arrest cracks long enough for biological remodeling processes to eliminate accumulated cracks (fatigue cracks). In the case of large structural bone allografts, which do not remodel, arrested cracks could continue propagating after a period of dormancy and lead to catastrophic fractures.

Many soft tissues exhibit an initially low stiffness, which rapidly increases near the fracture strain, known respectively as the "toe" and "heel" regions of the stress-strain curve (Chapter 5). This behavior has a direct and profound effect on the resistance of these tissues to tearing and fracture. Highly compliant materials stretch extensively at a crack tip and allow the sharp flaw to become blunt. Furthermore, the low shear modulus of the material limits the communication of the local stress conditions near the blunted crack tip to the bulk where other flaws may exist. Thus, while stiff tissues tend to increase their toughness through irreversible deformation processes, soft tissues prefer to elastically blunt to forestall the onset of tearing or fracture.

324 Fracture mechanics

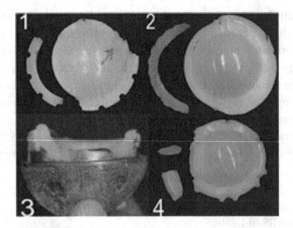

Figure 9.20

Four fractured XL-UHMWPE acetabular liners. All show a similar crescent-shaped fracture along the unsupported rim (shown in perspective in 3), and all fractures initiated at a stress concentration.

9.9 Case study: fracture of highly crosslinked acetabular liners

Four different implants were revised at four separate institutions, and it was found that each had catastrophically fractured in a roughly similar manner (Furmanski *et al.*, 2009). These retrievals are shown in Figure 9.20. None of the implants was surgically malpositioned or was associated with an unusual patient history, so the failures were assumed to be representative of the clinical performance of those designs. Further, since each design was different but failed in a similar fashion, a failure analysis was conducted with the aim of establishing that each occurred in an analytically predictable way, ultimately utilizing fracture mechanics based arguments. Since all four of the liners contained a relatively sharp notch near the location where fracture initiated (radius <0.2 mm), fracture and fatigue were certainly considered a risk.

A three-dimensional finite element analysis of a direct rim loading boundary condition was conducted, utilizing the experimentally determined elastic-plastic deformation response in tension, and no unloading was analyzed. The distribution of maximum principal stresses (Figure 9.21) in each case peaked in the region where the fracture initiated, with a value of 9–40 MPa.

To explain whether these failures were indeed driven by fracture at this loading condition, it was assumed that a crack of negligible size had initiated at the observed initiation location, and was subject to the simulated loading. In this case, the total crack length was assumed then to be the depth of the rim notch. For a 2 mm deep notch, and a fracture toughness of $K_c \sim 0.7$ MPa$\sqrt{m}$, the critical tensile stress for fracture was computed to be 8.8 MPa. Therefore, it was concluded that this loading condition could drive fracture if cracks had initiated in those locations in each of the four cases.

There were a number of limitations in this analysis. No actual argument regarding crack initiation was made, as the modeling employed did not compute the *J*-integral,

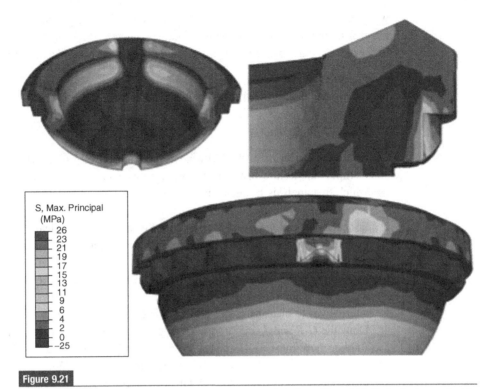

Figure 9.21

Maximum principal stress for a 500 N distributed load incident on the rim in Case 1 (Figure 9.20). The tensile stress peaks at the root of the rim notch and is also substantial at the sharp corner underneath the rim. Initiated cracks were found at both these locations in this case.

and actual data on values of J_c for UHMWPE in the literature vary widely. Since cracks were observed at those locations, it was a safe assumption, but it was not known how long it took those cracks to initiate, and if that was therefore the most critical step for determining the fatigue life of the part. A more detailed analysis treating the polymer as viscous and computing $J(t)$ and t_i would provide more tangible information on the likelihood of these fractures occurring in the future.

9.10 Summary

Fracture mechanics demonstrates how materials and components can fail far below their ultimate tensile strength due to the presence of flaws or cracks. The local stress near the tip of a crack is easily pushed beyond the strength of the material, and the local failure at the crack may result in an instability causing catastrophic fracture. The mathematical theories for predicting and therefore preventing this outcome depend on the extent to which the material can deform inelastically prior to local failure. Modified methods in linear-elastic fracture mechanics are useful for many problems involving metals, while elastic-plastic fracture mechanics better describes the phenomena in more ductile materials. Time-dependent fracture mechanics is necessary of viscous deformations that may occur during prolonged service, as is likely in many natural and synthetic polymers.

9.11 Problems for consideration

Checking for understanding

1. Complete the terms in the Griffith energy balance for a plate with a central crack. Compute the failure stress calculation for this geometry. Comment on the extent to which this analysis holds for pyrolytic carbon, a titanium alloy, a ductile stainless steel, and a polymer.
2. For each crack propagation Mode, give an example of a medical device and location in that device where a crack would propagate in that Mode. Give two examples of mixed-mode loading.
3. Explain why it is important to use sufficiently thick specimens when measuring K_{Ic}. What if the device you are analyzing itself is not thick? Are there nonlinear alternatives to measuring K_{Ic}, and what is given up if you employ those instead?
4. Draw the G-R and J_R curve for a ductile metal and for a non-toughening ceramic.
5. Explain how the CTOD relates to the Dugdale model, and the J-integral. What are some experimental methods for measuring CTOD? What is the advantage or disadvantage in measuring CTOD instead of employing other methods in modified LEFM or EPFM?

For further exploration

6. Even proponents of EPFM admit that it has serious drawbacks. List as many of these limitations as possible, and compare them to the limitations inherent to modified LEFM and CTOD methods. What would you propose to your company if you had to pick one approach for an elastic-plastic material? What details of the loading or application would weigh on your decision?
7. For a given crosslinked polymer, draw three different $J(t)$ curves for a subcritical initial J ($J_0 < Jc$) at a constant applied load. How does the crack initiation time depend on the applied load (hint: assume J is proportional to P^2).
8. For a given constant load P, plot the different $J(t)$ responses for the following normalized material properties for these material formulations:
   Material 1: $E = 1, \tau_0 = 1, d = 10, J_c = 1$
   Material 2: $E = 1, \tau_0 = 1, d = 8, J_c = 0.5$
   Material 3: $E = 0.7, \tau_0 = 1, d = 8, J_c = 0.5$
   Comment on the observed effect on the crack initiation time for each material property.
9. Your company can choose any of the materials in problem 8 for a given total hip replacement that is to be used by professional athletes, and which contains stress concentrations in regions of tensile loading. Which material do you choose? In this

case, assume that Material 1 is already in FDA approved use for a similar implant. What if the cost for material is 10 times that of Material 1, and Material 3 is 5 times that of Material 1? Recall that you can vary the design with the choice of material.

9.12 References

Anderson, T.L. (2004). *Fracture Mechanics*. New York: CRC Press.

Atkins, A.G., and Mai, Y.W. (1985). *Elastic and Plastic Fracture*. Chichester, West Sussex, UK: Ellis Horwood Limited.

Barenblatt, G.I. (1962). The mathematical theory of equilibrium cracks in brittle fracture. *Advances in Applied Mechanics*, 7, 55–129.

Burdekin, F.M., and Stone, D.E.W. (1966). The crack opening displacement approach to fracture mechanics in yielding materials. *Journal of Strain Analysis*, 1, 145–53.

Dugdale, D.S. (1960). Yielding of steel sheets containing slits. *Journal of the Mechanics and Physics of Solids*, 8(2), 100–104.

Evans, A.G., Ahmad, Z.B., Gilbert, G.B., and Beaumont, P.W.R. (1986). Mechanisms of toughening in rubber-toughened polymers. *Acta Metallurgica*, 34 (1) 79–87.

Furmanski, J., Anderson, M., Bal, S., Greenwald, A.S., Penenberg, B., Ries, M.D., and Pruitt, L. (2009). Clinical fracture of cross-linked UHMWPE acetabular liners. *Biomaterials*, 30(29), 5572–82.

Griffith, A.A. (1921). The phenomena of rupture and flow in solids. *Philosophical Transactions of the Royal Society of London*, A221, 163–97.

Hahn, G.T., and Rosenfield, A.R. (1965). Local yielding and extension of a crack under plane stress. *Acta Metallurgica*, 293–306.

Hutchinson, J.W. (1968). Singular behavior at the end of a tensile crack in a hardening material. *Journal of the Mechanics and Physics of Solids*, 16, 13–31.

Inglis, C.E. (1913). Stresses in a plate due to the presence of cracks and sharp corners. *Transactions of the Institute of Naval Architects*, 55, 219–41.

Irwin, G.R. (1948). Fracture dynamics. In *Proceedings of the ASM Symposium on Fracturing of Metals*, Cleveland, OH: American Society for Metals, pp. 147–65.

Irwin, G.R. (1956). Onset of fast crack propagation in high strength steel and aluminum alloys. *Sagamore Research Conference Proceedings*, 2, 361–64.

Kanninen, M.F., and Poplar, C.H. (1985). *Advanced Fracture Mechanics*. New York: Oxford University Press.

Kumar, V., deLorenzi, H.G., Andrews, W.R., Shih, C.F., German, M.D., and Mowbray, D.F. (1981). Estimation technique for the prediction of elastic-plastic fracture of structural components of nuclear systems. *4th Semiannual Report to EPRI, Contract No. RP1237-I*. Schenectedy, NY: General Electric Company.

Pearson, R., and Pruitt, L. (1999). Fatigue crack propagation in polymer blends. In *Polymer Blends: Formulation and Performance*, ed. D.R. Paul and C.B. Bucknall. New York: Wiley, Volume 27, pp. 269–99.

Rice, J.R., and Rosengren, G.F. (1968). Plane strain deformation near a crack tip in a power-law-hardening material. *Journal of the Mechanics and Physics of Solids*, 16, 1–12.

Ritchie, R.O. (1988). Mechanisms of fatigue crack propagation in metals, ceramics and composites: role of crack tip shielding. *Materials Science and Engineering A*, 103, 15–28.

Ritchie, R.O. (1999). Mechanisms of fatigue-crack propagation in ductile and brittle solids. *International Journal of Fracture*, 100(1), 55–83.

Ritchie, R.O., Gilbert, C.J., and McNaney, J.M. (2000). Mechanics and mechanisms of fatigue damage and crack growth in advanced materials. *International Journal of Solids and Structures*, 37(1–2), 311–29.

Schapery, R.A. (1975). Theory of crack initiation and growth in viscoelastic media. 1. *Theoretical Development. International Journal of Fracture*, 11(1), 141–59.

Schapery, R.A. (1975). Theory of crack initiation and growth in viscoelastic media. 2. Approximate methods of analysis. *International Journal of Fracture*, 11(3): 369–88.

Shih, C.F. (1981). Relationship between the J-integral and the crack opening displacement for stationary and extending cracks. *Journal of the Mechanics and Physics of Solids*, 29, 305–26.

Suresh, S. (1998). *Fatigue of Materials*, 2d edn. Cambridge, UK: Cambridge University Press.

Wells, A.A. (1963). Application of fracture mechanics at and beyond general yielding. *British Welding Journal*, 10, 563–70.

Williams, J.G. (1984). *Fracture Mechanics of Polymers*. Chichester, West Sussex, UK: Ellis Horwood Limited.

Wnuk, M.P. (1971). Subcritical growth of fracture (Inelastic fatigue). *International Journal of Fracture*, 7, 383–407.

10 Fatigue

Inquiry

How would you safely design a femoral component of a total hip replacement to prevent fatigue fracture of the stem without causing stress shielding to the adjacent bone?

The inquiry posed above represents a realistic design challenge that one might face in the field of orthopedics. Stress shielding is defined as bone loss that is owed to insufficient stress transfer from the implant to the neighboring bone. Stress shielding can be minimized through a combination of material selection and device design that provides improved compliance match to the tissue. Device design is a multifactorial process and it is equally important to ensure that the femoral stem is able to safely provide the requisite fatigue properties for successful performance of the implant. For a total hip replacement to function for two decades, the femoral component should offer a minimum of 20 million fatigue cycles to provide the estimated one million walking cycles per year for an average person. This fatigue life is further complicated by the anatomy, weight, health, and physical activity of the specific patient. The performance *in vivo* also depends on factors such as surgical placement and tissue healing as well as immunological response to the material. In order to design a fatigue-resistant implant, one would want to know the magnitude and nature of the expected stresses in the implant. For example, the implant may experience a combination of tensile, compressive, and bending stresses that are cyclic in nature. At a minimum, the peak stress ranges in the femoral stem should be below the **endurance limit** for the material and **mean stresses** should be incorporated into the life calculations. Further, the design should mitigate the use of sharp corners or other stress concentrations that can lead to fatigue crack propagation and fracture; if these elements exist within the design, then the fatigue life should be estimated conservatively using **fracture mechanics methodologies**.

10.1 Overview

Fatigue damage and associated failures are of critical concern to the medical device community as sustained loading of structural implants is inevitable. There are numerous cases of fatigue failures in medical devices owing to poor material selection, unsuitable sterilization methods, deficiencies in manufacturing processes, careless designs with inherent stress concentrations or crack initiating sites, and insufficient stress analysis given the complex multiaxial stresses of the body. The case example at the end of this chapter highlights a fatigue failure owing to the coupling of a poor component design and a material with inferior fatigue resistance.

Fatigue failures typically ensue in implants or medical devices as a result of accumulated damage or growth of a defect to a critical dimension in the biomaterial during service *in vivo*. The majority of fatigue failures occur well below the yield strength of the material utilized in the implant. Hence, designing a component to resist fatigue fracture that results from the nucleation and growth of a crack over millions of load variations is far more complicated than designing a device to prevent static fracture or yield. Fatigue damage can be mitigated through design by minimizing stress concentrations and surface roughness of the components. Notches serve to locally concentrate stresses and are often the nucleation site of critical defects. Intrusions or extrusions at the surface can initiate flaws that lead to a fracture process. Understanding the fatigue fracture behavior of biomaterials is important when designing components involving both sustained conditions and stress concentrations; the fast fracture criterion of the implant is dictated by the critical flaw size concomitant with the fracture toughness or fracture resistance of the biomaterial. Moreover, the **micromechanisms** of deformation and fracture are highly dependent upon the material microstructure.

Metals generally offer high fracture toughness and resistance to crack propagation; however, these materials are highly susceptible to stress corrosion cracking. The grain boundaries, dislocation slip systems, and precipitates within an alloy system can serve as an impedance to crack advance or they can serve as an initiation site for the nucleation of defects (Hertzberg, 1996; Suresh, 1998). Some metals, such as steels and nickel alloys, have an endurance limit that assures an "infinite" life for the material. The endurance limit is typically defined as stress range that provides at least 10 million fatigue cycles and it plateaus at that level so that any stress below the endurance limit is insufficient for crack growth. Most metals, however, do not have such a limit and decreasing stress amplitude merely results in an extended life rather than an infinite life. This is the case with titanium alloys and many cobalt-chromium alloy systems.

Many crystalline ceramics are known for their exceptional wear resistance and this is attributed to their high elastic modulus and yield strength; however, the intrinsic resistance to plastic deformation renders ceramic systems as brittle with very low fracture toughness. Accordingly, ceramics are unable to offer much resistance to fatigue crack growth. Toughening mechanisms such as crack bridging, microcracking, and phase transformations can be utilized to mitigate crack tip stresses, but fatigue resistance in ceramics remains limited by their low fracture toughness (Evans, 1980; Ritchie, 1999; Ritchie *et al.*, 2000).

The fatigue behavior of polymers is extremely sensitive to molecular variables such as molecular weight, molecular weight distribution, crystallinity, chain entanglement density, crosslink density, and the presence of fillers or reinforcements (Kausch and Williams, 1986; Kinloch and Guild, 1996; Pearson and Pruitt, 1999; Pruitt, 2000; Sauer and Hara, 1990). Additionally, many polymers are viscoelastic and are sensitive to frequency, waveform, and service temperature (Ferry, 1961; Hertzberg and Manson, 1980; Schapery, 1975). All of these factors are of considerable interest and practicality for the safe design of structural medical devices subjected to sustained loading.

10.2 Learning objectives

This chapter provides a broad overview of fatigue characterization and device design when biomaterials are expected to undergo sustained loading *in vivo*. The basic factors contributing to fatigue life and device design as well as the specific micromechanistic behavior associated with metals, ceramics, and polymers are discussed. The design philosophy of total life and defect-tolerant life predictions are presented, and the benefits and limitations of these philosophies are discussed. The key objectives of this chapter are that at the completion of the chapter the student will be able to:

1. list the key factors that contribute to fatigue life of a material
2. define the assumptions used in the total life design philosophy
3. use S-N plots and the Basquin equation to predict fatigue life
4. apply the Palmgren-Miner's rule to determine fatigue life for variable amplitude loading
5. employ the Coffin-Manson equation to predict low-cycle or high-cycle fatigue conditions
6. define the assumptions used in defect-tolerant design philosophy
7. describe the stress intensity factor and calculate the critical flaw size for a material
8. use *da/dN* plots and the Paris equation to predict fatigue life
9. discuss the benefits and limitations of total life and defect-tolerant methodologies
10. give an explanation of how the fatigue fracture behavior of a material is linked to its extrinsic and intrinsic damage mechanisms
11. elucidate the reasons why ductile systems are more prone to cyclic damage, while brittle systems are more prone to quasi static fracture
12. explain the micromechanisms of crack growth in metals, ceramics, and polymers

10.3 Fatigue terminology

The fatigue resistance of an implant generally depends on many factors, including the micromechanisms of deformation or fracture for the material, the component design, the stresses imposed on the device, and the environment in which the implant is utilized. In order to develop the basic equations utilized for fatigue design, it is useful to first define the primary mechanical variables associated with the sustained (cyclic) loading of a component. These factors include mean stress, peak stress, load ratio, waveform, frequency, and amplitude variation (**spectrum loading**). The basic terminology associated with sustained loading under constant amplitude conditions is illustrated

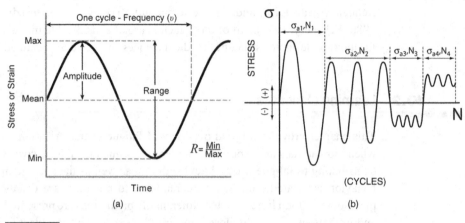

Figure 10.1

Basic definitions of stress or strain range, stress or strain amplitude and load ratio, R, for cyclic conditions imposed using a (a) sinusoidal waveform and (b) spectrum (variable amplitude) loading.

in Figure 10.1(a). The same terminology is utilized when the amplitude of loading is variable and the primary difference is that the parameters are then defined for specific intervals or cycles, N_i, of constant amplitude loading as shown in Figure 10.1(b).

In fatigue loading, the **stress range** is typically the dominant factor in the progression of fatigue damage, and failure, especially in metals, and is defined as:

$$\Delta\sigma = \sigma_{max} - \sigma_{min} \tag{10.1}$$

and correspondingly the **stress amplitude** and the **mean stress** of the loading cycle are defined as:

$$\sigma_a = \frac{\sigma_{max} - \sigma_{min}}{2} \quad \sigma_m = \frac{\sigma_{max} + \sigma_{min}}{2} \tag{10.2}$$

where σ_{max} is the maximum stress and σ_{min} is the minimum stress of the fatigue cycle or load reversal. The mean stress is an important parameter in component design as this captures the average level of stress and provides a direct measure of the nature of the stress state. For example, the mean stress is zero for fully reversed loading, but the average stress increases dramatically for fully tensile loading. For most materials, an increase in mean stress for an equal stress amplitude results in a shortened fatigue life. The ratio of the minimum stress normalized by the maximum stress is defined as the stress ratio or **R-ratio**:

$$R = \frac{\sigma_{min}}{\sigma_{max}}. \tag{10.3}$$

The R-ratio (R) also captures the stress state during loading. For example, in fully reversed loading, $R = -1$ whereas for tensile loading, R ranges from 0 to 1(static limit) and for fully compressive loading, $R > 1$. For materials that are dominated by strain-based mechanisms, such as many mineralized tissues or metals undergoing plastic deformation, the strain amplitude becomes a dominating factor in the fatigue life. One can define the strain range, strain amplitude, and the strain ratio for cyclic strain conditions using the same definitions above where strains are merely substituted for stress in Equations (10.1)–(10.3). It is important to understand such factors as mean stress, peak stress, and stress amplitude and their role in the fatigue behavior of specific materials.

Characterization of the fatigue behavior of an engineering biomaterial captures its inherent resistance to damage under sustained loading conditions and examines the micromechanisms that ensue in the nucleation and growth of a flaw. Designing for fatigue resistance is an intricate process that involves many factors, including the material utilized and its inherent fatigue resistance as well as geometry, surface finish, and tolerance of the component in the manufacturing process. One critical aspect of the design process is the decision as to whether the component's fatigue life will be dominated by the *initiation* or *propagation* process of a critical flaw. The **total life** design methodology assumes that the component is initially free of any flaws that are sufficiently sized for growth or ideally that the component is "defect-free." This methodology is based on the notion that fatigue failure is a consequence of crack nucleation and subsequent growth to a critical size and that the majority of the life is spent in the nucleation (initiation) phase. This design philosophy is distinct from the **defect-tolerant** approach in which the fatigue life of a component is based on the number of loading cycles needed to propagate an existing crack to a critical dimension for the material. The initial size of the flaw is assumed to correspond to the resolution of an inspection test. The critical dimension of the flaw is directly correlated to the fracture toughness of the material. The defect-tolerant philosophy is more commonly employed in safety-critical applications such as heart valve design.

Another facet of fatigue design is that most structural materials either cyclically soften or harden under the sustained loading. The specific response of a material is owed to its micromechanism of deformation, such as whether an alloy has the propensity to strain harden due to the interaction of dislocations. Figure 10.2 shows how the cyclic stress-strain loop continuously evolves with the stress amplitude either (a) increasing or (b) decreasing to maintain a fixed plastic strain range ($\Delta\varepsilon_{pl}$). For a material that cyclically hardens, the stress amplitude must be increased to maintain the fixed strain range, whereas for a softening material the stress range is continuously decreasing to maintain the specified strain amplitude. This phenomenon results in a cyclic yield stress that is either greater than or less than its monotonic equivalent. In metal systems, the yield stress saturates under sustained loading and the alloy may either soften or harden depending on its dislocation micromechanics. Polymeric materials always cyclically soften due to the nature of polymer chain slip and their tendency for hysteretic heating. The recurring stress strain behavior and concomitant cyclic yield stress should be known for any material that is subjected to sustained loading.

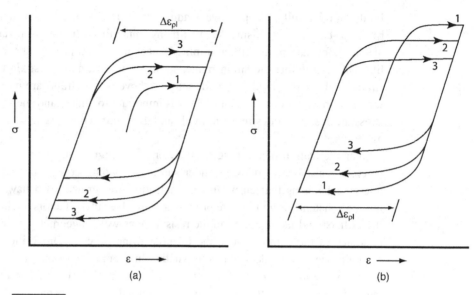

Figure 10.2

Evolution of cyclic stress-strain loops with the stress amplitude either (a) increasing or (b) decreasing to maintain a fixed plastic strain range. For a material that cyclically hardens, the amplitude must be increased to maintain the fixed strain range, whereas for a softening material the stress range is continuously decreasing to maintain the specified strain amplitude.

10.4 Total life philosophy

10.4.1 Stress-based loading

The total life philosophy for fatigue life prediction is founded on the notion that the component is initially defect free and that the life of the device is based on the initiation of a **coherent flaw** capable of growth and its subsequent extension into a critical crack size. Additionally, it is presumed in the stress-based total life design philosophy that the component will undergo high cycle fatigue conditions (discussed below) with little initial plastic strain involved in the fatigue process. Damage is accumulated in the component until a coherent flaw reaches a critical length and then failure occurs rapidly.

The fatigue characterization of a material based on the total life philosophy is based on either a *stress-based* test that examines the conditions for failure for a range of stress amplitudes and mean stress, or a *strain-based* test that examines the fatigue behavior under cyclic strain amplitudes. The former often utilizes an un-notched sample subjected to rotating or axial cyclic loading to determine the high cycle fatigue behavior (Figure 10.3(a, c)). In these tests, unnotched specimens are subjected to constant stress amplitude until failure occurs. Such tests are repeated over a range of stress amplitudes to generate stress amplitude-cycle (S-N) fatigue plots (Basquin, 1910). On the other hand, strain-based testing is generally utilized for materials with fracture criteria controlled by

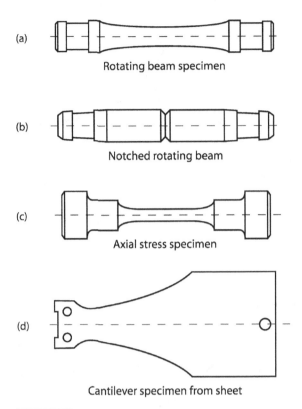

Figure 10.3

Schematic illustration of typical specimens utilized in total life tests: (a) un-notched rotating beam specimen, (b) notched rotating beam specimen, (c) un-notched axial stress specimen, and (d) cantilever specimen.

critical strain or for conditions where cyclic plastic strain is expected in service (Coffin, 1954; Manson, 1954; Morrow, 1968). Strain-based tests provide insight into the **low-cycle fatigue** process and often utilize notched samples to generate cyclic plastic strain conditions in the material (Figure 10.3(b,d)). Morrow (Morrow, 1968) outlined a strain-based testing methodology; however, cyclic plastic strains can be difficult to measure, and much of the design for fatigue resistance is centered upon stress-based tests. The testing methodology for the total life stress-based response of a material is provided in the ASTM Standards E466-E468 (American Society for Testing and Materials).

As mentioned above, in the stress-based total life tests, the un-notched samples are cycled until failure over a range of increasingly smaller values of stress amplitude to generate an (S-N) curve as shown in Figure 10.4. Due to the statistical nature of fatigue and the number of measurements needed to generate a S-N plot for a material, the number of specimens needed for this type of fatigue test is extensive. For a rotating test, a constant bending moment is applied to the specimen and cycled continuously until failure. The number of cycles or load reversals is determined for each stress level to generate stress (S) versus cycles to failure (N_f) data. The first test is performed just below

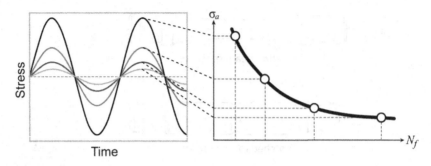

Figure 10.4

Schematic of typical S-N plot on linear-logarithmic scale showing the stress amplitude (σ_a) and concomitant cycles to failure for that loading range.

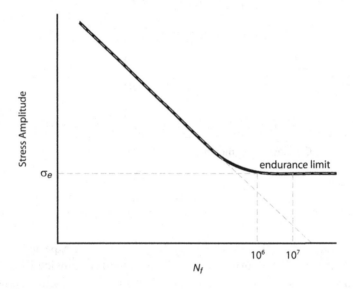

Figure 10.5

Schematic of typical S-N plot on linear-logarithmic scale showing the endurance limit (σ_e) that exists for certain metals such as ferrous alloys (steels). For materials without an endurance limit, such as titanium alloys, the curve continues to decrease after one million cycles and the fracture strength at 10 million cycles is often used for design purposes.

the monotonic or static strength for the material, and the subsequent tests are performed at a continuously decreasing stress level. This methodology is continued until either a stress level is reached for which anything below this value no longer results in failure (endurance limit, σ_e) or the tests are continued until a fatigue strength (the ordinate on the S-N plot) for a specified number of cycles is determined. Ferrous alloys (steels) have an endurance limit that occurs at one million cycles while most other alloy systems do not. An endurance limit is generally defined as the cyclic stress level that enables infinite life in the material (Figure 10.5). For systems without an endurance limit, the **fatigue**

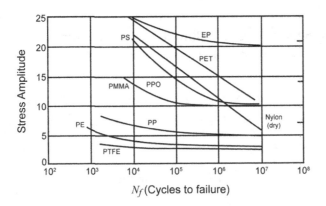

Figure 10.6

S-N fatigue data for several polymer systems. Note that PET and nylon do not have endurance limits.

strength for a given number of cycles (usually a minimum of 10 million cycles) is used instead as a material property for component design. The endurance limit is a critical material property for component design. The assumption is that if the device is subjected to a stress value below its endurance limit, then the device is safe from fatigue failure. The endurance limit of most steels is 35–50% of the ultimate tensile strength. The endurance limit can be affected by several factors such as surface finish, stress concentrations, heat treatment, environment, and component design (Marin, 1962). Often the component endurance limit is taken as the measured endurance limit under idealized conditions, and modifying factors are multiplied through to account for the various effects (Shigley and Mischke, 1989).

The relationship between the stress amplitude and the number of cycles to failure is known as the **Basquin equation** (Basquin, 1910) and is given as:

$$\sigma_a = \sigma'_f (N_f)^b \quad (10.4)$$

where σ_a is the stress amplitude, σ'_f is the fatigue strength coefficient and is comparable to the true fracture strength for the material, N_f is the number of cycles (or load reversals) to failure, and b is the Basquin exponent. The Basquin exponent is determined from the slope taken from the plot of stress amplitude versus the number of cycles to failure on a linear-log plot (Figure 10.5). The typical range of b for most metal and polymer systems is between -0.05 and -0.10. Figure 10.6 shows the S-N plots for several polymer systems used in medical devices. It can be seen in this figure that nylon and polyethylene terephthalate (PET) do not exhibit an endurance limit. On the other hand, polymers such as polyethylene (PE), polypropylene oxide (PPO), polystyrene (PS), polytetrafluoroethylene (PTFE), polypropylene (PP), polymethylmethacrylate (PMMA), and epoxy (EP) clearly exhibit an endurance limit below which failure does not occur in less than 10^7 cycles (Pearson and Pruitt, 1999).

Example 10.1 Determining the cycles to failure for a spinal implant

Consider the artificial spinal disk from Example 6.3. The stresses on the implant as shown in Figure 6.11 are:

$$\sigma_x = -2.2 \text{ MPa}, \sigma_y = -1.1 \text{ MPa}, \sigma_z = -0.58 \text{ MPa}$$
$$\tau_{xy} = -0.57 \text{ MPa}, \tau_{xz} = -0.79 \text{ MPa}, \tau_{yz} = -0.33 \text{ MPa}$$

and the principal stresses:

$$\sigma_1 = -0.25 \text{ MPa}, \sigma_2 = -0.86 \text{ MPa}, \sigma_3 = -2.77 \text{ MPa}$$

The Tresca yield criterion (Equation 8.4) was employed in Example 8.1 to determine the effective stress on the system:

$$\sigma_{\text{Tresca}} = 2.52 \text{ MPa}.$$

Using the Basquin equation (Equation 10.4), determine the number of loading cycles that can be sustained before failure occurs. The Basquin exponent for the material is -0.1.

Solution Since this implant is subjected to multiaxial loads, the first task is to determine the stress amplitude for the system. The largest stress range in this system is determined by the principal stress difference, $\sigma_1 - \sigma_3 = \sigma_{\text{Tres}} = 2.52$ MPa.

One can assume that the effective stress range serves as the effective stress amplitude in this case. The Basquin exponent for the material is -0.1 and recall that the yield strength is 8 MPa. If the yield stress is employed as the failure stress, then the Basquin equation takes the form:

$$\sigma_a = \sigma'_f (N_f)^b \Rightarrow 2.52 = 8(N_f)^{-.1}$$

$$N_f = \left(\frac{\sigma_a}{\sigma'_f}\right)^{\frac{1}{b}} = \left(\frac{2.52}{8}\right)^{-\frac{1}{0.1}} = 103{,}966.$$

Hence, for the effective stress range on the spinal implant, the number of cycles to failure is predicted to be 103,966 loading cycles.

The use of the Basquin equation assumes that the mean stress is zero – that is, that the specimen or component is undergoing **fully reversed loading** and that each cycle represents two reversals. If the mean stress is not zero, then these effects must be considered in predicting the life of the component. The mean stress has a dramatic effect on the fatigue behavior of a material, and this is schematically illustrated in Figure 10.7. As the mean stress of a fatigue cycle is increased, the number of cycles to failure and the endurance limit (if it exists) is decreased substantially. Most alloys and polymer systems are extremely sensitive to the mean stress of the sustained loading cycle.

10.4 Total life philosophy

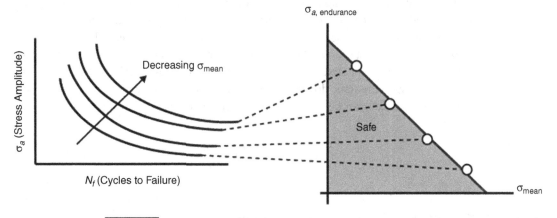

Figure 10.7

The effect of mean stress on the fatigue life of a material and the concomitant linearly decreasing endurance limit plotted as a function of mean stress.

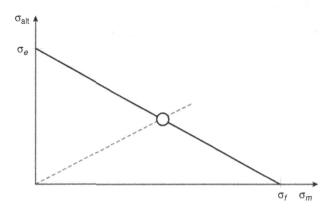

Figure 10.8

Illustration of the Goodman relationship between endurance limit and mean stress. The dotted line that initiates at the origin (zero mean stress) is used to determine the factor of safety for the material.

Figure 10.7 illustrates that the endurance limit decreases proportionately with increasing mean stress. This behavior is commonly modeled using the Goodman line that describes the linear proportionality between the endurance limit for a given alternating stress and the mean stress of the loading cycle. Figure 10.8 illustrates the Goodman relationship between endurance limit and mean stress. The dotted line that initiates at the origin (zero mean stress) is used to determine the factor of safety for the material.

The Goodman relationship uses the intercepts on the axes to develop the linear equation:

$$\frac{\sigma_a}{\sigma_e} + \frac{\sigma_m}{\sigma_f} = 1 \qquad (10.5(a))$$

where σ_e represents the endurance limit at zero mean stress, which can be written as $\sigma_a|_{\sigma_m=0}$, and σ_f represents the failure stress that is defined as the ultimate strength of the material, σ_{UTS}. The equation can also be modified to use the yield stress as the failure criteria under static conditions. The cyclic portion of the equation represents the endurance limit as compared to the cyclic driving force (stress amplitude) and the static portion represents the static driving force (mean stress) as compared to the static failure stress. The equation can be rewritten in the form commonly known as the Goodman equation:

$$\sigma_a = \sigma_a|_{\sigma_m=0}\left(1 - \frac{\sigma_m}{\sigma_{UTS}}\right). \qquad (10.5(b))$$

Example 10.2 Total life approximation for a femoral stem

A new low-modulus alloy is being considered for the femoral stem of a hip implant. Fully reversed loading tests are performed on un-notched specimens at a frequency of 20 Hz and in Hank's solution at 37°C. The cycles to failure for these specimens are obtained for a range of stress amplitudes.

Stress amplitude, σ_a(MPa)	Cycles to failure (N_f)
379	8000
345	13,000
276	53,000
207	306,000
172	1,169,000

(i) Determine the coefficients and basic form of the Basquin equation for this alloy using the stress-life data.

Solution (i) The basic form of the Basquin equation is: $\sigma_a = \sigma'_f(N_f)^b$.
Using the data in the table, the slope, b, is found as:

$$b = \frac{\log \sigma_1 - \log \sigma_2}{\log N_1 - \log N_2} = \frac{\log(379) - \log(172)}{\log(8000)_1 - \log(1,169,000)} = -0.159$$

The strength coefficient is found at the intercept where $N_f = 1$. Using the Basquin equation, one finds that $\sigma'_f = 1745$ MPa.

(ii) If the stress amplitude on the implant is 200 MPa, how many cycles will it last?

Solution (ii) Using $b = -0.159$, and $\sigma'_f = 1745$ MPa, one finds that N_f is on the order of 300,000 cycles.

(iii) Is this number of cycles sufficient for a hip implant to last a decade or more? What happens to the validity of the analysis if the mean stress is not zero?

Solution (iii) The average person walks one million steps per year. At the operating stress level of 200 MPa, the implant would last only about 4 months. This is an insufficient lifetime for the femoral stem as hip implants nominally have a 90% success rate after nearly 15 years of service. If the mean stress is not zero, then the Goodman equation should be used and the fatigue life will be shortened according to the level of mean stress applied on the component.

Example 10.3 Total life approximation using the Goodman equation

Fatigue tests are typically conducted in fully reversed loading, with no mean stress. Many structures, such as total joint replacements, experience conditions much closer to repeated loading, where the stress ranges from 0 up to the maximum, and the mean stress is approximately half the maximum stress ($R = 0$). Consider an orthopedic alloy with an endurance limit, $\sigma_e = 547$ MPa and a fracture strength, $\sigma_f = 1050$ MPa. How is the maximum cyclic stress allowable for infinite life determined?

Solution Because the mean stress is not zero, the Goodman equation (Equation 10.5) is employed for this analysis. Thus, employing Equation (10.5), write:

$$\frac{\sigma_a}{547} + \frac{\sigma_m}{1250} = 1$$

and thus

$$\sigma_a(1050/547) + \sigma_m = 1050 \text{ MPa.}$$

In repeated loading for this case, $\sigma_a = \sigma_m$, and hence $\sigma_a\,(3.29 + 1) = 1050$. Solving for σ_a we find that the maximum stress amplitude that can be sustained is $1050/4.29 = 291$ MPa, which is 50% of the endurance limit for fully reversed loading. The maximum allowable stress is twice the stress amplitude, 582 MPa, and is slightly greater than the endurance limit for fully reversed loading. Many devices do not experience constant amplitude loading but rather undergo continuous variations in stress or strain. In spectrum loading, the accumulated damage summed over the various amplitudes of loading in service is commonly used to estimate the lifetime of a device. In this method, the number of cycles at a given stress or strain range, n_i, is compared to the number of cycles to failure at that same amplitude, N_i, to estimate fractional lifetime, as shown in Figure 10.9.

The Palmgren-Miner accumulated damage model is commonly employed to estimate the amount of damage and fractional lifetimes used in each increment of

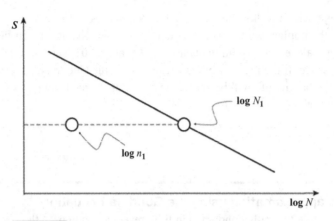

Figure 10.9

Illustration of the concept of fractional lifetime using the number of cycles, n_i, sustained at a given stress amplitude (S), as compared to the number of cycles to failure, N_i, at that same stress amplitude. This ratio serves as a measure of damage and provides an estimate of the fraction of lifetime used in the block of loading cycles.

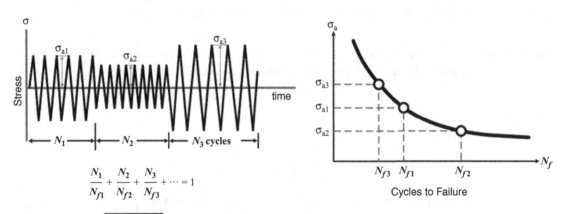

Figure 10.10

Illustration of incremental damage and Palmgren-Miner's rule (adapted from Dowling, 2007).

stress range (Miner, 1945; Palmgren, 1924). The incremental damage, d, is defined as:

$$d_i = \frac{N_i}{N_{f_i}} \tag{10.6}$$

and the total damage, D, is defined as the summation of incremental damage:

$$D = \sum \frac{N_i}{N_{f_i}}. \tag{10.7}$$

When the total damage sums to 1.0, then failure of the component is predicted (Figure 10.10).

Example 10.4 Fatigue life approximation using Palmgren-Miner's rule for variable amplitude loading

Consider a metallic implant that undergoes a series of variations in hourly loading as shown schematically in Figure 10.10. For this loading block, the variations in load are 6 reversals at $\sigma_{a1} = 290$ MPa, 10 reversals at $\sigma_{a2} = 200$ MPa, and 5 reversals at $\sigma_{a3} = 400$ MPa. The form of the Basquin equation for this alloy is $\sigma_a = 1758(N_f)^{-0.098}$.

(i) How many loading blocks can be sustained before fracture?

Solution (i) To solve variable loading problems, it is easiest to use a table format for the data so that damage increments can be readily calculated.

Stress amplitude, σ_a (MPa)	Cycles to failure (N_f)	Cycles at this amplitude (N)	Damage, $d = N/N_f$
400	136,000	5	3.67×10^{-5}
290	1,540,000	6	3.89×10^{-6}
200	4.29×10^9	10	2.33×10^{-9}

The total damage in one loading block is $\sum N/N_f = 4.04 \times 10^{-5}$; the number of loading blocks that can be sustained is 24,757.

(ii) Is this a sufficient number of cycles for a fracture fixation device that must last 6 months?

Solution (ii) The device offers 24,757 safe loading blocks. There are 24 hours in a day. The device offers 1,031 days of service. This is more than sufficient to support loads for 6 months.

10.5 Strain-based loading

Strain-based tests are often utilized when the structural component is likely to experience fluctuations in displacement or strain and are often utilized for components that are expected to undergo localized plastic strain, as may be the case for designs with notches or stress concentrations. The majority of strain-based fatigue tests are performed using fully reversed applied strain conditions and are often used when materials are expected to have some level of plastic strain in their applications. Cyclic strain data are often represented in a manner that is analogous to the S-N characterization used in stress-based testing. The total strain amplitude, $\Delta\varepsilon_a$, can be divided into the elastic, $\Delta\varepsilon_{a,el}$, and plastic, $\Delta\varepsilon_{a,pl}$, strain amplitude components:

$$\Delta\varepsilon_a = \Delta\varepsilon_{a,el} + \Delta\varepsilon_{a,pl} \qquad (10.8)$$

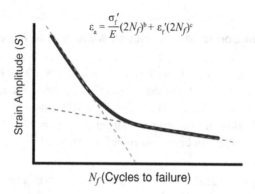

Figure 10.11

Illustration of the strain-based total life philosophy that incorporates both elastic and plastic strain amplitudes.

where the elastic strain amplitude component is obtained by normalizing the Basquin equation from the stress-based approach:

$$\Delta\varepsilon_{a,el} = \frac{\sigma'_f}{E}(2N_f)^b \tag{10.9}$$

and the plastic strain amplitude is given:

$$\Delta\varepsilon_{a,pl} = \varepsilon'_f(2N_f)^c. \tag{10.10}$$

The total strain amplitude of the fatigue cycle is plotted against the number of cycles or load reversals to failure, and is found by summing the elastic and plastic components:

$$\varepsilon_a = \frac{\sigma'_f}{E}(2N_f)^b + \varepsilon'_f(2N_f)^c. \tag{10.11}$$

Here, ε_a is the total strain amplitude, σ'_f is the strength coefficient, ε'_f is the ductility coefficient, $2N_f$ is the number of load reversals to failure, and b and c are material constants. The first term on the right side of equation represents the elastic component of the strain amplitude, while the second term signifies the plastic component of the strain amplitude. For strain-based tests, the strain amplitude is plotted as a function of number of cycles to failure, as shown in Figure 10.11.

Tests dominated by small amounts of cyclic plastic strain are designated **high-cycle fatigue**, while those with high plastic strains are termed **low-cycle fatigue**. The former occurs with a large number of cycles to failure and the latter is identified by the relatively small number of cycles to failure. It should be noted that most designs for structural components, with the exception of a balloon-expandable stent, avoid sustained plastic strain. However, local yielding can occur near stress concentrations or during rare overloading events. The number of cycles that marks the transition between

10.5 Strain-based loading

low-cycle fatigue and high-cycle fatigue is found by equating the elastic and plastic strain amplitudes and solving for N_t:

$$\frac{\sigma'_f}{E}(2N_t)^b = \varepsilon'_f(2N_t)^c \qquad (10.12)$$

and

$$2N_t = \left(\frac{\varepsilon'_f E}{\sigma'_f}\right)^{\frac{1}{b-c}}. \qquad (10.13)$$

Example 10.5 Determining critical strain for low-cycle fatigue in a tibial plateau

Consider the tibial plateau presented in Example 6.5 (likewise in Example 8.2).

This implant is made of UHMWPE in a constraining frame of CoCr and loaded with a force of 30 kN that is evenly spread across the tibial plateau. UHMWPE has a yield strength of 22 MPa and an elastic modulus of 1 GPa. The exponent for low-cycle fatigue, c, is -0.45.

Applying 3D Hooke's Law and geometric constraints provides the stresses and strains (Example 6.5):

$$\sigma_x = \sigma_y = 16.7 \text{ MPa}; \sigma_z = 25 \text{ MPa}$$

$$\tau_{xz} = \tau_{xy} = \tau_{yz} = 0 \text{ MPa}$$

$$\varepsilon_x = \varepsilon_y = 0; \varepsilon_z = 0.012$$

The effective von Mises stress, found in Example 8.2, is:

$$\sigma_{\mathit{eff}} = 8.3 \text{ MPa}$$

What is the critical strain amplitude needed for this device to fail in 50,000 load reversals (low-cycle fatigue condition)?

Solution The yield strain that occurs under uniaxial yielding is found by normalizing the yield strength with the elastic modulus and the strain to failure is measured under uniaxial tensile testing as 450%.

$$\varepsilon_y = \frac{\sigma_y}{E} = \frac{22}{1000} = 0.02$$

$$\varepsilon'_f = 4.5$$

Using Equation (10.10) for low-cycle fatigue, the cyclic plastic strain amplitude can be found:

$$\Delta\varepsilon_{a,pl} = \varepsilon'_f(2N_f)^c = 4.5(50{,}000)^{-.45} = 0.035.$$

Hence, a plastic strain amplitude of 3.5% results in failure after 50,000 cycles (low-cycle fatigue). Given that most orthopedic devices employed in knee replacement must last upwards of 15 million cycles, this strain level is inappropriate.

10.6 Marin factors

Fatigue test data are typically generated to standardized specimens under highly controlled and accelerated conditions defined by ASTM protocols. In this manner, large amounts of repeatable data can be accumulated for long-term failure conditions in a reasonable amount of time but in an idealized manner. However, actual components may differ radically in shape, size, loading configuration, and manufacturing process from the test specimens. Virtually all of these "real-world" distinctions from the test specimen condition result in a reduced fatigue life and endurance limit, and must be accounted for if an accurate prediction of fatigue performance is required for a given component.

A variety of common endurance limit-altering factors were studied and tabulated by Joseph Marin for use with metals (Marin, 1962), and have since been expanded and tabulated (Shigley and Mischke, 1989). Endurance limiting factors include a surface roughness effect, a part size effect, a loading mode effect (axial vs. bending), and a variety of less used factors for temperature, data reliability, or operating environment. The surface roughness of a component is a concern because it represents a population of flaws that nucleate fatigue cracks in service, and test specimens are typically mirror polished to reduce this effect. Larger components have a higher probability of containing an internal defect that would degrade the fatigue performance, and likewise, axial loading stresses more of a part cross-section than bending or torsional loading, in which the majority of the stress is experienced near the surface of the part. An example of how to apply these factors is given in Example 10.4, but for a more expanded treatment the reader may refer to any number of machine design textbooks (Norton, 2008; Shigley and Mischke, 1989, 1989).

It is clearly worth considering the differences between test specimens and manufactured parts before directly applying fatigue data in a prediction of fatigue performance. These factors are highly efficacious for component design when many types and sizes might be made from a relatively limited pool of materials. These factors may have limited validity for non-ferrous metals (Norton, 2008) or other less conventional structural materials. The development of such factors may not make sense for a unique material/application combination. Particularly for biomedical applications, where a device might contain materials from which it is difficult to form test specimens, or could include special coatings or material treatments unique to the manufacturing process of a component, it may be more straightforward to test the as-manufactured component in an *in vitro* configuration mimicking service in the body than to develop an extensive body of fatigue-modifying factors. The concept to take away is that fatigue data from accelerated

10.6 Marin factors

and idealized tests are often precise but not accurate depictions of the fatigue of the same material in a component, and special effort is recommended to estimate the error incurred in the use of such test data in failure-critical applications.

Example 10.6 Endurance limit error using Marin factors for a total hip replacement

A stainless steel total hip replacement stem is to be manufactured using the least amount of material possible to mitigate stress shielding in the adjacent bone. The steel stem uses a proprietary grade of corrosion-resistant steel that has been internally tested in uniaxial tension experiments according to standard protocols. The ultimate tensile strength obtained from the test specimens is 1470 MPa, and the endurance limit is 900 MPa.

The implants and specimens are machined from forged and heat-treated stock. The implants are then conventionally machined, while the test specimens are then ground and polished to a mirror finish. The test specimen (Figure 10.3) is loaded in fully reversed bending and has a diameter of 0.3 inches. The stress in the stem is primarily due to bending, and the stem has a circular cross-section of 0.5 inches at the location of interest. Using the provided tables and Marin factors, find the corrected endurance limit for the hip stem, accounting for size and surface effects.

The Marin factors are applied as coefficients to the endurance limit and are written as $\sigma_{\text{e-part}} = C_{\text{surf}} * C_{\text{size}} * \sigma_{\text{e-specimen}}$.

Surface roughness effect

The surface factor is computed as a function of the ultimate tensile strength

$$C_{\text{surf}} = A(\sigma_{ult})^b$$

where A and b are given as (Shigley and Mischke, 1989).

Table 10.1 Marin Factor coefficients

Surface finish	A (MPa)	b (unitless)
Ground	1.58	−0.85
Machined	4.51	−0.265
Hot-rolled	57.7	−0.718
As-forged	272	−0.995

With the data in Table 10.1, the machined component is found to have a surface factor of

$$C_{\text{surf}} = 4.51 * 1470^{-0.265} = 0.652$$

Thus, the surface roughness resulting from conventional machining (which is quite smooth to the touch) reduces the endurance limit by 35% alone, as compared to the value obtained from laboratory tests on specimens with mirror finish. It is worth noting that if the stem was left in the as-forged condition, $C_{\text{surf}} = 0.192$.

Size effect

The effect of the increased size of the component is accounted for using the following rule (Shigley and Mischke, 1989):

For $d < 0.3$ inches (7.62 mm); $C_{size} = 1$

For $0.3 < d < 10$ in (7.62 mm $< d <$ 254 mm); $C_{size} = 0.869 d^{-0.097}$

Else: $C_{size} = 0.6$

Thus, for the stem $C_{size} = 0.869(0.5)^{-0.097} = 0.929$, which is a much smaller effect than that from surface roughness.

Since the component and test specimen experience a similar load state, we are neglecting loading mode. Thus, altogether we find the endurance limit for the component (part) is $\sigma_{e\text{-part}} = 0.652*.929*\sigma_{e\text{-specimen}} = 0.606*\sigma_{e\text{-specimen}} = 545$ MPa.

A 40% drop in the endurance limit could represent a substantial loss of safety margin, and therefore we can see that it is critical to account for any difference between test specimens and components unless they can be dramatically overdesigned for safety.

10.7 Defect-tolerant philosophy

10.7.1 Fracture mechanics concepts

The defect-tolerant philosophy is based on the implicit assumption that structural components are intrinsically flawed and that the fatigue life is based on propagation of an initial flaw to a critical size. Fracture mechanics is used to characterize the propagation of fatigue cracks in materials that are capable of sustaining a large amount of sub-critical crack growth prior to fracture. The defect-tolerant approach is used in safety-critical applications where defect-tolerant life estimates are essential, such as heart valve designs.

In order to characterize the intrinsic resistance to crack propagation under cyclic loading conditions, specimens that contain notches and pre-cracks are used to determine that rate of crack advance under different combinations of stress amplitude and flaw size. The stress intensity factor, K, derived from linear elastic fracture mechanics is the parameter used to describe the magnitude of the stresses, strains, and displacements ahead of the crack tip. This parameter is used for linear elastic materials and for materials that conform to the conditions of small-scale yielding (local plastic effects ahead of the crack tip are limited). The linear elastic solution for the normal stress in the opening mode of loading is written as a function of distance, r, and angle, θ, away from the crack tip:

$$\sigma_{yy} = \frac{K_1}{\sqrt{2\pi r}} \cdot \cos\frac{\theta}{2} \left\{ 1 + \sin\frac{\theta}{2} \sin\frac{3\theta}{2} \right\} \qquad (10.14)$$

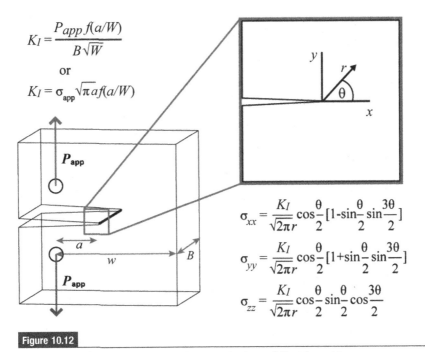

Figure 10.12

Illustration of the compact tension specimen and the associated stress fields at the crack tip.

where K_I is the Mode I (opening mode) stress intensity factor. The stress intensity parameter incorporates the boundary conditions of the cracked body and is a function of loading, crack length, and geometry (Anderson, 2004). The stress intensity factor can be found for a wide range of specimen types and is used to scale the effect of the far-field load, crack length, and geometry of the flawed component. The most common specimen type used in characterization of fatigue crack propagation resistance is the compact tension geometry, as shown in Figure 10.12.

The applied load can be applied to a specimen in one of three primary modes, as depicted in Figure 10.13. Mode I is the opening mode, in which the crack tip experiences predominantly tensile stresses; Mode II is the shear mode where the crack tip experiences primarily in-plane stresses; and Mode III is the out-of-plane shear mode where the crack tip experiences a torsional stress state. The most conservative mode (the harshest conditions for a propagating crack tip) is Mode I. This is the most common mode of loading in fatigue crack propagation tests. Any loading scenario that combines these modes of loading is termed "mixed-mode" crack propagation.

10.7.2 Fatigue crack propagation

Linear elastic fracture mechanics provides a conservative design approach in comparison to the total life methodology for predicting the life of a cracked structural component

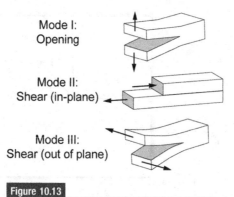

Figure 10.13

Illustration of the three primary modes of loading at a propagating crack tip.

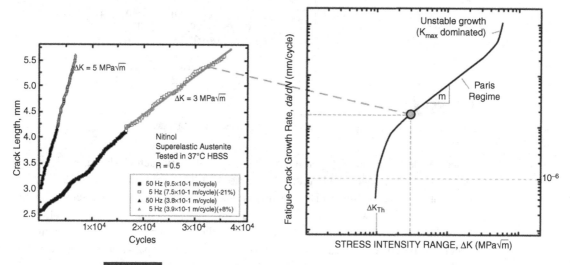

Figure 10.14

Illustration of linear crack growth as a function of loading cycles and how this crack velocity (slope) is used to generate data on a da/dN versus ΔK plot. The data on the left are shown for a Nitinol (Ni-Ti) shape memory alloy.

under cyclic loading conditions. Although the fracture micromechanisms vary for metals, polymers, and ceramics, the fatigue crack propagation behaviors of these materials share many similar attributes at the macroscopic scale. The velocity of a moving fatigue crack subjected to constant stress amplitude loading is determined from the change in crack length, a, as a function of the number of loading cycles, N. This velocity represents the fatigue crack growth per loading cycle, da/dN, and is found from experimentally generated curves where a is plotted as a function of N (Figure 10.14). For constant amplitude loading, the rate of crack growth increases as the crack grows longer, since the stress intensity is a function of the crack length (Equation 10.14). Fatigue crack propagation resistance is typically presented with the crack velocity as a function of stress intensity on a logarithmic scale.

10.7 Defect-tolerant philosophy

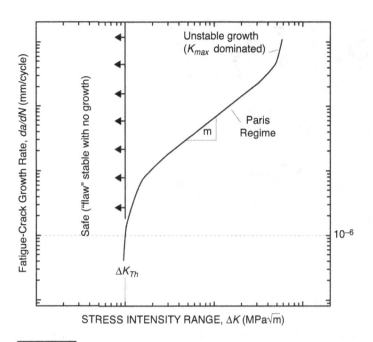

Figure 10.15

Illustration of the sigmoidal fatigue crack propagation plot on log-log scale. This schematic depicts the three primary regimes of crack growth: near-threshold, linear, and fast fracture where peak intensity drives the fracture process.

Paris (Paris, 1964) suggested that the stress intensity factor range, $\Delta K = K_{max} - K_{min}$, which itself captures the far-field stress, crack length, and geometry, should be the characteristic driving parameter for fatigue crack propagation. This is known as the Paris law, and it states that da/dN scales with ΔK through the power law relationship:

$$\frac{da}{dN} = C \cdot \Delta K^m \tag{10.15}$$

where C and m are material constants. The Paris equation is valid for intermediate ΔK levels spanning crack propagation rates from 10^{-6} to 10^{-4} mm/cycle.

Figure 10.15 schematically illustrates the sigmoidal curve that captures the crack growth rate as a function of stress intensity range. The plot illustrates three distinct regions: the slow crack growth or threshold (referred to as near-threshold in figure caption) regime, the intermediate crack growth or Paris regime, and the rapid crack growth or fast fracture regime. While the Paris regime is most often used for life prediction, the fatigue threshold is key for designing against the inception of crack growth when components are expected to have long service lifetimes or when intermittent inspections may be difficult. For safe design, engineers use the fatigue threshold value for estimates of allowable stresses that do not enable growth of a flaw. Such data, however, are difficult to generate and are generally not available for all biomaterials used in medical implants as the near-threshold stress intensity range is defined as a crack velocity of 10^{-7} m/cycle

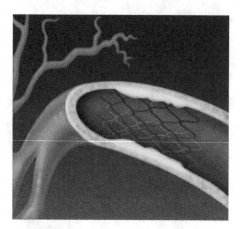

Figure 10.16

Schematic of a metallic stent used to restore blood flow in a vessel.

(10^{-10} mm/cycle) or less. Thus, it is more common that the linear crack growth (Paris) regime is used as a conservative measure to predict the life of a component.

The Paris law is commonly employed for fatigue life prediction of components that have known stress concentrations and which are used in safety-critical designs. It is implied in this defect-tolerant approach that the device or component contains an initial defect or crack size, a_i. Initial flaw size is typically determined from **non-destructive evaluation** (NDE) techniques such as electron microscopy, X-ray spectroscopy, or ultrasound. In the event that no defect is found, an initial defect whose size is the limit of resolution of the NDE method is assumed to exist as a worst-case scenario. Assuming that the fatigue loading is performed under constant stress amplitude conditions, that the geometric factor, $f(\alpha)$, does not change within the limits of integration, and that fracture occurs when the crack reaches a critical value, a_c, one can integrate the Paris equation in order to predict the number of cycles to failure (fatigue life) of the component:

$$N_f = \frac{2}{(m-2)Cf(\alpha)^m(\Delta\sigma)^m \pi^{m/2}} \cdot \left[\frac{1}{a_i^{(m-2)/2}} - \frac{1}{a_c^{(m-2)/2}}\right] \text{ for } m \neq 2. \quad (10.16)$$

Example 10.7 Fatigue design of a Nitinol stent

Consider a stent device made of the shape memory alloy, Nitinol. The stent is used to restore blood flow to a vessel as shown in Figure 10.16, and must withstand physiological cyclic loads for the life of the device. The average heartbeat is 72 beats per minute. The typical strut thickness is 500 μm.

The fatigue crack propagation constants for Nitinol (tube) are $C = 2 \times 10^{-11}$ (MPa$\sqrt{m}$)^m, $m = 4.2$, and $\Delta K_{th} = 2.5$ MPa$\sqrt{m}$ as seen in Figure 10.15. The geometric parameter, $F(a/W)$ (note that $F(a/W)$ is the same as $F(\alpha)$) for the strut is known from finite element analysis to be 0.624. The maximum allowable flaw size in a medical grade Nitinol alloy is 39 μm as per ASTM F2063. The physiological stress range on the implant is 294 MPa.

How can the defect-tolerant approach be used to safely design a stent that will resist fatigue crack growth and that will offer structural integrity for the life of the patient?

Solution Because of the critical nature of the application and the fact that the device undergoes 72 beats per minute for the life of the patient, it is essential that any flaw present is incapable of propagating. For this reason, the threshold stress intensity factor is used to determine the critical flaw size that initiates the onset of crack growth (Figure 10.15). Using the general form of the stress intensity equation shown in Equation (10.11), we can rearrange and solve for the critical crack length as:

$$\Delta K_{th} = \Delta\sigma\sqrt{\pi a} \cdot F\left(\frac{a}{W}\right);$$

$$a_{cr} = \frac{1}{\pi}\left[\frac{\Delta K_{th}^2}{F\left(\frac{a}{W}\right)^2 \Delta\sigma^2}\right] = 59 \text{ μm}$$

Because the critical flaw size for the inception of crack growth is 59 μm and the maximum allowable flaw size for the Nitinol alloy is 39 μm, the device is safe against fatigue crack growth.

Example 10.8 Fatigue crack propagation in a flawed hip implant

Consider the fracture of a hip stem that was traced back to its laser etching. Years ago, laser etching was used on the lateral side of the stem to mark a serial number. The material removed by the laser etching is essentially a small edge-notched crack that is 1 mm deep on the side of the stem as shown in Figure 10.17. It is estimated that the tensile bending stress on the stem is approximately 90 MPa for a typical active male weighing 200 pounds.

The femoral stem is made of a CoCr alloy with a fracture toughness of 9.5 MPa, and fatigue crack propagation constants $C = 6 \times 10^{-11}$ (MPa$\sqrt{m}$)^m and $m = 4$. The form of the stress intensity factor for a single edge-notched geometry is $K_I = 1.10\sigma\sqrt{a}\sqrt{\pi}$.

What is the critical flaw size for this alloy? How many fatigue cycles will this system last? Is this acceptable for a hip implant?

Solution Using the general form of the stress intensity range, we can rearrange and solve for the critical crack length as:

$$\Delta K = 1.12\Delta\sigma\sqrt{\pi a}$$

$$a_{cr} = \frac{1}{\pi}\left[\frac{K_{IC}^2}{1.12^2 \Delta\sigma^2}\right] = 2.8 \text{ mm}.$$

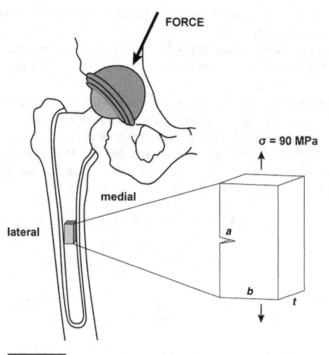

Figure 10.17

Schematic of a metallic femoral stem with an embedded flaw.

The number of cycles to failure can be determined using the integration of the Paris equation (Equation 10.16) from an initial flaw size of 1×10^{-3} m to 2.8×10^{-3} m, and assuming that the implant will undergo 2 million cycles per year, we find that the component will last 7 years:

$$N_f = \frac{2}{(m-2)Cf(\alpha)^m(\Delta\sigma)^m \pi^{m/2}} \cdot \left[\frac{1}{a_i^{(m-2)/2}} - \frac{1}{a_c^{(m-2)/2}}\right]$$

$$= \frac{1}{6 \times 10^{-11}(1.12)^4(90)^4\pi^2} \cdot \left[\frac{1}{(1 \times 10^{-3})^2} - \frac{1}{(2.8 \times 10^{-3})^2}\right]$$

$$= 14.2 \times 10^6 \text{cycles} = 7.1 \text{ years}$$

The average hip implant lasts 15–20 years. The laser etching in this case results in a fatigue life that is more than halved due to the initial stress concentration.

10.7.3 Fatigue behavior of structural materials

As with quasi-static loading there are two classifications of toughening mechanisms that are utilized in resisting crack growth under varying load conditions. **Intrinsic**

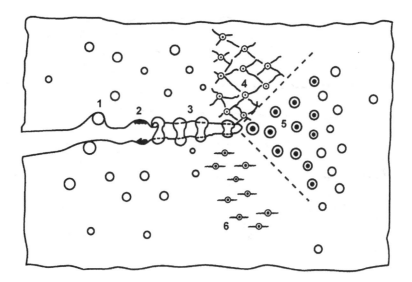

Figure 10.18

Schematic showing crack shielding mechanisms that can exist in fatigue crack propagation (1) crack deflection; (2) particle tearing; (3) bridging by fibers or second-phase particles; (4) shear banding; (5) plastic void growth or phase transformation; and (6) multiple craze formation or microcracking.

toughening mechanisms involve fundamental material properties that enhance the plastic deformation or ductility of the material and are controlled by the microstructure (heat treating, grain size, precipitate distribution, etc.). As the level of plasticity or ductility of the material is increased, the intrinsic toughness is enhanced. **Extrinsic toughening** involves mechanisms that are activated during the propagation of the crack and include: (i) *crack deflection* in which the crack is forced out of plane as it advances and the effective crack tip stress intensity is reduced; (ii) *contact shielding* in which mechanisms work behind the advancing crack and typically produce traction or closure in the crack wake through particle or fiber bridging; and (iii) *zone shielding* in which growth mitigating processes such as shear banding at second-phase additions, phase transformation, void formation around particles, multiple crazes around reinforcing phases, or microcracking are activated within the crack tip process zone (Evans, 1980; Pearson and Pruitt, 1999; Ritchie, 1999; Ritchie *et al.*, 2000; Suresh, 1998). These mechanisms play a critical role in the near-threshold and fast fracture regime of fatigue crack growth. The extrinsic crack tip shielding effect has been expressed as (Ritchie, 1999; Suresh, 1998):

$$\Delta K_{\text{tip}} = \Delta K_a - K_s \quad (10.17)$$

where ΔK_a is the applied stress intensity range and K_s is the stress intensity factor due to shielding. Figure 10.18 shows a schematic illustration of several crack shielding mechanisms that operate in structural materials.

Ritchie (Nalla *et al.*, 2003; Ritchie, 1999) has extensively characterized the difference in fatigue behavior between ductile and brittle solids and attributes the disparity primarily

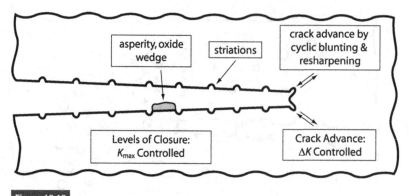

Figure 10.19

Illustration of crack advance mechanisms that are activated in the crack wake (K_{max} controlled) and ahead of the crack tip (ΔK controlled) in a ductile metal. Fatigue fracture surfaces in metals depict striations owing to the continual sharpening and blunting at the near tip zone (adapted from Ritchie, 1999).

in their dependence of crack growth phenomena on their respective intrinsic and extrinsic toughening processes. His work links the fatigue crack propagation behavior of a material under cyclic or static loading to the nature of the intrinsic and extrinsic fracture processes, such that cyclic phenomena drive a dependence on cyclic loading, and static mode phenomena are independent of load cycling and therefore are active even under static stresses.

The relative dependence of fatigue crack propagation under fluctuating loading on cyclic or static mode phenomena is characterized by the correlation (respectively) of crack growth to the range of the stress intensity factor, ΔK, or to the peak stress intensity factor, K_{max}. Ductile materials such as metals undergo fatigue and experience cyclic damage through unloading and crack tip resharpening. Therefore, the intrinsic driving force correlates primarily to ΔK, while extrinsic processes such as oxide wedging are quasi-static and correlate to K_{max} (Figure 10.19).

Unlike metals, the crack tip in brittle materials (ceramics) does not experience resharpening or other phenomena that are driven by cyclic loading. Thus, crack advance is not related directly to ΔK but rather is strongly correlated to the peak value of the stress intensity, K_{max}. Figure 10.20 shows that bridging mechanisms behind the crack tip in ceramics are controlled by ΔK; however, the intrinsic micromechanisms ahead of the crack tip are controlled by static mechanisms. Thus for ceramics, and other brittle materials that propagate cracks in a non-cyclic (static mode) manner, it is the peak stress intensity, K_{max}, that dictates the crack advance under global cyclic loading conditions.

Mathematically, this model couples the active cyclic and static fatigue crack propagation mechanisms in a synergistic way through the respective effects of their driving forces:

$$\frac{da}{dN} = C \Delta K^p K_{max}^q \qquad (10.18)$$

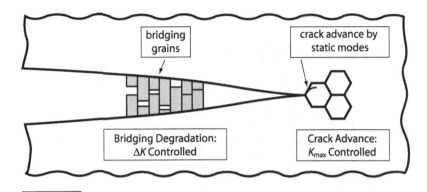

Figure 10.20

Illustration of crack advance mechanisms that are activated in the crack wake (ΔK controlled) and ahead of the crack tip (K_{max} controlled) in a brittle ceramic (adapted from Ritchie, 1999).

where p and q represent the cyclic and static exponents for the material. For materials that accumulate damage predominantly due to cyclic phenomena, $p \gg q$, whereas those in which static mode phenomena dominate have $q \gg p$. This argument also applies to brittle and ductile material behavior, where for the former case $q \gg p$ and crack propagation is practically only dependent on K_{max}, and the latter is driven by ΔK and collapses to the Paris equation (Ritchie, 1999).

This treatment essentially represents two opposing extremes of behavior (brittle vs. ductile), with many possible fatigue behaviors between that are a mixture of cyclic and static effects. For instance, the fatigue of intermetallic compounds, which are chemically both metal-like and ceramic-like, falls almost evenly between the behavior of metals and ceramics with $p \sim q$. The fatigue of polymers can likewise span the spectrum, depending on the material properties and extrinsic fracture phenomena that activate in the material.

10.7.4 Fatigue behavior of metals

The mechanism for crack propagation in most metals involves the incremental crack advance by a mechanism of localized plasticity and blunting during loading, and a resharpening processing upon unloading (Suresh, 1998). This mechanism results in striation markings observed in fracture surfaces of metals as illustrated in Figure 10.21. In metal systems, the stress in the neighborhood of the crack tip is governed by the stress intensity (in small-scale plasticity); however, the damage process is essentially cyclic, so the intrinsic micromechanisms of crack propagation in metals are dominated by ΔK under applied cyclic loading. The crack-wake mechanisms contributing to crack closure are essentially static mode and correlate to the peak stress intensity, K_{max} (Figure 10.19). The typical range of the cyclic fatigue exponent (the slope, m, in the Paris regime) for metal alloys is on the order of $m = 2$–4. This is considerably lower than values reported for ceramics (discussed below) and indicates that ductile metals are generally more

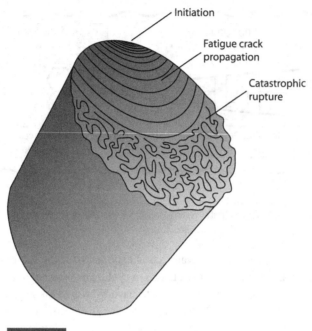

Figure 10.21

Schematic illustration of fatigue striations observed in the fracture surface of a metal.

tolerant of fatigue crack propagation as a class of materials, as the crack propagation rate accelerates after initiation at a much lower rate than in brittle materials. A comparison of fatigue crack propagation behavior of a range of medical alloys spanning the near-threshold regime through the fast fracture regime is shown in Figure 10.22.

Example 10.9 The effect of geometry on the critical flaw size

Consider the two hip implant geometries presented in Example 6.8. The design of the rectangular cross-section is altered with a laser etching that serves as an edge-notched crack that is 0.5 μm in depth. The circular implant is stamped with the lot number such that it serves as an edge-notched flaw that is 0.1 μm in depth. What will be the effect on the number of cycles to failure with all other factors considered equal, and propagation slope of $m = 4$? The form of the stress intensity factor for an edge-notched crack is:

$$K_I = 1.12\sigma^\infty \sqrt{\pi a} \Rightarrow \Delta K_I = 1.12\Delta\sigma^\infty \sqrt{\pi a}$$

Solution The Paris equation can be integrated to determine the number of cycles to failure (Equation 10.16):

$$N_f = \frac{2}{(m-2)Cf(\alpha)^m(\Delta\sigma)^m \pi^{m/2}} \cdot \left[\frac{1}{a_i^{(m-2)/2}} - \frac{1}{a_c^{(m-2)/2}}\right] \text{ for } m \neq 2$$

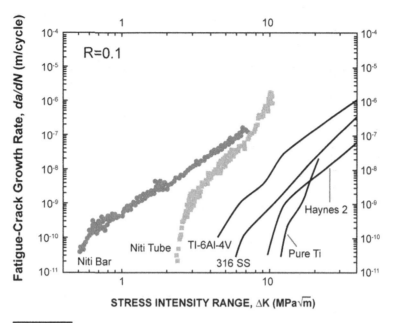

Figure 10.22

Fatigue crack growth rate as a function of stress intensity range for several medical grade alloys including stainless steel, Nitinol (Ni-Ti), Haynes 2, pure titanium, and Ti-6Al-4V. (After Ritchie, 1999.)

For all other factors considered equal, the ratio of the lifetimes can be written as:

$$\frac{N_f, \text{circ}}{N_f, \text{rect}} = \frac{a_{\text{Rect},i}^{(m-2)/2}}{a_{\text{Circ},i}^{(m-2)/2}} = \frac{a_{\text{Rect},i}^2}{a_{\text{Circ},i}^2} = \frac{0.5^2}{0.1^2} = 25$$

$$N_f, \text{circ} = 25 N_f, \text{rect}.$$

The number of fatigue cycles that can be sustained before failure is 25 times greater in the circular cross-section with the smaller flaw.

10.7.5 Fatigue behavior of ceramics

Ceramic systems are generally known to be highly sensitive to flaws and generally do not have high resistance to fatigue crack propagation. The Paris regime slope, m, for many ceramic systems is high and is often on the order of $m = 50$–100. Crack propagation in ceramics is highly sensitive to $K_{\max}$, and the growth of flaws in ceramics has been shown to proceed in a predominantly static mode (Ritchie, 1999). Because ceramics rely on extrinsic toughening mechanisms, the length of the flaw can have a dramatic effect on fracture resistance. Short cracks are not able to offer the same extent of resistance to crack advance as that same material with a longer crack (wake). Hence, as discussed

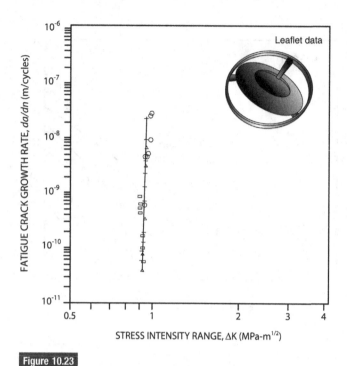

Figure 10.23

Fatigue crack growth rate as a function of stress intensity range for pyrolytic carbon. (After Ritchie, 1999.)

in the previous chapter, ceramics are known to have rising **R-curve** (resistance curve) behavior. The short crack phenomenon can strongly affect the near-threshold regime where the same stress intensity range may result in vastly different crack propagation rates depending on the crack length. Figure 10.23 shows the fatigue crack propagation rates as a function of stress intensity range for pyrolytic carbon used in heart valves (Ritchie, 1999). Note the steepness of the propagation slope in comparison to that for metals and that there is a very small region of stable crack propagation in brittle materials. As stated in the previous section, this is consistent with the micromechanism of fatigue crack growth driven by peak stress intensity (quasi-static fracture processes).

10.7.6 Fatigue behavior of polymers

Polymers are generally known to be susceptible to creep and strain rate effects; and thus it is important to understand the viscoelastic nature of polymers under cyclic loading conditions. The dissipation of energy in cyclic loading of polymers results in heat generation; the amount of temperature elevation depends strongly on frequency, deformation amplitude and the damping properties of the specific polymer. The energy dissipated per second is given as (Ferry, 1961; Schapery, 1975):

$$\dot{E} = \pi \nu J''(\nu, T, \sigma)\sigma^2 \qquad (10.19)$$

where ν is the frequency, J'' is the loss compliance, and σ is the peak stress of the fatigue cycle. The rate of change of temperature for adiabatic heating conditions in which the heat generated is transferred into temperature rise is given as:

$$\frac{dT}{dt} = \frac{\dot{E}}{\rho c_p} \qquad (10.20)$$

where ρ is the mass density and c_p is the heat capacity of the polymer. Due to this phenomenon, polymers can actually melt rather than fracture due to hysteretic heating (Hertzberg and Manson, 1980) and for this reason most polymer fatigue tests are not performed at frequencies above 5 Hz.

Another approach to handling the cyclic and static modes of crack advance in polymers is to couple the growth of the flaw due to cyclic loading conditions and due to creep but to consider them additive:

$$da = \left[\frac{\partial a}{\partial N}\right]_t dN + \frac{1}{f}\left[\frac{\partial a}{\partial N}\right]_N dN \qquad (10.21)$$

where the first term reflects the contribution to crack growth due to the cyclic component of the load (ignoring time effects, i.e., the Paris equation) and the second component gives the contribution of time-dependent creep component, which is dependent on the loading frequency (Hertzberg and Manson, 1980).

Polymeric materials can be highly dependent upon local stress states and the presence of flaws, and a seemingly ductile system can exhibit brittle behavior in the presence of a crack or stress concentration. An interesting polymer in this regard is ultra-high molecular weight polyethylene (UHMWPE), used in total joint replacements. There is a plethora of literature devoted to the mechanical characterization of UHMWPE (Kurtz, 2009; Kurtz et al., 1998; Pruitt, 2005; Rimnac and Pruitt, 2008); and yet it has been recently shown that this polymer is highly sensitive to the peak stress intensity of the fatigue cycle (Furmanski and Pruitt, 2007). In such a system, the polymer is ductile in bulk loading conditions yet is brittle in localized or concentrated stress states. In fact, the capacity of UHMWPE for creep may be sufficient to initiate flaws loaded under static conditions, and this polymer can behave like a time-dependent ceramic in the presence of a flaw or stress concentration (Furmanski and Pruitt, 2007). This finding has great implications for total joint replacement design ((Furmanski and Pruitt, 2009).

The fatigue behavior of polymers depends on many factors. In general, polymers with higher molecular weight, chain entanglement density, and crystallinity are more resistant to crack propagation while increased crosslinking decreases resistance to fatigue crack propagation (Hertzberg and Manson, 1980). Figure 10.24 presents compiled fatigue crack propagation data for UHMWPE subjected to varying degrees of crosslinking (radiation dose) and crystallinity (thermal treatments) and shows that the polymers with

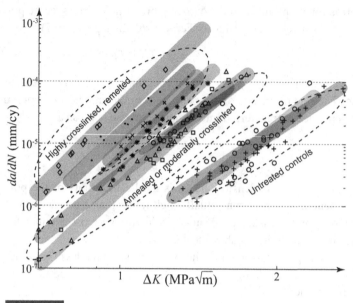

Figure 10.24

Fatigue crack growth rate as a function of stress intensity range for a range of clinically relevant formulations of UHMWPE. Untreated controls, with no crosslinking and greatest level of crystallinity, offer the greatest resistance to fatigue crack propagation (adapted from Atwood et al., 2009).

greatest crystallinity and least amount of crosslinking offer the best resistance to fatigue crack propagation (Atwood et al., 2009; Rimnac and Pruitt, 2008).

10.8 Case study: fatigue fractures in trapezoidal hip stems

In the early to mid-1970s, a hip stem design known as the Trapeziodal-28™ (T-28) was developed as an attempt to improve the stability and strength over previous stem designs. The Trapeziodal-28™ was so named because it had a trapezoidal cross-section in its stem and neck and its femoral head diameter was 28 mm. This hip implant design utilized medical grade 316L stainless steel in both the wrought and lightly worked condition, and its design helped to maximize bone cement penetration into adjacent bone upon insertion. In the 1970s, thousands of these T-28 designs were implanted; however, these implants failed at a rate that was four times that of other femoral stems (Rimnac et al., 1986). Rimnac and co-workers investigated the failure mechanisms of the T-28 design and found 21 out of 805 implanted devices exhibited fractures; 18 of these failures occurred as fractures in the femoral stem. The 18 stem failures were noted to generally occur in the top third (proximal region) of the femoral stem, while the remaining failures occurred in the neck region.

All devices were examined with both optical and scanning electron microscopy. The crack initiation sites were found to occur on the posterior corner on the medial side of the

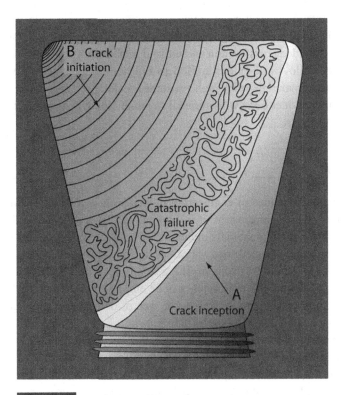

Figure 10.25

Fractography of the fractured trapezoidal T-28 stem. Flaw initiates at the posteromedial corner (A) and clamshell markings originate from the lateral side (B). (After Rimnac *et al.*, 1986.)

stem and clamshell markings associated with fatigue were noted on the fracture surfaces (Figure 10.25). Rimnac and co-workers concluded, based on their fractography, that the failures were initiated due to nucleation and growth of flaws under fatigue loading. They correlated the spacing on the clamshell markings to the rate of crack advance in the implants.

Using this information along with the stress intensity of a corner crack and the fatigue crack propagation data for 316L stainless steel, they were able to estimate the values of stress intensity range, ΔK, and the range of stress operating on the lateral side of the implant. They estimated that the lateral side cracks were subjected to cyclic tensile stresses in the range of 115–244 MPa. Using curved beam theory, Rimnac found that due to axial and bending stresses on the implant, initial stresses on the medial side were on the order of 600 MPa in compression and 550 MPa in tension on the lateral side.

The yield stress for 316L stainless steel is on the order of 550 MPa and hence plastic mechanisms and localized yielding would be expected; moreover, residual tensile stresses were found to be sufficient to nucleate crack growth on the medial side. This crack then enhances the stress levels on the lateral side of the device and a crack then originates on the lateral side as well. The merger of the two crack fronts under cyclic loading results in the fracture of the devices. Due to a preponderance of these types of failures, the femoral stem design evolved away from sharp corners found in the T-28 and moved to a rounded cross-section and utilized metals with enhanced fatigue resistance, such as CoCr alloys and Ti alloys. Modern hip stem designs generally utilize titanium alloys as these materials offer high resistance to fatigue crack inception and growth, and titanium provides bone with minimal stress shielding owing to its lower modulus as compared to steels.

10.9 Summary

The fatigue performance of engineering biomaterials is affected by many factors. For safe design of medical devices, it is imperative that appropriate test conditions are used to evaluate the biomaterials in laboratory settings. When possible, test conditions should mimic the expected load variations and physiological environment as well as the potential for stress corrosion cracking or any other *in vivo* degradation process of the implant. The total life philosophy can be employed to predict the life of the device if the structural component is likely to be free of defects and stress concentrations or if the component is likely to spend the majority of its lifetime in the initiation stage of crack growth. In contrast, the use of fracture mechanics should be used for safety critical fatigue designs and in flawed structural components likely to sustain a large degree of stable crack growth prior to fracture. The fatigue behavior of medical devices is complicated by the interplay of biochemical and mechanical environments *in vivo*. Safe design of biomaterial components requires evaluation under appropriate fatigue conditions which best mimic service conditions of the device.

10.10 Problems for consideration

Checking for understanding

1. Describe the different philosophies utilized in the fatigue design of medical implants.
2. Based on the above description of fatigue design philosophies, elucidate the process by which you would safely design a heart valve made of pyrolytic carbon. Would your design methodology change if instead you were designing an artificial tendon made of polyethylene? What, if anything, is different between these two situations?
3. Explain the sources and significances of statistical scatter in fatigue data. How might this affect the use of S-N or da/dN vs. ΔK plots for medical device design purposes?
4. The S-N curve for an alloy is characterized by the Basquin relationship:

$$\sigma_a = \sigma'_f(2N_f)^b$$

where σ_a is the stress amplitude, σ'_f is the strength coefficient, $2N_f$ is the number of load reversals to failure, and b is the Basquin exponent approximately equal to –0.09. When the stress amplitude is equal to the ultimate tensile strength of the

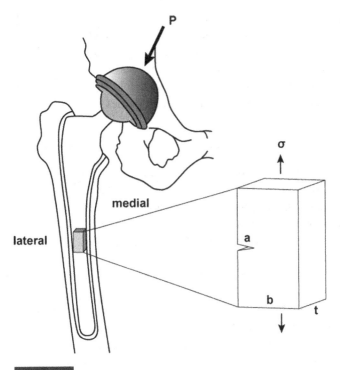

Figure 10.26

Schematic showing a side crack in a femoral stem.

material in this alloy, the fatigue life is 1100 cycles. If a sample spends 70% of its life subjected to an alternating stress level equal to its fatigue endurance limit, σ_e, 20% of its life subjected to $1.1\sigma_e$, and 10% at $1.2\sigma_e$, estimate the fatigue life using the Palmgren-Miner linear damage rule. Schematically illustrate the S-N behavior (log-log scale) for this material.

5. A medical device comprised of a surgical grade alloy is to be exposed to a cyclic tensile stress range of 100 MPa ($R = 0$). Prior to testing, it has been determined that the largest surface flaw is 2 μm in length. The fracture toughness of this material is 25 MPa√m and its yield strength is 210 MPa. The basic form of the Paris equation for this alloy is:

$$da/dN = 1 \times 10^{-10}(\Delta K)^3$$

(a) Estimate the fatigue life of this device.
(b) Is this an acceptable fatigue life for a cardiovascular or orthopedic device?

For further exploration

6. Consider a crack that propagates from the stress concentration due to laser etching of the implant as given above, as depicted in the image in Figure 10.26.

Your supervisor asks you to compare the fatigue performance of orthopedic materials. List the yield strength, elastic modulus, $\Delta K_{threshold}$, and K_{IC} for several common orthopedic metals in a table, and comment on how they might resist crack propagation and fracture. Are there any apparent trade-offs for performance in these materials? Be sure to include the worked condition (as wrought, cold rolled, hot forged, fully aged, etc.), for each entry in the table, as the material properties depend heavily on that state.

10.11 References

Anderson, T.L. (2004). *Fracture Mechanics*, 2d edn. New York: CRC Press.

Atwood, S., Van Citters, D.W., Furmanski, J., Ries, M.D., and Pruitt, L. (2009). Oxidative stability and fatigue behavior of below-melt annealed and remelted crosslinked UHMWPE. *Transactions of the Orthopaedic Research Society*, Las Vegas, NV.

Basquin, O.H. (1910). The exponential law of endurance tests. *Proceedings of the American Society for Testing and Materials*, 10, 625–30.

Coffin, L.F. (1954). A study of the effects of cyclic thermal stresses on a ductile metal. *Transactions of the American Society of Mechanical Engineers*, 76, 931–50.

Dowling, N.E. (2007). *Mechanical Behavior of Materials: Engineering Methods for Deformation, Fracture and Fatigue*, 3d edn. Upper Saddle River, NJ: Pearson-Prentice Hall.

Dugdale, D.S. (1960). Yielding of steel sheets containing slits. *Journal Mechanics and Physical Solids*, 8, 100–104.

Evans, A.G. (1980). Fatigue in ceramics. *International Journal of Fracture*, 16(6), 485–98.

Ferry, J.D. (1961). *Viscoelastic Properties of Polymers*. New York: Wiley.

Furmanski, J., and Pruitt, L. (2007). Peak stress intensity dictates fatigue crack propagation in UHMWPE. *Polymer*, 48, 3510–19.

Furmanski, J., and Pruitt, L. (2009). Polymeric biomaterials for use in load-bearing medical devices: The need for understanding structure-property-design relationships. *Journal of Metals*, September, 14–20.

Hertzberg, R.W. (1996). *Deformation and Fracture Mechanics of Engineering Materials*, 4th edn. New York: Wiley.

Hertzberg, R.W., and Manson, J.A. (1980). *Fatigue of Engineering Plastics*. New York: Academic Press.

Kausch, H.H., and Williams, J.G. (1986). Fracture and fatigue. In *Encyclopedia of Polymer Science and Engineering*, ed. J.I. Kroschwitz. New York: Wiley, pp. 333–41.

Kinloch, A.J., and Guild, F.J. (1996). Predictive modeling of the properties and toughness of rubber-toughened epoxies. In *Advances in Chemistry Series 252: Toughened Plastics II: Novel Approaches in Science and Engineering*, ed. K. Riew and A.J. Kinloch. Washington, DC.: American Chemical Society, 1–25.

Kurtz, S.M. (2009). *The UHMWPE Handbook: Principles and Clinical Applications in Total Joint Replacement*. New York: Elsevier Academic Press.

Kurtz, S.M., Pruitt, L., Jewett, C.W., Crawford, R.P., Crane, D.J., and Edidin, A.A. (1998). The yielding, plastic flow, and fracture behavior of ultra-high molecular weight polyethylene used in total joint replacements. *Biomaterials*, 19(21), 1989–2003.

Manson, S.S. (1954). Behavior of materials under conditions of thermal stress. *National Advisory Commission on Aeronautics: Report 1170*, Cleveland, OH: Lewis Flight Propulsion Laboratory.

Marin, J. (1962). *Mechanical Behavior of Engineering Materials*. Englewood Cliffs, NJ: Prentice-Hall.

Miner, M.A. (1945). Cumulative damage in fatigue. *Journal of Applied Mechanics*, 12, A159–64.

Morrow, J.D. (1968). *Fatigue Design Handbook – Advances in Engineering*. Warrendale, PA: Society of Automotive Engineers.

Nalla, R.K., Kinney, N.H., and Ritchie, R.O. (2003). Mechanistic fracture criteria for the failure of human cortical bone. *Nature Materials*, 2, 164–68.

Norton, R.L. (2008). *Machine Design*. Englewood Cliffs, NJ: Prentice-Hall.

Palmgren, A. (1924). Die lebensdauer von kugellagern (The service life of ball bearings). *Zeitschrift des Vereines Deutscher Ingenieure*, 68(14), 339–41.

Paris, P.C. (1964). The fracture mechanics approach to fatigue. *In Fatigue – An Interdisciplinary Approach*, Proceedings of the 10th Sagamore Army Materials Research Conference, ed. J.J. Burke. Syracuse, NY: Syracuse University Press, pp. 107–32.

Pearson, R., and Pruitt, L. (1999). Fatigue crack propagation in polymer blends. In *Polymer Blends: Formulation and Performance*, ed. D.R. Paul and C.B. Bucknall. New York: *Wiley*, 27, pp. 269–99.

Pruitt, L. (2000). Fatigue testing and behavior of plastics. In *Mechanical Testing and Evaluation*, Volume 8, ed. H. Kuhn and D. Medlin. Cleveland, OH: ASM International, pp. 758–67. Polymers. In *Comprehensive Structural Integrity*, ed. R.O. Ritchie and Y. Murakami. Oxford: Elsevier Science Limited.

Pruitt, L. (2005). Deformation, yielding, fracture and fatigue behavior of conventional and highly cross-linked ultra high molecular weight polyethylene. *Biomaterials*, 26 (8), 905–15.

Rimnac, C.M., and Pruitt, L. (2008). How do material properties influence wear and fracture mechanisms. *Journal of the American Academy of Orthopedic Surgeons*, 16(1), S94–S100.

Rimnac, C.M., Wright, T.M., Bartel, D.L., and Burstein, A.H. (1986). Failure analysis of a total hip femoral component: a fracture mechanics approach. In *Case Histories Involving Fatigue and Fracture*, ed. C.M. Hudson and T.P. Rich. Philadelphia: ASTM, STP 918, pp. 377–88.

Ritchie, R.O. (1999). Mechanisms of fatigue-crack propagation in ductile and brittle solids. *International Journal of Fracture*, 100(1), 55–83.

Ritchie, R.O., Gilbert, C.J., and McNaney, J.M. (2000). Mechanics and mechanisms of fatigue damage and crack growth in advanced materials. *International Journal of Solids and Structures*, 37, 11–329.

Sauer, J.A., and Hara, M. (1990). Effect of molecular variables on crazing and fatigue of polymers. In *Advances in Polymer Science 91/29*, ed. H.H. Kausch. Berlin: Springer-Verlag, pp. 71–118.

Schapery, R.A. (1975). Theory of crack initiation and growth in viscoelastic media. *Analysis of Continuous Growth. International Journal of Fracture*, 11(4), 549–62.

Shigley, J., and Mischke, C. (1989). *Mechanical Engineering Design*. New York: McGraw-Hill.

Suresh, S. (1998). *Fatigue of Materials*, 2d edn. Cambridge, UK: Cambridge University Press.

11 Friction, lubrication, and wear

Inquiry

You are designing a total knee replacement – a high-load device for which wear is a major concern. What would your ideal materials be for a low-wear device? Are there other concerns at play here, and how do they affect your material selection?

The inquiry posed above gives an idea of the issues that must be balanced in medical device design. A total knee replacement will see contact stresses of 40 MPa (Bartel *et al.*, 1986) and will undergo 1 million fatigue cycles per year. We have already discussed the dangers of stress shielding which can occur when using high stiffness materials. Selecting a material that can meet these demands and still maintain low amounts of wear is a challenge that must be tackled by a team of engineers and scientists.

11.1 Overview

When thinking about designing devices that will have contact with any other materials, it is necessary to consider issues of friction, lubrication, and wear, also known as the field of **tribology**. These are particularly critical in load-bearing medical device applications such as total joint replacements, where the generation of wear debris is a known issue that begins a cycle of immune response and the eventual need for device replacement. Frictionless contact would be ideal, but does not exist in the real world. It has also been shown that low-friction contact can still lead to appreciable wear (Stachiowak and Batchelor, 2005). In orthodontic implants such as braces, a smooth surface to the arch wires is very important – frictional forces are known to reduce the amount of force applied to the teeth by 50% or more (Bourauel *et al.*, 1998). The characterization of the interface between two surfaces, including the composition of the articulating materials, the surface finishes, and the lubricant (if any) will play a large role in determining the friction and wear behavior.

In the healthy body, there are numerous examples of lubricated articulation, from the contact between the eyelid and the sclera, to the femoral head moving against the acetabulum. In these examples, the body provides lubrication and these contacts can occur without negative consequences. Unhealthy contact, from grinding one's teeth

(Lavigne *et al.*, 2003) to bone-on-bone contact experienced by patients with osteoarthritis (Minas and Glowacki, 2003) will lead to pain and tissue damage. The design process for medical devices that replace these tissues will have to take into account material replacements that are not a true substitute for the natural tissues, as well as a degraded lubricant.

The problem of wear must be considered for any long-term implant. Wear can occur through strictly mechanical means, or through a combination of changes to the material's surface through chemical or thermal effects, followed by mechanical wear. This problem is of interest not simply because the loss of material can change the geometry and thus the dynamics of a device, but also because wear debris can set off an inflammatory cascade within the body, leading to implant rejection.

Testing biomaterials for friction and wear properties can be as simple as rubbing two materials together and taking measurements with basic sensors, or it can be as complex as developing a multi-station *in vitro* wear simulator for a hip replacement complete with load patterns that represent a single step repeated for millions of cycles. The majority of these tests can provide useful data, but it is important to remember the test parameters when cross-comparing results. The standardization of wear testing is ongoing.

This chapter touches upon basic definitions and useful analyses for the consideration of friction, wear, and lubrication in medical implants. The reader is encouraged to examine the listed references for more detail in these areas.

11.2 Learning objectives

This chapter provides discussions of friction, lubrication, and wear and the aspects of these topics that are relevant in the study of medical devices. Methods for testing friction and wear and measuring surface roughness are presented. After reading this chapter, a student should be able to:

1. define surface properties that affect friction and wear behavior, and calculate R_a and R_{RMS}
2. define static and kinetic friction
3. describe applications for Hertzian contact models
4. describe the different types of lubrication, and where they are commonly found, and how to read a Stribeck curve
5. compare the different types of wear, and describe where they are commonly found
6. identify areas of concern for friction/lubrication/wear *in vivo*
7. describe aspects of friction and wear that are specific to polymers
8. summarize basic test methods for surface characterization, friction testing, and wear testing
9. identify important design factors from the perspective of friction and wear concerns

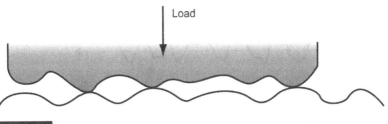

Figure 11.1

Two rough surfaces coming into contact, demonstrating the difference between real area of contact and apparent area of contact.

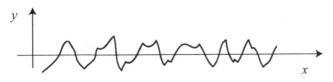

Figure 11.2

One-dimensional curve illustrating surface roughness.

11.3 Bulk and surface properties

When studying friction and wear behavior, there are certain bulk and surface properties that merit consideration. Bulk properties such as yield strength, hardness, and elastic modulus will have an effect on a material's response to high loads and moving contact. Surface properties such as surface energy will also play a role. Finally, the surface topography, or roughness, is important.

As two surfaces come into contact (illustrated in Figure 11.1), the actual contact comes at the **asperities**, small peaks on each imperfect surface. Asperities are commonly modeled as hemispherical elastic contacts. The **real area of contact**, A_r, is the term used to differentiate these asperity contacts from the more general contact area. Hence, a surface with high roughness will actually have a smaller A_r than a smoother surface. A_r can also change during sliding contact as the asperities undergo deformation and/or wear. In fact, for hard, rough surfaces under moderate load, the real area of contact may be only 15–20% of the apparent contact area (Bhushan, 2001).

A curve from a single roughness measurement is shown in Figure 11.2. From this curve, **average roughness** (R_a) and Root Mean Squared, or **RMS roughness** (R_{RMS}) can be determined. The formulas for these are given below:

$$R_a = \frac{1}{n} \sum_{i=1}^{n} |y_i| \qquad (11.1)$$

$$R_{RMS} = \sqrt{\frac{1}{n} \sum_{i=1}^{n} y_i^2} \qquad (11.2)$$

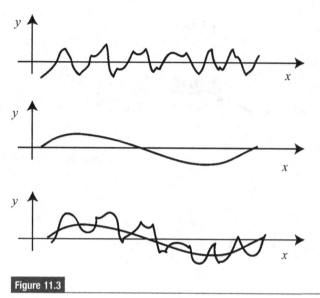

Figure 11.3

Curves indicating solely roughness, solely waviness, and the two overlaid.

where y_i is defined as the distance from the profile to the mean line, which in turn is defined such that the area between the profile and the line is equal above and below the mean line. One problem with R_a is that it can have the same value for different topographies due to the averaging effect. For this reason, it can be more useful to use R_{RMS}, which is weighted by the square of the peak heights. Either of these values should be considered in parallel with measurements of the surface waviness as shown in Figure 11.3 in order to gain a full picture of the surface topography.

Other parameters, such as the maximum peak-to-valley height, R_{max}, the skewness (which refers to the amount of valleys or peaks), and the kurtosis (a measure of peakedness) can also be determined, although these are less commonly used.

Surface roughness can be controlled through processing. For example, a machined surface such as in the acetabular cup of a total hip replacement will have measurable marks from the cutting tools. The additional cost of a polishing procedure must be weighed against the added benefit of a smoother surface. Generally, smoother surfaces do inflict less damage on bearing counterfaces during articulation. It is thought that rougher surfaces may use the valleys as lubricant reservoirs, which could help to reduce friction. However, these same valleys can be vulnerable locations for crack inceptions, so the fracture resistance of a material should also be taken into account when evaluating surface topography. The relative sizes of some surface features are given in Figure 11.4.

Example 11.1 Calculating roughness

Given the following 1D curve taken using a profilometer, calculate R_a and R_{RMS}.

Peak/valley	Distance from centerline (μm)
1	10
2	-4
3	6
4	4
5	-8
6	4
7	-7
8	8
9	-10
10	-3
11	5
12	-8
13	5
14	-6
15	8
16	-3
17	-10
18	7
19	-5
20	7

Solution   Calculate R_a using Equation (11.1). $R_a = \frac{128 \, \mu m}{20} = 6.4 \, \mu m$.

Calculate R_{RMS} using Equation (11.2). $R_{RMS} = \sqrt{\frac{16{,}384 \, \mu m^2}{20}} = 28.6 \, \mu m$.

11.4 Friction

Friction is the resistance to relative motion between two surfaces. It is normally expressed as the coefficient of friction, μ, which expresses the ratio between the lateral and normal force in a simple friction test as shown in Figure 11.5. It can be subdivided into **static friction**, or friction between non-moving surfaces, and **kinetic friction**, or friction between moving surfaces. The coefficients of friction for static and kinetic friction are known as μ_s and μ_k, respectively. It is known that the coefficient of friction is generally

374 Friction, lubrication, and wear

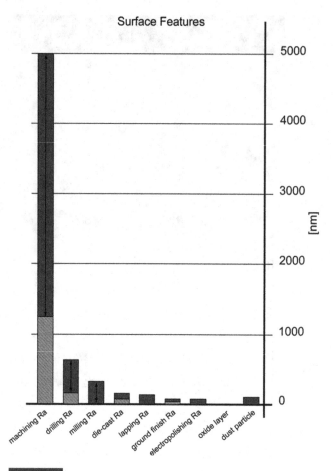

Figure 11.4

Relative sizes of surface features (Bhushan, 2002; Schafrik, 2005). Dark portions indicate range of expected values.

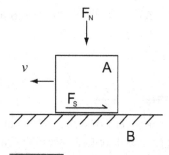

Figure 11.5

Classic friction definition with a sliding block. F_N is normal stress, while F_S is shear stress.

Table 11.1 Coefficient of friction for common materials pairs (Black, 2006; Kennedy et al., 2005; Wang and Wang, 2004)

Materials	Lubricant	μ
Tire/concrete	Dry	0.7
Tire/concrete	Wet	0.5
Steel/steel	Dry	0.50–0.57
Steel/PE	Dry	0.1
Steel/PTFE	Dry	0.05
Steel/ice	Water	0.01
Cartilage/cartilage	Synovial fluid	0.001–0.03
Cartilage/cartilage	Ringer's solution	0.005–0.01
CoCr/PE	Bovine serum	0.08
Al_2O_3/Al_2O_3	Ringer's solution	0.1–0.05

independent of apparent contact area, velocity, and surface roughness, also known as Amonton's Law. Exceptions to this include cases where surface effects dominate (for example, at the nano-scale) or if loads are high enough to cause large amounts of deformation in the materials. Friction can also be thought of as a combination of adhesion and deformation. The former is due to asperities that have become attached, and the latter is due to the need to slide asperities from one surface past the asperities of the opposing surface. Adhesion dominates between like materials (for example, metal sliding against metal) but deformation becomes important when unlike materials are paired (as in a metal sliding against a polymer).

For general cases, the coefficient of friction is defined as follows:

$$\mu = \frac{F_S}{F_N} \tag{11.3}$$

where F_S is shear force and F_N is normal force. Some basic values for μ are given in Table 11.1. Note the importance of lubricant in reducing the value for μ.

Figure 11.6 shows a standard curve from a friction test. It is clear that the static coefficient of friction is higher than the kinetic coefficient of friction, and in fact this is always true. For many material pairs the kinetic friction is constant. In certain cases, it will change over time, likely reflecting changes to the material surface. Friction can be measured experimentally, but rarely predicted.

11.5 Surface contact mechanics

Let us return to the moment when two asperities come into contact. If these asperities are thought of as two spheres with radii R_1 and R_2 under a load F, the resulting contact

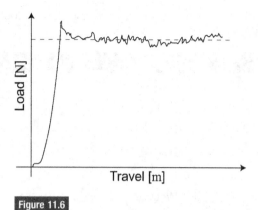

Figure 11.6

Curve resulting from a friction test. Shear load is plotted against distance traveled.

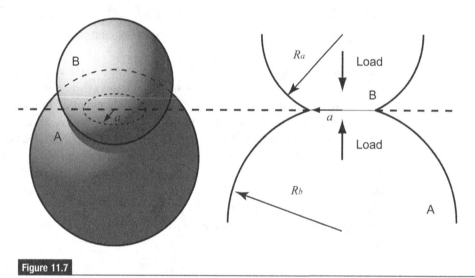

Figure 11.7

Contact between two elastic spheres. The contact area has radius a.

area will be a circle with radius a, and both spheres will deform slightly in order to create that contact area, as shown in Figure 11.7. The equations that govern this classical problem are given below, assuming that the contact is elastic and frictionless. This type of contact is known as Hertzian contact. The effective radius and elastic modulus are known as R^* and E^*, respectively. The maximum pressure, p_0, will be at the center of the contact area, and the pressure distribution is described as $p(r)$. The maximum shear stress occurs at the center of the contact area at a depth of approximately $0.5a$.

$$\frac{1}{E^*} = \frac{1-\upsilon_1^2}{E_1} + \frac{1-\upsilon_2^2}{E_2} \tag{11.4}$$

$$\frac{1}{R^*} = \frac{1}{R_1} + \frac{1}{R_2} \tag{11.5}$$

$$a^3 = \frac{3FR^*}{4E^*} \tag{11.6}$$

$$p_0 = \frac{3F}{2\pi a^2} \tag{11.7}$$

$$p(r) = p_0 \left(1 - \frac{r^2}{a^2}\right)^{\frac{1}{2}} \tag{11.8}$$

There are other solutions for contact problems, such as contact between a sphere and an elastic half-space, contact between a rigid cylindrical indenter or a conical indenter and a half-space, and contact between two cylinders with parallel or crossed axes. The solutions to these problems can be easily found in tribology textbooks (Bhushan, 2001; Thomas, 1999).

11.6 Lubrication

Lubrication is defined as the use of a film or layer to separate opposing surfaces during relative motion. The benefit of lubrication is that the separation of the surfaces can reduce the opportunity for damage. In conceptualizing lubrication options, it is constructive to think of them as ranging from fluid layers that keep the opposing surfaces entirely separate, to an intermediate lubricant layer that allows for some asperity contact, to boundary lubrication where a film is all that protects a surface from damage. Finally, all lubricant options are exhausted and direct asperity contact is all that is left. Lubricants can be solid, such as graphite; liquid, such as synovial fluid; or in some cases gas, as in air bearings. It is not easy to control lubrication in a medical device, since both the quality and quantity of the available lubricant will be dictated by the patient's physiology. Keeping this in mind, one tactic for improving lubrication in an implant is to select materials that are highly wettable. The different types of lubrication are summarized below.

Hydrodynamic lubrication is the most common type of lubrication. As long as there is sufficient lubricant and speed, this is difficult to avoid even in undesirable situations such as slipping while walking on wet brick. As seen in Figure 11.8, relative motion between the two surfaces draws the lubricant into the contact area. The lubricant then acts as a wedge and holds the opposing surfaces apart. The surface separation in this scheme is usually $10^{-3} - 10^{-4}$ cm. In hydrodynamic lubrication, the surfaces are completely separated with no asperity contact and thus this lubrication scheme is considered ideal in that the potential for surface damage is minimal.

In Figure 11.9, **elastohydrodynamic lubrication** is shown. Characterized by $10^{-4} - 10^{-5}$ cm separation, the normal load is transmitted through the lubricant, resulting in

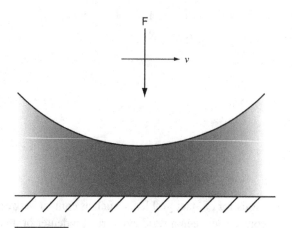

Figure 11.8

A schematic of two surfaces separated by hydrodynamic lubrication. Gray region denotes lubricant layer between asperity and opposing surface.

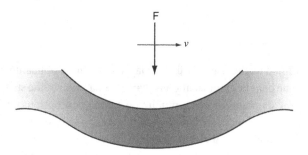

Figure 11.9

A schematic of two surfaces separated by elastohydrodynamic lubrication. Note that the lubricant layer (gray) transmits pressure and deforms the opposing surface.

elastic deformation of at least one of the surfaces. This process may result in localized fatigue over long-term use. Elastohydrodynamic lubrication is thought to be the main type of lubrication in total hip replacements during the walking phase (Wang and Wang, 2004). Because the film thickness in elastohydrodynamic lubrication is quite small, it is reasonable to question whether or not it is on the same order of magnitude as the surface roughness. This is characterized by a parameter, λ, which is defined as follows:

$$\lambda = \frac{h_0}{\sqrt{\left(R_{RMS,A}^2 + R_{RMS,B}^2\right)}} \qquad (11.9)$$

where h_0 is the film thickness in meters, and R_{RMS} is the RMS roughness for materials A and B. The parameter λ has been shown to correlate with the onset of surface damage (Stachiowak and Batchelor, 2005). In elastohydrodynamic lubrication, the pressure distribution across the contact zone is assumed to follow the Hertzian curve.

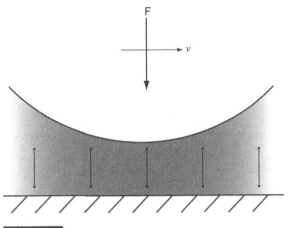

Figure 11.10

A schematic of two surfaces separated by squeeze film lubrication. In this case, the lubricant layer resists deformation and flow.

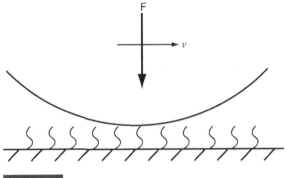

Figure 11.11

A schematic of two surfaces separated by boundary lubrication. The lubricant is bound to the opposing surface.

Squeeze film lubrication is illustrated in Figure 11.10. In this case, a highly viscous lubricant has its own elastic response to loading. This is seen during transient overloads, for example a sudden increase in load in the human knee (Murakami *et al.*, 1998).

Boundary lubrication, shown in Figure 11.11, is demonstrated when the lubricant is a solid or gel coating on one (or both) of the surfaces, as opposed to simply a low-shear interface as in the previous schemes. Boundary lubrication is found in high-pressure, low-velocity situations. In boundary lubrication, the surface separation is usually less than 10^{-5} cm. The asperities carry the majority of the transmitted load and the lubricant may undergo chemical changes due to changes in pressure and temperature. It is thought that lubrication in human joints is boundary during motion initiation, elastohydrodynamic during motion, and transitions to squeeze film lubrication in extreme loading situations (Ateshian and Mow, 2005; Lee *et al.*, 2006). Under pressure, the synovial fluid will partially penetrate articular cartilage, leaving a concentrated gel made

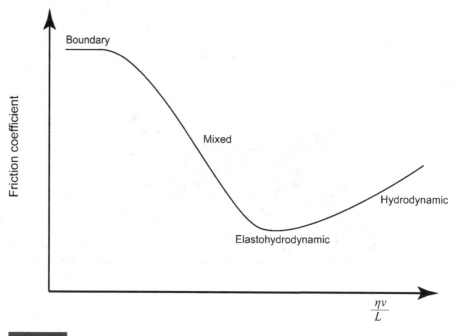

Figure 11.12

A representative Stribeck curve, showing the variation in friction coefficient with lubrication type.

of mucoprotein acid behind to act as a boundary lubricant (Wang and Wang, 2004). This is sometimes referred to as **mixed lubrication**, a situation in which a coating provides additional protection, but the standard lubrication mode is hydrodynamic or elastohydrodynamic.

The relationship between coefficient of friction and the dimensionless parameter $\frac{\eta v}{L}$ is unified in a Stribeck curve, where η is the lubricant viscosity, v is the relative motion between the two surfaces, and L is the load. Developed by Richard Stribeck in the 1920s, this curve provides a method for determining what type of lubrication is to be expected. A sample Stribeck curve is shown schematically in Figure 11.12. The highest friction coefficient is where boundary lubrication is expected. As speed and/or viscosity increase, or load decreases, a fluid film will form, lowering the coefficient of friction. During these changes to the system, the lubrication mode becomes mixed, with elastohydrodynamic lubrication at the minimum coefficient of friction. The subsequent increase in coefficient of friction during hydrodynamic lubrication is attributed to the increased drag due to the thicker fluid film which is found as the lubricant viscosity increases or the load decreases further (Hamrock et al., 2004).

Synovial fluid provides an interesting example of a lubricant. Firstly, it is **rheopectic**, meaning that it becomes less viscous during shear motion. In Figure 11.13, the decrease in viscosity of synovial fluid post-trauma is shown, and also the degradation of the synovial fluid as a result of disease. Studies have shown that the hyaluronic acid

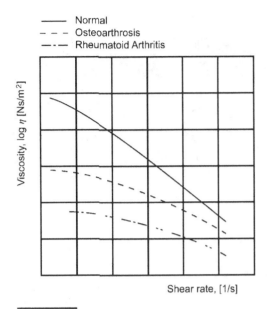

Figure 11.13

A schematic of synovial fluid viscosity as a function of shear rate. (After Dumbleton, 1981.)

molecules are depolymerized and that the concentration of hyaluronic acid decreases in the diseased state (Balasz *et al.*, 2005; Elsaid *et al.*, 2005; Fam *et al.*, 2007). It is likely that there will be higher rates of cartilage-on-cartilage contact with the lower viscosity synovial fluids, leading to joint damage.

11.7 Wear

Wear, or the removal of material from a surface, is problematic in medical devices for two reasons. First, although the material used in a medical device is likely inert, it is possible that the wear debris particles will still invoke an inflammation response as described in Chapter 1. This is not ideal, particularly for long-term implants. Second, the wear of a device leads to a change in the geometry due to the removal of material. This in turn can negatively affect the device function and reduce its useful life. Thus the study of wear, or how to reduce it, is of great importance to medical device designers. There are, in general, three stages of wear: running in, steady-state wear, and catastrophic wear. The standard types of wear are defined below.

Adhesive wear accounts for nearly half of all wear. As illustrated in Figure 11.14, fragments from one surface are pulled off and adhered to the second surface. These particles tend to be 10–100 μm in diameter. Adhesive wear volume, V, is described by the following equation:

$$V = \frac{kF_N x}{3p_{\text{soft}}} \quad (11.10)$$

Table 11.2 Typical k values (Komvopoulos, 1998)

Condition of interface	Metal/Metal		Metal/Polymer
	Similar	Dissimilar	
Clean	5×10^{-3}	2×10^{-4}	5×10^{-6}
Poorly lubricated	2×10^{-4}	2×10^{-4}	5×10^{-6}
Average lubrication	2×10^{-5}	2×10^{-5}	5×10^{-6}
Well lubricated	$2 \times 10^{-6} - 10^{-7}$	$2 \times 10^{-6} - 10^{-7}$	2×10^{-6}

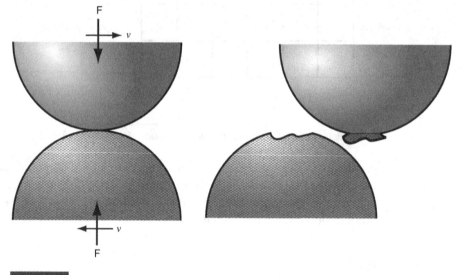

Figure 11.14

A schematic of adhesive wear. Material is removed from one surface through adhesion to the opposing surface.

where k is Archard's coefficient (typical values given in Table 11.2), p_{soft} is the Vickers hardness of the softer material (which is the one more likely to lose volume to wear debris), and x is the total sliding distance (Bhushan, 2001).

It is clear from Table 11.2 that there are drastic increases in k for failed lubrication in metal/metal pairings. The change in k for failed lubrication in metal/polymer pairings is less dramatic. Adhesive wear, although extremely common, is also difficult to eliminate. It is important to recognize that k is not a material property or even a constant, but a value that changes based on operating conditions and material couples.

Abrasive wear is the second most common type of wear and is shown in Figure 11.15. Abrasive wear occurs when hard asperities remove material from a softer surface, in processes characterized as plowing, cutting, or cracking. Abrasive wear particles are typically 2–3 orders of magnitude larger than adhesive wear particles. Abrasive wear can be harnessed in a beneficial fashion, in the form of polishing. It can be two-body (only

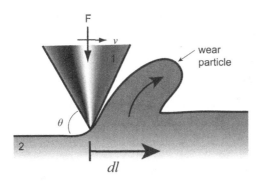

Figure 11.15

A schematic of abrasive wear. Hard particles remove material from a softer counterface.

the two counterfaces are involved) or three-body (two counterfaces and unconstrained grains of material). The formulas that describe volume of abrasive wear particles during sliding in the x-direction are as follows:

$$\text{for } p_2 < 0.8p_1, \quad \frac{dV}{dx} = \overline{\tan\theta}\frac{F}{\pi p_2} \tag{11.11}$$

$$\text{for } 0.8p_1 < p_2 < 1.25p_1, \quad \frac{dV}{dx} = \frac{\overline{\tan\theta}}{5.3p_2}\left(\frac{p_1}{p_2}\right)^{2.5} \tag{11.12}$$

$$\text{for } p_2 > 1.25p_1, \quad \frac{V}{x} = \frac{1}{p_2^7} \tag{11.13}$$

In these equations, θ refers to the sharpness of a single asperity (which is then averaged over the entire surface), F refers to the normal load, and p_1 and p_2 refer to the hardness of the asperity and counterface, respectively. Some strategies for avoiding abrasive wear include preventing the incorporation of hard grains and/or hard, rough counterfaces, making soft surfaces harder, and using elastomeric materials, which are resistant to abrasive wear because they do not deform in the same way as elastic-plastic materials. Abrasive wear is found in implants that utilize bone cement. Small particles of the bone cement can scratch the implant, as can bone and metallic particles that are created during the implant process (Davidson et al., 1994). It is also seen in dental implants, when replacement materials are used against enamel (Ghazalab et al., 2008; Hirano et al., 1998). Ghazalab and co-workers recommend harder materials (ceramic teeth versus composite resin teeth) when trying to improve stability by reducing abrasive wear.

Corrosive wear is depicted in Figure 11.16. In this situation, a brittle oxide corrosion layer is removed from its corresponding surface, resulting in pitting. Another type of corrosion wear involves the gradual removal of a coherent corrosion layer, leading to general topography change. Corrosive wear can be controlled by careful attention to lubricant and environment. Titanium is an example of a material that is susceptible to corrosive wear. Because it is so electrically active, it continually renews its oxide layer

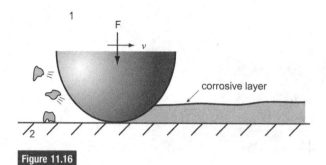

Figure 11.16

A schematic of corrosive wear. Asperities remove the protective oxide layer.

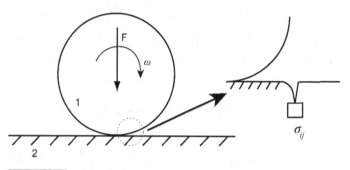

Figure 11.17

A schematic of contact fatigue. Stresses due to normal load F develop in subsurface cracks.

during corrosive wear, leading to more material removal. Thus, it is a poor choice for a bearing material, not because it wears readily due to abrasive or adhesive mechanisms, but because its oxide layer is so easily removed.

Contact fatigue, or rolling contact fatigue, refers to the process by which material removal occurs due to near-surface alternating stresses induced through rolling. Illustrated in Figure 11.17, this type of wear is common in bearings and gears. The contact can be conforming or non-conforming, and the near-surface stress field in elastic materials is often approximated using Hertzian contact theory. Contact fatigue will take place in addition to other wear mechanisms, so, for example, it is possible for a system to include both adhesive wear and contact fatigue.

Erosive wear involves the removal of material from a surface that is being bombarded by liquid or solid particles. The volume of material removed is a function of particle velocity, shape, mass, and the impingement angle. Erosive wear is maximized at 30–90° to the surface, depending on the material. This type of wear is commonly seen in situations with high-velocity flow such as valves, and is not generally an issue *in vivo*. Erosive wear is shown in Figure 11.18.

Fretting wear is the removal of material when two contacts under load are also experiencing minute reciprocating or vibrating motion. The sliding motion can range

11.7 Wear

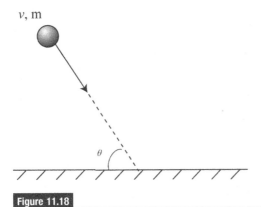

Figure 11.18

A schematic of erosive wear. High-velocity particle contact leads to the removal of material.

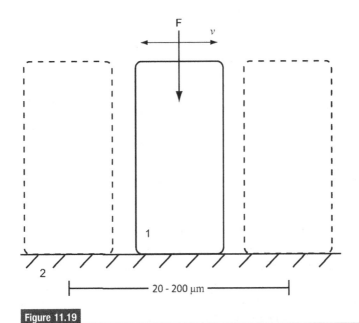

Figure 11.19

A schematic of fretting wear. Reciprocating micro-motion leads to material removal.

from 20–200 μm. Fretting wear is represented in Figure 11.19. The wear debris from fretting wear can contribute to further material degradation as abrasive wear particles. As an example of where fretting wear may occur, consider the manner in which modular hip stems for total hip replacement are assembled. As shown in Figure 11.20, the femoral head is slipped over a taper on the proximal end of the femoral stem. Although this is meant to be a tight fit, micro-motion between these two components can occur, and fretting wear is the result. Fretting is usually a combination of corrosive and abrasive wear.

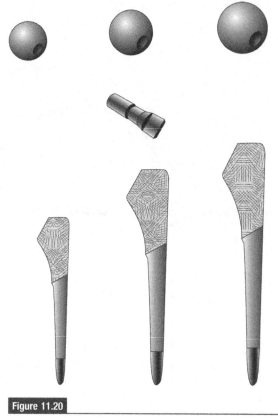

Figure 11.20

A modular hip stem system.

Delamination wear occurs under somewhat similar conditions to contact wear. During loaded rolling or sliding, subsurface stresses are developed in a material. In the presence of these stresses, a void or imperfection (as illustrated in Figure 11.21) will grow into a crack parallel to the surface, and ultimately, this piece of material will delaminate. Delamination wear is characterized by large, thin wear particles. It is more common in layered materials, but can be seen in homogenous objects as well.

11.8 Surface contact in biomaterials

The coefficient of friction in natural joints is very low, on the order of 0.002. The natural lubricity in the interface between articular cartilage and synovial fluid is so good that no synthetic option has come close to approximating it. As described in Chapter 5, articular cartilage exudes fluid during loading. Cartilage-on-cartilage contact is normal in healthy joints. It is only when the cartilage and synovial fluid become degraded due to disease or trauma that bone contact, and pain, come into existence (Dumbleton, 1981).

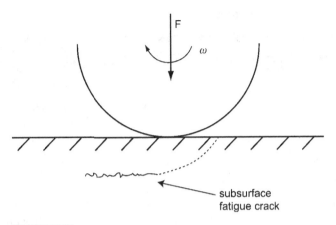

Figure 11.21

A schematic of delamination wear, in which a larger wear particle is removed due to subsurface crack propagation.

The relationships between patient physiology, cartilage degradation, and total joint replacement implant performance are very complex and not entirely understood. The study of "biotribology" is an emerging and exciting field, encompassing both positive aspects of wear (such as in the removal of debris during toothbrushing) and negative aspects of wear (as in the generation of wear debris in a total joint replacement) (Neu et al., 2008).

In polymers, the tribological behavior depends on molecular structure, crystallinity, molecular weight, and crosslink density, in addition to surface finish (Joyce, 2009). Adhesive or abrasive wear can be reduced through particle or fiber reinforcement such as carbon fillers, but there is a trade-off in the increased risk of delamination wear and the almost certain scratching of the counterface. The inclusion of solid lubricants can aid in reducing the coefficient of friction. As discussed earlier, the coefficient of friction should be independent of sliding velocity. For polymers this is only true if the temperature change due to increased sliding velocity is negligible. If the temperature range is large, or if it nears a transition temperature for the polymer, then there will be a more complex relationship between coefficient of friction and sliding velocity for that material. Finally, there is evidence that polymer microstructure can be modified through loaded sliding contact, making these materials even more susceptible to wear (Klapperich et al., 1999).

11.9 Friction and wear test methods

There are many methods for surface roughness characterization. The most common (profilometer and atomic force microscope, or AFM) are briefly described here, while the reader is referred to ASTM F2791–09: *Standard Guide for Assessment of Surface*

Table 11.3 Comparison of roughness measurement methods

Method	Quantitative information	Spatial Res (nm)	Vertical Res (nm)	Limitations
Stylus	Yes	1–100	0.1–1	Contact can damage sample
Specular reflection	No	10^5–10^6	0.1–1	Semi-quantitative Does not always correlate with stylus readings
Diffuse reflection	Limited	10^5–10^6	0.1–1	Smooth surfaces
Optical interference	Yes	500–1000	0.1–1	
Scanning tunneling Microscopy	Yes	0.1	0.01	Requires conducting surface Scans small areas
AFM	Yes	0.1–1	0.02	Scans small areas
Fluid/electrical	No			Semi-quantitative
Electron microscopy	Yes	5	50	Expensive instrumentation Requires conducting surface Scans small areas

Texture of Non-Porous Biomaterials in Two Dimensions, and ASME B46.1–2002: *Surface Texture, Surface Roughness, Waviness and Lay* for further details. In profilometry measurements, a diamond-tipped stylus is placed in contact with the material being characterized, and its vertical displacement is recorded as it is moved across the material surface under a specified load. The radius of curvature of the diamond tip limits the ability to detect surface features. The resulting curve is then analyzed as described in Section 13.3, or is combined with subsequent scans to form a surface map.

The AFM uses a sharp tip on a flexible cantilever to measure the ultra-small forces between the tip and the material surface. The forces cause the cantilever to deflect, and this deflection is measured and transformed into a surface roughness curve. An AFM can be used in contact mode, in which the force between the tip and surface is kept constant by changing the cantilever's position, or in non-contact mode, where the cantilever is oscillated close to its resonance frequency, and changes to this oscillation are measured and used to give information about the material's surface characteristics. During non-contact mode, there is a risk of the cantilever tip hitting the sample. To circumvent this issue, tapping mode can be used. In this mode, the cantilever is oscillated in such a way that the oscillation decreases when the tip approaches the sample, thus ensuring no damage to the sample by the tip. A comparison of these methods, along with less frequently used methods, is given in Table 11.3 (Bhushan, 2002; Thomas, 1999).

Friction measurements can be very simple. At the most basic, a moving solid is rubbed against a stationary solid. A normal load is provided using dead weights, and the lateral force is measured. The details of this test are laid out in ASTM G115–04: *Standard Guide for Measuring and Reporting Friction Coefficients*.

Wear testing, on the other hand, encompasses a wide variety of test methods, from unidirectional pin-on-disk testing to multi-station wear simulator testing. A challenge with wear testing is to balance replication of *in vivo* loading conditions and environment with the ability to perform multiple tests at an accelerated pace. No machines are able to fully duplicate the loading, motion, and environment in a human joint over 20 years. There are, however, machines that claim to do something close, and in a matter of months!

A pin-on-disk test is the most simple wear test, and is described in ASTM G99–05: *Standard Test Method for Wear Testing with a Pin-On-Disk Apparatus*. In this method, a pin of one material is loaded with dead weight and held against a rotating disk made of the counterface material. A lubricant may be used, and the resulting wear can be characterized through multiple means including surface characterization before and after testing, dimensional changes, and weight loss.

The choice of lubricant is not a trivial decision. The main options are saline, Ringer's solution (a solution that is isotonic with blood), and bovine serum (a portion of plasma from fetal bovine blood) (Furey and Burkhardt, 1997). Deionized (DI) water can be used but results in a transfer film between the polymer and the metallic counterface, while human synovial fluid would be ideal but is obviously difficult to obtain. Bovine serum is the most commonly used lubricant. It is generally diluted with DI water in order to maintain a protein concentration similar to that of synovial fluid, although the actual relevant concentration is by no means standardized (Liao *et al.*, 1999). A range of 20–30 mg/ml is considered acceptable (Joyce, 2009). Bovine serum degrades over time and must be changed over the duration of a long-term test in order to avoid contamination.

The more complex, but ostensibly more representative of *in vivo* loading, method for wear testing is a wear simulator. These machines allow the users to select the peak contact stress, a loading cycle that can match human gait, and sliding velocity. In both wear simulator testing, and pin-on-disk testing, it is common to use controls to measure lubricant absorption over the test duration. After wear testing in a simulator, a component can be characterized for weight loss, dimensional changes, and changes to surface features. Finally, wear debris can be gathered and analyzed for size and distribution. In friction and wear tests, the most important consideration should be matching the prospective use environment. Loads, frequencies, lubricants, and temperatures should be as close as possible to the actual application in order for the test to be truly relevant.

11.10 Design factors

When designing against friction and wear in a medical device, it is important to consider what elements are controllable. In general, the environment (such as temperature, pH, chemical makeup) is not, so materials must be selected that are compatible and even resistant to this environment. Some established methods for reducing tribological methods include the use of coatings, lubricants, and changes to surface geometry. Although the load may be predetermined, are there ways to change the geometry to mitigate the contact stress and motion? How long is this device intended to be used? Can the surface be modified through different processing techniques to make it more wear-resistant while still maintaining cohesion to the bulk material? Coatings are often described as a simple method for reducing friction, but there is a higher risk of delamination wear in this case. In total hip replacements, the lubricant layer changes throughout the walking cycle as a function of joint load and relative speed between the articulating surfaces. Because of this, the type of lubrication varies during the walking cycle. Wang and Wang describe a case in which the following effect was calculated: increased head diameter (of the femoral component) led to a thicker elastohydrodynamic lubricant layer. These results would seem to indicate that a large head diameter would improve lubrication. However, a large head diameter also increases the frictional torque on the implant during boundary lubrication, which is present during some portions of the walking cycle. Therefore, according to this study, the best path for implant designers is to optimize the head diameter and maximize the lubricant layer during elastohydrodynamic lubrication while minimizing frictional torque (Wang and Wang, 2004). These types of trade-offs must be carefully considered during the design process.

11.11 Case study: the use of composites in total joint replacements

In the early days of total joint replacement, surgeons were trying many options to find a long-lasting implant. After initial attempts at developing a metal-on-metal bearing system resulted in fears that the high friction between the two surfaces would loosen the implant and reduce its life, Sir John Charnley attempted to find a low-friction bearing pair (Kurtz, 2009). In the late 1950s, he decided to use polytetrafluoroethylene (PTFE), a well-known low-friction and inert material. After many design iterations in which he attempted to reduce contact stresses through increasing contact area and improve fixation through the use of different designs and adhesives, he arrived at a design that combined a PTFE acetabular cup with a cemented stainless steel femoral stem. Unfortunately, patients complaining of pain led to the discovery that these devices were experiencing such severe wear that over 90% of implants had to be revised within three years of implantation. Additionally, the wear debris from these implants caused a dramatic foreign body response. Charnley continued his attempts to use PTFE by developing a glass-filled PTFE composite that was subsequently implanted into patients. This material also exhibited poor wear properties. Finally, he accepted ultra-high molecular weight polyethylene (UHMWPE) as a replacement. Although UHMWPE has a higher coefficient of friction against its counterface materials, it has very good wear properties and in fact is still the polymer used in these implants today. Charnley's initial attempts were based on the idea that reducing friction would amount to reducing wear. In reality, wear is very much situational, and the coefficient of friction is only one part of the system that creates a wear response.

11.12 Summary

The friction and wear behavior of medical devices can be a determining factor in the overall device life. Both friction and wear are dependent on the entire system and so factors such as the material properties, surface roughness, and presence of a lubricant must all be considered. In this chapter, common lubrication schemes were discussed and their relevance to healthy joints was described. The different types of wear were outlined and methods for avoiding these when possible were suggested. Friction and wear can be difficult to measure in situations that are relevant to *in vivo* use, so care must be taken when applying the results of laboratory testing to predictive estimates of device performance.

11.13 Problems for consideration

Checking for understanding

1. Prove that R_a and R_{RMS} are the same for sinusoidal profiles with wavelengths λ and 2λ and amplitude A_0. What does this mean for surfaces where the only difference in surface roughness is the wavelength?
2. What is the expected contact area for initial asperity contact between UHMWPE and CoCr surfaces, assuming that the asperities can be modeled as hemispheres and that the mechanical properties are as follows: $E_{UHMWPE} = 1$ GPa, $\nu_{UHMWPE} = 0.4$, $E_{CoCr} = 214$ GPa, and $\nu_{CoCr} = 0.3$. Assume that the load is 50N and that the radii of both asperities are 100 μm.
3. What are the primary differences between adhesive and abrasive wear?
4. Describe a situation in the medical device field where you might see hydrodynamic lubrication.

For further exploration

5. Design a method for testing μ of cartilage samples.
6. Describe how you would minimize each type of wear for an ACL replacement.

11.14 References

American Society of Mechanical Engineers. (2002). *ASME B46.1–2002: Surface Texture, Surface Roughness, Waviness and Lay*. New York: ASME.

ASTM, International. (2004). *ASTM G115–04: Standard Guide for Measuring and Reporting Friction Coefficients*. West Conshohocken, PA: ASTM, International.

ASTM, International. (2005). *ASTM G99–05: Standard Test Method for Wear Testing with a Pin-On-Disk Apparatus*. West Conshohocken, PA: ASTM, International.

ASTM, International. (2009). *ASTM F2791–09: Standard Guide for Assessment of Surface Texture of Non-Porous Biomaterials in Two Dimensions*. West Conshohocken, PA: ASTM, International.

Ateshian, G.A., and Mow, V.C. (2005). Friction, lubrication and wear of articular cartilage and diarthrodial joints. In *Basic Orthopaedic Mechanics & Mechano-biology*, 3d edn., ed. V.C. Mow and R. Huiskes. Philadelphia, PA: Lippincott, Williams & Wilkins, pp. 447–94.

Balasz, E.A., Watson D., Duff, I.F., and Roseman S. (2005). Hyaluronic acid in synovial fluid. I. Molecular parameters of hyaluronic acid in normal and arthritic human fluids. *Arthritis & Rheumatism*, 10(4), 357–76.

Bartel, D.L., Bicknell, V.L., and Wright, T.M. (1986). The effect of conformity, thickness, and material on stresses in ultra-high molecular weight components for total joint replacement. *Journal of Bone & Joint Surgery*, 68, 1041–51.

Bhushan, B. (2001). *Modern Tribology Handbook*, Vol. I. Boca Raton, FL: CRC Press.

Bhushan, B. (2002). *Introduction to Tribology*. New York: Wiley.

Black, J. (2006). *Biological Performance of Materials: Fundamentals of Biocompatibility*, 4th edn. Boca Raton, FL: CRC Press.

Bourauel, C., Fries, T., Drescher, D., and Plietsch, R. (1998). Surface roughness of orthodontic wires via atomic force microscopy, laser specular reflectance, and profilometry. *European Journal of Orthodontics*, 20, 79–92.

Davidson, J.A., Poggie, R.A., and Mishra, A.K. (1994). Abrasive wear of ceramic, metal, and UHMWPE bearing surfaces from third-body bone, PMMA bone cement, and titanium debris. *Biomedical Materials Engineering*, 4(3), 213–29.

Dumbleton, J.H. (1981). *Tribology of Natural and Artificial Joints*. Tribology Series, 3. The Netherlands: Elsevier.

Elsaid, K.A., Jay, G.D., Warman, M.L., Rhee, D.K., and Chichester, C.O. (2005). Association of articular cartilage degradation and loss of boundary-lubricating ability of synovial fluid following injury and inflammatory arthritis. *Arthritis & Rheumatism*, 52(6), 1746–55.

Fam, H., Bryant, J.T., and Kontopoulou, M. (2007). Rheological properties of synovial fluids. *Biorheology*, 44(2), 59–74.

Furey, M.J., and Burkhardt, B.M. (1997). Biotribology: friction, wear and lubrication of natural synovial joints. *Lubrication Science*, 9(3), 255–71.

Ghazalab, M., Yanga, B., Ludwiga, K., and Kerna, M. (2008). Two-body wear of resin and ceramic denture teeth in comparison to human enamel. *Dental Materials*, 24(4), 502–7.

Hamrock, B. J., Schmid, S. R., and Jacobson, B. O. (2004). *Fundamentals of fluid film lubrication*, 2nd edn. New York, NY: Marcel Dekker.

Hirano, S., May, K., Wagner, W., and Hacker, C. (1998). In vitro wear of resin denture teeth. *Journal of Prosthetic Dentistry*, 79(2), 152–55.

Joyce, T. (2009). Biopolymer tribology. In *Polymer Tribology*, ed. S.K. Sinha and B.J. Briscoe. London: Imperial College Press, pp. 227–66.

11.14 References

Kennedy, F.E., Booser, E.R., and Wilcock, D.F. (2005). Tribology, lubrication and bearing design. In *The CRC Handbook of Mechanical Engineering*, 2d edn., ed. F. Kreith and D. Y. Goswami. Boca Raton, FL: CRC Press, pp. 3-129–3-169.

Klapperich, C., Komvopoulos, K., and Pruitt, L. (1999). Tribological properties and microstructure evolution of ultra-high molecular weight polyethylene. *Journal of Tribology*, 121(2), 394–402.

Komvopoulos, K. (1998). *Fundamentals of Tribology and Contact Mechanics*. Berkeley: University of California Press.

Kurtz, S.M. (2009). The origins of UHMWPE in total hip arthroplasty. In *UHMWPE Biomaterials Handbook*, ed. S.M. Kurtz. Burlington, MA: Elsevier, pp. 31–41.

Lavigne, G.J., Kato, T., Kolta, A., and Sessle, B.E. (2003). Neurobiological mechanisms involved in sleep bruxism. *Critical Reviews in Oral Biology & Medicine*, 14(1): 30–46.

Lee, T.Q., Fornalski, S., Sasaki, T., and Woo, S. L.-Y. (2006). Biomechanics of synovial joints. In *Cartilage Injury in the Athlete*, ed. R. Mirzayan. New York: Thieme Medical Publishers, Inc., pp. 10–24.

Liao, Y.S., Benya, P.D., and McKellop, H.A. (1999). Effect of protein lubrication on the wear properties of materials for prosthetic joints. *Journal of Biomedical Materials Research*, 48(4), 465–73.

Minas, T., and Glowacki, J. (2003). Cartilage repair and regeneration. In *Operative Arthroscopy*, 3d edn., ed. J.B. McGinty, S.S. Burkhart, R.W. Jackson, D.H. Johnson, and J.C. Richmond. Philadelphia, PA: Lippincott, Williams and Wilkins, pp. 127–38.

Murakami, T., Higaki, H., Sawae, Y., Ohtsuki, N., Moriyama, S., and Nakanishi, Y. (1998). Adaptive multimode lubrication in natural synovial joints and artificial joints. *Proceedings of the Institution of Mechanical Engineers – Part H*, 212(1), 23–35.

Neu, C.P., Komvopoulos, K., and Reddi, A.H. (2008). The interface of functional biotribology and regenerative medicine in synovial joints. *Tissue Engineering Part B*, 14(3), 235–47.

Schafrik, R.E. (2005). Unit manufacturing and assembly processes. In *The CRC Handbook of Mechanical Engineering*, 2d edn., ed. F. Kreith and D.Y. Goswami. Boca Raton, FL: CRC Press, pp. 13-3–13-62.

Stachiowak, G.W., and Batchelor, A.W. (2005). *Engineering Tribology*, 3d edn. Burlington, MA: Elsevier Butterworth-Heineman.

Thomas, T.R. (1999). *Rough Surfaces*, 2d edn. London: Imperial College Press.

Wang, C., and Wang, Y. (2004). Tribology of endoprostheses. In *Biomechanics and Biomaterials in Orthopedics*. London: Springer-Verlag, pp. 168–78.

Part III Case studies

12 Regulatory affairs and testing

A medical device company is developing a material for use in vascular grafts. What should the designers keep in mind from a regulatory standpoint as they move toward developing a prototype implant?

The regulatory aspects of medical device approval can be the least familiar part of the medical device development process, particularly for engineers and scientists. It can be very useful to think ahead to what types of tests will be instrumental in gaining approval from the Food and Drug Administration (FDA). This company will want to decide if it is planning to market the device itself, or merely supply the material to another company that will be the primary distributor of the implant. Second, the company might want to check the FDA database to see what guidance the FDA may provide for these implants. Third, the company will need to determine whether a device made from this material is going to be entirely novel, or whether it can be classified as substantially equivalent to currently approved devices (perhaps the more simple path to market). Further details regarding medical device classification and the approval process will be given in this chapter.

12.1 Historical perspective and overview

The Food and Drug Administration (FDA) regulates $1 trillion in products per year, which includes 1.5 million medical devices (Chang, 2003; Swann, 2003). The Center for Devices and Radiological Health (CDRH) within the FDA is responsible for evaluating approximately 4000 new medical device applications each year, as well as monitoring the medical devices already offered on the U.S. market (FDA, 2010). With just over 1200 employees, the CDRH works daily to engage with industry members and physicians in defining the balance between safety, efficacy, and health benefits of medical devices. In addition to the large numbers, the devices approved and monitored by the FDA are also quite diverse, ranging from specialties such as neurology to gastroenterology; from implants as complex as the total artificial heart to items as simple as a bandage.

Attention to regulatory affairs is a legal obligation of any medical device manufacturer, and thus, from the very beginning of a project, it is recommended that the approval process be considered. The history and organization of the FDA and the process for submitting a medical device for FDA approval will be discussed in detail in this chapter. An overview of basic testing standards will also be given.

12.2 Learning objectives

This chapter provides an overview of FDA legislative history and the medical device approval process. After reading this chapter, students should be able to:

1. list important legislation and surrounding events in the development of the current FDA
2. define the term "medical device"
3. classify basic medical devices as Class I, II, or III
4. summarize the path to FDA approval for various devices
5. explain the need for testing standards
6. describe the general information given in a testing standard
7. compare the regulatory approval processes for various countries

12.3 FDA legislative history

In the 19th century, individual states maintained control over domestically produced and distributed food and drugs, while federal authority was limited to imported items. These state-level regulations varied widely from state to state. Deceptive products, including **adulterated** and misbranded food and drugs, were rampant. By the late 19th century, advances in scientific techniques improved scientists' abilities to detect food and drug fraud. At this time, food safety was under the jurisdiction of Charles M. Wetherill in the U.S. Department of Agriculture. As the amount of deceit in the food and drug industries increased, legitimate manufacturers were becoming troubled by the damage to their reputations by association with less ethically minded colleagues. The uncontrolled market outraged consumers and the stage was set for regulation to improve.

The beginnings of the current FDA can be found in the Division of Chemistry (later changed to the Bureau of Chemistry). Dr. Harvey Wiley was appointed Chief Chemist in 1883. He earned bachelor's degrees from both Hanover College and Harvard University, and a medical degree from Indiana Medical College. Prior to joining the Division of Chemistry, he was a professor at Purdue University, publishing work in the adulteration of sugar with glucose. His arrival at the Division of Chemistry spurred additional research into the area of food and drug contamination. The most public of these experiments was called "Foods and Food Adulterants." This was a 5-year experiment beginning in 1902 in which groups of 12 healthy volunteers (known as the "Poison Squad") ate food additives incorporated into their three daily meals in order to determine their effects on health. Additives studied include borax, salicylic acid, sulfuric acid, sodium benzoate, and formaldehyde. Participants were checked weekly by doctors, and their progress was also followed closely by the press. Although the results of the experiment were not very useful scientifically, the enormous amount of publicity led to public support for a congressional hearing on the development of a federal law to prohibit adulteration and misbranding of food and drugs (Hilts, 2003).

The final push needed to write such a law came in the form of Upton Sinclair's *The Jungle*, which was published in 1906. Although this book was intended to highlight the social inequity of industrial labor and working conditions, it became known for exposing horrifying food safety transgressions, such as workers falling into tanks and being incorporated into meat products. The successive public outcry led to the Pure Food and Drug Act of 1906, signed by President Theodore Roosevelt. Upton Sinclair was notoriously underwhelmed by the effect of his own book. His intent was to encourage higher wages and improved living conditions for workers, and instead America received meat-packing regulations. He is quoted as saying, "I aimed at the public's heart, and by accident I hit it in the stomach" (Orton, 1907).

The Pure Food and Drug Act of 1906, administered by the Bureau of Chemistry (formerly the Division of Chemistry), had two major effects. First, it granted permission for federal inspection of meat products. Second, it prohibited the manufacture, sale, or transportation of adulterated food products and poisonous drugs. It introduced stricter labeling requirements such that all products were required to bear labels that clearly described the contents. If the manufacturer decided to include the product weight or volume, this amount had to be correct. Eleven habit-forming substances, including alcohol, heroin, and cocaine, were not considered illegal as long as their presence was noted on the label.

In 1927, the Bureau of Chemistry changed its name to Food, Drug and Insecticide Administration, which was shortened to the Food and Drug Administration in 1930. It was slowly becoming clear that the 1906 law was not stringent enough and that many harmful products could still be found on the market. The FDA assembled a grisly collection of products that showed ways in which the 1906 law failed to protect consumers. This collection included a false cure for diabetes, an eyelash dye that led to blindness, and a radium-containing tonic (Swann, 2003). In 1937, S.E. Massengill Co. marketed Elixir Sulfanilimide for sore throats. This product, which contained a sulfa drug and raspberry flavoring dissolved in diethylene glycol, was not subject to animal testing, and the 1906 law did not require it. Had it been tested, the scientists at Massengill would have found what was already commonly known at the time – that diethylene glycol is poisonous to humans. The prescription of this medication led to more than 100 deaths, many of them children who preferred the sulfa drug in its liquid form over the solid form due to its taste (Tobin and Walsh, 2008).

Again, public outcry helped propel a new law into being and in 1938, the Federal Food, Drug and Cosmetic Act was signed. This law replaced the Pure Food and Drug Act of 1906 and required that all drugs be pre-approved by the FDA and that they be labeled with directions for safe use. It also brought medical devices and cosmetics under FDA control. This Act allowed factory inspections and increased the agency's ability to enforce the new rules. The powers given to the FDA under this law were unprecedented.

In 1962, the Kefauver-Harris Amendments mandated that manufacturers must demonstrate drug efficacy as well as safety in order to gain FDA approval. These amendments followed on the heels of the Thalidomide disaster, in which a drug (never approved in the United States) was prescribed to pregnant women as a sedative and led to birth defects. Subsequent amendments enhanced FDA control over foods and drugs (Babiarz and Pisano, 2008).

The Dalkon Shield was an intrauterine device marketed in the United States from 1971 to 1974. The device was designed to prevent pregnancy, but because of poor design decisions it actually introduced bacteria into the uterus. In initial studies that tested the effectiveness of this device, it was in use for an average of 6 months over a 12-month period in 640 subjects. The results were published immediately afterward, obviously not allowing time for all pregnancies to be detected and accounted for. Additionally, subjects were encouraged to use contraceptive foam during the study, meaning that the results could not be attributed to the Dalkon Shield alone. Thus, the marketing for this device included claims that ranged from deliberate omissions to falsehoods regarding its use, effectiveness, and safety. Implanted in an estimated 3.6 million women worldwide, usage of the device led to infections, pregnancies, infertility, and deaths (Sobol, 1991). At that time, pre-market approval by the FDA was not required for medical devices, only for pharmaceuticals. This fact, however, was about to change.

In 1976, President Gerald Ford signed into law the Medical Device Amendments, which required that manufacturers follow strict quality control protocols and new medical devices were now subject to pre-market approval. At that point in time, it would have been difficult for the FDA to process and approve all medical devices already on the market. As a result, the majority of medical devices that were on the market when these amendments were adopted (known as "pre-amendment devices" or "pre-1976 devices") were approved, with pre-market approval only required for novel devices from this point forward ("post-amendment devices"). These amendments also established three classes of medical device, which will be further described later in this chapter.

The next major piece of legislation to affect medical device approval is the 1990 Safe Medical Devices Act. This law required that facilities that use medical devices report any adverse events involving these devices to the FDA. It also required that manufacturers conduct post-market surveillance on the more high-risk, permanently implanted devices and that methods be established for tracing and locating these devices to streamline product recalls (Babiarz and Pisano, 2008). This legislation followed on the heels of the Bjork-Shiley heart valve replacement, in which faulty heart valves were implanted into approximately 86,000 patients worldwide, with pre-market failures downplayed to the FDA during the approval process and manufacturing problems kept hidden (Blackstone, 2005). This case is discussed in detail in Chapter 14.

In 1997 and 2002, further wide-ranging reforms were instituted in the Food and Drug Administration Modernization Act (1997) and the Medical Device User Fee and Modernization Act (2002). The combined result of these was to accelerate the device review process and to regulate advertising of medical devices. Manufacturers now pay fees to the FDA for application review, and in return, the FDA agrees to review devices according to certain timelines. Finally, limited review by third party reviewers was allowed, in order to relieve some of the burden on the FDA (Babiarz and Pisano, 2008). The history of the FDA and associated drug and device tragedies are summarized in Figure 12.1.

A recent legal case has firmly cemented the FDA as the highest authority with regard to medical device safety. In *Riegel v. Medtronic (2008)*, a Medtronic catheter burst during an angioplasty conducted on Charles Riegel in 1996. The label expressly warned against the usage of this catheter for his condition, and the doctor inflated it beyond the rated burst pressure. Riegel and his wife sued the company, claiming that the design, labeling, and manufacture of the catheter violated New York common law. This was, at the time, a common method for suing medical device manufacturers. The case worked its way through district court up to the Supreme Court, with the final conclusion being that states can not impose liability on medical device manufacturers as long as the device is approved by the FDA. Essentially, this case gives the FDA the final word in medical device regulation, and it also gives consumers limited legal recourse when it comes to device malfunction. This case can only be applied to medical devices, and would not affect victims of pharmaceutical problems.

12.4 Medical device definitions and classifications

Medical devices can range in complexity from tongue depressors to long-term implants such as heart valves. The definition of a medical device is laid out in Section 201(h) of the Federal Food, Drug and Cosmetic Act as follows:

an instrument, apparatus, implement, machine, contrivance, implant, in vitro reagent, or other similar or related article, including a component part, or accessory which is:

- recognized in the official National Formulary, or the United States Pharmacopoeia, or any supplement to them,
- intended for use in the diagnosis of disease or other conditions, or in the cure, mitigation, treatment, or prevention of disease, in man or other animals, or
- intended to affect the structure or any function of the body of man or other animals, and which does not achieve any of its primary intended purposes through chemical action within or on the body of man or other animals and which is not dependent upon being metabolized for the achievement of any of its primary intended purposes.

If a product meets this definition, it is subject to pre- and post-market controls and regulated by the FDA. Note that this definition excludes other products regulated by the FDA such as drugs or blood products. It is possible for a medical device and a drug to be combined, as in a drug-eluting stent. In this case, the Office of Combination Products will help to assign a primary product center such as CDRH that will then oversee its regulation.

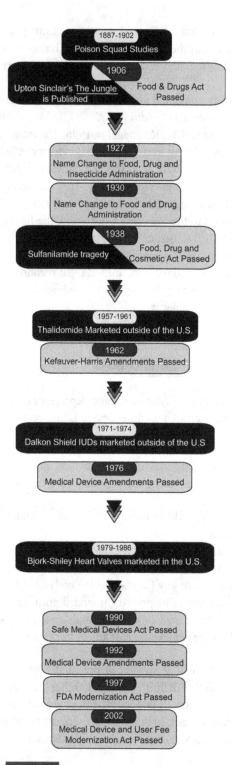

Figure 12.1

Timeline of FDA legislative history and relevant device or pharmaceutical tragedies.

12.4 Medical device definitions and classifications

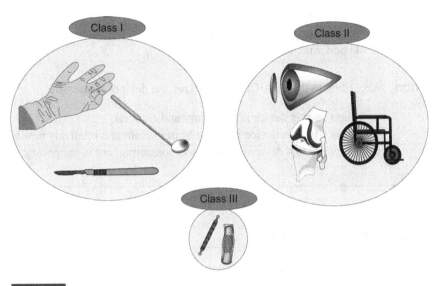

Figure 12.2

Some examples of device organized by class: Class I (examination gloves, dental mirror, scalpel); Class II (contact lens, total knee replacement, wheelchair); and Class III (synthetic ligament, coronary stent).

Device classifications are defined in the Medical Device Amendments of 1976. Based on intended use, indications for use, and risk, devices are divided into three classes. When determining device classification, a manufacturer must consider what the device is intended to treat (body system affected) and how it will be used (degree of invasiveness, duration of contact). Also, the prospective patient demographic (age, weight, activity level, etc.) is important. Class I devices are typically low-risk, short-term, and/or used to treat illnesses that are not life-threatening, and include items such as bandages or tongue depressors. The majority of Class I devices (93%) are exempt from FDA scrutiny and thus have a relatively simple path to market. Class II devices have moderate risk, such as certain types of needles or dental cement. Class III devices are the highest risk and/or are used for life-threatening diseases, such as replacement heart valves or coronary stents. Long-term implants are typically Class II or Class III. Class III devices must undergo clinical testing before FDA approval, regardless of similarities to pre-1976 devices. The CDRH maintains a large database that manufacturers can use to determine device classification if there are predicate devices. Forty-six percent of devices are Class I, 47% are Class II, and 7% are Class III. Some samples of devices and their classifications are given in Figure 12.2.

Example 12.1 Device classification

Using the FDA/CDRH database, find the device classes for the following implants after making a guess based on the guidelines above:

a. external ankle brace
b. prosthetic intervertebral disc
c. hypodermic needle

Solution According to the FDA/CDRH database, the devices classes are:

a. Class I – this device is temporary and external.
b. Class III – this device is a long-term implant and relatively new to market.
c. Class II – this device can be blood-contacting, but is temporary.

12.5 CDRH organization

The FDA is the agency responsible for ensuring the safety and/or efficacy of food (except for meat, poultry, and some dairy products, which are overseen by the U.S. Department of Agriculture), drugs, medical devices, biologics (such as vaccines and tissue or blood transplants), animal food and drugs, cosmetics (after they are released to the marketplace), and radiation-emitting products. The FDA divides these responsibilities among six centers as given in Figure 12.3, one of which is the Center for Devices and Radiological Health (CDRH). The CDRH regulates medical devices as wide-ranging as latex gloves and artificial hearts, MRI machines and hospital beds. It also oversees radiation-emitting devices that are not related to medicine, such as microwave ovens and cellular telephones.

Under the Medical Device Amendments of 1976, all new medical devices are required to undergo clinical testing before market, but all devices that were legally marketed before these amendments were written are able to submit an application for review without clinical trials. New devices, if substantially equivalent to pre-1976 devices and if they are classified as lower-risk devices, are also able to submit an application for review without clinical trials. As clinical trials are both expensive and time-consuming, there is an enormous benefit to introducing devices that are similar to pre-1976 ones.

12.5.1 Types of submissions

When deciding what type of application to use in submitting a medical device for approval, a company must first classify the device and decide if it is substantially equivalent to an existing FDA-approved device. Specifically, the existing device (called the **predicate device** for the device for which approval is being sought) must be either a pre-1976 device or a 510(k)-approved device itself. The question of substantial equivalence determines whether the device requires **Pre-Market Notification** (510(k)) or **Pre-Market Approval**. If the device is **substantially equivalent** (defined in the Safe

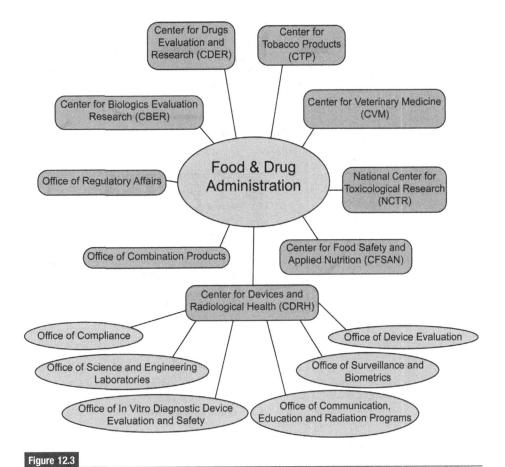

Figure 12.3

A schematic representation of the organization of the FDA.

Medical Devices Act as having the same intended use, having the same technological characteristics *or* having different technological characteristics but not raising new questions about safety and effectiveness) as its predicate device, then only Pre-Market Notification is required. In fact, the majority of Class I devices are exempt from even Pre-Market Notification by virtue of being either pre-1976, for veterinary use, or from a general list of exempt devices which is maintained by the FDA (this list includes devices such as stethoscopes, dental cement, and wheeled gurneys).

For example, consider the case of a Class I device that is substantially equivalent to another device that has been marketed since 1995. The path to market for this type of device is shown in Figure 12.4. This new device would be required to submit an application for Pre-Market Notification as well as pay the user fees (approximately $4000 for FY2010). The CDRH is required to review the submission within 90 days, with an average review time of 64 days for FY2008. In 2009, the CDRH approved nearly 3000 Pre-Market Notification submissions (FDA, 2010).

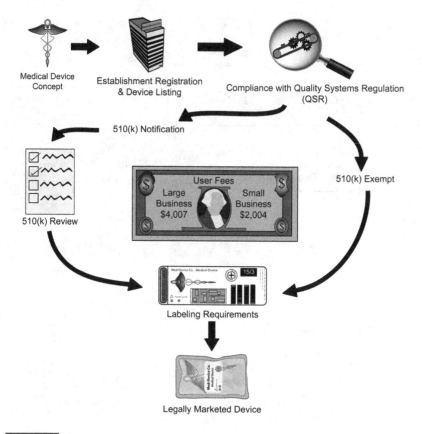

Figure 12.4

Path to market for a 510(k) device (Pre-Market Notification).

If, however, the device is completely novel *or* is a Class III device, the path to market can be more complex, as illustrated in Figure 12.5. Preclinical or laboratory testing is required, and possibly clinical testing as well. In the case where clinical testing poses a significant risk to patients, an **Investigational Device Exemption** (IDE) is required before the clinical tests can be done. The CDRH receives approximately 220 IDE applications per year, which it is required to review in 30 days or less. The clinical tests are overseen by local **Investigational Review Boards** (IRBs). After successful clinical testing, an application for Pre-Market Approval can be submitted. These applications are extremely detailed and can be upwards of 2000 pages. The user fees for Pre-Market Approval are nearly $218,000 for FY2010 and the CDRH is obligated to review the application within 180 days. In 2009, the average review time for PMA applications was 145 days (FDA, 2010). The costs for Pre-Market Notification and Pre-Market Approval submissions are reduced for small businesses (defined as having less than $100 million in gross receipts or sales in the most recent tax year) to $2,004 and $54,447 respectively for FY2010.

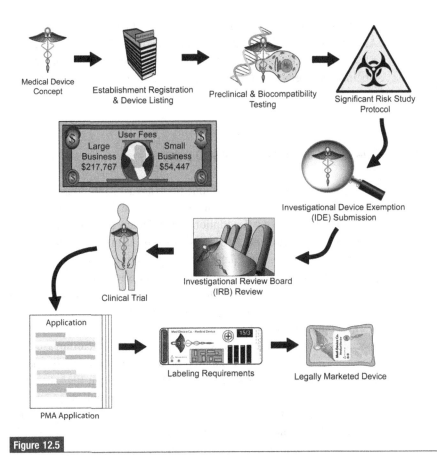

Figure 12.5

Path to market for a novel device seeking Pre-Market Approval.

Example 12.2 Path to market

Describe the path to market for a total hip replacement with an acetabular cup made of a novel polymer and a hip stem that is made of a ceramic material.

Solution Although many ceramic/polymer hip replacements are Class II devices, the fact that this one has a component made of a novel polymer means that it would likely be labeled a Class III device. Once the manufacturer confirms Establishment Registration and Device Listing with the FDA, biocompatibility testing and preclinical mechanical testing on the materials and components should be done. For the polymer, this could include a full mechanical characterization (tensile, compressive, torsional, fatigue, fracture, impact, friction) and we will assume that the ceramic component has already been tested, although that data will need to be assembled. *In vitro* wear tests should be run with the polymer and ceramic components. Once all of this has taken place, the manufacturer should put together a Significant Risk Study Protocol and apply for an

Investigational Device Exemption in order to do clinical testing under IRB supervision. After a clinical trial demonstrating success of the device, the manufacturer will submit a PMA application to the CDRH. The manufacturer will demonstrate its ability to meet labeling requirements and subsequent to receiving approval, will market its device.

12.5.2 FDA guidance in the submission process

In their applications, device manufacturers are asked to submit any data that might assist the CDRH in evaluating safety and efficacy of the device. In almost all cases, companies are required to turn in a description of the device, a proposed label, indications for use, material safety data, and a risk analysis. For each risk mentioned, the company must propose a type of test and define criteria to demonstrate whether or not the device has passed the test. If the device is in a well-established category, such as vascular grafts or total knee replacements, the FDA will put together "Guidance Documents" to provide more explicit, though non-binding, recommendations.

The applications go to the reviewers within the Office of Device Evaluation that specialize in the device area, such as orthopedic or cardiovascular implants. A lead reviewer is assigned to each submission and consults with clinicians, statisticians, and engineers if necessary. As detailed in MDUFMA, the FDA has a specific amount of time to review each submission and if information is lacking, the reviewers "stop the clock" and return the submission to the manufacturer for further details.

The FDA does not normally specify how a test should be done, but it does recognize many international testing standards, such as those developed by ASTM International (formerly the American Society for Testing and Materials). If a manufacturer chooses to do a test in a manner different from a published standard, it must also include a thorough explanation of the test method used and, if it is similar to a standard, what the differences were. In this case, the manufacturer would be required to develop and validate that its test gives results with repeatability. Contrary to popular belief, the FDA does not conduct any of the tests or take any data for the medical devices that it reviews. It is up to the medical device company to do the correct tests and submit the necessary results. If the CDRH still has unresolved questions, they can request additional tests or explanations. The reason that the FDA avoids giving specific test requirements is due to the 1997 law that requires that the FDA allow companies to use the "least burdensome" approach for demonstrating device safety and efficacy. Mandating specific test standards could potentially increase the burden on medical device companies if they had historical data using non-standard tests.

12.6 Anatomy of a testing standard

The purpose of a testing standard is to provide a framework for measuring, recording, and reporting material properties or other performance gauges. A formal definition is given by the International Organization for Standardization, also known as ISO (the name derives from the Greek word for "equal," *isos*): "Standards are documented agreements containing technical specifications or other precise criteria to be used consistently as rules, guidelines or definitions of characteristics, to ensure that materials, products, process and services are fit for their purpose" (Chang, 2003). Ideally, if a testing standard is followed, the same results should be achieved regardless of testing facility and equipment used. In this way, data can be compared across institutions and over time. Items that are commonly specified in a standard include:

- Equipment (grips, dimensions, loads, and measurement accuracy)
- Environment (temperature, humidity, stability)
- Test specimens (shape, size, number, method of manufacture, measurement locations, conditioning)
- Test method (test conditions, procedure)
- Analysis (calculations, corrections)
- Report (what details need to be shared)
- What constitutes a repeatable and reproducible result

For example, ASTM D638: *Standard Test Method for Tensile Properties of Plastics* explains that it is intended for use on samples between 1 mm and 14 mm thick. It cautions against using the results for applications in which the time scales and/or environment are largely different from those of testing. It requires testing equipment with a moving crosshead and self-aligning grips and mentions some tips for cloth or abrasive surfaces to hold the sample if the grips have smooth plates. Eight different dimensions for the test samples are given, including gage length, sample length, and distance between grips. For an isotropic material, a minimum of five samples are required. The temperature (73.4 +/− 3.6°F) and humidity (50 +/− 5%) are specified and the procedures, including measurement of the sample dimensions before, during, and after the test, are given. The calculations necessary to find tensile strength, percent elongation, elastic modulus, secant modulus, and Poisson's ratio are included. For reference, the standard includes the results of these tests from six to ten laboratories for various materials. Finally, the standard states the details that should be included in the report and acceptable standard deviations between samples and tests. Sample dimensions and configurations for some common standardized tests can be found in Appendix C.

12.7 Development of testing standards

A typical standards development process might occur as follows, beginning with the need for a standard. First, it should be determined if an acceptable standard already exists from among the numerous **standards development organizations** (SDOs) such as ISO, ASTM, or the American National Standards Institute (ANSI). Next, a technical committee is set up that ensures the input of stakeholders. In the development of standards relating to medical devices, these stakeholders would include representatives from regulatory bodies, industry members, academics, and clinicians. A first draft of the standard is developed and comments from the SDO's members are solicited. The committee then works to resolve the comments and revise the draft, leading to a second-level review. The standard can then be approved through the accepted method of the SDO. Some organizations use a consensus-building approach in which all parties must agree (such as ASTM International), while others (for example, ISO) ensure adoption of the standard through a majority vote. Finally, the standard is published, with built-in intervals for review and revision as necessary.

There are many benefits to this consensus-building approach. Firstly, there is increased understanding and communication between all parties. Standards development meetings provide a useful forum for open discussion between all stakeholders. It also takes the responsibility for standards development out of the hands of the government, while allowing FDA employees to still take part. Finally, this type of approach ensures that the standards development process is informed by the most up-to-date research and results. These organizations have large memberships (over 33,000 members from more than 135 countries in ASTM, and 157 countries represented by ISO, for example (ASTM, 2008; ISO, 2008) and are able to easily tap into the current research trends. Although these bodies traditionally write standards only for existing materials and devices, having members on the forefront of research allows these standards to be written with possible future developments in mind. One drawback of the consensus-building style is that due to the large membership and the emphasis on reaching a consensus, the standards development process cannot always keep pace with technological advances. There can be a significant lag between the introduction of a technology and the creation of a consensus-driven standard that evaluates it.

12.8 International regulatory bodies

Brief summaries of the regulatory processes for the European Union (EU), Canada, Japan, and Australia are given below, to highlight some similarities and differences between these and the regulatory processes in the United States. Also, efforts to bring these regulatory processes into closer alignment are discussed.

In the EU, medical devices are regulated by the Directorate General for Health and Consumers. The directives set out by this Directorate are then applied by agencies within the member countries of the EU such as the Medicines and Healthcare products Regulatory Agency (MHRA) in the United Kingdom. Medical devices are grouped into three categories, although Class II has two subcategories, so there are effectively four categories. Manufacturers must register with a physical address in Europe, and are required to submit Declarations of Conformity to the EU regulations, as well as the details of how they conform. Class I devices are exempt, and need only satisfy safety, performance, and labeling requirements. Once a device is approved, it receives a CE mark (Conformité Européenne, or European Conformity) from a Notified Body (a group that has fulfilled the requirements to carry out conformity assessment) and is then permitted on the market (European Commission, 2010; Medicines and Healthcare Products Regulatory Agency, 2008).

Medical devices in Canada are regulated by the Therapeutic Product Directorate (TPD). There are four device categories, with Classes III and IV subject to regulatory scrutiny. Class II devices require manufacturer declaration of device safety and effectiveness, while Class I is exempt after meeting general safety, effectiveness, and labeling requirements. Manufacturers must register with the TPD for an Establishment License, and approved devices are given a Device License as permission to enter the market. In 2007–2008, TPD saw more than 7000 applications for Class II, III, and IV devices (Health Canada, 2010; Therapeutic Products Directorate, 2008).

In Japan, the Pharmaceutical and Medical Safety Bureau within the Ministry of Health, Labour and Welfare is responsible for medical device regulation. There are three device classes: Class I and certain Class II devices are eligible for regional approval, but novel Class II devices and Class III devices must obtain central government approval. Manufacturers or medical device distributors must have a Japanese business license. When approved for market, Class I devices receive pre-market notification (todokede), Class II devices receive pre-market certification (ninsho), and Class III devices receive pre-market approval (shounin) (Pharmaceutical and Medical Safety Bureau, 2010).

Australia's Therapeutic Goods Association (TGA) requires that manufacturers have an Enterprise Identification Number. There are four device classes, based on risk, and the devices must conform to Australian standards, which are developed by the government. Once a device is approved, it is given a number within the Australian Registrar of Therapeutic Goods (ARTG). As of May 2008, there were more than 54,000 devices in the ARTG (Therapeutic Goods Association, 2010; Therapeutic Goods Association, 2010).

The Global Harmonization Task force is made up of the members described above: the United States, the European Union, Canada, Japan, and Australia. This group, beginning in 1992, has worked toward the goal of unifying the various regulatory processes to enhance global public health. There could be many benefits of having both greater medical device availability and sharing of adverse event reports such as recalls (Chang, 2003).

12.9 Case study: examining a 510(k) approval

The approval of and subsequent FDA internal review relating to the ReGen Menaflex Collagen Scaffold (MCS) gives a rare window into some of the weaknesses of the current medical device approval process. The MCS is a curved resorbable collagen scaffold meant to be used in the meniscus, which is located in the knee (Dichiara, 2008). In 2004, the FDA recommended a Class III designation for the MCS instead of Class II, as the device was meant to fully replace meniscal tissue instead of simply repairing it. In response to this decision, ReGen began the application process for pre-market approval. The company ran a two-year randomized clinical trial comparing patients with a partial menisectomy to patients with a partial menisectomy and the MCS implanted. Results showed no benefit with the MCS (FDA, 2009). During this time, the FDA began to approve devices called surgical meshes, which reinforce natural tissue. Given the results of the clinical trial, ReGen removed its PMA application and decided instead to submit a 510(k) application using surgical meshes as predicate devices (Hines et al., 2010). This submission and the one following in 2007 were rejected on the basis that the MCS was Not Significantly Equivalent (NSE) to the predicate devices listed in the applications. However, a third 510(k) application was eventually deemed Significantly Equivalent (SE) and the MCS was approved for market (Hines et al., 2010).

Numerous deviations from standard protocol were documented throughout the approval process, including a shortened review timeline and intervention by the Commissioner's office on behalf of the manufacturer. ReGen itself was described as aggressive, and the reviewers for the third and final 510(k) mentioned persistent pressure from members of Congress relating to this application (FDA, 2009). ReGen also claimed that it was receiving unfair treatment. It refused to submit additional clinical data because it alleged that clinical data are not traditionally required for 510(k) submissions. The many deviations from the standard protocol were poorly documented, and reasons for the deviations frequently were not recorded (FDA, 2009).

Aside from the communication and protocol problems described above, there was a further issue. As explained earlier in this chapter, if a manufacturer is seeking approval for a device using a 510(k), then it must demonstrate that the device is SE to a pre-1976 device, which means that it has the same intended use, same technological characteristics, *or* different technological characteristics and equivalent safety and efficacy. If Device B uses Device A as a predicate device, and Device C uses Device B as a predicate device, it is simple to see how in a few iterations, a device can receive FDA approval while having only minor similarities to the original pre-1976 device (Device A). This incremental modification to the supposed basis for approved devices is "predicate creep." In the case of the MCS, the application cited 11 predicate devices, all surgical meshes which were predicate creep on their own due to having different intended uses (Dichiara, 2008). It shared some characteristics with each and ostensibly those similarities qualified them as predicate devices, but none of these surgical meshes was used where loads were expected to be as high as in the knee, making them poor predicates.

The combination of the issues with the selected predicates, the outside pressures on the reviewers, and the documented rejection of the review panel's conclusions by the FDA Commissioner and Center Director (Goode, 2008), led to a highly flawed approval for the MCS device in 2009. In a review of this process, an FDA report outlined suggestions that could prevent a similar event from re-occurring. Among these suggestions were the following: adhere more strongly to existing procedures, review and revise current procedures for resolving inter-Center science-based differences of opinion, fully investigate and document claims of unfair treatment, fully document alternate procedures if used, and create an office to handle company-Center disputes at the Center level so that the Commissioner's office does not become involved. It was unusual that these issues at the FDA came to light, but this case received national press attention in addition to the internal review (Harris and Halbfinger, 2009; Mundy, 2009). The approval of the MCS without obvious clinical benefit or strong relationship to predicate devices demonstrates how the current approval process can be exploited.

12.10 Summary

The legislative history of the FDA demonstrates that drug and device regulation increases in strictness only following disasters with high human cost. This reactive posture may be the result of the balance between encouraging innovation and protecting public health, although many of the pharmaceutical and device disasters described in this chapter were preventable even given the level of scientific analysis available at the time. The role of the FDA continues to evolve with leaps in technology and heightened attention to the need for more uniform international approval systems.

In the United States, medical devices are regulated by the CDRH. Devices are grouped into three classes based on intended use, indications for use, and device risk. The path to market has added requirements for the higher-risk categories. Guidance documents are provided for some well-established implant types, but generally the FDA/CDRH does not require specific types of testing for pre-market notification or approval applications. It is recommended that device manufacturers conduct their tests according to international standards, such as those developed by ASTM or ISO. These standards development organizations use the consensus of large memberships to generate testing standards that reflect currently accepted methods for deriving data, defining performance, or assuring quality.

Regulatory approval procedures in other countries hold many similarities to those in the United States, with the primary differences being found in the amount of assessment that is conducted by third parties, and the reliance on specific standards for safety and efficacy. In the future, there is hope that international approval procedures and post-market surveillance will converge into a uniform system that will provide increased protection and safety to consumers worldwide.

12.11 Problems for consideration

Checking for understanding

1. For a device of your choosing, describe how it would have achieved FDA approval in 1920, 1950, 1980, and now.
2. What is the recommended standard used for verifying biocompatibility of a heart valve, and what information does it give?
3. Describe the path to market for an artificial ligament made of novel polymer fibers.

For further exploration

4. What are the device classes for the following implants? Use the FDA database after making your own guesses. (a) renal stent, (b) cemented metal/polymer hip implant, (c) powered hospital bed, (d) vascular graft, (e) resorbable suture.
5. Using any available FDA Guidance Documents, summarize the materials and mechanical testing the FDA recommends for intervertebral fusion devices.
6. Propose a modification to the current FDA approval process that would increase public health, public safety, or the number of devices approved per year without decreasing the other two options.

12.12 References

ASTM, International. (2008). *2008 Annual Report*. West Conshohocken, PA: ASTM, International.

ASTM, International. (2010). *D638–10: Standard Test Method for Tensile Testing of Polymers*. West Conshohocken, PA: ASTM, International.

Babiarz, J.C., and Pisano, D.J. (2008). Overview of FDA and drug development. In *FDA Regulatory Affairs: A Guide for Prescription Drugs, Medical Devices and Biologics*, 2d edn., ed. D.J. Pisano and D.S. Mantus. New York: Informa Healthcare USA, Inc., pp. 1–32.

Blackstone, E.H. (2005). Could it happen again?: The Bjork-Shiley Convexo-Concave heart valve story. *Circulation*, 111: 2717–19.

Chang, M. (2003). *Medical Device Regulations: Global Overview and Guiding Principles*. Geneva, Switzerland: World Health Organization.

Dichiara, J. (2008). *510(k) Summary for Collagen Scaffold (CS) – K082079*. Rockville, MD: Food and Drug Administration.

European Commission. (2010). http://ec.europa.eu/consumers/sectors/medical-devices/index_en.htm, retrieved July, 2010.

Federal Food, Drug, and Cosmetic Act of 1938. (1938). Pub.L. No. 75–717, 52 Stat. 1040.

Food and Drug Administration. (2009). Review of the Regen Menaflex: departures from processes, procedures, and practices leave the basis for a review decision in question, preliminary report. Rockville, MD: Food and Drug Administration.

Food and Drug Administration. (2010). *Monthly device approvals*. www.fda.gov. Rockville, MD: Food and Drug Administration. Retrieved July, 2010.

Food and Drug Administration Modernization Act of 1997. (1997). Pub.L. No. 105–115, 111 Stat. 2296 (codified in scattered sections of 21 U.S.C.).

Goode, J. S. (2008). 510(k) memorandum regarding ReGen Collagen Scaffold (CS), August 14, 2008. Rockville, MD. Food and Drug Administration.

Harris, G., and Halbfinger, D.M. (2009). F.D.A. reveals it fell to push by lawmakers. *New York Times*. September 24.

Health Canada (2010). http://www.hc-sc.gc.ca/dhp-mps/md-im/index-eng.php. Retrieved July, 2010.

Hilts, P.J. (2003). *Protecting America's Health: The FDA, Business and One Hundred Years of Regulation*. New York: Alfred A. Knopf.

Hines, J.Z., Lurie, P., Yu, E., and Wolfe, S. (2010). Left to their own devices: breakdowns in United States medical device premarket review. *Public Library of Science Medicine*, 7(7): e1000280.

International Organization for Standardization. (2008). *2008 Annual Report*. Geneva, Switzerland: International Organization for Standardization.

Medical Device Amendments of 1976. (1976). 90 Stat. 539 (codified at 21 U.S.C. § 301 *et seq*.).

Medical Device Amendments of 1992. (1992). Pub.L. No. 102–300, 106 Stat. 238 (codified in scattered sections of 21 U.S.C.)

Medical Device and User Fee Modernization Act of 2002. (2002). Pub.L. No. 107–250, 116 Stat. 1588 (codified in scattered sections of 21 U.S.C.).

Medicines and Healthcare Products Regulatory Agency. (2008). Medicines & Medical Devices Regulation: *What you need to know*. London: MHRA.

Mundy, A. (2009). Political lobbying drove FDA process. *Wall Street Journal*. March 6.

Orton, J.F. (1907). *The Jungle* – its purpose and its author. *The Public*, ed. L.F. Post, 9(462), p. 1071.

Pharmaceutical and Medical Safety Bureau. (2010). www.mhlw.go.jp/english/index.html, retrieved July, 2010.

Riegel v. Medtronic, Inc., 552 U.S. 312 (2008).

Sobol, R.B. (1991). *Bending the Law: The Story of the Dalkon Shield Bankruptcy*. Chicago: University of Chicago Press.

Swann, J.P. (2003). History of the FDA. In *The Food and Drug Administration*, ed. M.A. Hickman. Hauppauge, NY: Nova Science Publishers, Inc., pp. 9–16.

The Kefauver-Harris Amendments. (1962). Pub.L. No. 87–781, 76 Stat. 780 (codified in scattered sections of 21 U.S.C.).

The Pure Food and Drugs Act of 1906. (1906). Pub.L. No. 59–384, 34 Stat. 768 (repealed 1938).

The Safe Medical Devices Act. (1990). Pub.L. 101–629, 104, Stat 4511 (codified in scattered section of 21 U.S.C.).

Therapeutic Goods Association. (2010). Australian Regulatory Guides for Medical Devices, Version 1.0 April 2010, Symonston, Australia: TGA.

Therapeutic Goods Association. (2010). http://www.tga.gov.au/devices/devices.htm, retrieved July, 2010.

Therapeutic Products Directorate. (2008). *Annual Report 2007–2008*. Ottowa, Ontario: TPD.

Tobin, J.J., and Walsh, G. (2008). *Medical Product Regulatory Affairs: Pharmaceuticals, Diagnostics, Medical Devices*. Weinheim, Germany: Wiley-VCH Verlag GmbH & Co.

13 Orthopedics

Inquiry

How have we arrived at the modern-day implants used in orthopedics? What engineering design challenges do we face in orthopedic medical devices?

Contemporary implants have evolved from original attempts to restore function in articulation of a joint or stabilization of a fracture. However, many early designers did not fully understand the functionality of the musculoskeletal system or the behavior of materials in the body, and this often resulted in premature failure of the device. Much of our modern-day success is built upon learning from these earlier failures. Many design challenges remain in the development of orthopedic implants such as bone loss due to stress shielding, **osteolysis** owing to particulate debris, fractures associated with stress-corrosion mechanisms, biocompatibility, as well as patient and surgical factors. What other factors can you think of that contribute to the structural longevity of an orthopedic implant?

13.1 Historical perspective and overview

Orthopedic implants are medical devices involving the musculoskeletal system and are generally used to restore function to synovial joints, bone, and the spine. Archaeological evidence suggests that orthopedic procedures were performed in a number of ancient civilizations; yet early surgical methods tended toward amputation rather than repair. Orthopedic surgery did not progress significantly until antiseptic surgical techniques were developed in the late 19th century. Much of the current field of orthopedics, fixation methods, and medical devices was founded upon early hip **arthoplasty**.

The historical evolution of total hip replacement designs offers great insight into the advancements of materials processing, improved understanding of physiological conditions, and orthopedic bearing combinations that offer long-term structural integrity. The hip was the first major joint reconstruction attempted, and the design progression of hip arthroplasty provided the groundwork for other total joint arthroplasty procedures including the knee, shoulder, and spine. In the late 19th century, a German surgeon by the name of Thomas Gluck introduced the use of ivory along with other grouting agents for hip reconstruction using a ball and socket design (Pospula, 2004). In this time frame, the hip restoration method commonly utilized **intramedullary** rods; however, many of these initial prototypes failed due to extrusion of the implant through the bone within months of implantation (Noble, 1999). A glass hip socket was introduced by Smith-Peterson in 1923, but this implant was limited by its strength and inability to tolerate the stresses in the hip joint (Praminik et al., 2005).

In the 1920s, **vanadium steel** was employed as the first medical grade alloy for orthopedic applications. This steel alloy offered strength but faired poorly in the body because of tissue incompatibility. Early metallic implants were prone to premature failures associated with corrosion-induced degradation and concomitant biological reactions to corrosion particulate debris; however, this lack of success also provided a platform for biocompatibility studies. In the mid-20th century, 316 and 316 L stainless steels were endorsed by the American College of Surgeons (Verdonschot, 2005). Within the following decade, the American Society for Testing Methods (ASTM) developed a standard for the composition and properties of surgical grade stainless steel. Cobalt alloys were developed in the 1930s and offered exceptional strength, tribological properties, and corrosion resistance to the orthopedic device industry. Once these superior alloys were introduced to the orthopedics community, they were continually improved and within a few decades they replaced stainless steel in total joint arthroplasty.

In 1938, an English surgeon by the name of Phillip Wiles bolted stainless steel femoral components to the upper end of the femur in total hip replacements, resulting in the first metal on metal design. Failure of these devices ensued due to the lack of bone fixation and resulted in further evolution of this design. In 1948, the bolted hip prostheses were replaced by a ball and socket design as shown in Figure 13.1 (Pramanik et al., 2005; Wiles, 1952). This bolted design was susceptible to corrosion and also stress concentrations associated with the short stem; however, this implant served as the foundation for successful design evolution. In 1951, the McKee-Farrar prosthesis (named after its surgical inventors) implemented the use of a long femoral stem and an undercut spherical head to reduce impingement into the aforementioned Wiles ball and socket design (McKee and Watson-Farrar, 1966). This successful design served as the foundation of many modern total hip prostheses.

The first *low-friction* hip arthroplasty was developed in 1959 by a British surgeon, Sir John Charnley, using a polytetrafluoroethylene (PTFE) cup affixed to the acetabulum and a metal (stainless steel) femoral head (Charnley, 1961). It was hypothesized that the low coefficient of friction and chemical inertness of PTFE would enable the polymer to perform well as a bearing surface; however, these devices typically failed within 2–3 years due to extensive creep, wear, and foreign body response, and for these reasons Sir Charnley exchanged the bearing material to an ultra-high molecular weight polyethylene (UHMWPE) with exceptional energetic toughness (Charnley, 1979). Having undergone extensive improvements in processing, structural development, and manufacturing over the last 50 years, UHMWPE remains the choice bearing material for articulation against CoCr or ceramics in total joint arthroplasty (Bellare and Pruitt, 2004). In fact, the basic

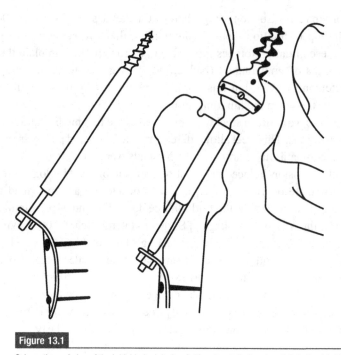

Figure 13.1

Schematic rendering of the initial bolted design (left) and early ball and socket design (right).

design and materials used in the Charnley model have served as the foundation for modern day hip arthroplasty (Figure 13.2). A contemporary total hip joint replacement comprises several elements, including those of the femur (femoral stem, neck, and head) and the acetabulum (acetabular shell and liner). The fixation between the implant and the bone can be achieved with press fit methods, acrylic bone cement, or porous coatings in the proximal end of the femoral stem (Kurtz, 2009a). If additional stability is necessary, cables, screws, or rods may also be used in the fixation and repair process of the joint. The functional requirements of these structural constituents are discussed below.

Materials have changed significantly in the historic evolution of total hip replacements and other orthopedic devices from the initial use of ivory, glass, and grouting agents to contemporary alloys and polymers that provide resistance to corrosion, fatigue, and wear. Figure 13.3 provides a basic timeline (1890–1971) showing the evolution of hip implant designs and materials selection. Early metallic implants were prone to corrosion and were later replaced by cobalt alloys with exceptional strength, tribological properties, and corrosion resistance. The cobalt alloys performed well as the bearing surface of the femoral head but suffered from a propensity for stress shielding and concomitant bone loss around the femoral stem (Pramanik *et al.*, 2005; Sullivan *et al.*, 1992). Many of these earlier femoral stems were also susceptible to loosening due to poor tissue integration and interface failures (Cook *et al.*, 1991; Maloney *et al.*, 1990). In the early eighties, titanium alloys were introduced into orthopedics (Thomas, 1990). Titanium provides

13.1 Historical perspective and overview

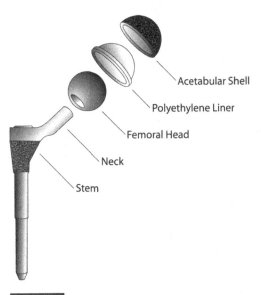

Figure 13.2
Schematic rendering showing the constituents of a contemporary total hip replacement.

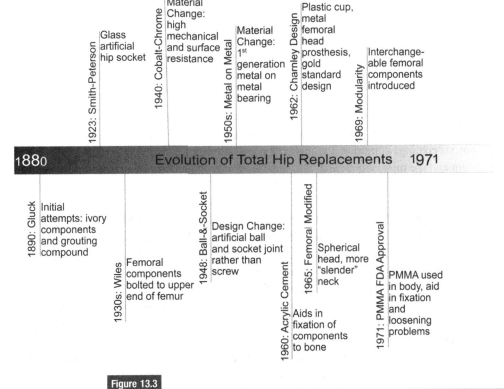

Figure 13.3
Illustration of the evolution in design and materials of total joint replacements. (Image courtesy of AJ Almaguer.)

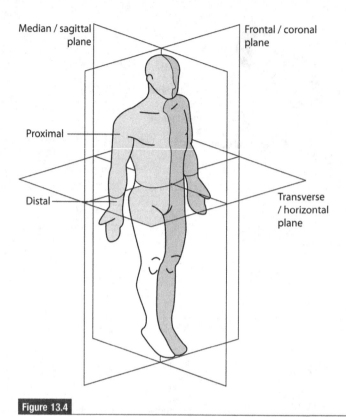

Figure 13.4

Planes of the body illustrated to facilitate an understanding of anatomy or device placement.

strength and fatigue resistance as well as improved compliance match and integration with bone tissue. More recent developments in the total hip implant design involve porous coatings comprising titanium or cobalt-chromium beads sintered on the femoral stem and the metal cup behind the acetabular liner to enhance bone integration, as well as the use of crosslinked forms of UHMWPE for improved wear resistance of the hip implant.

Many great advances in orthopedic device designs occurred between the mid-20th century and the late 1970s. Since this time, the research field has placed great emphasis on the understanding of intricate relationships that exist among biomechanical demands, material selection, device design and manufacture, biocompatibility, *in vivo* environment, as well as patient and surgical factors. This chapter provides an overview of medical devices used in the field of orthopedics with emphasis on total hip and knee arthroplasty, fracture fixation, and spinal implants. The basic factors contributing to device design and performance are discussed. A basic review of the anatomy for the hip, knee, and spine are presented. To facilitate clinical terminology used in the medical industry, Figure 13.4 provides an illustration of the primary body planes (cross-sections) used when describing the anatomical positioning (**frontal/coronal; median/sagittal; transverse/horizontal**; distal/proximal). The chapter ends with case studies representative of clinical challenges and looks forward to the future of implant development in this field.

13.2 Learning objectives

This chapter provides an overview of medical devices used in the field of orthopedics with emphasis on total hip and knee arthroplasty, fracture fixation, and spinal implants. The key learning objectives are that at the completion of this chapter the student will be able to:

1. describe the historical evolution of the primary orthopedic implants including total joint replacements, fracture fixation devices, and spinal implants
2. illustrate the basic anatomy of the hip, knee, and spine as well as provide the basic statistics of osteoarthritis and joint revision in the United States
3. explain the clinical conditions that lend themselves to total joint replacement
4. illustrate the design evolution and the general implant designs used in total joint reconstruction, spinal implants and fixation, and fracture fixation of bone
5. list the engineering materials that are utilized in orthopedic implants
6. specify the necessary structural properties needed for function in total joint replacements, fracture fixation components, and spinal implants
7. portray the engineering challenges, design constraints, and primary factors that contribute to the success of a medical device used in orthopedics
8. discuss case examples in orthopedics where an implant has failed prematurely due to corrosion, fatigue, wear, clinical complication, poor material selection, or inappropriate design

13.3 Total joint replacements

Total joint arthroplasty (TJA) is a highly successful surgical treatment for a number of degenerative joint disorders including osteoarthritis and rheumatoid arthritis. Total joint replacements are widespread and highly successful in restoring function to the synovial joints of the body. There are nearly 750,000 total joint replacements performed in the United States annually (AAOS, 2010). The most commonly replaced joints are the hip and knee, followed by the spine, shoulder, elbow, and ankle. The hip is the simplest of these articulating junctions and comprises a **conforming** ball and socket joint with a wide range of motion including translation and rotation. Because of the high conformity and cross-shearing motion that exist in the hip joint, a hip replacement experiences relatively low **contact stresses** (2–5 MPa). Due to its relatively simple (ball and socket) anatomy, the hip can rely on the joint conformity to provide much of its stability. On the other hand, the knee has a more complex anatomy with relatively low conformity and relies on surrounding ligament structure for stability. Knee articulation involves rolling, sliding, and rotation motion due to flexion and extension and results in an implant that is subjected to high contact stresses (20–40 MPa) and mostly uniaxial motion with some degree of rotation. Shoulders and most of the other synovial joints fall somewhere between in terms of conformity, motion, and stress levels (Bartel et al., 1986).

The failure rate of all total joint replacement types is about 10% and most of these complications are owed to loosening associated with osteolysis. The hip has an estimated *in vivo* success rate of 90% at 15 years, while that of the knee and shoulder is on the order of 10 years and 7 years, respectively. The percentage of procedures where the primary joint replacement is replaced or revised is estimated at 12% of all total joint surgeries performed (Kurtz, 2009b). With the aging baby boomers and the increased propensity for obesity in the United States, it has been estimated that by 2030 there could be as many as 4 million joint replacement procedures performed annually (Kurtz, 2009b). From the perspective of healthcare costs savings associated with implants with extended service life, there is a great interest in understanding the factors that contribute to the success and longevity of total joint replacements.

13.4 Total hip arthroplasty

13.4.1 Anatomy of the hip

The hip joint is classified as a **synovial joint** because it is freely movable and contains an open space between the articulating bones that is filled with a viscous lubricant known as synovial fluid. This synovial fluid is contained within a capsule (synovial membrane) that holds mostly blood filtrate with glycoprotein molecules that allow it to act as a good, low-friction lubricant for the healthy joint. The anatomy of a healthy hip joint is shown in Figure 13.5. As the joint is compressed during walking or other activities, the synovial fluid is able to squeeze into and out of the cartilage covering the surrounding bones, thoroughly lubricating the cartilage surfaces.

Cartilage surrounding a synovial joint such as the hip is known as articular cartilage and comprises water (60–80 wt.%), type II collagen (10–20 wt.%), proteoglycans (4–7 wt.%), and chondrocyte cells as the balance (Mow and Hayes, 1991). Articular cartilage is commonly modeled as **poroelastic** because of the combined effects of the elastic collagen matrix and the fluid flow through the pores in this matrix, and in this regard has time-dependent mechanical behavior similar to a viscoelastic solid. The collagen fibers are responsible for the elastic mechanical behavior and the fluid flow through the porous matrix provides a viscous behavior that may dampen loads in the joint.

13.4.2 Causes for total hip replacements

The need for total joint replacement results from damage to the articular cartilage associated with disease, trauma, or normal "wear and tear" as shown in Figure 13.6.

13.4 Total hip arthroplasty

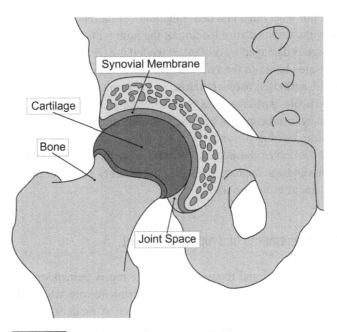

Figure 13.5

Illustration of a frontal cross-section of a healthy hip joint.

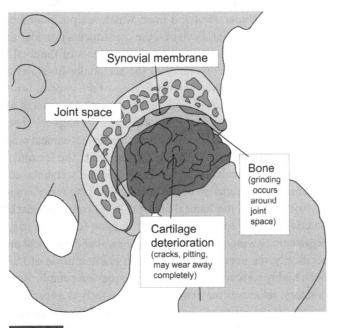

Figure 13.6

Illustration that shows the effect of osteoarthritis on the articulating surfaces of the hip joint.

Cartilage is not innervated and hence the lubrication mechanisms operating in healthy tissue can withstand repetitive loading of the joint without causing any pain. However, if the cartilage is damaged by disease, extended use, or trauma, intense pain may occur as the innervated bones of the synovial joint come into direct contact upon articulation. **Rheumatoid arthritis**, **osteoarthritis**, and **trauma** are the three primary sources of damage in articular cartilage (AAOS, 2010). Rheumatoid arthritis is a systemic disease in which the synovial capsules become inflamed, produce too much synovial fluid, and damage a large number of synovial joints in the body. Osteoarthritis is a degenerative disease that facilitates the degradation and wear-through of cartilage and ultimately results in painful bone contact. Trauma is the result of physical injury that damages the cartilage or fractures the hip joint.

13.4.3 Functional requirements of a hip replacement

Maintaining the functional requirements in the hip is tremendously important for the long-term performance of the device. The hip must restore articulation at the hip joint and provide stability within the pelvis and femur. A modern total hip replacement is considered to be a highly successful procedure when it restores function and pain-free living to a patient for ten years or more. The components of a modern total hip replacement include the femoral elements comprising the head, Morse taper (head-neck junction), neck, and stem that fixate into the femur as well as the acetabular components encompassing the acetabular shell and liner, which are positioned into the pelvis for restored articulation of the joint. Each of these constituents provides different attributes and uniquely contributes to the functional requirements of the total hip replacement, such as articulation and wear resistance of the acetabular liner and femoral head and fatigue resistance of the stem. Mechanical fixation of the implant is necessary for long-term stability and can be obtained with press fitting, bone cement along the femoral stem and backside of the acetabular cup, or porous coatings for tissue ingrowth. The materials used in these components must offer suitable mechanical properties in order to meet the functional requirements of the hip replacement. The femoral head provides the ball portion of the joint articulation and moves within the acetabular cup (socket portion of the joint). The primary function of the bearing pair is to restore range of motion and joint stability with minimal mechanical failures. These bearings must be highly resistant to wear (material loss due to sliding contact), fatigue (cyclic damage), and corrosion (material degradation owing to combined mechanical and chemical attack).

More specifically, the **acetabular cup**, which is often made of UHMWPE or CoCr alloys, must be highly resistant to wear in order to prevent complications with **osteolysis** and inflammatory response; the **femoral head**, which is generally a CoCr alloy or ceramic, must resist deformation, wear, and corrosion; the neck-head junction (**Morse taper**) must provide appropriate machining tolerance to prevent crevice corrosion; the **femoral neck** must be designed to sustain elevated bending stresses and fatigue loading; and the **femoral stem** must resist fatigue and corrosion, and should offer

Table 13.1 Table of functional properties of components in total hip replacements

Component	Functional requirement	Materials utilized	Modern THR
Femoral head	Hardness Wear resistance Corrosion resistance	CoCr alloys Zirconia or alumina	
Morse taper (Neck-head junction)	Corrosion resistance Flexural strength	CoCr alloys	
Neck	Flexural strength Corrosion resistance Fatigue resistance	CoCr alloys	
Femoral stem	Fatigue resistance Corrosion resistance Compliance match to bone	Titanium alloys CoCr alloys	
Porous coating	Integration with bone Interface strength	CoCr or Titanium beads Hydroxyapatite Bioactive glass	
Cement	Interdigitation with bone Interface strength Fatigue strength	PMMA	
Acetabular cup (liner)	Wear resistance	UHMWPE CoCr alloys	
Acetabular shell (backing)	Compressive strength Integration with bone	Titanium alloys CoCr alloys	

compliance match to bone to minimize stress shielding, and for this reason is commonly manufactured from a titanium alloy. Porous coatings should provide bone integration to provide interface strength and stability, and these coatings may employ titanium beads, hydroxyapatite, or **bioactive glass** to assist in tissue ingrowth. Similarly, **bone cement** (if used) should provide interdigitation with adjacent bone tissue for optimal fixation. The functional requirements of components and concomitant material groups used in modern total hip arthroplasty are summarized in Table 13.1.

The ball and socket articulation mechanism of the hip results in repetitive loading of the joint constituents during normal daily activity. Gait analysis shows that loads on the hip can be twice that of body weight forces during normal walking and four to six times body weight forces during running (Tencer and Johnson, 1994). While these loads

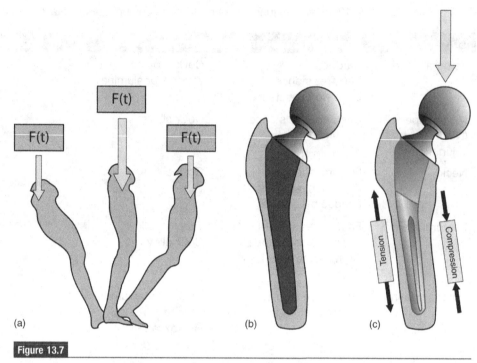

Figure 13.7

Forces develop in the hip through everyday activities. The (a) heel-strike gait results in varying forces. The (b) hip replacement resolves these forces into (c) complex cyclic stresses (fatigue loading) with tensile stresses developing on one side of the femoral stem and compressive stresses developing on the other side.

are not sufficiently high to break an implant in a single cycle, the repetitions of these loads over time may cause fatigue and wear damage to the materials or their interfaces and can contribute to implant loosening (Cook *et al.*, 1991). The forces in the hip result in a combined cyclic stress state of tension, compression, and bending of the femoral component as is illustrated in Figure 13.7. These stresses are cyclic in nature and result in fatigue loading of the stem. For this reason, the fatigue properties of the stem material and the stresses developed owing to implant design are of critical concern. Recall that the premature failures of the trapezoidal stem design (T-28) resulted from high cyclic stresses associated with the geometry that led to initiation as well propagation of a flaw until the device fractured through a fatigue failure of the alloy (see case study in Chapter 10).

A wide range of motion including translation and rotation exists at the articulating junction between the femoral head and the cup serving as the **acetabulum**. This ball and socket mechanism results in high conformity and the articulating surface of the hip replacement experiences relatively low contact stresses (2–5 MPa). Nevertheless, a large contact area and cross-shearing motion exist in the hip joint. This conforming biomechanical structural feature of the hip requires a highly wear-resistant bearing pair, as the bearing surfaces of the hip are susceptible to adhesive, abrasive, and third body

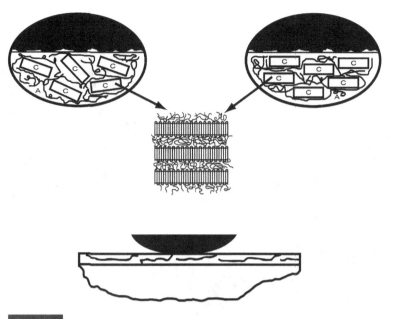

Figure 13.8

Schematic of wear mechanism that results in lamellar alignment in UHMWPE at the articulating surface. (After Klapperich *et al.*, 1999.)

wear. A mechanism of damage associated with the wear process is the generation of a plasticity-induced damage layer at the articulating surface which leads to lamellar alignment and loss of structural integrity as the implant undergoes a cross-shear articulation (Edidin *et al.*, 1999; Klapperich *et al.*, 1999; Kurtz *et al.*, 1999). This mechanistic process is illustrated in Figure 13.8.

The primary clinical concern in hip arthroplasty is loosening associated with wear-induced osteolyisis. Although all of the materials used in hip arthroplasty are biocompatible when in bulk form, problems arise when tiny particles of metal, UHMWPE, or cement are generated through contact loading and concomitant wear, interface failures, or corrosion processes. Particulates that are less than a micron in length are most damaging, as macrophages recognize them as foreign bodies and ingest the particles (Cook *et al.*, 1991). Through a complex mechanism that is not completely understood, the macrophages that have ingested particulate debris then signal osteoclasts, via cytokines, to begin eating away at the surrounding bone tissue. This bone loss is known as osteolysis, and it leads to partial loosening of the implant, followed by relative motion which causes more wear particles, and a vicious cycle of wear and loosening continues until the implant causes pain or loss of function. Mechanical loosening of the implant is the ultimate cause of failure. Ideally, the device should not cause any significant bone loss owing to stress shielding or particulate-induced periprosthetic osteolysis.

In order to overcome the clinical challenge of osteolysis associated with wear of the bearing surface, the orthopedics community has moved toward highly crosslinked forms

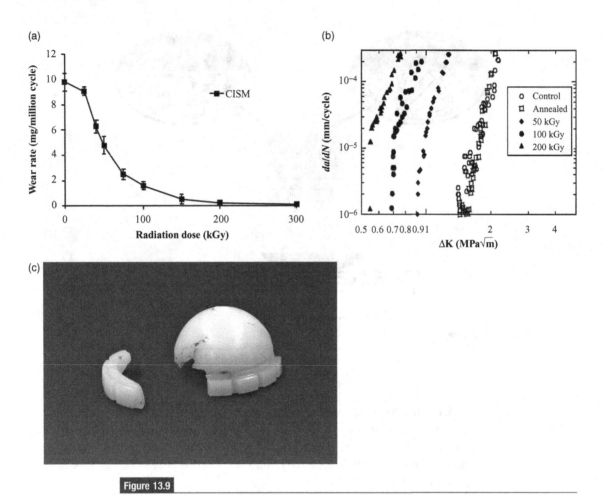

Figure 13.9

Illustration showing (a) wear rate and (b) fatigue crack propagation resistance as a function of crosslinking (radiation dose). (c) Optical micrographs showing a fracture in a highly crosslinked acetabular liner. (After Furmanski and Pruitt, 2009.)

of UHMWPE that offer exceptional wear resistance (Kurtz and Ong, 2009). Crosslinking is achieved with radiation and subsequent thermal treatments that annihilate free radicals that could otherwise render the polymer susceptible to oxidation. Figure 13.9(a) shows the effect of crosslinking (radiation dose) on the wear resistance of UHMWPE, and it is clear that a radiation dosage of about 20 MRad results in a material that is virtually without wear. A challenge for the orthopedic manufacturers, however, is that the crosslinking comes at the expense of fatigue and fracture resistance. Figure 13.9(b) shows the degradation of fatigue crack propagation resistance as a function of crosslinking (radiation dose); it is clear that the reduction in fatigue properties is concomitant with increasing crosslink density (Baker *et al.*, 2003; Furmanski and Pruitt, 2007;

Furmanski et al., 2009; Pruitt, 2005; Rimnac and Pruitt, 2008). Similar degradation has been reported for other mechanical properties such as ultimate strength, ductility, and fracture toughness (Kurtz and Ong, 2009; Pruitt, 2005; Ries and Pruitt, 2005). In fact, as is shown in Figure 13.9(c), there is evidence of fractures in retrieved highly crosslinked acetabular liners that have stress concentrations (Furmanski et al., 2009). Ultimately, the exceptional wear properties come at the expense of fracture properties and caution must be used when employing highly crosslinked UHMWPE in devices where stress concentrations or high cyclic contact stresses are expected (Furmanski and Pruitt, 2009).

In the past few decades there has been a trend toward modularity using designs that include metal-on-metal conical taper connections. Modularity has a number of benefits, including the customized fitting for a patient without exorbitant costs and the ability to replace only the femoral head while leaving a well-fixed femoral stem in place. The majority of modular implants use a conical press-fit fixation known as a Morse taper to achieve a junction between the femoral head and neck (Figure 13.11). The Morse taper makes use of mechanical design of a cone within a cone comprising a male portion (**trunnion** of the femoral shaft) and female portion (**bore** of the femoral head) with uniform tapers. In this way, the conical femoral taper compresses the walls in the bore of the femoral head as it is tapped into place by the surgeon and the mechanical stresses provide the fixation. The Morse taper connects two different materials and enables unique combinations in modular implants, such as the coupling of ceramic femoral heads with titanium stems. The stresses developed in the taper junction can cause catastrophic fractures of the ceramic femoral head if stresses exceed the strength level or fracture toughness of the ceramic employed. This failure mode is likely when a ceramic head is coupled with a femoral stem produced by a different manufacturer (with different taper angle) and is more apt to occur in a revision procedure where surgeons may couple a new ceramic head on the older femoral stem. An additional challenge in using the Morse taper junction is that the dissimilarity of materials can also lead to **fretting** or **corrosion** at the mechanical junction, and the surgeon must take care to make sure that joining surfaces are clean and dry before tapping the bore of the femoral head onto the stem trunnion. These junctions are also prone to high bending stresses as well as corrosion and fretting at the conical interface in the presence of any geometric mismatch (Figure 13.10). An example of fracture at the Morse taper owing to coupled effects of bending stresses and corrosion is presented at the end of this chapter.

It is clear that the design of a total hip replacement is not trivial and there are many factors to consider, including (i) the patient's health (joint anatomy, body weight, and physical activity level); (ii) the quality of bone available for fixation (level of bone porosity or osteoporosis); (iii) device design (choice of modularity, stem geometry, neck length, head size, and acetabular cup type); (iv) materials selection (choice of alloy system used in the femoral stem and acetabular shell, metal or ceramic bearing for the femoral head, and UHMWPE grade (degree of crosslinking) or metal as the counter bearing to the femoral head); and (v) fixation method (use of bone cement, porous coating, or bioactive surface treatment).

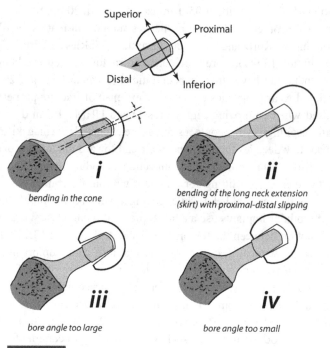

Figure 13.10

Schematic illustration of the Morse taper and various scenarios of loading in the superior-inferior and proximal-distal orientations that can lead to enhanced stresses and corrosion conditions in a modular implant (i) bending in the cone, (ii) bending in the long neck extension, (iii) bore angle too large, and (iv) bore angle too small. (After AAOS, 2010.)

13.4.4 Contemporary hip implants

Total hip replacement procedures are used to restore pain-free articulating function to a joint that has had extensive damage due to disease or trauma. The issues of biocompatibility and mechanical stability have evolved considerably in total hip replacements over the past century, with many devices offering pain-free lifestyles to patients for 20 years or more. Engineers as well as surgeons have a much better understanding of the biomechanics of the hip and the appropriate material properties necessary for providing resistance to wear, fatigue, fracture, and corrosion. Yet, implant design remains a highly interdisciplinary and multifactorial problem and thus it is not surprising that modern implants share the same challenges that have been present since the beginning of hip arthroplasty. The gait of a restored joint and concomitant lubrication schemes are not well understood in comparison to that of a healthy joint. To this day laboratory predictions of device performance often do not fully capture clinical outcomes of implants. Patients, depending on individual activity level and joint anatomy, may be more prone to implant dislocation or dissociation. Such clinical episodes put implant materials at risk for overloads and bearing contact that is not anticipated in the general device design or structural analysis of the implant of a general device. Moreover, minor shifts in

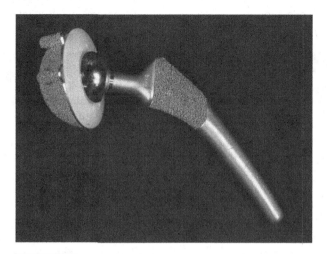

Figure 13.11

Contemporary total hip replacement showing the porous coating in the proximal portion of the femoral stem and the metal backing of the acetabular cup.

processing of a material to improve a property (such as wear resistance) often come at the expense of a different property (such as fracture resistance), and designs that have worked well for decades may be at risk for unpredicted failures.

There are many hip implant designs in use today, but the majority of these devices share the basic components employed in the innovative low-friction Charnley hip of 1962 as described above and shown in Figure 13.2. The ball and socket mechanism of the total hip replacement is designed to articulate with as little friction as possible and to mimic the motion and function of the healthy hip joint. One major shift from the earliest implant designs is the use of porous coatings on the metal backing of the acetabular cup and the proximal portion of the femoral stem (Figure 13.11). These coatings provide exceptional mechanical fixation and integration with bone tissue, and minimize the need for acrylic bone cements that can be prone to fatigue or interface fractures. Total hip replacements are used most frequently as treatment for osteoarthritis but can also be used for treatment of rheumatoid arthritis, developmental deformities, and some posttraumatic conditions including hip fracture. The number of fractures in the United States doubled from 1960 to 1980, and it is expected that there will be three times as many incidences in the year 2050 as there are today (Kurtz, 2009b).

13.4.5 Materials used in total hip arthroplasty

Although the materials today are similar to those introduced in the Charnley hip (a metal femoral ball, a plastic acetabular cup, and acrylic bone cement fixation) in the 1960s, a number of important improvements and modifications have been made in the last

Table 13.2 Material properties of alloys, polymers, ceramics, and coatings employed in contemporary total hip replacements

Material	Elastic modulus (GPa)	Yield strength (MPa)	Ultimate strength (MPa)	Endurance limit (MPa)	Fracture toughness (MPa√m)
Ti-6Al-4V	110	800	965	414	44–66
CoCr (cast)	210	450	655	310	60–70
CoCr (forged)	210	896	1207	414	70–100
CoCr (HIP)	210	825	1380	620	110–120
316L SS	200	700	965	345	70–120
UHMWPE	1	20	30	16	2–4
PMMA	3	–	35	34	1–2
Hydroxyapatite	15–18	–	100 (tensile) 500 (compress)	–	1
Bioactive glass	15	–	57 (tensile) 500–800 (compress)	–	1
Alumina	380–400	–	260 (tensile) 4500 (compress)	–	2–5
Zirconia	190–207	–	250 (tensile) 2500 (compress)	–	8–10

50 years. Owing to their high corrosion resistance and surface hardness, CoCr alloys are used for femoral heads, liners in all-metal prostheses, and femoral stems in contemporary implants. Differences in CoCr alloy properties depend on the manufacturing process employed. Wrought CoCr is currently preferred for use in hip arthroplasty as these alloys generally have lower surface roughness and lower friction, and wrought F1537–95 CoCr has been shown to have improved wear performance over cast F75–92 alloy (Noble, 1999). The titanium alloy Ti-6Al-4V is used primarily for the femoral stem but not the femoral head (owing to poor wear or tribological properties). The elastic modulus of the titanium alloy is better for evenly distributing loads between the prosthesis and the shaft of the femur and helps prevent atrophy of adjacent bone (stress shielding). Table 13.2 provides key material properties for the alloys used in total hip arthroplasty.

UHMWPE remains the polymer of choice for acetabular cups (as well as tibial inserts and **glenoid** components in knees and shoulders, respectively). The UHMWPE material properties listed in Table 13.2 are representative values; however, these figures can vary greatly depending on molecular weight, density, crystallinity, crosslinking, and processing conditions of the polymer (Kurtz and Ong, 2009). Crosslinking limits plasticity and greatly affects the wear processes as well as the fatigue-fracture properties of the polymer; for this reason, care must be taken in the design process to fully mitigate stress concentrations in order to prevent inadvertent fractures (Furmanski and Pruitt, 2009; Furmanski *et al.*, 2009).

Acrylic bone cement (**PMMA**) has been used as a load-bearing grout between the bone and the implant since the early days of hip replacements because of its self-curing characteristics that allow it to be press-fit into the appropriate space before it hardens. It has been shown that bone cement achieves its greatest mechanical fixation through bone penetration into a porous structure (osteoporitic bone), and this interdigitation is accompanied by an increased interface fracture toughness between bone and the inplant (Graham *et al.*, 2000). Cemented hip arthroplasties are still used in patients with porous bone structure and are used in a significant percentage of modern knee and shoulder replacement procedures. Other materials used in the fixation process include titanium or CoCr beads that are sintered to the proximal end of the femoral stem. Metal beads facilitate and encourage bone integration (Pawlowski, 2008). Other successful surface coatings for fixation include the use of bioactive ceramics, such as hydroxyapatite or bioactive glasses, with chemical structures similar to bone that encourage integration of the prostheses into osseous tissue (Pawlowski, 2008). The basic properties of the fixation materials are given in Table 13.2.

Ceramics offer exceptional wear resistance and are employed in the femoral head of some total hip replacement designs (usually targeting the more active patient). Femoral heads that utilize ceramics generally make use of zirconia or alumina, and both of these materials provide excellent tribological properties (low coefficient of friction, high hardness, and a very low wear rate). The ceramic bearing (head) generally articulates against an UHMWPE acetabular cup, and this bearing pair generally provides exceptional wear resistance. One challenge of ceramics, as discussed previously (Chapter 4), is that they have very low fracture toughness (Table 13.2). Any flaws or stress concentrations (such as a geometric mismatch at the Morse taper) will render the material susceptible to fracture, and great care must be taken in the processing and inspection of this component.

13.4.6 Design concerns in total hip arthroplasty

The design of a total hip replacement is a multifactorial process that combines the need to balance functional requirements of the hip, anatomical needs of the patient, material selection, and manufacturability. The primary concerns that exist for the design of total hip replacement include the anatomical fit and long-term restoration of the function (articulation) of the hip; the minimization of stress shielding and periprosthetic wear-induced osteolysis; the fatigue, wear, and corrosion resistance of the materials utilized; fixation of the implant with the tissue via press-fit, porous coatings, or bone cement; geometric tolerance and prevention of fretting; sterilization methodology; and shelf life. The basic factors that contribute to the success of a medical device were outlined in Chapter 1 and illustrated in Figure 1.2. However, this process can be further elucidated

Table 13.3 Engineering design process for development of a hip replacement

Phases	Elements	Examples
1. Identify problem	How do I design a ___that will ___?	How do I design a hip replacement that will offer pain-free articulation?
2. Identify criteria and constraints	Specify design requirements and constraints	Restore motion in the hip joint Suitable for a small framed female athlete with osteoporosis Offer functionality for 15 years Costs less than $7000 Unique anatomy in hip
3. Brainstorm solutions	Identify a number of solutions (usually done quickly with informal sketches, stickies®, etc.)	Metal on metal Ceramic on polymer Composites Ceramic on ceramic Metal on polymer Hydroxyapatite coatings Bone cement fixation Bioactive glass coatings Porous coatings Modular system (variable stem length, neck length, femoral head size, acetabular cup size, acetabular shell) Custom implants
4. Generate ideas	Narrow selection and develop 2–3 ideas (technical drawings, general data)	Metal on metal Ceramic on polymer Bone cement fixation Modular system (variable stem length, neck length, femoral head size, acetabular cup size, acetabular shell) Custom implant (specific to patient anatomy)
5. Explore	Discuss pros and cons on designs	Discuss the specificities of designs with respect to the criteria and constraints (above)
6. Select approach	Choose best design	Custom design (patient anatomy) Ceramic on polymer with bone cement fixation
7. Build prototype	Build device or full-size model Assess performance	Confirm the device design offers 15 years of service with no wear (15 million cycles with minimal material loss)
8. Refine design	Identify changes that need to be made (constraints/requirements)	Confirm anatomical fit and performance for active patient

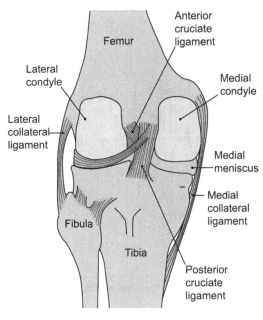

Figure 13.12

Anatomy of the healthy knee showing the stabilizing ligaments.

using a basic engineering design analysis, and a simple example for a hip replacement is exemplified in table form (Table 13.3).

13.5 Total knee arthroplasty

13.5.1 Anatomy of the knee

The knee is the largest synovial joint in the body and provides articulation between the femur and tibia as well as the patella (knee cap). The primary movement of the knee joint is flexion and extension in the sagittal plane with a small amount of rotation. These motions induce a small area of contact between the femoral condyles and the tibial plateaus due to the constraint provided by the ligaments. The knee achieves much of its stability from the large ligaments that attach the femur and tibia and fibula. As with the hip, the articular cartilage provides lubrication and smooth articulation between the long bones in a healthy knee joint. The synovial membrane (capsule) covering the knee facilitates the containment of the lubricating joint fluid. In comparison to the hip, the knee is much less conforming. The motion in the knee is also more complex than the hip, including rolling, sliding, and some rotation due to flexion and extension of the joint.

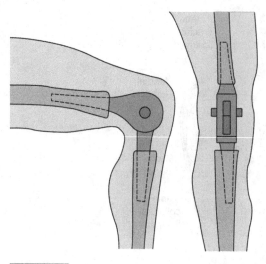

Figure 13.13

Illustration of the basic hinge mechanism design first utilized in total knee reconstruction.

13.5.2 Clinical reasons for knee replacement

It is estimated that there are more than 130,000 primary knee replacement surgeries in the United States annually, and as with the hip, it is estimated that 10% will require a revision within the coming decade (AAOS, 2010). The causes for knee replacement are quite similar to those described for the hip and are typically associated with some form of osteoarthritis, rheumatoid disease, and trauma. Osteoarthritis can lead to roughened surfaces, wear-through of the cartilage, and loss of lubrication. Rheumatoid arthritis is accompanied by inflammation of the synovial membrane and is often a systemic disease. Misalignment of the knee is another cause for knee replacement. The normal knee has its load-bearing axis down the middle of the leg through the hip, knee, and ankle; **varus** (bow-legged) knees cause more stress on the medial compartment of the knee, while **valgus** (knock-kneed) alignment causes more stress across the lateral compartment of the knee.

13.5.3 Historical evolution and contemporary knee design

In the late 19th century, a Greek surgeon by the name of Thoeodorus Gluck designed and implanted the first total knee replacement comprising ivory and plaster of Paris. The earliest implants were based on Gluck's initial design and utilized a hinge mechanism to replicate the kinematics of the knee (Shetty *et al.*, 2003). Figure 13.13 schematically illustrates the basic hinge mechanism employed in the earliest knee designs.

Figure 13.14

Example of a total knee reconstruction in the mid- 20th century utilizing the hinge design. (After Shetty *et al.*, 2003.)

Within the first half of the 20th century the hinge designs dominated the field, and by 1950, the fixed-hinge knee design became commonplace (Figure 13.14). At that time it was believed that the hinge mechanism replicated the kinematics of the knee joint; however, the metal-on-metal articulation in this design did not offer long-term structural integrity and was prone to mechanical failures. It was later realized that these hinge designs were incapable of replicating the internal rotational motion known to occur in the knee (Shetty *et al.*, 2003).

A major improvement to the hinge design was the invention of the "polycentric knee" developed by Gunston, who recognized that the motion in the knee comprised rolling, sliding, and rotation (Gunston, 1971). He postulated that during flexion of the knee the femur rocks (rolls) and glides (slides) in a posterior direction atop the tibial condyles about successive centers of rotation (Figure 13.15(a)). Gunston referred to this motion as a rotation about a "polycentric pathway" and named his invention accordingly. The polycentric knee comprises two metallic femoral runners that articulate in UHMWPE (also referred to as polyethylene) tracks in the tibial condyles (Figure 13.15 (b)). The alignment of these four primary compartments was a clinical challenge for orthopedic surgeons. Short-term results looked promising; however, in the long term the device suffered from complications with loosening, subsidence of components, poor range of motion, and instability (Jones *et al.*, 1981).

In the following decade, **unicondylar** and **bi-unicondylar** implants were developed and were available in a range of sizes to be implanted over the tibia (Insall and Walker, 1976). Figure 13.16 illustrates the designs associated with these early implants. Uni-condylar knee arthroplasty (where only one condyle was damaged enough to be replaced

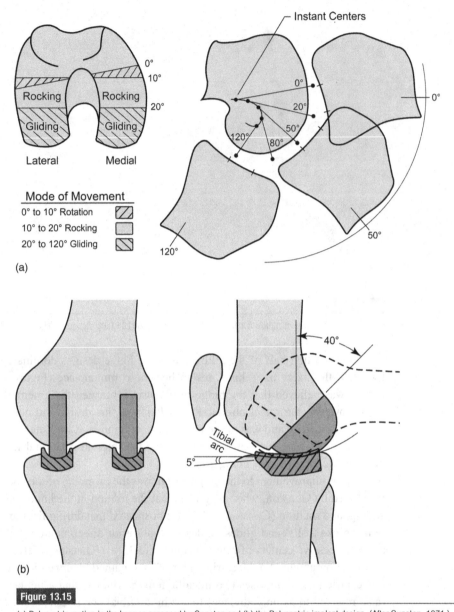

Figure 13.15

(a) Polycentric motion in the knee as proposed by Gunston and (b) the Polycentric implant design. (After Gunston, 1971.)

while the other retained the natural bone and cartilage) was found to provide postoperative functional advantage over total knee arthroplasty owing to the retaining of cruciate ligaments in the unicondylar procedure (Banks *et al.*, 2005). Unicompartmental designs were popular for patients with unicompartmental arthritis and were better suited for the less-active patients. However, many of the early unicondylar and bi-unicondylar

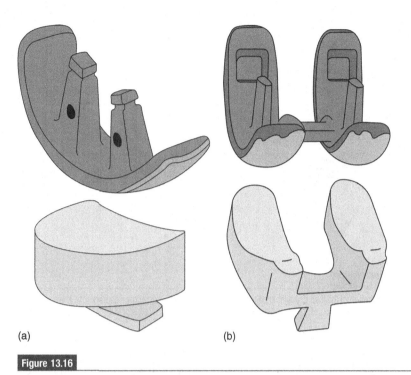

Figure 13.16

Unicompartmental (a) and bi-unicompartmental (b) implant designs that employed metal-on-polymer articulation. (After Insall and Walker, 1976.)

implants were difficult to work with surgically and were prone to loosening owing to their short stem design. Patient selection, implant design, and surgical technical features constituted the keys to successful results.

In the early 1970s, the clinical orthopedic research field was in search of the perfect condylar knee. In 1971, Ranawat and Insall implanted the first duocondylar knee at the Hospital for Special Surgery in New York (Ranawat and Sculco, 1985). Ranawat, Insall, and Walker introduced the "total condylar knee" in 1974, which had multiple-radii curvature on its articulating surfaces, more resembling the natural human knee (Insall *et al.*, 1982). The features of this design included multiple radii of curvature and replaced the trochlear groove and patella. In 1978, Insall and Burstein developed another version of the total condylar knee that was posterior-stabilized and comprised a metal backing to support the tibial insert (Bartel *et al.*, 1982; Insall *et al.*, 1982). This design has been one of the most successful condylar knee designs and has served as the prototype for the most common total knees in use today. Figure 13.17 shows a contemporary knee design founded upon the principles of the total condylar knee designed by Insall and Burstein.

The inability to fully replicate the kinematics of the knee joint resulted in a number of anatomical variations on total joint replacements. Some designs were highly conforming

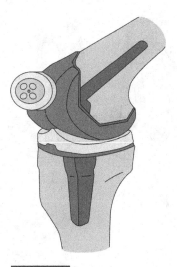

Figure 13.17

Illustration of the total condylar knee used as the foundation in many contemporary total knee replacement designs. (After Insall *et al.*, 1982.)

(more like the hip) but resulted in poor joint articulation and concentrated stresses at the point of contact (as shown in Figure 8.29, in Chapter 8). On the other extreme end of the design spectrum, non-conforming implants were introduced but were prone to instability through the flexion-extension cycle of the knee. The articulation of the non-conforming knee is illustrated in Figure 13.18.

The issue of stability was addressed in a number of ways. One method was to develop a total joint replacement that could retain the posterior cruciate ligament (Figure 13.19(a)), while another utilized a substitute for the ligament in the form of a cam or post (Figure 13.19(b)). One of the primary complications of the cam or post that substitutes for the posterior cruciate ligament is its susceptibility to fatigue fracture as the components articulate through extension-flexion (Bolanos *et al.*, 1998).

Mobile bearing designs were introduced in 1976 to prolong the life of the implants by lowering the stresses present in the polyethylene (Callaghan *et al.*, 2000). A computer-directed precision-grind finishing process was introduced to make the contacting surfaces more conforming. A bearing was added to facilitate rotation between the tibial tray and the polyethylene insert to lower the shear stresses and allow for a more natural range of motion (Figure 13.20).

The primary challenge in the mobile bearing design is the propensity for wear at the articulating surface of the primary bearing pair owing to enhanced conformity and also at the backside interface between UHMWPE and its underlying alloy in the rotating bearing. As with other implant designs, increased levels of particulate debris are accompanied by a greater likelihood for bone loss around the implant due to osteolysis.

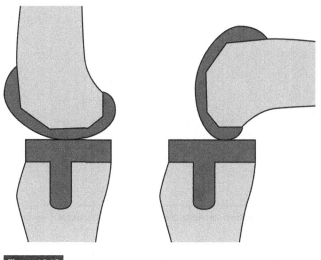

Figure 13.18

Illustration of the non-conforming design showing tendency toward instability during flexion-extension of the knee.

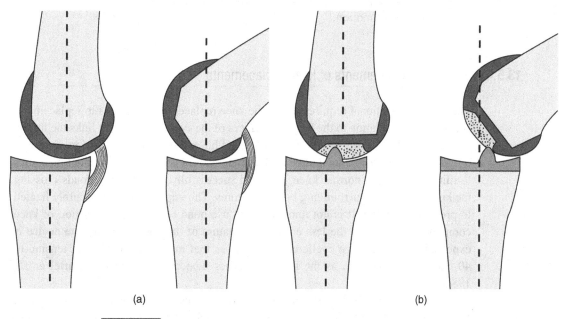

Figure 13.19

(a) Posterior cruciate ligament-retaining design in extension (left) and flexion (right); and (b) posterior cruciate ligament-substituting design in extension (left) and flexion (right).

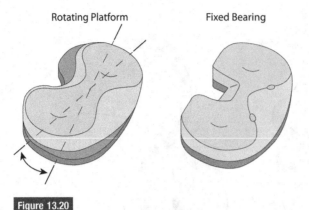

Figure 13.20

Schematic showing the difference between the rotating platform and fixed bearing design. (After Callaghan *et al.*, 2000.)

There are a number of contemporary total knee implant designs available, yet many of these share resemblance to the natural knee design developed by Insall and co-workers (Insall *et al.*, 1982) and shown in Figure 13.17 above. Today there exist a wide range of clinical procedures and medical devices that a surgeon can uniquely choose based on the patient's needs. Contemporary condylar knee designs utilize improved fixation methods, reduced particulate owing to wear or corrosion, better kinematics, and an enhanced range of motion.

13.5.4 Functional requirements of knee replacements

The primary functional requirement in the knee replacement, like the hip replacement, is to restore articulation to the bearing surfaces of the synovial joint that links the femur, tibia, and fibula. The implant must be biocompatible and resistant to corrosion, wear, and fatigue. The design of a total knee replacement should provide appropriate kinematics to mimic those of a normal knee joint and successfully transfer large loads crossing the knee joint to the surrounding bone structures. The implant must be securely fixated to prevent loosening. It is not uncommon to use bone cement in the fixation of knee components. Owing to the less conforming nature of the knee anatomy, the device is expected to sustain large cyclic contact stresses that may span 10 MPa in tension to 40 MPa in compression as the knee undergoes flexion and extension (Bartel *et al.*, 1986).

A source of failure in early metal-on-polymer knee replacements was subsurface fatigue fracture (delamination) of the UHMWPE caused by high contact stresses from the non-conforming surfaces and in some instances, oxidation of the polymer owing to poor sterilization methodology (Kurtz, 2009c). Non-conforming surfaces create large Hertz contact stress, resulting in a large subsurface von Mises stress. The large cyclic shear stresses can lead to fracture and delamination, resulting in liberation of sheets of UHMWPE that can lead to premature failure of the implant. To address the problem, a metal backing was added to the polyethylene tibial insert. The metal backing allows the force to be more evenly distributed in the component and lowers the overall stresses. However, the metal backing may increase stress shielding of the tibia, and also reduces the thickness of the polyethylene, which results in higher Hertz contact stress (Bartel *et al.*, 1986). In addition to metal backing, manufacturers also moved to more conforming designs to mitigated contact stresses, however, this led to transfer of the force to the bone-implant interface (and lead to loosening). Cobalt-chrome alloys replaced stainless steel to increase strength and wear resistance of the bearing couple. The materials used in total knee replacements are chosen to offer exceptional fatigue, wear, and corrosion resistance. A successful total knee replacement re-establishes articulation and pain-free living to a patient for at least ten years.

The design of a total knee replacement is multifactorial and depends on (i) patient's anatomy, health, activity level; the quality of bone available for fixation; (ii) geometric requirements of the femoral condyle, tibial plateau, and patellar button system; (iii) system design which is dependent upon whether the system is constrained or unconstrained, whether the cruciate ligament is retained or substituted, and whether the system is metal-on-metal or metal-on-polymer (at the bearing coupling); (iv) selection of materials utilized, including the choice of alloy system, grade of UHMWPE (degree of crosslinking and process conditions); and (v) fixation method such as porous coating, bioactive treatment, or bone cement. The materials used in these components must offer suitable mechanical properties in order to meet the functional requirements of the knee replacements. The functional requirements of the components of a total knee replacement are summarized in Table 13.4.

13.5.5 Materials used in contemporary total knee arthoplasty

As discussed above, much of the materials development employed in total knee reconstruction followed directly from the lessons learned from total hip arthroplasty. Table 13.4 provides representative properties for the materials used in total knee replacements. CoCr alloys offer exceptional corrosion and wear resistance and are used for femoral components that articulate against UHMWPE tibial plateaus, which may or may not be metal backed. The metal backing and stem attachments for fixation often employ titanium alloys for better compliance match and integration with bone tissue. UHMWPE remains the polymer of choice for the tibial inserts. The functionality of the knee joint mandates exceptional fatigue resistance; however, the exceptional wear resistance offered by highly crosslinked versions of UHMWPE comes at the expense of fatigue crack propagation and fracture resistance (Rimnac and Pruitt, 2008).

Because fatigue resistance is such an important functional property for the bearing surfaces in knee arthroplasty (owing to the high cyclic contact stresses), a number of

Table 13.4 Functional properties of components in modern total knee replacements

Component	Functional requirement	Materials utilized
Femoral condyles	Hardness Wear resistance Corrosion resistance	CoCr alloys
Tibial plateau (insert)	Fatigue resistance Wear resistance Creep resistance Corrosion resistance	UHMWPE CoCr alloys
Pattelar button (knee cap)	Wear resistance Fatigue resistance	UHMWPE
Metal backing (stem)	Fatigue resistance Corrosion resistance Compliance match to bone	Titanium alloys
Cement	Interdigitation with bone Interface strength Fatigue strength	PMMA

material developments have specifically attempted to improve the mechanical integrity of UHMWPE. Conceptually this remains a valid material design pursuit; however, historically these material enhancements have often resulted in poor clinical performance. One such example was the development of carbon fiber-reinforced UHMWPE (Poly II). This material was marketed to offer improved mechanical performance over conventional UHMWPE (Wright et al., 1988a) with the assumption that increasing the strength of the bearing material would improve the life of the implant. In laboratory studies, the composite offered a 25% increase in monotonic compressive strength and an order of magnitude improvement in wear resistance over the conventional UHMWPE; it was thought that the composite would be more resistant to delamination failure mechanisms typical of the knee as well as wear processes in the hip (Wright et al., 1988b). Regrettably, Poly II^{TM} fared poorly in clinical practice owing to poor interface strength (adhesion) between the fiber and the matrix as well as enhanced contact stresses associated with the increased elastic modulus of the composite structure. The consequence of increased contact stresses and fiber debonding negated any improvements in mechanical properties, and early failures employing these materials were often accompanied by osteolysis and chronic inflammation of the synovial capsule due to the carbon fiber particulate debris (Wright et al., 1988a, b). Figure 13.21(a) shows a typical fracture surface with fiber debonding in the UHMWPE matrix. Poly II was removed from the orthopedics market shortly after its implementation, and later research studies revealed that the fatigue crack propagations rates were nearly a decade higher for the carbon fiber reinforced UHMWPE over that observed in conventional UHMWPE (Kurtz, 2009d).

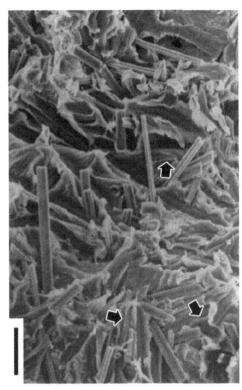
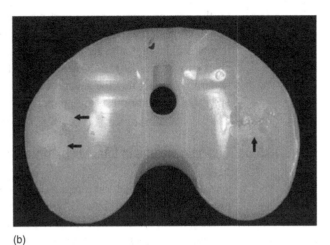

Figure 13.21

(a) Typical fracture surface showing fiber debonding in the UHMWPE matrix of Poly II™ and (b) delamination in Hylamer™ and sterilized by gamma radiation in air. (After Wright et al., 1988a.)

Another well-known modification to UHMWPE known as Hylamer™ was presented in the early 1990s (Bellare and Kurtz, 2009). This polymer system utilized hot isostatic pressing to create an extended-chain morphology with enhanced crystallinity and lamellae thickness. The greater crystallinity (70%) results in a twofold increase in elastic modulus as compared to conventional UHMWPE, and an increase in elastic modulus intensifies Hertzian contact stresses (Wright and Bartel, 1986). However, the enhanced crystallinity also provides substantially greater resistance to fatigue crack propagation (Baker et al., 1998; Pruitt, 2005) and it was thought that this enhancement would offset the increased contact stresses and provide improved resistance to delamination processes in total knee arthroplasty. Unfortunately, Hylamer™ did not fare well in clinical applications. The primary complication with Hylamer™ was attributed to an enhanced susceptibility to oxidative degradation owing to its extended-chain morphology (Bellare and Kurtz, 2009). Hylamer™ was introduced into the market at the same time that manufacturers sterilized devices using gamma radiation in air, and this caused an oxidative degradation reaction in Hylamer™. Clinical studies revealed pitting and delamination on the bearing surface after short periods of service (Figure 13.21(b)). Fatigue crack propagation studies on Hylamer™ performed after sterilization using gamma radiation in air resulted in a 50% reduction in the stress intensity range needed for inception of crack growth (Baker *et al.*, 1998). Clinical studies such as those in Hylamer and Poly II underpin the fact that the function of medical devices is complicated by many factors and often laboratory tests are not predictive of the implant performance *in vivo*.

13.5.6 Design concerns in total knee arthroplasty

The design of a total knee replacement depends on numerous parameters, including processing the need to balance functional requirements of the knee, anatomical needs of the patient, material selection, and manufacturability. As with other medical device designs, there are design pathways (such as that shown in Table 13.3) that are employed in creating implants that are both functional and cost-effective. The concerns that exist for total knee replacement designs are quite similar to those for a hip replacement and include anatomical fit and restored articulation of the joint; resistance to fatigue fracture owing to high cyclic contact stresses; the minimization of stress shielding and wear-induced osteolysis; corrosion resistance; fixation; manufacturability, sterilization methodology; and shelf life. Unlike a hip replacement, the kinematics of a revised knee are quite often altered from that of a healthy joint. In total knee reconstruction, the anterior cruciate ligament and menisci are removed; and for these reasons, there remains an ongoing design challenge to fully restore the kinematics and function of a healthy knee joint using engineered materials.

13.6 Fracture fixation

13.6.1 Background

It is estimated that more than 6 million people each year sustain skeletal fractures in the United States alone (AAOS, 2010). More than 15% of fractures heal abnormally, exhibit delayed union, or result in non-unions, and hence the design of treatment strategies for fracture healing remains of utmost importance (Einhorn, 1995). The term **fracture fixation** broadly encompasses any material system or device that is used to facilitate healing of a fractured bone, and this includes both external methods (casts, braces) as well as internal fixation techniques (wires, screws, plates). The development of internal fracture fixation is considered to be one of the most significant advances in the field of orthopedics (AAOS, 2010). Internal fixation allows shorter hospital stays, enables individuals to return to function earlier, and reduces the incidence of improper healing. The topic of fracture fixation is extensive and the reader is referred elsewhere to have a broad overview of the field (Bray, 1992). This text will focus on internal fracture fixation systems as they are utilized in long bones and the spine. The first attempt at internal

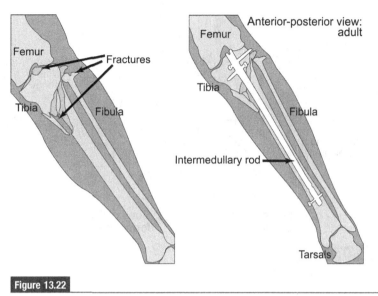

Figure 13.22

Internal fracture fixation process for a fractured tibia. (Image courtesy of Dr. Michael Ries.)

fixation dates back to the use of an iron pin in the mid-1800s; however, these systems fared poorly in the body owing to corrosion of the iron and concomitant inflammatory response in the body (Peltier, 1990). Subsequently, a surgeon by the name of Lister introduced open reduction and performed the first sterile surgery in which he stitched a fractured patella together with a silver wire (Peltier, 1993). The use of plates, screws, and wires in the fixation process was first documented in the late 19th century. The early internal fracture fixation methods were often complicated by infection, inflammatory response to the metals, and a poor understanding of the biomechanics of bone repair. In the last 60 years there has been a great improvement in the understanding of the fracture healing process, resulting in a number of successful clinical fixation techniques (Peltier, 1990). A typical internal fracture fixation process for a fractured tibia is shown in Figure 13.22.

13.6.2 Anatomy of fracture healing

As fracture fixation applies to the repair of any bone fracture in the body, the relevant anatomy is extensive. Skeletal regeneration of fractured bone tissue involves a spatially rich cascade of biological processes, and fracture healing is generally characterized by three distinct but overlapping phases involving inflammation, repair, and remodeling. Fracture of bone tissue results in the tearing of the bone **periosteum**, damage to the bone marrow, and interruption of the local vasculature. The trauma to the tissue triggers a pathway of inflammatory events and the formation of a hematoma. As the necrotic

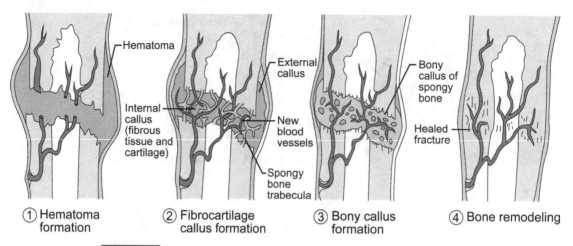

Figure 13.23
Illustration of the bone healing process.

tissue is resorbed, the dissemination of cells around the fracture site leads to the formation of a soft granulation tissue that bridges and supports the fractured bone ends. The formation of this tissue, known as the fracture callus, signals the onset of the reparative phase. During the early reparative phase, a soft (**fibrocartilage**) callus is formed with fibrous tissue growth and cartilage proliferation within the fracture gap. Concomitant with these processes is the expression of genes that regulate **ossification** (bone formation via cartilage mineralization) and primary bone deposition. The subsequent hard (bony) callus phase is accompanied by the formation of woven bone. Finally, during the remodeling stage, the callus is completely mineralized and the primary woven bone starts to remodel into a mature lamellar bone structure. Figure 13.23 illustrates the basic processes of bone healing that involve the formation of a (1) hematoma at the fracture site, (2) fibrocartilage callus, (3) bony callus, and (4) bone remodeling.

13.6.3 Clinical reasons for fracture fixation

Fracture fixation implants are necessitated by the various fractures that occur within the skeletal system. These fractures need to be set with proper alignment and must ensure that the bone carries sufficient load for healing. There are *six* basic classifications associated with fracture types (Peltier, 1990; Tencer and Johnson, 1994). A bone that breaks into many small pieces is termed a (i) **comminuted fracture** and this type of fracture is commonplace in people who have **osteoporosis**. A (ii) **compression fracture** refers to bone that has been crushed and it is also common in osteoporotic bones. If the fractured bone is depressed inward, it is referred to as a (iii) **depressed fracture**, and this is observed in skull fractures. Fractured bones that have been pushed into each other during the fracture process are known as (iv) **impacted fractures** and are commonly

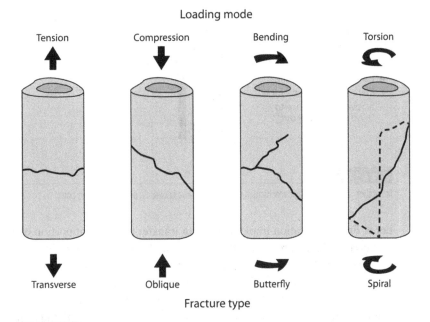

Figure 13.24

Fractures characterized by the types of loads that generate them. (After Tencer and Johnson, 1994.)

seen in falls involving an outstretched arm. Sports injuries often involve a torsion stress that results in a failure plane that is oblique to the long axis of the bone, and these fractures are known as (v) **spiral fractures**. A bone that is broken incompletely is referred to as a (vi) **greenstick fracture** and is more prevalent in children. Fractures can also be characterized by the stress state that generates them, as is shown in Figure 13.24. A **transverse fracture** occurs under tensile loading, **oblique fractures** occur in the presence of compressive stresses, **butterfly fractures** are associated with bending stresses, and **spiral fractures** involve torsion loads (Tencer and Johnson, 1994).

13.6.4 Functional requirements of fracture fixation devices

The primary purpose of fracture fixation implants is to provide a functional platform for bone healing. During fracture fixation or bone surgery, different metal devices can be used to provide immobilization of the bone fragments leading to faster patient mobility. This process may involve pinning, screwing, tying, reinforcing, or plating the various fractures using individual components (pins, screws, wires, rods, or plates) or any combination thereof (see below). Implants are used to properly align the fractured bones so that they heal in the appropriate orientation. Fracture fixation devices provide mechanical stability throughout the healing processes, including the initial inflammatory process, the formation of soft callus and subsequent bony callus, and finally remodeling of the

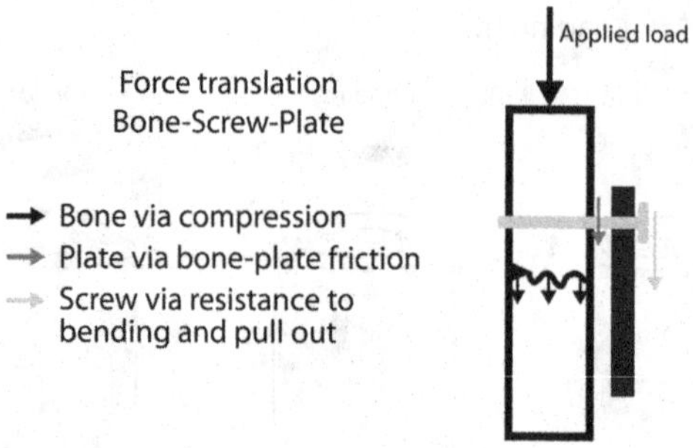

Force translation Bone-Screw-Plate

→ Bone via compression
→ Plate via bone-plate friction
→ Screw via resistance to bending and pull out

Applied load

Figure 13.25

Illustration showing the basic mechanism of force translated between a plate, screw, and bone.

bone. Fracture fixation implants have to transfer a sufficient portion of load to bone so that cell-mediated processes of recovery can be activated, and for this reason, *compliance match to bone* is an important factor. The implants must offer excellent mechanical properties, including *tensile strength*, *compressive strength*, *flexural strength*, and *torsion strength*. Additionally, the implants ought to be *fatigue resistant*, to withstand the cyclic loading associated with the musculoskeletal system. Of equal importance is the *corrosion resistance* and for this reason CoCr alloys, stainless steels, and titanium alloys are utilized in fracture fixation devices. Figure 13.25 shows the basic mechanism by which forces are translated between a plate, screw, and bone.

13.6.5 Fracture fixation designs

The primary constituents utilized in internal fracture fixation include pins, wires, plates, screws, and rods. **Pins** are used to fasten or attach fragmented bones together and are commonly used in fractures that are too small to be fixed with screws. Pins may be removed or left in place after the fracture heals. **Wires** are generally employed like metal sutures to thread or tie fractured bones back together. **Plates** serve as internal splints and are screwed in place to hold the fractured bone together. These devices may be removed after healing or left in place, depending upon the nature of the fracture. **Screws** are the most commonly used implants for internal fixation of fractured bone and these may be used independently or in combination with the other fixation implants. There are two basic types of screws available. Cortical screws are designed for compact **diaphyseal** (cortical) bone, whereas cancellous screws are designed for the **metaphyseal** (trabecular) bone. The former typically have a larger thread diameter and pitch and a greater difference between major and minor (shaft) diameters in comparison to cortical screws, providing more surface area for bone attachment. These screws are intended for use in metaphyseal fixation, where bone is softer. **Cannulated** screw sizes vary from 4.0–7.5 mm and cancellous screws are typically between 4.0–6.5 mm. **Rods** are sometimes referred to as intramedullary rods (or nails) and are as a rule employed in

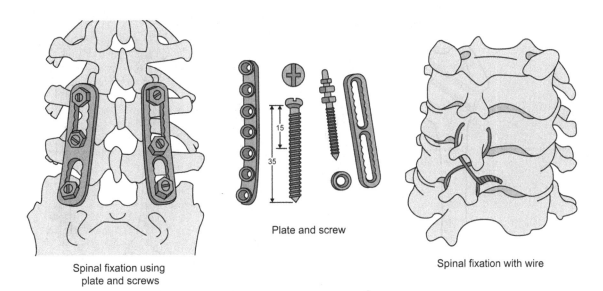

Figure 13.26

Illustrates an example of a fracture fixation process in a spine using plate and screws (left, center) as well as the method that employs wire (right) for fixation. (After Mazel, 2008.)

the repair of long bones; they are typically inserted through the center of the long axis of the bone. Like the other implants, the rods may be left in place once the fracture is repaired. Figure 13.26 provides an example of a fracture fixation process in a spine using plate and screws as well as a method that employs wire for fixation. Figure 13.27 shows the use of cannulated screws for femoral neck repair; intramedullary rods for fracture in long bones, such as the femur; and the **Knowles pin** used in small fractures (after Peltier, 1990).

The basic mechanism utilized for a dynamic compression plate is shown is Figure 13.28. This fracture fixation implant is designed to compress the fracture. It makes use of screws to exert force on holes in plate. The force between the screw and plate moves the bone until the screw sits properly and compressive forces are transmitted across the fracture (Tencer and Johnson, 1994).

13.6.6 Materials used in fracture fixation

The evolution of materials employed in internal fracture has closely mirrored the alloy development and implementation utilized in total joint arthroplasty (described above). The earliest implants attempted the use of iron but suffered from premature failures due to immunological response and chronic inflammation in the patient. In the late 1930s, vanadium steel was introduced in plates and screws for bone fixation; however, these metals were also predisposed to corrosion. By 1940, scientists recognized a need for

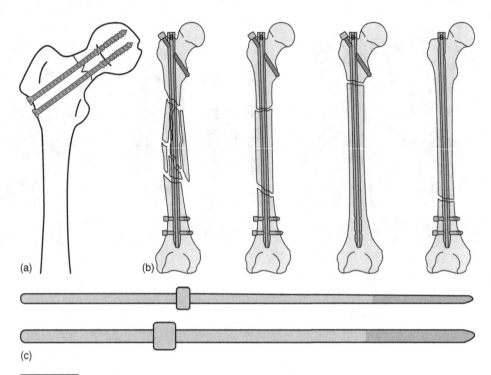

Figure 13.27

Schematic shows the use of cannulated screws for (a) femoral neck repair; (b) intramedullary rods for fracture in long bones, such as the femur; and (c) the Knowles pin used in small fractures. (After Peltier, 1990.)

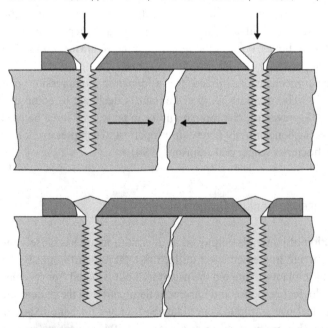

Figure 13.28

Illustration of a typical dynamic compression plate used in fracture healing. (After Tencer and Johnson, 1994.)

a material that offered resistance to corrosion and fatigue fracture. By the mid-20th century, stainless steels were commonplace in internal fixation devices. The following two decades realized the introduction of cobalt chromium alloys and titanium alloys, which have become a successful mainstay in the fracture fixation industry. Today the primary materials used in fracture fixation are medical grade 316L stainless steels, CoCr alloys, and titanium alloys (the same alloys used in total joint arthroplasty).

13.6.7 Design concerns in fracture fixation devices

The primary concerns today with internal fracture fixation designs involve optimization of mechanical properties, load transfer to bone, and corrosion resistance. Fracture fixation devices can be implemented for short-term bone healing or they may be implanted with the intent that they will last the lifetime of the patient. In the latter case, not only are the foregoing properties essential, but also the resistance to damage under cyclic loading (fatigue loading) becomes critical in long-term devices.

13.7 Spinal implants

13.7.1 Background

There are two primary categories of medical devices used in the spine, and this includes spinal fixation and disk replacement. It is estimated that more than 800,000 Americans are afflicted with **scoliosis** and suffer from related back pain, loss of motion, and spinal fracture (Kuptniratsaikul, 2008). In the late 19th century, a surgeon by the name of Wilkins performed the first spinal surgery to stabilize a fractured spine (Omeis *et al.*, 2004). In the years that followed, the methods for stabilization of spine disorders and fractures were further developed. A physician by the name of Hadra first used a wire technique to secure and stabilize the **spinous processes** associated with a cervical fracture (Hadra, 1891) and later another surgeon named Fritz Lange implemented the use of steel rods in combination with wire to stabilize the spine (Lange, 1910). Wire methods of fixation were primarily utilized through the first half of the 20th century, and an example of this configuration is shown in Figure 13.26 above. One of the drawbacks of wiring was that it failed to provide the support necessary to resist twisting in the transverse and coronal planes. To compensate for this problem, patients required an external cast or brace system in addition to the surgical procedure. In the 1940s, facet screws for spinal stabilization were introduced, and a decade later the use of the pedicle as a fixation point was implemented (Boucher, 1959).

In 1911, two surgeons, Fred Albee and Russell Hibbs, using different techniques, performed the first spinal fusion procedure independently. Hibbs used what he referred to as a "feathered fusion," in which he connected the lamina to the cortical bone with pieces of bone from adjacent vertebrae, while Albee used **autologous** tibial grafts on both sides of the spine much like the steel rods in Lange's technique (Bick, 1964). The bone grafts were used to stabilize adjacent vertebrae in the spine by facilitating intervertebral bone growth; however, this technique resulted in a spine that lacked range of motion and which was susceptible to degenerative diseases. Albee and Hibbs attempted to use bone grafts as a scaffold to promote the fusion of the spinal bones, but new methods were required to further stabilize and fuse the spine.

In the early 1960s, Roy-Camille introduced a screw-plate construction for the stabilization of a spine (Mazel, 2008). This device is illustrated in Figure 13.26 above. Roy-Camille applied plates and screws previously used in the repair of long bone fractures to specifically address posterior spinal fixation through the pedicles, the strongest part of the vertebra. These plates focused on anterior cervical fixation in order to increase the chances of bone fusion (Kinnard et al., 1986). However, disadvantages of the unconstrained system such as screw backout drove the development of restricted and constrained backout plates that locked the screws in place. Unfortunately, restricted systems also decreased bone fusion and were limited in use due to anatomical complexities. Lateral screw rod systems were created to better accommodate individual differences seen among patients. Semi-constrained dynamic plates were also developed to allow rotational and translation motion, reaping advantages of the plate system while still allowing for settling of plates to promote bone strengthening.

In contemporary spinal fixation designs, the screws are usually inserted into an appropriate pedicle and the rod is then fitted to contour the individual anatomy and is connected to the screws (Omeis et al., 2004). An advantage of the screw-rod system is that it allows a solid stabilized construct for the **occipito-cervical** junction as well as for the **cervical** and upper **thoracic spine**. Positioning of screws can be tailored to the individual needs of the patient and the screw-rod systems facilitate bone growth as they leave and leave most of the musculature undisturbed in the procedure. Today, a combination of wires, plates, and rods are used in multi-construct systems depending on the condition of individual spines.

Spinal fusion has been used to treat **disk arthrosis** and collapse, and essentially any disorder of spinal instability that contributes to chronic pain (West et al., 1991). Regrettably, the successful **arthrodesis** (fusion) of two or more spinal segments is generally concomitant with reduced mobility and function of the spine. The motivation of using an implant to replace an intervetebral disk is to remove the damaged or diseased disk while preserving motion in the spine. However, the long-term clinical success of spinal arthroplasty remains uncertain. Many of the initial designs utilized in disk replacement failed as a result of their inability to withstand the complex stresses of the spine. In the 1960s, Fernström surgically implanted the first nucleus prosthesis (Kurtz and Edidin, 2006). The implant comprised a stainless steel ball to restore intervertebral height and motion while preserving the **annulus** and vertebral endplates (Freeman and Davenport, 2006). As the field of biomechanics and joint kinematics has evolved, a number of designs that better mimic the viscoelastic and biomechanical processes of the intervertebral disk have been introduced into the orthopedics community.

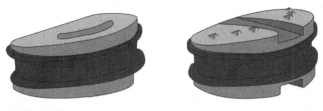

Figure 13.29

AcroFlex disks with solid core and contoured endplates. The disk on the right has "teeth" to provide better fixation to bone.

A major advancement in spinal disk replacement technology was the AcroFlex™ design developed and introduced in the 1970s by a spine surgeon by the name of Steffe. The novelty of this design was that the implant replicated the anatomy of a human vertebral disk with a rubber disk embedded between two titanium endplates (Figure 13.29). It was believed that this design would prevent subsidence of the implant that had been a prior complication in the ball bearing implants (Fraser and Raymond, 2004). An anatomical disk replacement device seems like a logical design; however, in order for the rubber core to be mounted it was necessary to **vulcanize** it to the titanium plates. It was at the titanium endplate-rubber core interface where failures occurred owing to repetitive shear forces induced by physiological loading and kinematics of the spine. An additional complication in the rubber implant was the generation of particulate debris that resulted in an inflammatory response and propensity for osteolysis in patients (Freeman and Davenport, 2006). After a short time on the market, AcroFlex™ was recalled and the implant was revised using a silicone elastomer in place of the rubber core. The modified device received much criticism for its use of silicone at a time when this material was under scrutiny during the Dow Corning litigation involving silicone breast implants, and the revised AcroFlex™ device was never implemented in clinical practice (Kurtz and Edidin, 2006). While this implant was not clinically successful, it provided the spinal implant community with a foundation of knowledge that would lead to development of disk replacements with multiple components that allow for motion without large deformation of the implant.

Two surgeons, Schellnack and Buttner-Janz, engineered the SB Charité™ intervetebral disk in the early 1980s (Buttner-Janz *et al.*, 1989). The design utilized materials that had shown success in total hip arthroplasty with two metal endplates made of a cobalt-chromium and a mobile core comprising ultra-high molecular weight polyethylene (UHMWPE). A modified form of this device remains in clinical use today with nearly 10,000 Charité™ devices implanted worldwide (Blumenthal and McAfee, 2005).

Although there has been extensive development in the intervetebral disk replacement field, the spine community remains in search of the archetypal implant that can withstand the complex loading and kinematics of the human spine and that can restore normal spinal curvature, alignment, and sagittal balance (Blumenthal and McAfee, 2005).

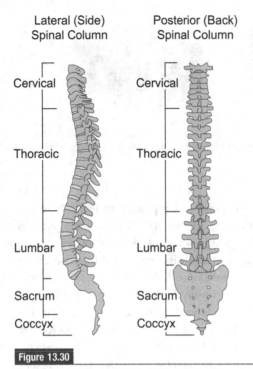

Figure 13.30

Illustration showing the anatomy of a healthy spine.

13.7.2 Anatomy of the spine

The anatomy of the spine is quite complex and its skeletal structure offers mechanical strength, flexibility, and stability. The spinal column protects the spinal cord and is connected to the central nervous system. The spinal column extends from the skull to the pelvis and comprises 33 individual bones termed vertebrae that are stacked upon each other (Figure 13.30). There are four primary regions of the spine, each with its own unique curvature; these segments are termed cervical, thoracic, lumbar, and sacrum. There are 7 cervical vertebrae, 12 thoracic vertebrae, and 5 lumbar vertebrae.

The first seven vertebrae (C1–C7) comprise the cervical spine and enable rotation of the neck and movement of the head. The first cervical vertebra (C1) is termed the **atlas** and the second (C2) is known as the **axis**, and these facilitate the nodding of the head. The next 12 vertebrae make up the thoracic spine (T1–T12). The thoracic region houses the rib cage and assists with the motion of bending, twisting, and side extension. Five vertebrae are found in the lumbar spine (L1–L5) and these provide support of body weight in the action of sitting or standing. Following the lumbar spine is the sacrum that comprises 4–5 fused bones and serves as the foundation of the pelvis. The tailbone (coccyx) serves as the base of the vertebral column and contains 3–5 fused bones. The coccyx and sacral curve together support and balance the upright body. The normal spine is capable of bending, reaching, and twisting.

13.7 Spinal implants

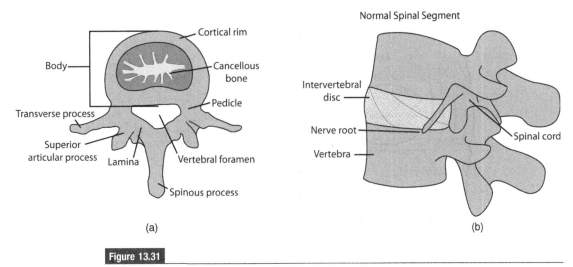

Figure 13.31

Illustration of (a) an individual vertebra and (b) the intervetebral disk between vertebra.

There are several important constituents in an individual vertebra (Figure 13.31(a)). The body of the vertebra serves as the primary region of load-bearing and serves as the foundation for the fibrous disks which separate each of the vertebrae. The **spinal canal** serves as the central pathway for the spinal cord (spinal nerves) and is covered by a bony protective layer termed the **lamina**. The spinous process is orthogonal to the matched **transverse processes** and together they provide attachment for back muscles and soft tissue. The vertebrae are separated by intervertebral **disks** that serve as shock absorbers between the bones (Figure 13.31(b)). Each disk has a tough outer layer termed the **annulus** that protects its soft viscoelastic interior known as the **nucleus** (Cramer and Darby, 1995).

This can be further divided into the **nucleus pulposa**, the **annulus fibrosus**, and the **lamellae** (Figure 13.32). The viscoelastic, gel-like center of the disk that is high in water content is termed the nucleus pulposa, and this structural element supports the elevated pressures of the upright spine. There are a number of functions of the nucleus pulposa and these include bearing the axial loads of the spine, providing a pivot point for torsional movement of the lower body, and serving as the "glue" between the vertebrae. The function of the annulus fibrosa is to serve as the pressure vessel wall for its highly pressurized contents – the nucleus pulposa. For this reason, the annulus fibrosus has a substantially greater content of collagen and is much more fibrous than the nucleus pulposa. The outermost layer of the intervetebral disk is known as the annulus, and this comprises concentric sheets of collagen that are arranged to optimize their strength and to assist in the containment and structural protection of the nucleus pulposa.

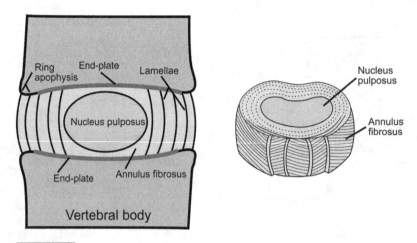

Figure 13.32

Schematic illustrating the primary components of the intervertebral disk.

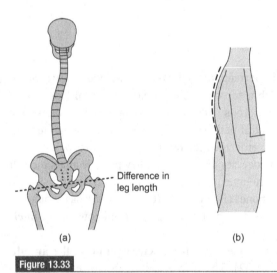

Figure 13.33

Illustration that shows a spine that is afflicted with (a) scoliosis and (b) kyphosis. Note curvature is in different planes.

13.7.3 Clinical reasons for spinal implants

It is estimated that over 26 million Americans of working age suffer from lower back pain and up to 80% of the population will experience back pain at some point in their lives (Lahiri, 2005). Further, approximately 816,000 Americans have scoliosis causing pain, loss of motion, and spinal fracture (Kuptniratsaikul, 2008). Other diseases such as **ankylosing spondilosis** can lead to spinal deformities such as **kyphosis** that require surgical intervention (Kuptniratsaikul, 2008). Figure 13.33 shows two common spine

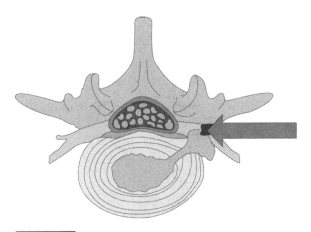

Figure 13.34

Illustration of a herniated (ruptured) intervertebral disk (denoted by arrow).

disorders – a spine that is afflicted with scoliosis, and one with kyphosis. One method of restoring some of the spinal function is through a procedure known as spinal fusion, stabilizing the spine by fusing two or more vertebrae. Arthrodesis is achieved through the use of spinal instrumentation described above. One of the most common causes of spine pain is degenerative disease of the intervertebral disks in which the structure becomes more fibrous and dehydrated, resulting in loss of disk height and mechanical function (Kurtz and Edidin, 2006). Figure 13.34 shows an intervetebral disk that has **herniated**. The rupture of an intervertebral disk results in loss of containment of its viscoelastic gelatinous substance comprising the nucleus. As this material leaks out through a tear in the annulus, it can lead to nerve impingement and severe pain. Such clinical cases may be candidates for intervetebral disk replacement.

13.7.4 Functional requirements of spinal implants

The complex biomechanics and anatomy of the spine make the design of implants extraordinarily difficult. From an anatomical perspective, there is the proximity to the nerve bundles along the spinal column that has little tolerance for impingement of any sort. As with the orthopedic devices described above, the implants must be biocompatible and offer resistance to both wear and fatigue damage. Additionally, however, these devices must withstand the rigorous loading that is expected in the spine. There are 6 degrees of freedom in the kinematics of the spine, including flexion-extension, lateral flexion, and axial rotation (Figure 13.35). This means that the implants must offer exceptional mechanical properties including tensile and compressive axial strength, flexural strength, and torsional strength.

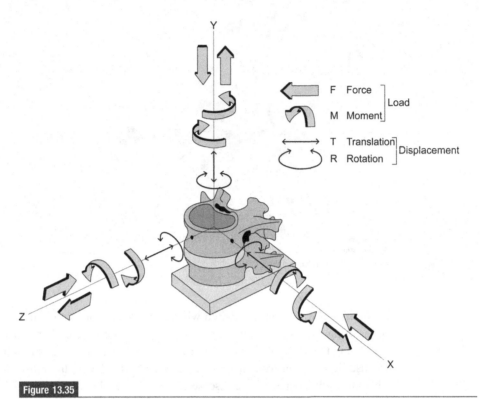

Figure 13.35

Illustration depicting the complex biomechanics of the spine.

13.7.5 Spinal implant designs

The Charite™ intervertebral disk is the most widely used lumbar interbody implant currently available. The design utilizes a biconvex UHMWPE insert that acts as a mobile bearing within endplates made of a cobalt-chromium alloy. The design of the implant has changed somewhat since its implementation, and the most recent designs have shown good clinical results after four years *in vivo* (Blumenthal and McAfee, 2005). Figure 13.36 shows the design evolution in the Charite™ intervertebral disk replacement.

The Pro-Disc™ intervertebral implant also utilizes a metal-on-plastic bearing coupling with UHMWPE and CoCr alloys. The 8–10 year post-operative data for this design have looked quite promising (Tropiano *et al.*, 2005). The contemporary device design known as the Pro-Disc II is currently being evaluated in a multi-institutional study investigating its safety and efficacy (Delamarter *et al.*, 2005). Disk replacement technology has improved significantly in the past few decades and yet even successful implants such as the Charité and ProDisc are far from perfect. Concerns with subsidence and migration as well as polymer wear remain in both designs (Ooij and Kurtz, 2007).

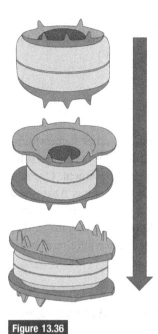

Figure 13.36

Illustration showing the design evolution of the Charite™ implant. (After Kurtz and Edidin, 2006.)

Metal-on-metal designs have also fared well in the intervetebral disk replacement market and may serve as a solution to the polymer debris generated in the UHMWPE-CoCr articulation. The Maverick™ implant utilizes CoCr alloy articulation with a fixed posterior center of rotation to better mimic the kinematics of the intervertebral motion segment and larger endplates to enhance stability of the implant. The Flexicore™ device uses a ball and socket CoCr articulation with dome-shaped endplates, a rotational stop, and a tension bearing to prevent separation or dislocation of the endplates. The Flexicore™ is under evaluation in a multi-center clinical trial (Errico, 2005). The device industry continues to improve its designs with the ultimate goal of restoring function and mobility to the spine.

13.7.6 Materials used in spinal implants

Material selection for spinal implant technology has evolved considerably from its inception more than a century ago. Prior to the implementation of stainless steel into the device industry in the 1930s, there were few material options that could withstand the complex biomechanical demands of the body. The alloys utilized in the spine require exceptional resistance to corrosion and to the cyclic stresses associated with flexion, extension, and lateral bending. As with the orthopedic devices described above, cobalt-chromium alloys have provided a platform for structurally sound implant design, and titanium has greatly facilitated osseointegration. Similarly, UHMWPE has fared well as an intervetebral insert. A complete overview of contemporary materials used in the spine is provided elsewhere (Kurtz and Edidin, 2006).

13.7.7 Design concerns in spinal implants

Intervertebral implants are more complicated than other successful joint implants such as the hip and knee, and designers face many obstacles in spinal fixation technology and the development of an ideal intervertebral disk replacement. One significant challenge is the complex biomechanics and kinematics of the spine. Anatomically, the loads and strains vary greatly depending upon position in the spine, and the proximity of the vertebrae to the spinal cord must be considered in the designs of the various implants. Future devices implemented in the spine must offer long-term stability, balance, and restored mobility.

13.8 Engineering challenges and design constraints of orthopedic implants

There are numerous challenges in the design of orthopedic implants used in total joint arthroplasty, fracture fixation, and spinal devices. Foremost, the implant must restore function to the intended anatomical location within the musculoskeletal system without any adverse reaction to the patient. Sterility of devices must be well controlled by the manufacturers, and the device must meet FDA specifications. Clinicians need appropriate instruction concomitant with specific surgical requirements of the implant. Patient factors such as health, activity level, and body mass index need to be incorporated into the indications or contraindications of the device. Materials utilized in the implants must offer the appropriate mechanical properties for the specific device design. Such properties include strength under multiaxial loading conditions, compliance match to the tissue, as well as resistance to corrosion, wear, fatigue damage, and fracture. In developing new material-design combinations, it is critical to review historical or clinical performance of the intended materials and designs, as previous failures serve as a foundation for learning and can prevent inadvertent device failures in the future.

13.9 Case studies

13.9.1 Delamination of a highly crosslinked UHMWPE acetabular cup (clinical failure)

A crosslinked UHMWPE liner and CoCr femoral head that had only been implanted for 5 months was retrieved after dislocating five times (Patten *et al.*, 2010). The damage to the liner included an area of delamination substantially larger than any previously reported surface damage for cross linked UHMWPE used in the hip. Areas of delamination decrease the conformity further and can lead to accelerated failure of the implant. This retrieved hip implant was investigated to determine the source of its early surface damage.

Roughening of CoCr femoral heads can occur with routine *in vivo* use; however, clinical dislocations may cause more severe damage to the femoral head because of contact between the femoral head and metal acetabular shell. In this study, it was hypothesized that the damage to the liner was due to titanium transfer deposits and roughening of the femoral head from contact with the rim of the acetabular shell during the dislocations. The retrieval was analyzed using optical microscopy, scanning electron microscopy (SEM), energy dispersive X-ray spectroscopy, and profilometry.

A prevalent region of delamination was found on the liner-bearing surface. Pitting, scratching, and abrasion were also prominent over the bearing surface (Figure 13.37). The Co-Cr femoral head exhibited surface damage that was attributed to its contact with the titanium acetabular shell during the clinical dislocations. This was corroborated by the EDX analysis that revealed regions of titanium deposited on the CoCr surface.

From this case report, clinical implications can be drawn relevant to three areas: wear of crosslinked ultra-high, fatigue behavior of crosslinked ultra-high, and dislocation damage to metal heads. Although crosslinked UHMWPE has demonstrated improved adhesive and abrasive wear, it has sacrificed resistance to fatigue crack propagation in the bulk material and is more susceptible to fracture in the presence of flaws than conventional UHMWPE. Metallic deposits on ceramic heads have been shown to increase the damage and wear to crosslinked UHMWPE liners, and close monitoring has been recommended following clinical dislocation of ceramic femoral heads. This case illustrates that similar monitoring is appropriate for metallic femoral heads.

13.9.2 Corrosion-induced fracture of a double modular hip prosthesis (design failure)

A cementless ProFemur® Z hip stem with a long neutral neck and ceramic-on-ceramic articulation that had been implanted in a 30-year-old man with a BMI of 29 fractured after two years *in vivo* when he suffered a fall on his right hip (Atwood *et al.*, 2010). The failure is shown in Figure 13.38. The Profemur® Z design is based on double-modular devices used extensively in Europe, and was first introduced to the United States in 2002 after approval by the FDA. The stem and neck are both manufactured from an orthopedic titanium alloy, and are offered in a range of sizes to accommodate a range of leg length and femoral anteversion adjustments.

The scanning electron microscopy uncovered several fracture processes associated with the chronology of failure events: a dark rough area near the initiation site indicative of an initial sharp crack owing to crevice corrosion; a lighter, smoother area associated

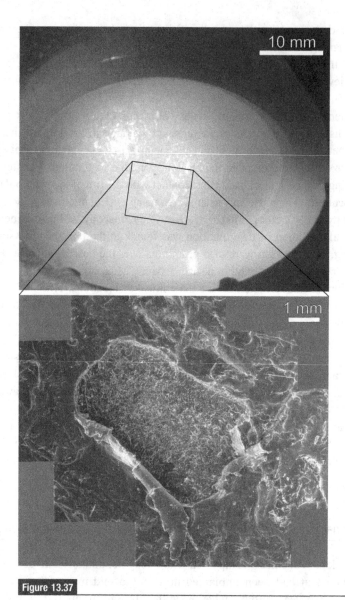

Figure 13.37

Figure (SEM) showing delamination associated with clinical dislocation. (After Patten *et al.*, 2010.)

with catastrophic fracture; and a region of tearing in the opposite diagonal corner is where the component was finally fractured (Figure 13.39).

Stem-neck design (modularity) may contribute to susceptibility to corrosion-induced fracture. Ti6Al4V alloys are generally protected against corrosion *in vivo* by a passive oxide layer on the surface; however, fretting occurred at the modular junction in the crevice of the neck-stem interface and wore away the protective oxide film. This design

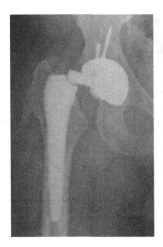
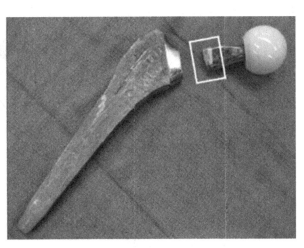

Figure 13.38

(Left) Anterior-posterior radiograph showing fracture of the modular neck. (Right) The retrieval with fracture of the modular neck component about 2 mm below the edge of the stem.

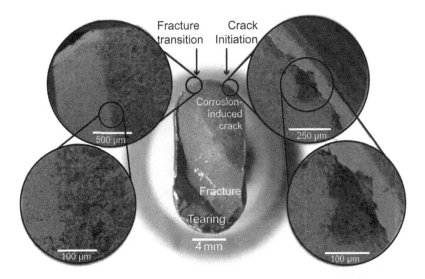

Figure 13.39

Initiation site occurred at the lateral anterior corner occurred on the tensile side of bending (right) and transitioned to catastrophic fracture (left). (After Atwood *et al.*, 2010.)

exacerbated the corrosion process, resulting in pitting and crack nucleation at this site. The bending stresses and likelihood for failure are greater in longer neck designs because the force scales with neck length (Figure 13.40). The progression of events leading to the fracture in this implant is illustrated in Figure 13.41.

466 Orthopedics

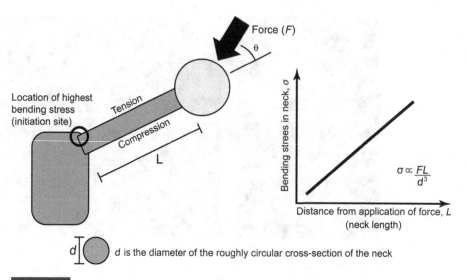

Figure 13.40

Schematic showing that bending stresses at the neck-stem junction increase linearly with neck length. Long neck lengths result in increased stresses that may contribute to corrosion-induced fracture at the neck-stem junction. (After Atwood et al., 2010.)

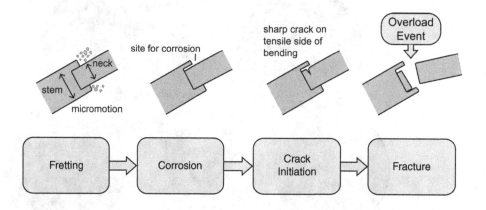

Figure 13.41

Schematic showing progression of events leading to the fracture in this implant.

The fracture of this modular neck is a representative case of the trade-off in modular implant design: modularity enables the surgeon to better optimize the patient's biomechanics and anatomy, but additional mechanical junctions can lead to corrosion processes, resulting in fracture. This case shows that there is a risk of fracture at the stem-neck junction when using double modular devices, particularly with long neck lengths in heavy patients. Furthermore, given the occurrence of fretting due to modularity, the high stresses at the neck-stem junction, and the cyclic nature of loading on hip prostheses, the modular neck component is also at risk for fatigue fracture without the occurrence of a traumatic event.

13.9.3 Sulzer recall (manufacturing failure)

In September 2000, Sulzer Orthopedics (Austin, Texas) began receiving adverse reports from surgeons who were getting complaints of leg pain from their patients recovering from recent total hip replacements procedures utilizing the Sulzer Inter-Op™ cup (Blumenfeld and Bargar, 2006). Following a two-month investigation of patient records, surgical techniques, and the product itself, the device was recalled in December 2000 by Sulzer Orthopedics. A letter was issued to patients and doctors from Gary Sabins, the president of Sulzer Medica Inc., and 31,000 Inter-Op acetabular shells were recalled back to the company. The company recalled lot numbers of its Inter-Op acetabular shells that had been manufactured between 1997 and 1999, and of these recalled devices, 17,500 had already been implanted. No other implant manufactured by Sulzer, other than the Inter-Op acetabular shell, was affected.

In 1997, Sulzer who at the time represented 11% of the device market, brought a previously outsourced manufacturing process in-house. It was determined that a small amount of mineral oil-based lubricant had leaked into the machine coolant during the manufacturing process and the lubricant remained as residue on the shell. Once the problem was identified, the manufacturing process was modified and validated to prevent a recurrence of the problem.

This retained oily residue on the defective implants inhibited the osseointegration necessary for fixating the shell in the bone of the pelvis. Designed so that the acetabular component snaps into place and bone grows into it, the oily residue interfered with the bone bonding and integration. This lack of fixation resulted in the formation of scar tissue and a loosened implant. Patients with the defective device reported pain in the groin and leg to their surgeons, and many of these patients received a new implant (Figure 13.42). Not all patients were adversely affected. Only 2,400 implants were removed and revised as of 2005. Sulzer agreed to pay a $1 billion settlement in a class action lawsuit filed in 2002. Patients who had undergone a revision surgery were awarded $160,000. After the lawsuit settled, Sulzer Orthopedics changed its name to Centerpulse Orthopedics.

13.9.4 Failure of pedicle screws used in spinal instrumentation (structural failure)

The pedicle screw is used to stabilize and immobilize the segments of the spine (Figure 13.43). Systems of plates, rods, and wires are anchored to the vertebra by pedicle

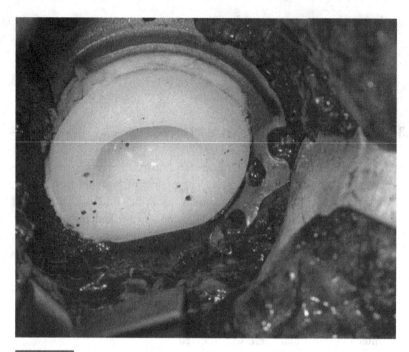

Figure 13.42

Image of the surgical procedure for an acetabular cup and polymer insert. (Image courtesy of Dr. Michael Ries.)

Figure 13.43

Fracture of the pedical screw near the plate-screw interface.

screws, drilled through the pedicles and into the vertebral body. Failure of the pedicle screw through screw loosening or screw bending or breaking can result in instability, failure of fusion, incorrect fusion, pain, or neurological damage if nerves are damaged. The pedicle represents the strongest point of attachment of the spine and thus significant forces can be applied to the spine without failure of the bone-metal junction.

The pedicle screw spinal fixation method involves the use of metallic screws drilled into juxtaposed vertebrae in the spine. A rigid plate is utilized to immobilize and align the vertebrae with screws embedded into the pedicles of the spine. Pedicle screws are highly susceptible to fracture, as the rigid fixation procedure hastens adjacent spinal segment degeneration and increases the stresses in the spine. The fracture of pedicle screws has been a frequent complication in spinal fixation systems. Failure commonly occurs in close proximity to the screw head owing to large stress concentrations at the plate-screw interface in the distal portion of the fusion device.

Failure analysis using scanning electron microscopy revealed fatigue striations and beach marks characteristic of fatigue failure (Chen *et al.*, 2005). Fatigue studies performed on commercially available pedicle screw systems demonstrated failures in less than 30,000 cycles for peak loads in the spine. These *in situ* tests emphasize the importance of post-operative stabilization of the spine. Stabilization with better bone integration or enhanced flexibility in the design may prevent device failures by reducing local stress concentrations in the device.

13.9.5 Failure of AcroFlex lumbar disk replacements (material failure)

In 1987, a spine surgeon by the name of Arthur Steffee developed the AcroFlex spinal disk replacement. The monolithic design utilized two titanium endplates fused with a polyolefin rubber core as described previously (Figure 13.29). A year later, he implanted this design into six of his patients. After three years, four of his six patients had fair or better results; however, one implant fractured and the other was removed due to chronic pain. The FDA banned the use of the rubber material used in the AcroFlex due to concern that the material was carcinogenic. Steffee was forced to suspend his clinical trials and a silicone core was used in the next iteration of his device. However, silicone was quickly out of favor as a biomaterial owing to the controversy over its alleged complications and inflammatory response in silicone breast implants. In the final iteration of the AcroFlex design, the core was substituted with a hexene-based elastomer and titanium beads were added to the polymeric side of the titanium plates to improve the integrity of the mechanical interface between the fused endplates and the elastomeric core.

In 1998, a clinical study was implemented with 11 patients, and another 17 were added over the next two years (Fraser and Raymond, 2004). Sixteen out of the 28 patients had complications owing primarily to tearing or other mechanical failures of the polyolefin core. An inflammatory response to the polymeric material was found in three patients.

Although laboratory tests comprising uniaxial cyclic loading and shear tests indicated that the AcroFlex had a high resistance to fatigue loading, the tests did not predict the clinical outcome for the implant design. With advancements in spine biomechanics it is now known that the device experienced a complex stress state comprising simultaneous shear, axial, and bending loads in the body. The elastomeric material used in the AcroFlex devices behaved very differently under mixed mode conditions, and the majority of the implants failed due to multiaxial fatigue loading within two years of service *in vivo*. About 30% of the patients with AcroFlex devices underwent revision surgery within one year of implantation.

The AcroFlex spinal implant evolved through three major material changes in design, yet none of these implants made it past clinical trials into the mainstream spinal industry. To date, clinicians and researchers have yet to develop a biomaterial or implant system that functions as well as the healthy intervertebral disk.

13.10 Summary

This chapter provided a broad overview of orthopedic implants, including total hip and knee arthroplasty, fracture fixation, and spinal implants. The historical development of the various implant designs was presented with emphasis on materials selection and functionality in the contemporary devices. The factors contributing to device performance and case studies that look at failures owing to complications with clinical factors, structural analysis, material performance, design, and manufacturability were examined.

13.11 Looking forward in orthopedic implants

As one looks forward in the field of orthopedic implants it is clear that there are a number of emerging fields that will contribute to the long-term restoration of function in the body. These include but are not limited to genomics and its use in presenting disease, tissue engineering and its application in tissue repair, improvements in imaging and robotic surgery, and the development of minimally invasive methods.

13.12 Problems for consideration

Checking for understanding

1. List five variables that contribute to the longevity of an orthopedic implant.
2. Describe how the conformity differs between a healthy hip and knee joint. Is this same level of conformity maintained in total joint replacements in comparison to a healthy joint? Explain.

3. Explain the primary design concerns in fracture fixation. How might this technology change in the future?

For further exploration

4. What are the attributes of a healthy intervetebral disk? Why has there been no engineered substitute that can fully restore function to the spine?
5. Discuss the benefits and trade-offs in modular implant design.

13.13 References

American Academy of Orthopedic Surgeons (AAOS). (2010). www.aaos.org.

Atwood, S., Patten, E., Bozic, K.J., Pruitt, L., and Ries, M.D. (2010). Corrosion-induced fracture of a double-modular hip prosthesis: a case study. *Journal of Bone and Joint Surgery*, 92, 1522–25.

Baker, D., Bellare, A., and Pruitt, L. (2003). The effects of degree of crosslinking on the fatigue crack initiation and propagation resistance of orthopedic grade polyethylene. *Journal of Biomedical Materials Research*, 66A, 146–54.

Baker, D., Coughlin, D., and Pruitt, L. (1998). The effect of accelerated aging and sterilization method on the fatigue crack propagation resistance of Hylamer-M and GUR4150HP at Body Temperature. *Transactions of the Society for Biomaterials*, 21, 122.

Banks, S.A., Benjamin, J., Fregly, J., Boniforti, F., Reinschmidt, C., and Romagnoli, S. (2005). Comparing in vivo kinematics of unicondylar and bi-unicondylar knee replacements. *Knee Surgery and Sports Trauma Arthroscopy*, 13, 551–56.

Bartel, D.L., Bicknell, V.L., and Wright, T.M. (1986). The effect of conformity, thickness and material on stresses in ultra-high molecular weight components for total joint replacement. *Journal of Bone and Joint Surgery*, 68A: 1041–51.

Bartel, D.L., Burstein, A.H., Santavicca, E.A., and Insall, J.N. (1982). Performance of the tibial component in total knee replacement. *Journal of Bone and Joint Surgery*, 64-A, 1026–33.

Bellare, A., and Kurtz, S.M. (2009). High pressure crystallized UHMWPEs. In *The UHMWPE Handbook*, 2d edn., ed. S.M. Kurtz. London: Elsevier, pp. 277–90.

Bellare, A., and Pruitt, L. (2004). Advances in UHMWPE: structure-property-processing inter-relationships. In *Joint Replacements and Bone Resorption*, ed. A. Shanbhag H. Rubash, and J. Jacobs. New York: Marcel Dekker.

Bick, E.M. (1964). An essay on the history of spine fusion operations. *Clinical Orthopaedics and Related Research*, 35, 9–15.

Blumenfeld, T.J., and Bargar, W.L. (2006). Early aseptic loosening of a modern acetabular component secondary to a change in manufacturing. *Journal of Arthroplasty*, 21(5): 689–95.

Blumenthal, S., and McAfee, P. (2005). A prospective, randomized, multicenter food and drug administration Investigational Device Exemptions study of lumbar total disc replacement with the Charite Artificial Disc versus lumbar fusion: Part I: Evaluation of clinical outcomes. *Spine*, 30(14), 1565–75.

Bolanos, A.A., Wayne, A., Colizza, A., McCann, P., Gotlin, R., Wootten, M., Kahn, B., and Insall, J. (1998). A comparison of isokinetic strength testing and gait analysis in patients with posterior cruciate-retaining and substituting knee arthroplasties. *Journal of Arthroplasty*, 13(8), 906–15.

Boucher, H.H. (1959). A method of spinal fusion. *Journal of Bone and Joint Surgery*, 41B, 248–59.

Bray, T.J. (1992). *Techniques in Fracture Fixation*. New York: Gower Medical Publishers.

Buttner-Janz, K., Schellnack, K., and Zippel, H. (1989). Biomechanics of the SB Charite® lumbar intervertebral disc endoprosthesis. *International Orthopaedics*, 13, 173–76.

Callaghan, J., Insall, J.N., Greenwald, A.S., Seth, A., Dennis, D.A., Komistek, R., Murray, D.W., Bourne, R.B., Rorabeck, C.H., and Dorr, L.D. (2000). Mobile-bearing knee replacement: concepts and results. *Journal of Bone and Joint Surgery (American)*, 82, 1020–41.

Charnley, J. (1961). Arthroplasty of the hip: A new operation. *Lancet*, I, 1129–32.

Charnley, J. (1979). *Low Friction Arthroplasty of the Hip: Theory and Practice*. Berlin: Springer Verlag.

Chen, W-J., Cheng, C.K., Jao, S-H.E., Chueh, S-C., and Wang, C-C. (2005). Failure analysis of broken pedicle screws on spinal instrumentation. *Medical Engineering and Physics*, 27, 487–96.

Cook, S.D., McCluskey, L.C., and Martir, P.C. (1991). Inflammatory response in retrieved noncemented porous-coated materials. *Clinical Orthopaedics*, 264, 209–22.

Cramer, G.D., and Darby, S.A. (1995). *Basic and Clinical Anatomy of the Spine, Spinal Cord and SNS*. New York: Elsevier.

Delamarter, R.B., Bae, H.W., and Pardhan, B.B. (2005). Clinical results of ProDisc-II® lumbar total disc replacement: Report from the United States clinical trial. *Orthopedic Clinics of North America*, 36, 301–13.

Edidin, A., Pruitt, L., Jewett, C.W., Crawford, R.P., Crane, D.J., Roberts, D., and Kurtz, S.M. (1999). Plasticity-induced damage layer is precursor to wear in radiation-crosslinked UHMWPE. *Journal of Arthroplasty*, 14(5), 616–27.

Einhorn, T. (1995). The enhancement of fracture healing. *Journal of Bone and Joint Surgery (American)*, 77, 940–56.

Errico, T.J. (2005). Lumbar disc arthroplasty. *Clinical Orthopaedics and Related Research*, 435, 106–17.

Fraser, R., and Raymond, R. (2004). AcroFlex design and results. *Spine Journal*, 4(6), S245–S251.

Freeman, B., and Davenport, J. (2006). Total disc replacement in the lumbar spine: a systematic review of the literature. *European Spine Journal*, 15(3), S439–S447.

Furmanski, J., Anderson, M., Bal, S., Greenwald, S., Halley, D., Penenberg, B., Ries, M., and Pruitt, L. (2009). Clinical rim fracture of cross-linked acetabular liners. *Biomaterials*, 30, 5572–82.

Furmanski, J., and Pruitt, L. (2007). Peak stress intensity dictates fatigue crack propagation in UHMWPE. *Polymer*, 48, 3512–519.

Furmanski, J., and Pruitt, L. (2009). Polymeric biomaterials for use in load-bearing medical devices: The need for understanding structure-property-design relationships. *Journal of Metals*, September, 14–20.

Graham, J., Pruitt, L., Ries, M., and Gundiah, N. (2000). Fracture and fatigue properties of acrylic bone cement. *Journal of Arthroplasty*, 15(8), 1028–35.

Gunston, F.H. (1971). Polycentric knee arthroplasty: Prosthetic simulation of normal knee movement. *Journal of Bone and Joint Surgery*, 53B, 272–77.

Hadra, B.E. (1891). Wiring the spinous processes in Pott's Disease. *Journal of Bone and Joint Surgery*, 20, 6–210.

Insall, J.N., and Walker, P. (1976). Unicondylar knee replacement. *Clinical Orthopaedics*, 120, 83–85.

Insall, J.N., Lachiewicz, P.F., and Burstein, A.H. (1982). The posterior stabilized condylar prosthesis: A modification of the total condylar design, two to four year clinical experience. *Journal of Bone and Joint Surgery*, 64A: 1317–23.

Jones, W.T., Bryan, R.S., Peterson, L.F., and Ilstrup, D.M. (1981). Unicompartmental knee arthroplasty using polycentric and geometric hemicomponents. *Journal of Bone and Joint Surgery (American)*, 63(6), 946–54.

Kinnard, P., Ghibely, A., Gordon, D., Trias, A. and Basora, J. (1986). Roy-Camille plates in unstable spinal conditions. A preliminary report. *Spine*, 11(2), 131–5.

Klapperich, K., Komvopoulos, K., and Pruitt, L. (1999). Tribological properties and microstructure evolution of ultra-high molecular weight polyethylene. *Journal of Tribology*, 121(2), 394–402.

Kuptniratsaikul, S. (2008). Role of metal in spine stabilization. *Metals, Materials and Minerals Bulletin*, 6: 32–40.

Kurtz, S.M. (2009a). The origins of UHMWPE in total hip arthroplasty. In *The UHMWPE Handbook*, 2d edn., ed. S.M. Kurtz. London: Elsevier, pp. 31–42.

Kurtz, S.M. (2009b). The clinical performance of UHMWPE in hip replacements. In *The UHMWPE Handbook*, 2d edn., ed. S.M. Kurtz. London: Elsevier, pp. 43–54.

Kurtz, S.M. (2009c). *In vivo* oxidation of UHMWPE. In *The UHMWPE Handbook*, 2d edn., ed. S.M. Kurtz. London: Elsevier, pp. 325–40.

Kurtz, S.M. (2009d). Composite UHMWPE biomaterials and fibers. In *The UHMWPE Handbook*, 2d edn., ed. S.M. Kurtz. London: Elsevier, pp. 249–58.

Kurtz, S., and Edidin, A.A. (2006). *Spine Technology Handbook*. Burlington, MA: Elsevier Academic Press.

Kurtz, S.M., and Ong, K. (2009). Contemporary total hip arthroplasty: hard-on-hard bearings and highly crosslinked UHMWPE. In *The UHMWPE Handbook*, 2d edn., ed. S.M. Kurtz. London: Elsevier, pp. 55–80.

Kurtz, S, Pruitt, L., Jewett, C.W., Foulds, J.R., and Edidin, A.A. (1999). Radiation and chemical crosslinking promote strain hardening behavior and molecular alignment in UHMWPE during multi-axial conditions. *Biomaterials*, 20(16), 1449–62.

Lahiri, S. (2005). Estimation of net costs for prevention of occupation low back pain: Three case studies from the US. *American Journal of Industrial Medicine*, 48(6), 530–41.

Lange, F. (1910). Support for the spondylitic spine by means of buried steel bars, attached to the vertebrae. *Journal of Bone and Joint Surgery*, S2–8:344–61.

Link, H. (2002). History, design and biomechanics of the LINK SB Charité artificial disc. *European Spine Journal*, 11(2), S98–S105.

Maloney, W.J., Jasty, M., and Rosenberg, A. (1990). Bone lysis in well-fixed cemented femoral components. *Journal of Bone and Joint Surgery (British)*, 72B, 966–70.

Mazel, C. (2008). The contribution of Raymond Roy-Camille to spine surgery. *Argospine News and Journal*, 19, 97–102.

McKee, G.K., and Watson-Farrar, J. (1966). Replacement of arthritic hips by the McKee-Farrar Prosthesis. *Journal of Bone and Joint Surgery (British)*, 48B, 245–59.

Mow, V.C., and Hayes, W.C. (1991). *Basic Orthopedic Biomechanics*. New York: Raven Press.

Noble, P.C. (1999). Biomechanics of revision hip replacement. In *Revision Total Hip Arthroplasty*, ed. J.V. Bono, J.C. McCarthy, T.S. Thornhill, B.E. Bierbaum, and R.H. Turner. New York: Springer-Verlag, pp. 135–41.

Omeis, I., Demattia, J.A., Hillard, V.H., Murali, R., and Das, K. (2004). History of instrumentation for stabilization of the subaxial cervical spine. *Neurosurgical Focus*, 16, S418–S421.

Ooij, A., and Kurtz, S. (2007). Polyethylene wear debris and long-term clinical failure of the Charite disc prosthesis: a study of 4 patients. *Spine*, 32(2), 223–29.

Patten, E., Atwood, S., Van Citters, D.W., Jewett, B.A., Pruitt, L., and Ries, M.D. (2010). Delamination of a highly crosslinked UHMWPE acetabular liner associated with titanium deposits on the CoCr head following dislocation: A case report. *Journal of Bone and Joint Surgery (British)*, 92B, 1306–11.

Pawlowski, L. (2008). *The Science and Engineering of Thermal Spray Coating*, 2d edn. Chichester, UK: Wiley.

Peltier, L.F. (1990). *Fractures: A History and Iconography of their Treatment*. Novato, CA: Norman Publishing.

Peltier, L.F. (1993). *Orthopedics: A History and Iconography of their Treatment*. Novato, CA: Norman Publishing.

Pospula, W. (2004). Total hip replacement: Past, present and future. *Kuwait Medical Journal*, 36(4), 250–55.

Pramanik, S., Agarwal, A.K., and Rai, K.N. (2005). Chronology of total hip joint replacement and materials development. *Trends in Biomaterials and Artificial Organs*, 19(1), 15–26.

Pruitt, L. (2005). Deformation, yielding, fracture and fatigue behavior of conventional and highly cross-linked ultra high molecular weight polyethylene. *Biomaterials*, 26(8), 905–15.

Ranawat, C.S., and Sculco, J.P. (1985). History of the development of total knee prosthesis at the Hospital for Special Surgery. In *Total Condylar Knee Arthroplasty*, ed. C.S. Ranawat. New York: Springer-Verlag.

Ries, M.D., and Pruitt, L. (2005). Effect of crosslinking on the microstructure and mechanical properties of UHMWPE. *Clinical Orthopedics and Related Research*, 440, 149–56.

Rimnac, C.M., Baldini, T.H., Wright, T.M., Saum, K.A., and Sanford, W.M. (1996). In vitro chemical and mechanical degradation of Hylamer ultra-high molecular weight polyethylene. *Transactions of the Orthopedic Research Society*, 21, 481.

Rimnac, C.M., and Pruitt, L. (2008). How do material properties influence wear and fracture mechanisms? *Journal of the American Academy of Orthopedic Surgeons*, 16(1), S94–S100.

Shetty, A.A., Tindall, A., Ting, P., and Heatley, F.W. (2003). The evolution of total knee arthroplasty. Part II: the hinged knee replacement and the semi-constrained knee replacement. *Current Orthopaedics*, 17, 403–7.

Sullivan, P., Scott, K., and Johnston, R. (1992). Current concepts in hip joint replacement. *Iowa Medicine*, 276, 468–69.

Tencer, A., and Johnson, K. (1994). *Biomechanics in Orthopedic Trauma*. London: Martin Dunitz.

Thomas, K.A. (1990). Biomechanics and biomaterials of hip implants. *Current Opinions in Orthopaedics*, 1, 28–37.

Tropiano, P., Huang, R.C., Girardi, F.P., Cammisa, F.P., and Marnay, T. (2005). Lumbar total disc replacement. Seven to eleven-year follow-up. *Journal of Bone and Joint Surgery (American)*, 87, 490–96.

Verdonschot, N. (2005). Implant choice: stem design philosophy. In *The Well-Cemented Total Hip Arthroplasty: Theory and Practice*, ed. S. Breush and H. Malchau. Berlin: Springer-Verlag.

West, J.L., Bradford, D.S, and Olgilvie, J.W. (1991). Results of spinal arthrodesis with pedicle screw-plate fixation. *Journal of Bone and Joint Surgery (American)*, 8, 1179–84.

Wiles, P. (1952). Surgery and the surgeon. *Proceedings of the Royal Society of Medicine*, 45, 493–96.

Wright, T.M., and Bartel, D.L. (1986). The problem of surface damage in polyethylene total knee components. *Clinical Orthopaedics*, 205: 67–74.

Wright, T.M., Astion, D.J., Bansal, M., Rimnac, C.M., Green, T., Insall, J.N., and Robinson, R.P. (1988a). Failure of carbon fiber-reinforced polyethylene total knee-replacement components. A report of two cases. *Journal of Bone and Joint Surgery (American)*, 70, 926–29.

Wright, T.M., Rimnac, C.M., Faris, P.M., and Bansal, M. (1988b). Analysis of surface damage in retrieved carbon fiber-reinforced and plain polyethylene tibial components from posterior stabilized total knee replacements. *Journal of Bone and Joint Surgery (American)*, 70, 1312–19.

14 Cardiovascular devices

Inquiry

How would you design a novel mechanical heart valve using only synthetic materials?

In the development of medical implants, designers and engineers frequently attempt to replicate both the geometry and function of the natural tissue requiring replacement. This is not true in the case of mechanical heart valves, which utilize designs that mimic the function of natural heart valves but not necessarily the geometry. Perhaps this is because the most important aspect of heart valve design – that is, to provide a long-term method for regulating fluid flow without damaging red blood cells – can be achieved in a number of ways. When considering the lifetime of the device, another challenge presents itself: the average adult heart rate is approximately 72 beats per minute and this translates to nearly 38 million fatigue cycles per year. What are other methods, beyond what is already on the market, to creatively design an implant that can withstand high-cycle fatigue while regulating blood flow without damaging red blood cells?

14.1 Historical perspective and overview

Cardiovascular disease can refer to any set of symptoms that affects the **cardiovascular system** (namely, the heart and blood vessels). Some people are born with cardiovascular disease and others develop it over a lifetime. In this textbook, the focus is on aspects of cardiovascular disease that can be handled through the use of a structural or load-bearing implant. This includes **coronary heart disease** (a lack of circulation to the heart and surrounding tissue), **congestive heart failure** (the inability of the heart to pump sufficient blood through the body), and **valvular disease** (heart valves experiencing regurgitation or incompetency). The electrical stimulation of the heart muscle through the use of a pacemaker will not be discussed, but the reader is encouraged to read the text by Al-ahmed and co-workers (2010) for more information on that topic. The most common type of cardiovascular disease is coronary artery disease, in which the arteries that supply oxygenated blood to the heart become partially blocked by the formation of **plaque** (the build-up of fibrous tissue, fatty lipids, and calcifications that occurs on the intimal side of vascular wall). This reduction of the arterial **lumen**, or opening, due to plaque accumulation is one of the characteristics of **atherosclerosis** (Ross, 1999).

Cardiovascular disease is the leading cause of death in the United States, where it was responsible for 1 of every 6 deaths in 2006, according to the American Heart Association (Lloyd-Jones et al., 2010). The same report shows that from 1996 to 2006, the number of inpatient procedures related to cardiovascular disease increased by 33%, from 5.4 million to 7.2 million. This trend demonstrates that the combined effects of cardiovascular disease, including obesity and diabetes, are creating an increased need for treatments – whether through implanted devices, pharmaceuticals, or lifestyle changes. The total direct and indirect costs of cardiovascular disease in the United States are estimated to be $503.2 billion in 2010 and it is thought that over 81 million American adults have one or more types of cardiovascular disease (Lloyd-Jones et al., 2010).

The implants (short term and long term) that are commonly used to treat coronary heart disease include stents, vascular grafts, and mechanical or biological heart valves. Also used are intra-aortic balloon pumps, ventricle assist devices, mitral valve clips, balloon catheters, septal occluders, and the total artificial heart. The primary requirements of cardiovascular implants are the same as for most other devices: biocompatibility, the restoration of function, durability, and the ability to be sterilized. Because these implants are blood-contacting, they are also required to be non-irritating to surrounding tissue, blood-compatible, or non-thrombogenic (no tendency to form clots) and generally resistant to platelet adhesion and blood absorption. Finally, they must emulate the function of their natural counterparts and be of a reasonable size and weight for implantation.

The evolution of materials selection for vascular grafts provides an interesting example of cardiovascular implant design. In the early 1900s, doctors experimented with autologous transplants, meaning that the vascular graft came from the patient's own body (Bhat, 2005). From these implants, research turned toward using hard tubes made of metals or polymers. These tubes were sometimes lined or polished for reduced thrombogenicity. As a wider variety of materials became available, in the mid-1900s, researchers began to test flexible tubes made of rubber and of woven fabrics such as nylon and polytetrafluoroethylene (PTFE). Other options, such as polyethylene terephthalate (PET), also known as polyester, were introduced in the 1960s. PTFE and PET are also commonly known by their tradenames, Teflon® and Dacron®. At around this time, **heterografts** (also known as **xenografts**, or grafts from another species) were developed using bovine vasculature. Expanded PTFE (ePTFE) was introduced in the 1970s, and quickly became a leading material that is still in use today (Ku and Allen, 2000). Another current option for vascular grafts is human **umbilical vein**. This parallel advancement of solutions based on natural tissues and synthetic materials is somewhat unique to cardiovascular device development. Current designs for vascular grafts are discussed further in Section 14.4.3.

14.2 Learning objectives

This chapter will discuss structural cardiovascular implants made from both synthetic and natural tissues. The focus will be on stents, grafts, and heart valves, although other implants will be briefly touched upon as well. Case studies for valve, graft, and stent failures will be given. The chapter will conclude with a look forward to the future of cardiovascular implant development. After reading this chapter, students should be able to:

1. name and delineate the function of major components of the cardiovascular anatomy
2. describe the clinical reasons that necessitate stents, grafts, and heart valves
3. provide a synopsis of the historical development of stents, grafts, and heart valves
4. summarize the functional requirements for stents, grafts, and heart valves
5. list natural and synthetic materials currently used in stents, grafts, and heart valves
6. portray design concerns for stents, grafts, and heart valves
7. explain uses of catheters, intra-aortic balloon pumps, ventricular assist devices, total artificial hearts, and septal occluders
8. discuss cases in cardiovascular device history where an implant has failed prematurely due to fatigue, wear, clinical complication, poor material selection, or inappropriate design and processing

14.3 Cardiovascular anatomy

For a complete description of cardiovascular anatomy, the reader should see Netter (2011). However, basic descriptions of the heart, vessels, and blood are given here.

The heart is a hollow muscular organ that is responsible for pumping blood through the body. In adults, it weighs between 250 and 350 g. It is 12–13 cm high and 7–8 cm at its widest point. The heart has three layers – the **endocardium**, which lines the blood-contacting surfaces; the **myocardium**, which is composed of cardiac muscle; and the **epicardium**. The epicardium is the inner layer of the double-walled sac that encloses the heart, known as the **pericardium**. The pericardium keeps the heart in the correct location and prevents it from over-filling with blood.

The heart has four chambers, as shown in Figure 14.1. The upper two chambers, called the **atria**, are thin-walled and are the holding areas for blood coming into the heart. The lower chambers, known as the **ventricles**, are thick-walled and pump the blood out of the heart. Each chamber has a capacity of approximately 120 mL. There are four heart valves that are used to direct and restrict blood flow: the tricuspid valve, the pulmonary valve, the mitral valve, and the aortic valve.

The aortic and pulmonic valves (also known as the semi-lunar valves), shown in greater detail in Figure 14.2, are made up of leaflets (or cusps) that have three layers of tissue: **fibrosa**, **ventricularis**, and **spongiosa**. Fibrosa is a dense connective tissue composed of crimped collagen fibers (mainly Type I, III, and V) arranged parallel to the free edge of the leaflet. It extends from the free edge to the valve sleeve. The arrangement of the collagen in the fibrosa layer is what gives the leaflet its strength in the circumferential direction and minimizes sagging in the center. Leaflets have greater compliance in the radial direction than in the circumferential direction, which allows for increased blood flow. Ventricularis is made of packed collagen with elastic fibers that

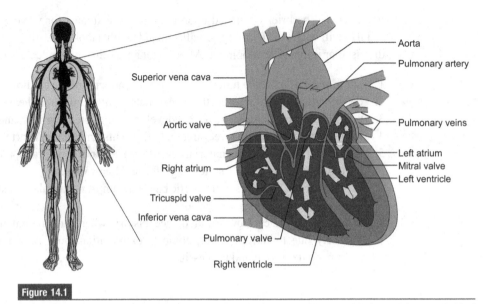

Figure 14.1

The human cardiovascular system and the heart. The atria hold blood coming into the heart, while the ventricles pump the blood back out of the heart. Flow between these chambers is controlled by the heart valves.

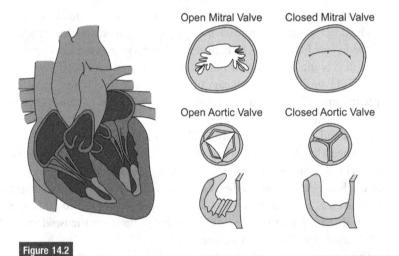

Figure 14.2

Schematics of the aortic, pulmonic, mitral, and tricuspid valves, showing their structure and support tissue.

are aligned radially and is found on the ventricular surface of the leaflet. It allows the leaflet to have minimal surface area during blood flow but stretch back during the closed phase. Spongiosa is a looser tissue made of collagen and proteoglycans and is located between the fibrosa and the ventricularis. This layer does not extend to the free edge and has little mechanical strength, showing enough compliance to accommodate the large shape changes the cusps undergo during each cycle. An endothelial layer covers the

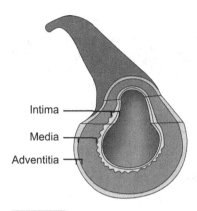

Figure 14.3

A cross-section of arterial tissue, showing the intima, media, and adventitia.

valve on both the ventricular and arterial sides. The mitral valve and the tricuspid valve have a similar biological makeup to the semi-lunar valves, but are larger and rely on support tissue called chordae tendineae, which are mainly collagen and connect to the papillary muscles in order to prevent backflow (Fong *et al.*, 2007; Fligner *et al.*, 2004; Schoen, 2005; Silver, 1994).

Blood is carried to and from the heart through blood vessels, which can be classified as **arteries** or **veins**. Arteries carry blood away from the heart, while veins carry blood toward the heart. Arteries and veins both have a trilaminate structure, but arteries in general have thicker walls to accommodate their higher mean pressures. A schematic cross-section of an artery is given in Figure 14.3. The layers of the artery are called the **intima**, the **media**, and the **adventitia**. At the innermost layer, the intima has endothelial cells and an elastic layer. The intima is thin and does not contribute significantly to the mechanical properties of the artery but interfaces directly with blood flow. It is known to both thicken and stiffen with age (Silver, 1994). The media has muscle fibers that are aligned circumferentially to assist with restricting blood flow when necessary. The ultrastructure of the elastin varies within the media: it is fenestrated at the interior and fibrous at the exterior. Finally, the adventitia is made up of fibroblasts with large-diameter collagen fibrils and proteoglycans.

Veins have a trilaminate structure similar to arteries, with one difference being that veins generally have thinner walls. In the lower limbs, veins also have valves to assist in controlling flow direction. The largest artery is the aorta, which branches into arteries, arterioles, and capillaries. The largest vein is the inferior vena cava, which is fed into by veins and venules. Some of the dimensions and relevant stresses in arteries and veins are given in Table 14.1.

Approximately half of the volume of human blood is made up of cells, with the other half being composed of water, albumin, fibrinogen, and ions (Silver, 1994). Blood cells include red blood cells, white blood cells, and platelets, illustrated in Figure 14.4. Red blood cells are disks with a concave center that are roughly 8 μm in diameter. These cells

Table 14.1 Dimensions and pressures in blood vessels (Bhat, 2005; Okamoto, 2008)

Vessel	Lumen diameter (mm)	Wall thickness (mm)	Wall tension (Pa)	Mean pressure (kPa)
Aorta	25	2	19,000	13.3
Artery	4	1	6000	12.0
Arteriole	0.03	0.02	50–120	8.0
Capillary	0.008	0.001	1.6	4.0
Venule	0.02	0.002	2.6	2.7
Vein	5	0.5	40	2.0
Vena Cava	30	1.5	2100	1.3

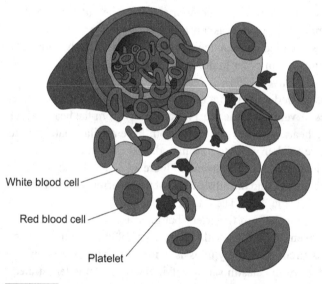

Figure 14.4

Red blood cells, white blood cells, and platelets emerging from a blood vessel.

contain hemoglobin and are responsible for transporting O_2 from the lungs to the body tissues and CO_2 exhaust back to the lungs. Trauma or changes in mechanical pressure can cause these cells to release hemoglobin, which in turn leads to a reduced ability to carry O_2. White blood cells are spherical and can be 7–22 μm in diameter. There is a low concentration of white blood cells in blood; it is higher in the lymphatic system. White blood cells defend the body against foreign matter and disease. Platelets are disks 2–4 μm in diameter. They contain growth factors that are released upon activation to limit bleeding.

Immediately after injury, blood vessels constrict and platelets adhere to the site in order to reduce blood loss through clot formation. A fixed clot is called a **thrombus**, while a floating clot is known as an **embolus**.

14.4 Load-bearing devices

14.4.1 Stents

A stent is used as a scaffold to maintain or increase the lumen of a blood vessel. The need for this implant arises when a patient is suffering from atherosclerosis. In atherosclerosis, fatty and mineral deposits as well as cells and scar tissue owing to inflammation can build up on the walls of the artery, reducing blood flow. Initially, balloon angioplasty alone was used to compress the plaque and increase the lumen. A balloon catheter could be inserted in the femoral artery and threaded to the site of the blockage. The balloon was then inflated repeatedly in order to open the area. Dr. Andreas Gruentzig piloted this technique in the mid-1970s, using double-lumen polyvinyl chloride (PVC) balloon catheters of his own invention (Meier, 2007). While this technique does restore blood flow, it can also critically damage the vessel, leading to scarring and **restenosis**, or re-closing of the vessel (Sigg and Hezi-Yamit, 2009). In the 1990s, studies showed that stents could help keep the artery open better and longer than balloon angioplasty alone (Fischman *et al.*, 1994; Sigwart *et al.*, 1997). Since that time, stents have evolved from being made of metal (also known as "bare metal" stents) to being coated with a drug-eluting polymer layer that is microns thick. The drugs employed in drug-eluting stents are used in order to reduce restenosis caused by the damage to the vessel during the stenting procedure. In 2006, there were 652,000 angioplasty procedures with stents, 76% of which used drug-eluting stents (Lloyd-Jones *et al.*, 2010). Stents are currently implanted using **percutaneous** procedures.

There are two main methods for stent delivery. This first method, shown in Figure 14.5, is commonly referred to as balloon-expanded stent delivery. In this method, the stent is slipped over the outside of a **balloon catheter**. When the balloon catheter is in the appropriate position, the balloon is inflated and the stent expands and plastically deforms in order to stay in position. The advantage of this type of stent is that the diameter can be made to fit the lumen size. The second technique for stent deployment involves a thermal effect and is known as self-expanding stent delivery, illustrated in Figure 14.6. A shape-memory metal, such as Nitinol, is formed into its final geometry and then cooled and compressed to fit within a catheter. Once the catheter is in the correct location, it is pulled back, leaving the stent in place. The stent expands to its original size once its temperature rises above its transitional temperature, as described in Chapter 2. When using this method, correct stent sizing is crucial, as is stent processing to produce the final shape and mechanical properties. Using a stent that is too large for the vessel, or one with extremely high radial force, can result in tissue injury or excessive stress on the vessel (White and Ramee, 2003).

484 Cardiovascular devices

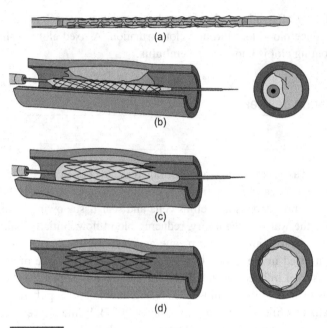

Figure 14.5

Balloon-expanded stent delivery. (a) The stent is loaded onto a catheter and (b) threaded to the location of the blockage. (c) The balloon catheter is inflated to press the stent against the vessel wall and then (d) withdrawn, leaving the stent to support the vessel.

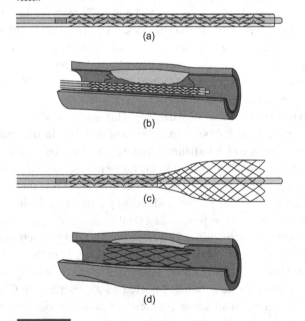

Figure 14.6

Self-expanding stent delivery. (a) The stent is compressed onto a catheter and a sheath is pulled over it. (b) The catheter is then threaded to the location of the blockage and (c) the sheath is retracted, allowing the stent to expand. (d) The catheter is retracted, leaving the stent to support the vessel.

Table 14.2 A representative sample of stents approved for use in United States

Name	Manufacturer	Materials	Drug
Multi-link Vision	Abbott Vascular	CoCr	None
Xience V™	(Abbott Park, IL)	CoCr, fluorinated copolymer	Everolimus
Express®	Boston Scientific Corp.	316L	None
VeriFLEX™	(Natick, MA)	316L	None
Taxus® Express®		316L, Translute SIBS	Paclitaxel
Driver®	Medtronic, Inc.	CoCr	None
Endeavor®	(Minneapolis, MN)	CoCr, phosphorlycholine	Zotarolimus
Cypher®	Cordis Corp. (Bridgewater, NJ)	316L, nonresorbable polymer (Parylene, PEVA, PBMA)	Sirolimus

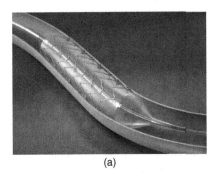
(a)

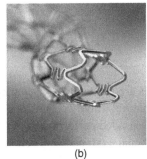
(b)

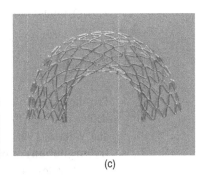
(c)

Figure 14.7

Representative stents currently marketed in the United States: (a) Abbot Multilink Vision (Image courtesy of Abbott Vascular. © 2010 Abbott Laboratories. All Rights Reserved.); (b) Cordis Cypher (Image courtesy of Cordis Corp., © Cordis Corporation 2010); and (c) Medtronic Endeavor (Image courtesy of Medtronic, Inc.).

Some current coronary stent designs are given in Table 14.2 with examples shown in Figure 14.7.

Stents are made from 316L stainless steel, CoCr alloys, and Ni-Ti alloys such as Nitinol. The polymer coatings used in the United States are non-resorbable, but stents with resorbable polymer coatings, such as the BioMatrix™ (BioSensors International Technologies, Singapore), are approved in other countries. This stent utilizes a polylactic acid (PLA) coating to deliver Biolimus, an immunosuppressant drug used to reduce restenosis.

The structural requirements of stents are that they exhibit radial/torsional flexibility (necessary during the insertion process), reliable expansion, radiopacity (for precise placement), and appropriate dimensions and radial force post-deployment. Designing the stent profile and leading edge to reduce tissue injury is a major concern, as injuries during placement can lead to thrombosis and **hyperplasia**, an increase in the number of cells that then block the vessel. Equally important is designing the stent for long-term durability with good fatigue and corrosion performance under physiological loading. The FDA recommends the following non-clinical tests for stent characterization (FDA, 2010b):

- Materials characterization (composition, shape memory and superelasticity, corrosion resistance)
- Stent dimensional and functional attributes (dimensions, recoil, radial stiffness, stress/strain, fatigue, accelerated durability, particle evaluation, radiopacity, crush and kink resistance, plus additional tests for specialized stents)
- Delivery system dimensional and functional attributes (dimensions, delivery/deployment/retraction, balloon burst pressure/fatigue/compliance/inflation and deflation time, catheter bond strength, tip pull, flexibility, kink resistance, torque strength, coating integrity, stent securement)
- Shelf life
- Biocompatibility

14.4.2 Heart valves

Prosthetic heart valves are needed to replace natural valves when there is a **congenital** defect or when the natural valve can no longer perform due to disease (Didisheim and Watson, 1996). Degraded performance of a heart valve is manifested as leakiness (**regurgitation**) or blockage (**stenosis**). There were 104,000 valve replacements in the United States in 2006. Also, over 77,000 deaths were attributed to valvular disease that year (Lloyd-Jones et al., 2010). Heart valves are currently implanted through open-heart surgery. The aortic and mitral valves are the most commonly replaced valves.

The first mechanical heart valve was implanted in 1952 by Dr. Charles Hufnagel. It was implanted in the descending aorta (this surgery occurred before the development of cardiac bypass in the mid-1950s). The use of tissue valves, or bioprosthetics, in the correct anatomical position followed not far behind in the 1960s (Yoganathan, 2000). Initial mechanical heart valve replacements utilized a caged ball design. In this arrangement, changes in fluid pressure due to the heart pumping are used to move a metal or polymer ball back and forth in a metal cage that is fixed in the valve location. This design was modeled after ball valves used in industry, and an early version, the Starr-Edwards Silastic Ball Valve, is shown in Figure 14.8. Although it was used for many years, the design was prone to damaging blood cells and required more work from the heart due to the necessary change in the course of the blood to move around the ball. The tilting disk design was invented in 1969 and initially used a disk made of Delrin (the tradename for polyoxymethylene) (Buvaneshwar et al., 2002). The disk floated on

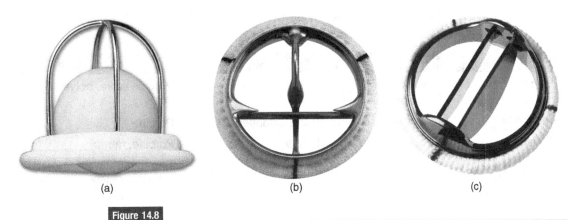

Figure 14.8

The evolution of heart valve design, 1952–present: (a) Starr-Edwards Mitral Ball Valve (Image courtesy of Edwards Lifesciences); (b) Medtronic Hall Valve (Image courtesy of Medtronic, Inc.); and (c) Regent[TM] Mechanical Heart Valve (Regent[TM] is a trademark of St. Jude Medical, Inc. Reprinted with permission of St. Jude Medical[TM], © 2010 All rights reserved).

two metal struts that allowed for opening and closing, restricting backflow while still allowing for central flow through the valve during pumping. The two problems with this design were strut fracture (described later in this chapter) and the fact that Delrin swelled slightly in the presence of blood or during autoclaving. This led to the replacement of Delrin by pyrolytic carbon (Yoganathan, 2000). Bileaflet valves were introduced in 1979 and remain the mechanical design most widely used today (Chandran *et al.*, 2003). Many current valves also have the capability to rotate within their housing in order to provide maximum blood flow. Because of the risk of clotting when a mechanical heart valve is used, a person who has had one of these implanted will take anti-coagulant medications for their remaining lifetime.

Silicon-alloyed, low-temperature isotropic pyrolytic carbon is very commonly used in mechanical heart valves (Dauskardt and Ritchie, 1999). The leaflets are formed by the co-deposition of silicon and carbon onto a polycrystalline graphite substrate in a fluidized bed at high temperatures (1000–1500°C) (Chandran *et al.*, 2003). This material demonstrates excellent biocompatibility and resistance to wear and fatigue in this application, as demonstrated by its high success rates over decades. Common materials for the sewing cuffs are polyester and PTFE fabrics, which have good biocompatibility and mechanical strength.

Tissue **heterograft** valves, either porcine or made from bovine pericardium, came into use when it became clear that cadaver **autografts** would not be able to provide the number of needed valve replacements. These heterografts are often crosslinked with gluteraldehyde for improved durability. Biological valves substantially reduce the risk of clotting and thus anti-coagulant therapies are not needed when these implants are utilized. However, tissue valves are vulnerable to calcification and do not last as long as their mechanical counterparts: 5–10 years as compared to the 20–30 years of expected lifetime from a mechanical valve (Didisheim and Watson, 1996).

Table 14.3 A representative sample of heart valves currently marketed in the United States

Device	Manufacturer	Type	Material
Hancock®	Medtronic, Inc. (Minneapolis, MN)	Tissue	Porcine valve, polyester
Freestyle®		Tissue (stentless)	Porcine
Mosaic®		Tissue	Porcine, fabric
Hall Easy-Fit®		Mechanical (tilting disk)	Titanium (no welds, tungsten-impregnated graphite with pyrolytic carbon coating, Teflon sewing ring
BioCor®	St. Jude Medical, Inc. (St. Paul, MN)	Tissue (stentless)	Porcine
SJM Masters		Mechanical (bileaflet)	Pyrolytic carbon over graphite substrate, sewing cuff
Regent™		Mechanical (bileaflet)	Pyrolytic carbon, sewing cuff
Perimount	Edwards LifeSciences, LLC (Irvine, CA)	Tissue	Bovine pericardium
ATS 3F®	ATS Medical, Inc. (Minneapolis, MN)	Tissue (stentless)	Bovine pericardium
Open Pivot®		Mechanical (bileaflet)	Pyrolytic carbon, graphite
Mitroflow®	CarboMedics, Inc. (Austin, TX)	Tissue	Bovine pericardium with PET and silicone support
Standard Aortic Valve		Mechanical (bileaflet)	Pyrolytic carbon, tungsten-loaded graphite, titanium, PET
On-X	On-X Life Technologies, Inc. (Austin, TX)	Mechanical (bileaflet)	Pyrolytic carbon, tungsten-loaded graphite, titanium, PTFE

Some current available heart valve replacements are shown in Table 14.3 and shown in Figure 14.8 and Figure 14.9.

Important requirements in valve design include: reduced turbulent flow, tissue ingrowth, and radiopacity. The reduction of turbulent flow is necessary to avoid trauma to the blood cells and to reduce the load on the heart (Chandran *et al.*, 2003). Tissue

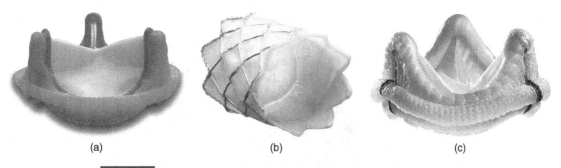

Figure 14.9

Representative tissue valves currently marketed in the United States: (a) Carpentier-Edwards Perimount Aortic Heart Valve (Image courtesy of Edwards Lifesciences); (b) Medtronic Melody Valve (Image courtesy of Medtronic, Inc.); and (c) Epic™ Supra Stented Tissue Valve with Linx™ AC Technology (Epic™ is a trademark of St. Jude Medical, Inc. Reprinted with permission of St. Jude Medical™, © 2010. All rights reserved).

ingrowth at the sewing cuff helps assure that the valve placement is secure. Finally, radiopacity is needed for post-surgical monitoring. Other issues that are of concern to valve designers are: strut fracture, pitting, and corrosion in mechanical valves; leaflet rupture in tissue valves; thrombosis; and availability of a range of sizes. The FDA requests the following types of characterization for mechanical heart valves (FDA, 2010a):

- **Pyrogenicity**
- Sterilization
- Pre-clinical *in vitro* assessment (valve samples, reference valves, biocompatibility, durability, fatigue, dynamic failure mode, cavitation, corrosion, flammability, hemodynamic performance, MRI safety, shelf life)
- Pre-clinical animal studies
- Clinical investigations

14.4.3 Vascular grafts

Vascular grafts are needed when a patient is suffering from severe atherosclerosis and whose effects cannot be relieved through balloon angioplasty or stenting. In extreme cases, when the coronary arteries are blocked, the heart does not receive adequate blood. This condition is known as angina and the resulting surgery is called **coronary artery bypass graft** (CABG) surgery. A person can have multiple bypass grafts in a single surgery. In 2006, 253,000 patients underwent 448,000 bypass procedures (Lloyd-Jones *et al.*, 2010). Grafts are also used when an **aneurysm** occurs. An aneurysm is a bulge in a blood vessel due to thinning and weakening of the vessel wall, as shown in Figure 14.10. They commonly occur in the aorta and at the base of the brain. In the aorta, an endovascular stent-graft can also be implanted to serve as an alternative conduit for blood flow.

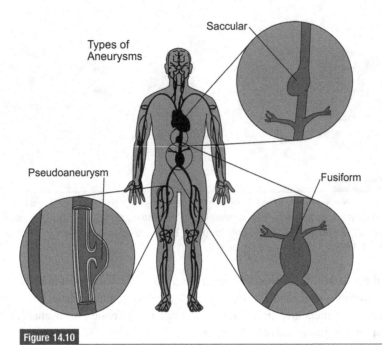

Figure 14.10

A schematic illustrating different types of aneurysms.

Modern vascular graft history shows that in 1949, the first graft surgery occurred using the patient's own **saphenous vein**. From 1950 to 1970, venous homografts were used, but this practice was abandoned due to a high rate of thrombosis and failure. In 1960, the umbilical vein homograft was introduced and is still used today (Didisheim and Watson, 1996).

Current available vascular grafts can be divided into natural and synthetic options. Natural options include the saphenous vein and the umbilical vein, which is sometimes enclosed in a polyester mesh to increase wall strength. These grafts show good durability and a lack of surgical complications. The success rate of these grafts is moderate: 80% success at one year, and 70% success at five years (Chandran et al., 2003). Unfortunately, 20–30% of patients requiring a graft do not have a suitable saphenous vein for the transplantation process (Didisheim and Watson, 1996). This fact, combined with a need for larger diameter replacements such as for the aorta, drives the development of synthetic grafts. These grafts, made of Dacron or expanded PTFE (ePTFE), can be knitted, crimped, or otherwise processed to maintain their structure. ePTFE has the necessary mechanical strength to perform well in this application and has been shown to have low occlusion due to thrombosis (Didisheim and Watson, 1996). PET fabrics have high stiffness and low extensibility (Ventre et al., 2009). Woven fabrics have lower porosity than knitted and do not require pre-clotting, while knitted grafts with velour are believed to have better tissue integration and to enhance endothelialization.

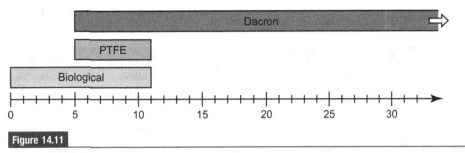

Figure 14.11

Recommended material selection for grafts by vessel size (mm).

In Dacron grafts, 10–45 μm pores are believed to support tissue ingrowth for better fixation (Bhat, 2005) and the hope is that the inner surface of these grafts will become covered with a pseudo-intimal layer that improves blood flow and reduces friction between the blood cells and the graft wall, an example of bioactive design. Large-diameter knitted Dacron grafts are often pre-impregnated with the patient's blood, albumin, or collagen in order to pre-clot the surface. There are also attempts to reduce coagulation altogether using surface treatments of anti-coagulant drugs.

As illustrated in Figure 14.11, it is suggested that 12–30 mm diameter vessels be replaced by Dacron grafts, 5–11 mm diameter vessels be replaced by Dacron, ePTFE or biological materials, and small grafts (<5 mm) be replaced solely by biological materials (Bhat, 2005). Thrombosis occurs readily in synthetic grafts of diameters less than 6 mm, and so these are not generally implanted below the knee or in CABG surgery (Chandran et al., 2003).

Some current synthetic vascular grafts and stent grafts are shown in Table 14.4 and Figure 14.12.

The requirements of this implant are lumen integrity, patency, flexibility, and durability. Some relevant internal pressures and wall strengths were given earlier in Table 14.1. Reducing surface roughness can improve resistance to thrombosis simply by reducing surface area. When vascular graft designs are submitted to the FDA for pre-market approval, the FDA suggests the following risks be characterized (FDA Guidance Document, 2000):

- Thrombosis
- Leakage
- Biocompatibility
- Graft disruption
- Seroma
- False aneurysm
- True aneurysm
- Sterilization
- Performance (biostability, leakage, strength, length, relaxed internal diameter, pressurized internal diameter, wall thickness, suture retention strength, kink diameter/radius)

Table 14.4 A representative sample of vascular grafts approved for use in the United States

Device	Manufacturer	Material	Available diameters (mm)
Excluder® (stent graft)	W.L. Gore & Associates, Inc. (Flagstaff, AZ)	ePTFE with Nitinol	19–29
Stretch® (vascular graft)		ePTFE	4–10
Propaten® (vascular graft)		ePTFE with heparin	5–8
Lifespan (vascular graft)	Edwards Lifesciences, LLC (Irvine, CA)	ePTFE	6–10
Ultramax™ (vascular graft)	Atrium Medical Corp. (Hudson, NH)	Gelatin-impregnated	6–24
Advanta™ (vascular graft)		Dacron	4–10
Flixene™ (vascular graft)		PTFE Trilaminate PTFE	6–8
Zenith Flex® (stent graft)	Cook Medical, Inc. (Bloomington, IN)	PTFE with stainless steel	22–36
Talent® (stent graft)	Medtronic, Inc. (Santa Rosa, CA)	PET with Nitinol	22–36
Flair® (stent graft)	Bard Peripheral Vascular, Inc. (Tempe, AZ)	ePTFE with Nitinol	6–9
Debakey® (vascular graft)		Woven PET	5–35
Dynaflo® (vascular graft)		Carbon-impregnated ePTFE	7–8

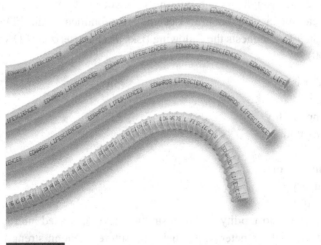

Figure 14.12

Vascular grafts showing a range of diameters. (Image courtesy of Edwards Lifesciences.)

Figure 14.13

A balloon catheter used in percutaneous transluminal angioplasty. (Image courtesy of Abbott Vascular. © 2010 Abbott Laboratories. All Rights Reserved.)

14.4.4 Other cardiovascular devices

There are many other cardiovascular implants. Some of the implants described below are not analyzed here because they are too complex, and references are given for more in-depth study. Others are summarized only, because they are not used in very large numbers.

Catheters (Figure 14.13) are placed nearly everywhere in the body during different medical procedures (Didisheim and Watson, 1996). These are long tubes of polyurethane, silicone, or low-density polyethylene used to administer fluids, withdraw samples, or insert devices. Sometimes catheters have a balloon at the end, which can be used for several procedures described above, such as angioplasty. They are short-term implants and so their main design concerns are flexibility, tractability, biocompatibility, and resistance to thrombosis. For more information on catheter design and materials selection, the text by Hamilton and Bodenham (2009) is an excellent reference.

The intra-aortic balloon pump (IABP), shown in Figure 14.14, was introduced in the late 1960s (Didisheim and Watson, 1996). This device is made up of a cylindrical balloon placed in the aorta and connected by catheter to a computer-controlled external pump that inflates and deflates the balloon to assist the heart. This is one of the most commonly used forms of circulator assistance. It supports patients undergoing CABG surgery and recovering from cardiac surgery. IABPs can be used to supplement heart function, allowing for heart recovery (Padera and Schoen, 2004). The text by Bolooki (1998) gives more information on IABPs.

A left ventricular assist device (LVAD), illustrated in Figure 14.15, is another mechanical method for assisting heart function. Using a pump that is implanted in the abdomen or chest cavity, this device pulls blood from the left ventricle and pushes it through the aorta. The pump is powered and controlled with an external unit. LVADs can be short-term implants, used to assist in surgical recovery, or long-term as a "bridge-to-transplant" or as an open-ended therapy, meaning that the patient will use the pump indefinitely or until a donor heart is made available for transplant. Similar to the IABP, evidence has shown that the use of LVADs can allow the heart enough time to heal on its own and eventually recover its full function. Challenges faced when developing this

494 Cardiovascular devices

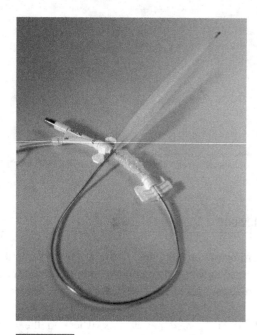

Figure 14.14

An intra-aortic balloon pump. (Image courtesy of Arrow International, a division of Teleflex, Inc.)

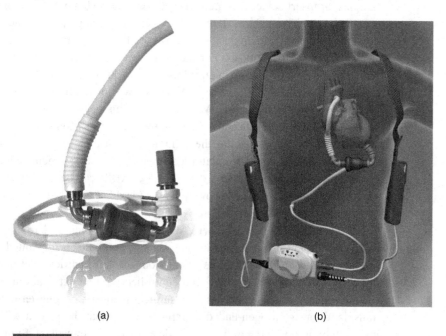

Figure 14.15

(a) A left ventricular assist device, (b) also shown schematically *in situ*. (Reprinted with permission from Thoratec Corporation.)

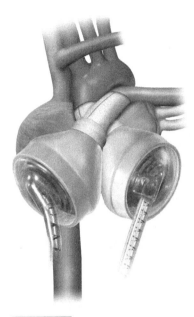

Figure 14.16

A total artificial heart (Courtesy syncardia.com).

device include clotting and durability, and also making one that is small enough for women and children to use. The durability issue is being approached through the use of pumps whose rotor is electromagnetically suspended, thus reducing wear. Ventricular assist devices can also target the right ventricle alone (RVAD) or both ventricles together (bi-VAD). For more information on LVADs, a good reference is Blitz and Fang (2010).

The total artificial heart (TAH) is rarely used, but is still an important device in the cardiovascular area. In 2008, there were 2,163 heart transplantations in the United States, while thousands more could have benefited from this surgery if donor hearts were available (Lloyd-Jones *et al.*, 2010). An artificial heart is distinct from VADs (which do not replace the heart) and cardiopulmonary bypass machines (which are mainly used for short-term applications). There are only two total artificial hearts approved for use by the FDA. The Syncardia temporary Total Artificial Heart (Syncardia Systems, Inc., Tucson, Arizona) was approved in 2004 and is used as a bridge-to-transplant. The AbioCor Replacement Heart (AbioMed, Inc., Danvers, Massachusetts) was approved under a Humanitarian Device Exemption in 2006 and is used only in patients with severe end-stage heart disease who are not eligible for a heart transplant. For more information on TAHs, Blitz and Fang (2010) provide a strong overview. The Syncardia TAH is shown in Figure 14.16.

Septal occluders, as seen in Figure 14.17, are used to block congenital heart defects (also known as septal defects) that are manifested as a hole between the left and right

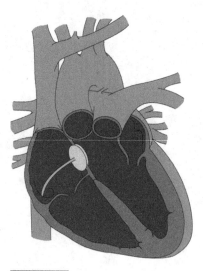

Figure 14.17
A schematic illustration of a septal occluder.

sides of the heart. These are among the most common birth defects and estimates of their occurrence in adults range from 650,000 to 1.3 million (Lloyd-Jones *et al.*, 2010). Of these defects, 20% close within the first year of life (Palacios, 2006). Septal occluders, implanted through a percutaneous procedure, are generally made of a Nitinol mesh with a fabric cover made of PTFE or PET. These devices experience very high success rates (Bass and Gruenstein, 2009).

14.5 Case studies

14.5.1 Bjork-Shiley Convexo-Concave heart valve (design and manufacturing failure)

The Bjork-Shiley Convexo-Concave mechanical heart valve was invented in 1969, and marketed by Shiley, Inc. It was a tilting disk valve made of Delrin and a cobalt-based alloy, illustrated in Figure 14.18. Because early tests showed problems with thrombosis, the next model had a convex-concave pyrolytic carbon disk that improved blood flow and reduced clotting. This model received FDA pre-market approval in 1979.

In pre-market testing, these devices showed strut fracture, but changes to the strut location were assumed to fix this issue (Blackstone, 2005) before it went to market, although no clinical testing was done. In November 1986, the device was recalled due to reports of strut fracture *in vivo*. When the strut fractured in an implanted device, the patient would die within minutes to hours (Baura, 2006). It was later found that the company was aware of manufacturing problems with the struts but continued to market

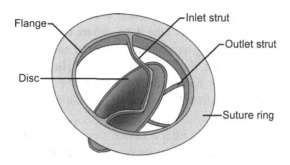

Figure 14.18

The Bjork-Shiley Convexo-Concave Heart Valve. This device experienced strut fractures and was eventually recalled.

the product. During manufacturing, company inspectors had noticed poor welds at the struts, but decided to have the welded joints ground down and polished to look smooth instead of discarding the faulty valves. Additionally, the design was changed again to incorporate a wider opening angle. Shiley, Inc. also later admitted that the company had re-welded earlier versions of the design with a 60° opening angle into a 70° opening angle instead of making new valves. By the time the FDA acted to require Shiley to notify all patients with implanted valves, approximately 86,000 had been implanted worldwide.

In failure studies after the recall, it was found that the large stresses during the closing of the valve put high bending stresses on the valve struts. In this case, the failures were due to design problems and poor manufacturing. The design problems are linked to inadequate testing of the engineering changes as they were made. For this disaster to have been avoided, the new design should have been fully tested before implantation began and manufacturing protocols should have been strictly followed. At least half of the faulty valves were implanted before 1981, at which point only 12 of the eventual 274 fractures had been reported to the FDA (Piehler and Hughes, 2004). Looking back, it is apparent that in order to issue the recall early enough to save the bulk of valves from being implanted, the FDA would have had to act based on a very small number of reported fractures.

14.5.2 ANCURE™ Endograft System (design and regulatory failure)

The ANCURE™ Endograft System was a stent graft designed to repair abdominal aortic aneurysms (AAA), marketed as a Class III device by Endovascular Technologies, Inc. Aneurysms were described earlier and shown in Figure 14.10. This device was approved by the FDA in September 1999 (FDA, 1999) and withdrawn in March 2001 (FDA, 2001; U.S. Attorney, 2003). This case illustrates both a design failure and a regulatory failure.

Before implantation, the graft is housed within a catheter. This catheter is inserted through the femoral artery and moved to the AAA, where the graft sheath is retracted. The graft utilizes its self-expanding frame and small metal hooks to secure itself to the correct location in the aorta. Then, the catheter is retracted (Baura, 2006).

When the device was initially approved by the FDA, the clinical data showed no deaths associated with the ANCURE™ tube implant at <30 days and 6% at 12 months (Guidant, 1999).

As discussed in Chapter 12, even after a device receives pre-market approval, it is subject to post-market surveillance. If a device causes or contributes to death or serious injury, the manufacturer is required to file a Medical Device Report (MDR) with the FDA. This provision was put in place in response to the Bjork-Shiley recall described above (Baura, 2006; FDA, 1999).

The first problem that developed after the device was on the market is that the delivery catheter became lodged in the patient after the stent was deployed. There were two options for surgeons to rectify this problem. They could either convert to a surgery and remove the delivery system by cutting open the patient, or they could break off the handle of the delivery system and pull the catheters out piece by piece. The handle-breaking technique was untested and was partially devised by a sales representative, hoping to avoid surgical conversions that would need to be reported to the FDA.

Clearly the fact that the delivery system could become stuck in the patient was a design flaw. If this showed up in clinical testing before the PMA submission, it was not mentioned in the Safety and Effectiveness data summary (Guidant, 1999). The public might never have known about this problem at all were it not for the "Anonymous Seven," employees from Guidant who felt compelled to send a letter to both Guidant management and the FDA, charging that Guidant was covering up these problems by failing to submit MDRs and that it was incorrectly labeling the device by not including instructions for the handle-breaking technique with the device documentation. During the time that the ANCURE™ stent graft was on the market, Guidant submitted only 172 MDRs, when in actuality, nearly 3000 events occurred which would have required one of these reports (Baura, 2006). In total, the delivery system failed in more than 30% of cases and contributed to 12 deaths (U.S. Attorney, 2003). Endovascular Technologies, Inc. paid $92.4 million after being found guilty on nine charges of introducing misbranded devices into interstate commerce and one count of misleading the FDA.

14.5.3 Nitinol stent strut failure (structural failure)

As described earlier in this chapter, stents are metal frames used to support arteries and maintain the lumen in order to improve blood flow. Nitinol stents are used in various locations in the body, including the superficial femoral, carotid, and renal arteries (Pelton et al., 2008). One concern associated with stents is that restenosis might occur, forcing additional procedures to improve blood flow. Another concern is that the stent might undergo strut fracture, shown schematically in Figure 14.19. Particularly in the

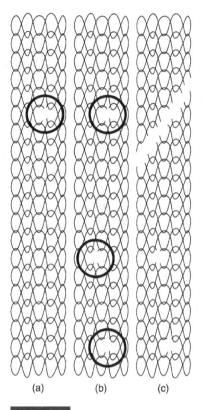

Figure 14.19

The different degrees of stent fracture: (a) single strut, (b) multiple strut, and (c) total separation.

superficial femoral artery, which undergoes a complex mixture of flexion, torsion, compression, and tension (Allie *et al.*, 2004), the mechanical strength of stents is deeply challenged.

Stent fracture can be classified in a number of ways. Some scales use numbers, starting with I (meaning single strut fracture) and ending with IV (associated with total stent separation) (Allie *et al.*, 2004). Other scales describe stent fractures as minor (single strut fracture), moderate (>1 strut fracture), or severe (complete separation of stent segments) (Scheinert *et al.*, 2005). The assessment methods can also differ. A stringent assessment protocol is necessary to catch single strut fractures. In current practice, it is common to use flat plane X-ray detection or **fluoroscopy**, a technique that uses X-rays to create a continuous, real-time image. Other techniques, such as angiography or CT angiography, are further variants of X-ray and fluoroscopy (Allie *et al.*, 2004; Duda *et al.*, 2002).

Allie and co-workers (2004) found stent fracture in 65.4% of nitinol stents, mainly non-coil designs. Approximately 80% of these were Type I and II fractures, yet were still associated with restenosis. The stent fractures were found to correlate with overlapping stenting, as was also seen by Duda *et al*. (2002), although the overall nitinol stent fracture rate was lower in this study, at 18.1%. Similar results were found by Scheinert and co-workers (2005), who showed that 24.5% of nitinol stents in the superficial femoral artery demonstrated fracture. Furthermore, in that study, the fracture rates for single stents, two stents, and three or more stents were 16.7%, 41.2%, and 59.0%, respectively. Stent fracture was also associated with reduced **patency**, or the fraction of blood vessels that allow blood flow past the procedure location.

Pelton *et al*. (2008) used a combination of finite element analysis and *in vitro* deformation testing to develop the correct conditions for stent subcomponent fatigue testing. This group was investigating the contributions of pulsatile fatigue to the stent fractures observed in nitinol stents in the superior femoral artery. The results of the subcomponent fatigue testing showed good agreement with the clinical data, and demonstrated that pulsatile fatigue was not the main cause of stent fracture. It was recommended that the other deformations (flexion, torsion, compression, tension) be similarly studied.

Concerns about stent fracture are twofold. First, a self-expanding metal stent that has a strut fracture may continue to expand and allow the fractured strut to push into tissue, possibly creating continuous damage. Second, the studies cited above demonstrate a strong relationship between stent strut fracture, increased occlusion, and reduced patency. Manufacturers should adopt improved testing of these stents, including testing at the boundaries or even exceeding physiological loading, as suggested by Pelton and co-workers, and testing that mirrors conditions such as overlapping stents, as proposed by the FDA (FDA, 2010b).

14.6 Looking forward

There are many interesting cardiovascular devices on the horizon. Some, such as bare metal stents coated with a biodegradable drug-eluting polymer, are already approved for use in other areas of the world, and it is likely that they will be approved in the United States soon (BioMatrix stent, Biosensors Interventional Technologies Pte Ltd., Singapore). Other fully absorbable stents will doubtless follow. Increased percutaneous surgeries, such as percutaneous valve replacement, open the door for novel device designs that can be passed through a catheter. Finally, tissue engineering approaches could generate new solutions for septal defects and aneurysms. An interesting challenge is the development of heart valves that can grow with the patient, enabling children with artificial heart valves to go longer or possibly indefinitely without needing surgery for the implantation of a larger valve.

14.7 Summary

This chapter provides an overview of cardiovascular devices with a focus on those that have primarily structural requirements: stents, grafts, and heart valves. The historical development of these implants is described and current designs are presented along with structural requirements and relevant physiological loads. FDA guidance for device approval is given wherever possible. Additional interesting cardiovascular devices are briefly touched upon, followed by case studies demonstrating design, materials, manufacturing, and regulatory failures.

14.8 Problems for consideration

Checking for understanding

1. What are the primary considerations when designing cardiovascular implants?
2. How does the tissue microstructure of the vessel wall contribute to its mechanical behavior?
3. What are the tissue elements that enable a healthy heart valve to function (open and close)?
4. What is the wall tension in an artery that is 5 mm in diameter and 1 mm thick when the systolic blood pressure is 120 mm Hg?
5. If you have a catheter made of PET with $S_y = 57$ MPa, diameter $= 6$ mm, and wall thickness $= 0.1$ mm, what pressure will cause it to yield?

For further exploration

6. How could you determine the radial stiffness of a stent? Would it be different if the stent were Nitinol as opposed to stainless steel?
7. Propose a novel stent design for the superficial femoral artery that would be less vulnerable to strut fracture than current designs.

14.9 References

Allie, D.E., Hebert, C.J., and Walker, C.M. (2004). Nitinol stent fractures in the SFA. *Endovascular Today*, July/August, 22–29.

Bass, J.L., and Gruenstein, D.H. (2009). Cardiac septal defects: treatment via the Amplatzer family of devices. In *Handbook of Cardiac Anatomy, Physiology and Devices*, 2d edn., ed. P.A. Iaizzo. New York: Springer, pp. 571–82.

Baura, G.D. (2006). *Engineering Ethics: An Industrial Perspective*. Burlington, MA: Elsevier Academic Press.

Bhat, S.V. (2005). *Biomaterials*. Middlesex, UK: Alpha Science.

Blackstone, E.H. (2005). Could it happen again?: the Björk-Shiley Convexo-Concave Heart Valve story. *Circulation*, 111, 2717–19.

Buvaneshwar, G.S., Ramani, A.V., and Chandran, K.B. (2002). Polymeric occluders in tilting disk heart valve prostheses. In *Polymeric Biomaterials*, 2d edn., ed. S. Dumitriu. New York: Marcel Dekker, pp. 589–610.

Chandran, K.B., Burg, K.J.L., and Shalaby, S.W. (2003). Soft tissue replacements. In *Biomaterials: Principles and Applications*, ed. J.B. Park and J.D. Bronzino. Boca Raton, FL: CRC Press, pp. 141–63.

Dauskardt, R.H., and Ritchie, R.O. (1999). Pyrolitic carbon coatings. In *An Introduction to Bioceramics*, ed. L.L. Hench and J. Wilson. River Edge, NJ: World Scientific Publishing, pp. 261–80.

Didisheim, P., and Watson, J.T. (1996). Cardiovascular applications. In *Biomaterials Science: An Introduction to Materials in Medicine*, ed. B.D. Ratner, A.S. Hoffman, F.J. Schoen, and J.E. Lemons. San Diego, CA: Elsevier Academic Press, pp. 283–95.

Duda, S.H., Pusich, B., Richter, G., Landwehr, P., Oliva, V.L., Tielbeek, A., ... Bérégi, J.P. (2002). Sirolimus-eluting stents for the treatment of obstructive superficial femoral artery disease: six-month results. *Circulation*, 106, 1505–9.

Fischman, D.L., Leon, M.B., Baim, D.S., Schatz, R.A., Savage, M.P., Penn, I., ... Goldberg, S. (1994). A randomized comparison of coronary-stent placement and balloon angioplasty in the treatment of coronary artery disease. *New England Journal of Medicine*, 331(8), 496–501.

Fligner, C.L., Reichenbach, D.D., and Otto, C.M. (2004). Pathology and etiology of valvular heart disease. In *Valvular Heart Disease*, 2d edn., ed. C.M. Otto, pp. 18–50. Philadelphia, PA: Elsevier.

Fong, P.F., Park, J., and Breuer, C.K. (2007). Heart valves. In *Principles of Tissue Engineering*, 3d edn., ed. R.P. Lanza, R.S. Langer, and J. Vacanti. Burlington, MA: Elsevier Academic Press, pp. 585–600.

Food and Drug Administration. (1999). P990017 Premarket Application for Ancure Tube System, Ancure Bifurcated System, Ancure Iliac Balloon Catheter. September 28, 1999. Office of Device Evaluation, Center for Devices and Radiological Health, Food and Drug Administration. www.fda.gov, retrieved September 06, 2010.

Food and Drug Administration. (2000). Guidance Document for Vascular Prostheses 510(k) *Submissions*. Rockville, MD: FDA.

Food and Drug Administration. (2001). FDA Public Health Notification: Problems with Endovascular Grafts for Treatment of Aortic Abdominal Aneurysm (AAA). April 27, 2001. Center for Devices and Radiological Health, Food and Drug Administration. www.fda.gov, retrieved September 06, 2010.

Food and Drug Administration. (2010a). Heart Valves – Investigative Device Exemption (IDE) and Premarket Approval (PMA) Applications. Rockville, MD: FDA.

Food and Drug Administration. (2010b). Non-Clinical Engineering Tests and Recommended Labeling for Intravascular Stents and Associated Delivery Systems. Rockville, MD: FDA.

Guidant Corporation. (1999). Ancure Summary of Safety and Effectiveness Data. www.fda.gov, retrieved September 06, 2010.

Ku, D.N., and Allen, R.C. (2000). Vascular grafts. In *The Biomedical Engineering Handbook*, 2d edn., Volume 2, ed. J.D. Bronzino. Boca Raton, FL: CRC Press, pp. 128-1–128-8.

Lloyd-Jones, D., Adams, R.J., Brown, T.M., Carnethon, M., Dai, S., De Simone, G., ... Wylie-Rosett, J. (2010). Heart disease and stroke statistics 2010 update: A report from the American Heart Association. *Circulation*, 121, e46–e215.

Meier, B. (2007). Percutaneous coronary intervention. In *Textbook of Cardiovascular Medicine*, Volume 355, ed. R.M. Califf, E.N. Prystowsky, J.D. Thomas, and P.D. Thompson. Philadelphia, PA: Lippincott, Williams & Wilkins, pp. 1258–72.

Okamoto, R.J. (2008). Blood vessel mechanics. In *Encyclopedia of Biomaterials and Biomedical Engineering*, 2d edn., Volume 1, ed. G.E. Wnek and G.L. Bowlin. New York: Informa Healthcare, pp. 392–402.

Padera, Jr., R.F., and Schoen, F.J. (2004). Cardiovascular medical devices. In *Biomaterials Science: An Introduction to Materials in Medicine*, 2d edn., ed. B.D. Ratner, A.S. Hoffman, F.J. Schoen, and J.E. Lemons. San Diego, CA: Elsevier Academic Press, pp. 470–93.

Palacios, I.F. (2006). Closure of interatrial communications using the CardioSEAL and STARFlex devices. In *Percutaneous Device Closure of the Atrial Septum*, ed. S.J.D. Brecker. London: Informa UK, pp. 121–40.

Pelton, A.R., Schroeder, V., Mitchell, M.R., Gong, X.-Y., Barney, M., and Robertson, S.W. (2008). Fatigue and durability of nitinol stents. *Journal of the Mechanical Behavior of Biomedical Materials I*, 1(2), 153–64.

Piehler, H.R., and Hughes, A.A. (2004). The role of post-market surveillance in the medical device risk management system. *Engineering Systems Symposium*, Cambridge, MA, March 29–31, 2004.

Ross, R. (1999). Atherosclerosis – an inflammatory disease. *New England Journal of Medicine*, 340 (2), 115–26.

Scheinert, D., Scheinert, S., Sax, J., Piorkowski, C., Braunlich, S., Ulrich, M., ... Schmidt, A. (2005). Prevalence and clinical impact of stent fractures after femoropopliteal stenting. *Journal of the American College of Cardiology*, 45(2), 312–15.

Schoen, F.J. (2005) Cardiac valves and valvular pathology: update on function, disease, repair and replacement. *Cardiovascular Pathology*, 14, 189–94.

Sigg, D.C., and Hezi-Yamit, A. (2009). Cardiac and vascular receptors and signal transduction. In *Handbook of Cardiac Anatomy, Physiology and Devices*, 2d edn., ed. P.A. Iaizzo. New York: Springer, pp. 191–218.

Sigwart, U., Puel, J., Velimir, M., Joffre, F., and Kappenberg, L. (1997). Intravascular stents to prevent occlusion and restenosis after transluminal angioplasty. *New England Journal of Medicine*, 316(12): 701–6.

Silver, F.H. (1994). Cardiovascular implants. In *Biomaterials, Medical Devices, and Tissue Engineering: An Integrated Approach*. London, UK: Chapman and Hall, pp. 153–93.

United States Attorney's Office, Northern District of California. (2003). Endovascular Press Release. June 12, 2003. *U.S.* Department of Justice. www.justice.gov, retrieved September 06, 2010.

Ventre, M., Netti, P.A., Urciuolo, F., and Ambrosio, L. (2009). Soft tissues characteristics and strategies for their replacement and regeneration. In *Strategies in Regenerative Medicine: Integrating Biology with Materials Design*, ed. M. Santin. pp.16–50. New York: Springer.

White, C.J., and Ramee, S.R. (2003). Aortoiliac artery angioplasty and stenting. In *Peripheral Vascular Stenting for Cardiologists*, ed. R. Heuser. London: Martin Dunitz Ltd., pp. 39–52.

Yoganathan, A.P. (2000). Cardiac valve prostheses. In *The Biomedical Engineering Handbook*, 2d edn., Volume 2, ed. J.D. Bronzino. Boca Raton, FL: CRC Press, pp. 127-1–127-23.

Bibliography

(2009). *Central Venous Catheters*, ed. H. Hamilton and A. Bodenham. Chichester, UK: John Wiley & Sons, Ltd.

(2010). *Pacemakers and Implantable Cardioverter Defibrillators: An Expert's Manual*, ed. A. Al-ahmad, K.A. Ellenbogen, A. Natale, P.J. Wang. Minneapolis, MN: CardioText Publishing.

Blitz, A., and Fang, J.C. (2010). Ventricular assist devices and total artificial hearts. In *Device Therapy in Heart Failure*, ed. W.H. Maisel. New York: Humana Press, pp. 339–72.

Bolooki, H. (1998). *Clinical Application of the Intra Aortic Balloon Pump*, 3d rev. edn. Armonk, NY: Futura Publishing.

Netter, F.H. (2011). *Atlas of Human Anatomy*, 5th edn. Philadelphia, PA: Saunders Elsevier.

15 Oral and maxillofacial devices

Shikha Gupta

Inquiry

How can the design of dental implants be improved to permit more efficient implant osseointegration?

Dental implants are artificial tooth roots that serve as a foundation for dental prostheses, such as crowns and fixed dentures. Modern dental implants are placed into the sockets of missing teeth, and, like the natural tooth root, must bear and transfer loads between the tooth (or prosthesis), and the alveolar bone surrounding the socket. However, for the implant to effectively bear weight and transfer load, it must osseointegrate, or achieve rigid fixation through intimate contact with the bone. Though efforts to improve the osseointegrative potential of implants have driven much of the research around dental implants for the last 50 years, osseointegration of implants in regions of low bone density remains problematic. What new design and/or material innovations might be explored to further promote osseointegration of dental implants in these areas?

15.1 Overview

Like cardiovascular devices, total joint replacements of the **appendicular** skeleton, and soft tissue implants, the demand for oral and **maxillofacial** implants has increased steadily over the last decade with the aging of the global population (Misch, 2005). The longer lifespan and more active lifestyles of this older demographic have also placed more stringent demands on the longevity and performance of these implants.

Oral implants, or implants placed through the oral cavity, typically refer to dental implants, which are used in the treatment of **edentulism**, or tooth loss. Dental implants are artificial tooth roots onto which dental restorations, such as crowns, are attached. Perhaps due to the accessibility of the oral cavity, dental implants have one of the longest recorded histories amongst medical devices, dating back over 4000 years to the ancient Chinese, who hammered bamboo pegs into the sockets of missing teeth. Despite their millennia long history, dental implants did not achieve clinical success until the 1940s. Though myriad attempts at using **homoplastic**, **heteroplastic**, and **alloplastic** materials were made, the absence of biologically compatible, sterilizable, and corrosion-resistant materials that were able to withstand static and dynamic **occlusal** loads precluded progress in implant design. Once the biocompatibility of cobalt-chromium-molybdenum alloys was discovered in 1939, however, rapid innovations in dental implant design ensued (Venable *et al.*, 1937; Venable and Stuck, 1941). These efforts were further bolstered by the discovery of the **osseointegrative** ability of titanium in 1952, which was the seminal discovery of modern oral implantology (Branemark *et al.*, 1969; Branemark, 1983).

Today, there is one basic implant type, available in hundreds of different designs that may be optimized for the anatomy of individual patients. Though most are composed of titanium and titanium alloys, several newer devices have adopted zirconium oxide, despite past failures with ceramics as dental materials (Andreiotelli *et al.*, 2009). Nearly one million dental implants are now placed yearly, and the number continues to rise with improvements in surgical technique, treatment planning, and implant longevity (Misch, 2008).

Maxillofacial implants are implants associated with restoring form and/or function to the **maxilla** (upper jaw) and the face. Maxillofacial implants include soft-tissue replacements used in facial augmentation, such as cheek and chin implants, and temporomandibular joint (TMJ) replacements. This chapter will focus on **TMJ** replacements, which are the primary load-bearing maxillofacial implant. The TMJ is an articulating joint that shares many features common to the knee, hip, and shoulder joints, and TMJ replacements, like other joint replacements, are modular devices that serve to restore both movement and load-bearing capacity of the joint.

Unlike dental implants, the history of TMJ replacements is barely a century old. Despite its modern evolution, the development of TMJ implants has been rife with problems. While the TMJ's small size, complex anatomy, and proximity to sensitive **craniofacial** features contributed to these problems, the primary culprit was the lack of basic research on TMJ mechanics, and the surrounding clinical controversy over whether the TMJ was even a load-bearing joint. The decades-long debate stifled attempts to comprehensively characterize the loading environment around the TMJ. While the 1940s and 1950s saw an experimental period where many different materials and designs were utilized, most met with limited clinical success, largely because these early devices were selected based on incomplete knowledge of the static and dynamics loads experienced by the TMJ. This knowledge gap also led to the disastrous adoption of **Proplast**-Teflon and **Silastic** as TMJ implant materials. The performance of devices composed of these materials was evaluated with an incomplete understanding of the biomechanics of the TMJ, and their mechanical deficiencies eventually led to painful biological failures that were widespread (Dolwick and Aufdemorte, 1985; Lagrotteria *et al.*, 1986). The failures finally prompted the maxillofacial community to establish more rigorous guidelines for ensuring the safety and efficacy of new devices, and only a few new TMJ implants have received FDA approval since the 1990s. Today, a few thousand TMJ partial and total joint replacements are placed each year, and modern TMJ implants, which now utilize the same materials employed in other total joint replacements, are meeting more success than their predecessors.

15.2 Learning objectives

This chapter will discuss the historical development and design evolution of load-bearing oral and maxillofacial implants. The focus will be on the two primary load-bearing implants: dental implants and temporomandibular joint replacements. The chapter will conclude with a look forward to the future of dental and TMJ implant development. By the end of this chapter, students should be able to:

1. name and describe the function of major components of the tooth and TMJ anatomy
2. describe the clinical reasons that necessitate dental implants and TMJ replacements
3. provide a synopsis of the historical development of dental implants and TMJ implants
4. summarize the functional requirements for dental implants and TMJ implants
5. list natural and synthetic materials currently used in dental implants and TMJ replacements
6. assess design concerns for dental implants and TMJ replacements
7. discuss cases in oral and maxillofacial device history where an implant has failed prematurely due to corrosion, fatigue, wear, clinical complication, poor material selection, or inappropriate design

15.3 Oral and maxillofacial anatomy

15.3.1 Tooth anatomy

Though humans are born toothless, they will possess two sets of teeth during their lifetime. The first 20 primary teeth, or **deciduous** teeth, develop by 2 to 3 years of age. These are replaced by 28 adult teeth by adolescence, and, after the eruption of the third molars, a total of 32 adult teeth. Eight of the adult teeth are incisors, four are canines, eight are **bicuspid** pre-molars, and 12 are molars (Scheid, 2007).

Each tooth is divided into two anatomically distinct regions – the crown and the root (Figure 15.1). The crown, which sits almost entirely above the gum line, is exposed to the oral cavity and is the portion of the tooth that is covered with enamel. The shape of the crown varies greatly depending on the type of tooth, from the sharp, straight-edged incisors to the long, pointed canines to the bicuspid pre-molars. Irrespective of the shape, every tooth crown is comprised of three layers – an outer layer of enamel, a middle layer of dentin, and an interior made of pulp. The thickness of the outer enamel layer is non-uniform, varying across tooth type and along the cross-section of individual teeth. The thickest portion is at the cusps, while the thinnest section lies at the base of the crown. Below the enamel is a thicker, more uniform layer of primary dentin. This dentin provides a protective cover for the underlying pulp chamber. The chamber, which contains coronal pulp, is composed of blood vessels, nerves, connective tissues, and cells, including fibroblasts and a peripheral layer of odontoblasts that is responsible for

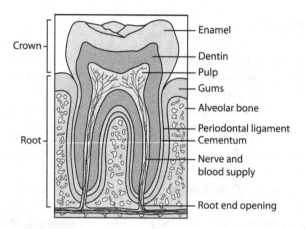

Figure 15.1

Schematic of the tooth crown and tooth root anatomy.

dentin formation. The pulp is imperative since it also provides nutrients and moisture to the surrounding mineralized tissues.

Unlike the crown, the tooth root is nestled below the gum margin. The tooth root may actually refer to one or more individual roots, which, like the number of cusps in the crown, varies with tooth type. At the center of each individual root is a root canal, which is a long, narrow extension of the pulp chamber. The root canal contains radicular pulp, which is similar in composition to coronal pulp. The pulp ends at the apical foreman, where nerves and blood vessels exit the root and enter the **periodontal** tissues. Surrounding the root canal is a layer of secondary dentin. This dentin has a composition and structure comparable to primary dentin, but, unlike primary dentin, it forms only after the root has fully developed and the tooth erupted completely. Rather than enamel, a thin layer of cementum, averaging several hundred microns in thickness, sheaths the secondary dentin (Stamfelj *et al.*, 2008). Surrounding the cementum are the periodontal tissues – the gums, the periodontal ligaments, and alveolar bone. Embedded at the periphery of the cementum are the periodontal and the gingival ligaments. The periodontal ligaments anchor the roots to the alveolar bone of the jaw in several locations, including the root apex and the cementoenamel junction. The **gingival** ligaments attach the root to the gums, separating the underlying alveolar bone from the tooth root, and protecting the periodontal ligaments and alveolar process. The alveolar processes are the bony ridges that contain the sockets of the teeth. The alveolar bone forms the borders of the upper jaw and the lower jaws. The coronal aspect of the alveolar bone, or the portion nearest the gum margin, is known as the alveolar crest. The upper arch of the jaw is the maxilla, while the lower jaw bone is the **mandible** (Figure 15.2), and both bones are comprised of both cortical and cancellous compartments.

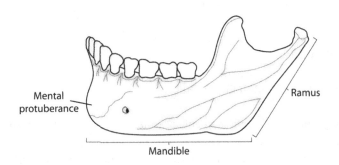

Figure 15.2

Schematic of the mandibular bone.

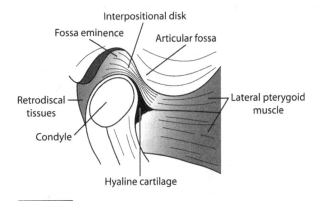

Figure 15.3

Schematic of the anatomy of the temporomandibular joint.

15.3.2 Temporomandibular joint anatomy

The TMJ, commonly known as the jaw joint, is a complex synovial joint that connects the upper part of the temporal bone of the skull to the mandible. The joint is bilateral, and movement of the jaw is guided by two mutually linked TMJs, one on either side of the face. In each joint, the posterior mandibular condyle articulates incongruently with a concavity, called the **glenoid fossa**, on the temporal bone (Figure 15.3). Anterior to the fossa is a convex region known as the articular eminence. The condyle and fossa are covered with a thin layer of hyaline cartilage. Between them lies a saddle-shaped fibrocartilaginous disk, which, much like the meniscus in the knee, acts as a stress absorber and directs synovial fluid flow. This **interpositional** disk is thinnest in the central, intermediate region and thickens in the anterior and posterior bands. Anteriorly, the disk is contiguous with the **pterygoid** muscles, which also attach to the neck of the mandibular condyle and aid in joint movement. Posteriorly, the disk is contiguous with the vascular, innervated **retrodiscal** tissues, which wrap around the condyles and supply nutrients to the disk (Scheid, 2007).

15.4 Dental implants

15.4.1 Clinical reasons for dental implantation

The primary function of teeth is to enhance the efficacy of **mastication**, or chewing of food. The long, sharp canine teeth tear up food, while the wide, flat molars grind and mash up food. Since mastication physically breaks down food, it facilitates enzymatic degradation by increasing the surface area of the food and also enables swallowing. Healthy teeth are thus important for both food intake and proper digestion. Unfortunately, tooth loss is common in both the adolescent and adult population. In the United States, more than 12 million people are completely edentulous in either the maxillary or mandibular arch, and over 18 million, 70% of whom are over 65, have no teeth in both arches (Misch, 2008). Though less than a few percent of children are missing permanent teeth, over 40% of adults aged 30–50 have lost at least one tooth to trauma or disease, and nearly 44 million people are missing at least one molar tooth. The leading causes of tooth loss in the United States are **dental caries** and **periodontitis**.

Dental caries, more commonly known as cavities, are regions of tooth enamel and dentin that have undergone acid-induced demineralization. These acids, particularly lactic acid, are produced by the billions of bacteria housed in the human mouth. The bacteria survive by consuming the sugars found in the food that remains in our mouth after we eat. Some bacteria, including *mutans streptococci* and *lactobacilli*, produce acids as a waste product of sugar digestion. Most of these acids do not contact or diffuse into the tooth enamel, but instead get diluted and washed away with saliva. However, dental **plaque**, comprised of salivary proteins, glycoproteins, food debris, and different bacterial colonies, shield the acids produced by these colonies from the diluting, buffering, and washing effects of saliva. Since plaque is immobilized on the surface of the tooth, concentrated acids in the plaque can react with the tooth enamel, solubilizing its hydroxyapatite crystals and releasing calcium and phosphate ions and water into the oral cavity. The weaker, demineralized enamel permits even greater penetration by bacteria into the tooth interior. If left untreated, bacterial invasion will result in sufficient decay of the dentin to expose the nerves in the pulp, leading to oral pain and eventually, tooth extraction.

Periodontal disease is the inflammation and degradation of the periodontal tissues surrounding the tooth, including the gums, periodontal ligaments, and alveolar bone. **Gingivitis**, which is the precursor to periodontitis, is a plaque- and **tartar**-induced inflammation of gingiva, or the gums. Bacteria residing in plaque release toxins that initially elicit an inflammatory response in the gums adjacent to the plaque. If the plaque is not removed, the continued release of toxins results in chronic inflammation, with an accumulation of macrophages and monocytes around the plaque site. In the process of fighting infection, these inflammatory cells also start to break down the gingival tissue along the gum line, causing it to recede from the base of the crown and form a **sulcus**, or a deep, narrow groove. Left untreated, the plaque can infiltrate this gap between the tooth and gums, causing subgingival inflammation and further deepening of the sulcus. Eventually, the bacterial invasion may move far enough down the root to expose the deeper periodontal tissues, including the periodontal ligaments and alveolar bone, to disease. If enough inflammation-induced resorption of the gums, connective tissue, and alveolar bone occurs, the tooth will no longer remain anchored to the tooth socket, become mobile, and eventually detach.

Fortunately, edentulism (tooth loss) is not a fatal condition. However, replacement may be necessary for many reasons, especially if several teeth are lost. Edentulism compromises mastication of food. Occlusal forces, or biting forces, between 50–200N are typically generated in dentate individuals during chewing (de Las Casas *et al.*, 2007; Richter, 1995). This number decreases rapidly with tooth loss, falling by nearly 80% in completely **edentulous** individuals (Atkinson and Ralph, 1973). The reduction in chewing forces limits the type and quantity of food suitable for consumption, and may have secondary health consequences. Further, gaps in the **dentition** may result in speech defects. Teeth have a natural tendency to shift, the extent of which is limited by the presence of adjacent teeth. However, once a tooth is lost, the remaining teeth may drift into the space vacated by the missing tooth. The subsequent tilting or rotation of adjacent teeth can lead to **malocclusion**, or misalignment of the opposing teeth of the maxillary and mandibular arches. In addition to reducing chewing efficiency, malocclusion may generate super-physiologic loads on the TMJ, causing joint inflammation and/or degeneration. Most important, tooth loss is known to have deleterious consequences on the underlying periodontal tissues, particularly the alveolar bone. Since bone is a mechanosensitive tissue, the reduction in compressive and tensile loads exerted by the tooth and periodontal ligaments, respectively, results in bone atrophy in the vicinity of edentulous regions. Bone resorption occurs in both the trabecular and compact bone compartments and it is manifested as a decrease in both the width and height of the alveolar ridge (Misch, 2005). This weaker bone is more susceptible to fracture. Further, as bone is lost, the gingival covering becomes thinner and weaker, and several extraoral facial changes associated with aging may accelerate, including the formation of wrinkles around the mouth and jowls (sagging chin). Thus, implants are also needed to restore and maintain oral and facial aesthetic.

Presently, there are three major types of dental prosthesis – removable dentures, fixed bridges, and dental implants. There are over 40 million fully or partially edentulous people in the United States. Though many still opt for removable dentures, the percentage using dental implants has been increasing dramatically over the last two decades, particular for single tooth replacements, from less than 100,000 in the early 1980s to over 1 million in 2005 (Misch, 2008).

Briefly, removable dentures, both complete and partial, generally consist of two primary components – **pontics**, or artificial teeth, and a denture base, onto which the pontics are attached. The denture base is a pink acrylic plate, custom-made to fit the

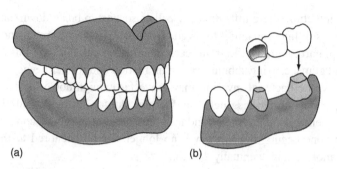

Figure 15.4

Illustration of (a) a complete removable denture and (b) a fixed partial bridge.

gums of each patient, while the pontics, which provide tooth facsimiles, are constructed of porcelain or, more recently, acrylic (Figure 15.4(a)). On the other hand, fixed partial bridges permanently attach artificial teeth to those adjacent to the missing tooth (teeth). Typically, pontics are fused to crowns for the adjacent **abutment** teeth with a metal bridge. The size of the abutment teeth is first reduced by removing the enamel and dentin in order to create room for the crowns, as well as to maintain natural alignment and correct contact with the other surrounding teeth. The bridge is then aligned and the crowns cemented to the abutment teeth (Figure 15.4(b)).

Completely removable dentures remain the most common prostheses used for fully edentulous individuals due to their low cost, ease of fabrication, nearly painless insertion, and minimal time to complete treatment. However, since dentures are soft-tissue borne prostheses, they only produce about 25% of the occlusal forces generated by natural teeth (Haraldson *et al.*, 1979; Kapur and Garrett, 1984; Michael *et al.*, 1990; Ogata and Satoh, 1995). This value is further reduced with long-term denture use. Even with adhesives, complete dentures are plagued by instability, are prone to clicking sounds while speaking, and slip out of place while eating. Further, all dentures require daily cleaning to maintain oral hygiene. The resorption of the alveolar bone may also accelerate with use of soft-tissue borne implants such as dentures. Since masticatory stresses are transmitted to the oral mucosa (gums) rather than the underlying bone, alveolar resorption continues, often at faster rates than those observed for edentulous individuals. In partial dentures, failure of the teeth adjacent to the prosthesis due to decay or bone resorption is also a concern – more than 60% of abutment teeth require treatment after 5 years, and within 10 years almost 44% of patients lose at least one abutment tooth (Misch, 2008).

Though fixed bridges generate greater chewing forces than dentures, they are notoriously difficult to clean, making them susceptible to plaque buildup. Not only do the abutment teeth have to be trimmed to place the bridge, but, like removable partial dentures, they are at risk for decay. In light of the detrimental effects of removable dentures and fixed bridges on oral health, their high rate of failure, and their inability to satisfactorily restore function, the shift to dental implants in recent years comes as no surprise.

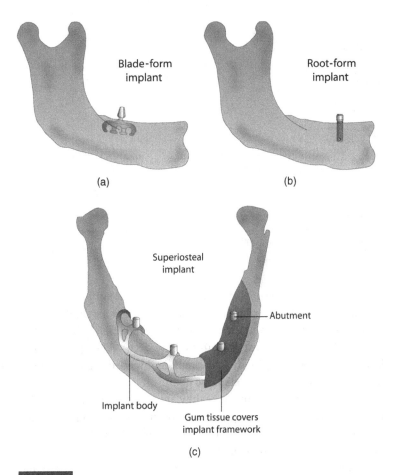

Figure 15.5

Examples of the three different types of dental implants: (a) blade-form, (b) root-form endosteal implants, (c) subperiosteal implants.

15.4.2 Types of dental implants

Three types of dental implants are utilized in modern dentistry – **subperiosteal**, blade (plate form) **endosteal**, and root-form endosteal (Figure 15.5). All three types of implants are composed of two main parts – the implant body, which is the portion of the implant placed within or atop the alveolar bone, and the abutment(s), which attach to the implant body and are **transmucosal**, extending through the gingiva. Dental prostheses, including crowns and fixed bridges, are secured to the abutments.

Endosteal implants are implants in which the implant body is embedded within bone. In blade-form endosteal implants, the implant body is a thin, blade-like metal framework that typically forms a single piece with the abutments (Figure 15.5(a)). Blade-form implants may be used in healthy or mild to moderately atrophied alveolar

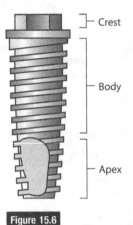

Figure 15.6

The three regions of a root-form implant.

bone, provided the vertical height of the ridge is sufficient. On the other hand, the implant body in modern root-form implants is either a cylinder or screw that resembles the natural tooth root. In screw-type implants, the implant body is further comprised of an **apex**, body, and a **crest module** (Figure 15.6). The apex is the portion embedded deepest in the bone, the body is the middle threaded segment, and the crest module is the implant neck, meant to retain the abutment and prevent rotation of the prosthesis.

Root-form implants require a healthy, intact vertical column of alveolar bone around the tooth root for support (Figure 15.5(b)). In endosteal implants, the implant body and abutment may be a single piece (nonsubmergible implant) or two separate components (**submergible**), depending on whether a one-stage or a two-stage surgery is to be performed. In a two-stage surgery, the implant body is first placed into the tooth socket. The wound is sutured closed to protect the implant from load. For a few months, the surrounding soft tissue is allowed to heal and the implant is given time to integrate with the bone. Since the implant body typically contains an internal thread for the placement of an abutment, the top of the implant body is usually covered with a short healing abutment to prevent the invasion of bacteria, bone, and soft tissue into these internal threads. During the second stage, the healing abutment is typically replaced with a permanent abutment, which is subsequently fitted with the desired restoration. In the single-stage surgery, both the implant and abutment are placed and allowed to heal simultaneously, while still precluding load-bearing of the implant for a few months.

Subperiosteal implants are dental prostheses placed just beneath the periosteum of the mandible or maxilla. They consist of a custom-made, patient-specific metal framework that sits below the gums but on top of the alveolar bone (Figure 15.5(c)), conforming to the shape of the jaw. The frame is fabricated with metal posts which protrude from the gums, and provide the equivalent of multiple tooth roots. Subperiosteal implants are typically utilized for completely edentulous arches, where there is insufficient alveolar bone to support endosteal implants, and are often accompanied by bone grafting.

Over the last two centuries, three primary motives have driven innovation in dental implants:

(1) Improving implant fixation and integration with surrounding bone
(2) Optimizing function within the existing bone, including mild, moderate, and severely atrophied bone, without the need for bone grafting
(3) Reducing implant-induced inflammation of the surrounding mucosa and subsequent bone resorption

15.4.3 Historical development of dental implants

Though blade-form and subperiosteal implants are innovations of the 20th century, the history of root-form dental implants dates back thousands of years. Archeological evidence from ancient burial sites indicates that the Egyptians placed hand-shaped prosthesis of ivory, seashell, or animal bones into the deceased prior to burial. Evidence of the first artificial teeth implanted into live patients comes from the Mayan civilization. While excavating in Honduras in 1931, the archeological team of Wilson and Dorothy Popenoe discovered a female mandible, dating back to about AD 600, in which tooth-shaped pieces of shell had been inserted into the sockets of the three lower incisors (Ring, 1995a). Though the implants were initially thought to have been placed after her death, radiographs taken several decades later revealed bony growth along the surface of the shells, indicating the implant must have been placed while the woman was alive. Through the 18th century, attempts to utilize homoplastic tooth replacements were common. Teeth were often forcibly taken from the poor and placed into freshly extracted teeth sockets of the rich. The success rate of the transplantations was very low, due both to non-sterile surgical technique and to the immune response experienced by the hosts.

The 19th century saw experimentation with several different metals as dental implants. In 1809, Dr. Maggiolo of France fabricated a tube-shaped 18 carat gold implant, which he placed into the socket immediately after tooth extraction. After allowing the tissue around the implant to heal for several weeks, Maggiolo attached a tooth to the gold implant. An identical procedure was attempted by Harris in 1887 with a lead-coated platinum post. Other experiments with silver and platinum also followed, but met with little success due to metal toxicity and, in some cases, material deformation (Ring, 1995b; Rudy et al., 2008). Rapid developments in dental implant design took place in the 20th century, particularly after the 1940s.

15.4.4 Subperiosteal implants

The first subperiosteal dental devices, implanted by Gustav Dahl and Drs. Gershkoff and Goldberg in the 1940s, were prefabricated metal substructures that were trimmed

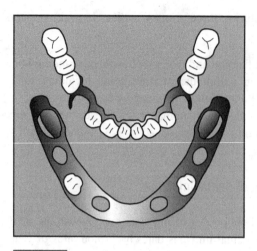

Figure 15.7

The earliest subperiosteal implant design.

to approximate the topography of the underlying bone and were screwed directly into the alveolar crest (Goldberg and Gershkoff, 1952). Later designs were customized by surgically opening the gingival tissue to expose the jaw, taking an impression of the bone with plastic, and casting a mold from the impression to fabricate the implant. With the advent of computed tomography and rapid prototyping, 3-dimensional models of the jaw morphology could be made non-invasively and used as a template to construct custom-fit implants.

The design of the metal in the subperiosteal framework continuously evolved from the 1950s to the 1980s. The earliest design was simply a thin strip of CoCr, a cobalt-chromium-molybdenum alloy, which rested primarily on the alveolar ridge (Figure 15.7) and was screwed into the bone. Not only did the implant place a significant amount of metal below the gums, but it concentrated occlusal loads onto an already atrophied bone crest, leaving sensitive alveolar nerves susceptible to damage (Linkow, 1990).

Since early implants did not conform to the shape of the jaw, they often displaced during mastication, eventually causing the screws fixing them to the alveolar bone to loosen and the implants to fail. Later designs included latticed meshes of both tantalum and CoCr. These too were formed from distorted plastic impressions of the jaw and contained significant amounts of metal. The abutments in these early implants were also short, bulky cylinders that were prone to infection. Subsequent designs, primarily composed of CoCr, experimented with many different strut configurations, some of which resulted in either excessively rigid or flexible implants that impeded jaw motion. Efforts to better distribute masticatory stresses around the alveolar ridge eventually lead to wider, expanded truss-like frameworks with strategically placed struts, and longer, tapered abutments that were often connected to one another with a thin metal bar to hold restorations more firmly in place (Figure 15.8). The addition of **fenestrations** in

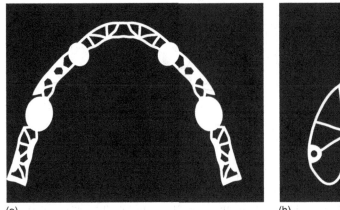
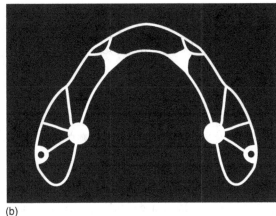

Figure 15.8

(a–b) Evolution of the subperiosteal implant strut configurations from many thick struts and large abutments to thinner, strategically placed struts and longer, narrow abutments.

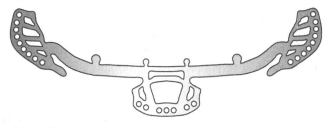

Figure 15.9

Typical tripodal subperiosteal implant design. The feet straddle the alveolar bone.

the peripheral struts also facilitated fixation of the implant (Bodine *et al.*, 1996; Linkow, 1990).

Once the techniques and materials used to make high-fidelity impressions of the alveolar bone were perfected, failure of subperiosteal implants was primarily attributed to chronically recurring inflammation, often originating at the site where the abutments project from the mucosa. The inflammation, which was accompanied by bone resorption, was present primarily around the posterior abutments, and could be caused by poor oral hygiene (Bodine, 1963; Bodine, 1974). In a few instances, severe bone resorption of the alveolar ridge was reported even in the absence of inflammation. In many cases, the loss led to intraoral exposure of the metal struts and/or pain (Bailey *et al.*, 1988). In the rare instances in which the implants fractured, the failure was a direct result of injury or trauma.

Modern subperiosteal implants are typically tripodal, and were introduced in 1984 by Leonard Linkow. The implants have three metal frame "feet," one on each side and one in the center, that sit over the mandibular or maxillary arch (Figure 15.9). The tripodal

Figure 15.10

Various early blade-form implant designs.

implants were originally composed of CoCr, but were gradually transitioned to titanium. Further, implants made of both materials were coated with different ceramics, including alumina or hydroxyapatite, in an effort to promote integration with the underlying alveolar ridge (Benson, 1972; Fettig and Kay, 1994; Kay et al., 1987).

Compared to its predecessors, the tripodal design had many advantages. It utilized less material and eliminated the need to expose the entire bony arch, both to determine the alveolar topography and to place the implant, simplifying the surgical procedure. Further, since the placement of the two side feet could be adjusted posteriorly to lie on the ramus region, preventing potential damage to regions in which alveolar nerves were too close to the bone crest, it could be used in patients otherwise contraindicated for dental implants. The most common causes of failure in modern subperiosteal implants are (1) improper surgical technique and (2) chronic inflammation around the abutments, generally associated with preexisting conditions, such as diabetes, or poor oral hygiene.

Subperiosteal bone implants are used infrequently today, particularly in comparison to endosteal implants. Most are placed in the completely edentulous mandibles. Though the reported success rate varies widely, a 60–80% survival rate at 10 years is commonly cited (Bodine et al., 1996; Mercier et al., 1981; Moore and Hansen, 2004).

15.4.5 Blade-form endosteal implants

The first blade-form implants were introduced by Dr. Leonard Linkow in the late 1960s (Babbush, 1972; Linkow, 1970). These were single-piece, wedge-shaped, vented CoCr blade implants that were placed by cutting a thin channel into the alveolar crest, slowly tapping the implants into the bone, and suturing the soft tissue around the posts. Linkow conceived of many different blade designs, including bow-shaped maxillary implants with elliptical vents, multi-toothed implants, and mandibular implants with spiral-shaped vents (Figure 15.10). By 1981, he had created a generic blade design whose geometry could be readily adapted during surgery for immediate use in many different anatomic sites. The blade width, curvature, and metal substructure design were determined by the size, location, and bone quality of the endentulous region. Submergible, two-stage blade vent systems also became available in the 1970s. Like subperiosteal implants, these early

Figure 15.11

Schematics of the ramus blade-form implant and the tripod blade-form ramus frame implant. (After Morimoto *et al.*, 1988.)

blade-form implants were primarily composed of CoCr. Early failures of these implants were often ascribed to improper implantation technique, including placement of implants into channels that had been drilled too wide to permit effective osseointegration, or placement of the implant blades too close to the alveolar crest, which lead to crestal inflammation and resorption. Blade designs with sharp corners and open apical vents were also less successful as they created stress concentrations around which excessive bone atrophy occurred (Linkow, 1990).

In the early 1970s, Drs. Ralph and Harold Roberts also introduced two different blade-form implants: the ramus blade and the ramus frame (Figure 15.11). The ramus blade consisted of a single abutment situated on one end of a curved, pick-axe-shaped blade – the blade was meant to be embedded into the ramus of the mandible. The ramus frame was similar to the tripodal subperiosteal implant, except that the feet were blade shaped, with the two peripheral feet meant to be embedded in the ramus and the central foot placed in the central alveolar ridge. Three short abutments protruding from the feet were connected via a continuous rail a few millimeters thick and about 5 mm wide. The ramus implants provided an alternative to the Linkow implants when not enough bone was present in the mandible for a thicker, tapered frame. Though both types of ramus implants were initially compromised of 316L stainless steel, later implants, like other blade implants, were composed of pure titanium, ceramic-coated titanium, and, for a short time, alumina (Kent *et al.*, 1982; Morimoto *et al.*, 1988).

Most blade-form implants are single-piece, with a provisional crown attached to the abutment posts immediately after the gums have been sutured. These implants are subject to immediate loading – not necessarily from direct occlusal loads, but other intraoral forces, including motion of the tongue, lateral movement of adjacent teeth, or passing contact with dentition from the opposing arch. Due to these loads, the implants experience micromotion, and the extent of this motion is an important determinant of the type, quantity, and quality of the **peri-implant** tissues, and consequently, the longevity of the implant.

Micromotion may preclude effective osseointegration and promote the formation of an avascular, collagenous layer at the tissue-implant interface instead (Cameron *et al.*, 1973; Pilliar *et al.*, 1986). This lack of osseointegration has been implicated in the failure of several blade-form implants. However, the formation of a fibrous pseudo-periodontal membrane has also been observed around implants that have survived successfully for more than 20 years *in vivo* (Cappuccilli *et al.*, 2004). In fact, Linkow, one of the pioneers in blade-form implantology, believed that this membrane played a role comparable to the periodontal ligaments, absorbing and transferring occlusal loads more effectively to the bone. Its function as a load-bearing tissue, however, remains unknown. Further, differences in the morphology and composition of the fibrous layer surrounding failed versus successful implants have been reported, with the membrane around failed implants comprised of a more inflamed and granulated fibrous tissue (Meenaghan *et al.*, 1974). Though a fibrous layer surrounds most blade-type implants, not all fail to osseointegrate – several animal and human studies have demonstrated direct bone-to-implant contact in blade-form devices (Smithloff *et al.*, 1975; Trisi *et al.*, 1993). Failure of osseointegrated implants has been linked to bone resorption around the shoulders of the blade, just below the transgingival region. This resorption, based on both experimental work and finite element models, occurs in the regions with the highest stress gradients, and has been attributed to persistent overload and microdamage of the bone. Much like in natural teeth, once sufficient resorption occurs, excessive implant mobility eventually leads to device failure.

15.4.6 Root-form endosteal implants

The first major advancement in the design of root-form implants in the 20th century came in 1913 – Dr. Edward Greenfield used an iridio-platinum wire to create a latticed, hollow-basket implant which was placed into a trough in the alveolar bone. The trough was created with a **trephine** (cylindrical blade) that was the same diameter as the implant. Greenfield believed that bone from the periphery of the basket and within its core would unite after growing into the interstitial spaces, anchoring the implant in place. At the top of the basket Greenfield soldered a gold plate that sat above the mucosa, with a slot machined into the upper surface of the plate to permit later attachment of a crown (Figure 15.12). Greenfield's was one of the first submerged implant bodies which also attempted to take advantage of the concept of implant anchoring via osseointegration (Greenfield, 1913). Though Greenfield claimed that the implants lasted for years, many of them failed due to the overload they experienced from the large bridge prosthesis Greenfield attached to them (Ring, 1995b).

Though Adams was the first to patent the design for a threaded implant body in 1937, the Strock brothers from Boston were the first to successfully implant a true screw-type root-form implant in 1939. The original implant, which was a one-stage implant composed of CoCr, was placed in both dogs and humans. Through their work with animal models, the Strocks were the first to provide histological evidence of bony ingrowth at the bone-implant interface (Strock, 1939). The brothers later went on to design and implant the first submergible, two-stage titanium screw implant in 1946 as well. Several of their implants functioned successfully for over 15 years and became the template for future implant innovators. The 1940s also saw the invention of helical wire implants

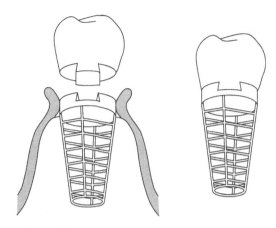

Figure 15.12

Schematic of Greenfield's hollow basket root-form implant. (After Greenfield, 1913.)

Figure 15.13

Schematic of Formaggini's spiral implant, which consisted of a wire coiled around itself.

(Figure 15.13) of stainless steel and tantalum. These implants, developed by Formaggini, were fabricated at the time of surgery by bending a wire back on itself to form a helical spiral. The two ends of the wire were soldered together to form an abutment post to receive the prosthesis. Formaggini's spiral implants met with limited success for several reasons. Since the implants were often hand-shaped, they did not typically conform well to the patient's anatomy. Further, bending the wires weakened them, making them more susceptible to fracture *in vivo*. Last, the spires near the alveolar crest were often infiltrated with fibrous tissue rather than bone. Several attempts at improving the spiral design by casting the metal implant, narrowing the spire width, and using a double helix were attempted, but none gained widespread clinical success (Linkow, 1990).

Figure 15.14

Schematic of Branemark's original titanium implant.

In 1958, Raphael Chercheve was the first to develop a "sleep away" implant that consisted of an internally threaded hollow screw that was meant to "sleep" submerged into the bone for several months to permit healing, after which the implant would be re-exposed and fit with a screw bearing prosthesis. Though the procedure was plagued by soft-tissue damage during the time of prosthetic insertion, it was the precursor to the Branemark implant.

The seminal development in dental implantology is attributed to Dr. Per Ingvar Branemark of Sweden, whose laboratory accidentally discovered the osseointegrative properties of titanium in 1952. After their initial discovery, Branemark's team conducted research for more than a decade on the biocompatibility of titanium in animal models, hitherto unprecedented. Due to the accessibility of the oral cavity, Branemark chose to focus his efforts on dental rather than orthopedic applications, and in 1965, he implanted his first set of titanium dental implants – four implants into the mandible of a 34-year-old man born with chin and jaw deformities (Branemark *et al.*, 1969). The implant was a submergible, two-stage straight-walled threaded implant (Figure 15.14). All the components, including the implant body, the abutment, and abutment screw, were made of pure titanium. In the 15 years following his first surgery, Branemark and his team continued to experiment with and optimize designs, primarily with tubular and screw titanium implants (Branemark *et al.*, 1977). After commercializing his designs in 1978, Branemark presented his clinical results from the previous 17 years at the Toronto Conference for Osseointegration in Clinical Dentistry in 1982 (Branemark, 1983). There, he also outlined different surgical techniques he felt were requisite for implant success. These included the use of a low-speed hand piece for bone drilling to minimize bone damage and the adoption of two-stage surgeries in which implant bodies was first allowed to heal and osseointegrate, submerged, for 3–6 months, before the implants received an abutment or prosthesis.

Contemporaneously, various self-tapping implant designs were fabricated by different researchers, including Linkow's vent-plant implants (Linkow, 1990). At that time, groups

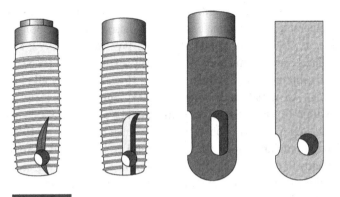

Figure 15.15

Different types of root-form implants with surfaces roughened through plasma spraying, sand-blasting, or bead sintering.

of researchers with the International Team of Oral Implantology (ITI) were also experimenting with hollow cylinder and hollow screw implant bodies. These were designed to minimize the amount of metal, and several studies reported a 10-year survival rate of over 90% (Buser *et al.*, 1997). These researchers also tried to improve osseointegration through the introduction of implants with surfaces roughened via plasma-spraying (Ring, 1995b). Though sand-blasting and bead sintering were later adopted in place of plasma-spraying (Albrektsson and Sennerby, 1991; Albrektsson *et al.*, 2008), roughened surfaces became commonplace, particularly in cylindrical bodies, which require a surface-based fixation mechanism (Figure 15.15).

Surface modifications with ceramic coatings were also attempted – alumina- and zirconia-coated root-form implants were introduced in 1975 (Cranin *et al.*, 1975), and further efforts to enhance bony ingrowth also led to implants with coatings of hydroxyapatite. However, since the coating was found to flake or crack during implant insertion, promoting greater plaque formation and enhancing bacterial retention, their clinical efficacy remains inconclusive (Buser *et al.*, 1991; Cooley *et al.*, 1992; Takeshita *et al.*, 1996; Taylor *et al.*, 1996).

The use of all-ceramic implant bodies dates back even further. Ceramics materials were attractive for two main reasons. First, most ceramics are white colored, so they mimic the natural look for teeth better than metals do. Second, ceramics provide an alternative for patients with metal allergies or sensitivity to titanium. Research in dental ceramics has focused on three different materials: polycrystalline aluminum oxide (alumina), single crystal aluminum oxide (sapphire), and yttria-stabilized zirconium oxide. Vitreous carbon implants also generated some excitement, albeit short-lived, as discussed in greater detail in the Case Studies section of this chapter.

The Crystalline Bone Screw (CBS), made of alumina, was first produced by the Swiss dentist S. Sandhaus in 1967, who believed that the physiochemical similarities between ceramic and bone would produce a stronger bone-to-implant interface (Linkow, 1990). Though pre-clinical experiments verifying Sandhaus's assertion did not exist at that time, later studies investigating ceramics as dental materials showed that they possess comparable, and in some cases, better, osseointegrative potential than titanium (Depprich *et al.*, 2008; Dubruille *et al.*, 1999; Hayashi *et al.*, 1992; Kohal *et al.*, 2006). After Sandhaus, several manufacturers, many of them in Europe, began producing alumina and sapphire dental implant systems. However, nearly all clinical trials proved disappointing. An initial study with the CBS found that six out of eight implants fractured within 6 years of service, while another observed a 23% success rate, with failure associated with loss of bone-implant contact (Brose *et al.*, 1988; Diedrich *et al.*, 1996). Though sapphire, which does not fracture as readily, had more favorable survival rates, especially in the mandible, the implants failed to osseointegrate during healing or lost integration, at rates significantly higher than those observed in titanium (Berge and Gronningsaeter, 2000; Koth *et al.*, 1988; Steflik *et al.*, 1995). With the recent withdrawal of the Bioceram (Kyotera, Japan) sapphire implant from the market, neither alumina nor sapphire implants bodies remain commercially available. On the other hand, there are presently five manufacturers of zirconia implants, such as Ceraroot and Incermed (Andreiotelli *et al.*, 2009). Many of these implants have been on the market for nearly a decade. However, there are surprisingly few investigations of their short or long-term *in vivo* performance. Although no randomized clinical trials exist, data from a few 1-year follow-ups suggest that the early survival rate of zirconia implants in the mandible is over 90%, and that most of the failed implants are lost due to implant mobility during the initial healing phase (Oliva *et al.*, 2007). Despite these initial promising results, recent reviews have cautioned against the widespread adoption of zirconia implants until more definitive evidence of their long-term safety and efficacy is available (Andreiotelli *et al.*, 2009).

In the last few decades, four basic geometries for root-form implant bodies have emerged: basket, hollow cylinder, solid cylinder, and screw, most of which are manufactured with titanium or titanium alloys, with either a smooth or roughened surface. Though cylindrical designs are more easily implanted, particularly into weaker bone, they are more susceptible to failure from occlusal overload since the shear strength of the bone-implant interface is the primary source of implant stability (Maniatopoulos *et al.*, 1986). Threaded screw type designs are now favored largely due to their greater bone-implant contact area, which may be manipulated through the thread pitch, profile, and depth, and the ease with which they may be relocated, if necessary, during surgery. Though the threads create stress concentration at the bone-implant interface, these too can be mitigated through thread design. Further, screw-type bodies may either be parallel walled or tapered to better accommodate different bone morphology. Like the body, the apex in root-form implants may also be straight or tapered, have flat side walls or grooves, and have vents of different geometries, all which optimize bony ingrowth and help resist torsional loads. The crest module, which creates a transition zone for load transfer between the abutment and the implant body, may be machined, roughened, or threaded, and may receive the abutment screw through an internal or external connection that may be rounded or hexagonal, which provides greater rotational resistance. Since the crest module is situated in the region with the largest stress gradients, the bone around the crest module is the most susceptible to resorption (Misch, 2008), and several design changes of the crest, including angling of the module walls, minimizing the length of the crest, and increasing the module diameter relative to the body and apex have been incorporated in an effort to curb implant-induced crestal bone atrophy.

To accommodate different jaw sizes, implants bodies may further be categorized by width, which generally varies from 3.5 mm to 5 mm, and by height, which typically ranges from 7 mm to 16 mm. Though optimal implant size is largely determined by the height and width of the underlying alveolar bone, the largest implant that may be accommodated may not always be the most advantageous. Wider implants may provide a greater surface area and decrease stresses surrounding the implant (Himmlova et al., 2004; Ivanoff et al., 1997), but do not necessarily enhance bone to implant contact, adjacent bone density, or implant survival (Brink et al., 2007; Ivanoff et al., 1997). When anatomically feasible, longer implant lengths (>9 mm) are preferred – the greater number of threads and a deeper apex provide greater surface area for osseointegration. However, alveolar ridge resorption and ill-positioned anatomic structures such as the mandibular canal and mental foreman may contraindicate the placement of standard length implants without bone augmentation techniques, such as bone grafts. Further, after a certain length, longer implants do not significantly reduce bone-implant stresses (Anitua and Orive, 2010). Thus, in some cases, short implants (<8.5 mm) may provide an attractive option. Despite the high rates of failure associated with early clinical results, improved surgical techniques have increased short implant survival to the levels cited for standard implants, and, with appropriate treatment planning, may be used in patients for whom longer implants are contraindicated (Romeo et al., 2010).

Like implant bodies, abutment designs have also transformed from short, bulky cylindrical posts plagued by bacterial infiltration and mucosal inflammation to longer, leaner, tapered geometries. In one-piece abutments, the lower portion of the abutment typically consists of a thin, threaded pin which screws directly into the internal threads of the implant body. In two-piece abutments, the portion that receives the prosthesis is anchored to the implant body via an abutment screw. The geometry of the abutment head varies depending whether the prosthesis will be retained via a screw, cement, or an attachment (Figure 15.16). Single tooth crowns are typically cemented to the abutment, while multiple crowns and larger prostheses are usually attached with screws or snap-in attachments (Misch, 2005).

Abutments are typically composed of the same materials as the implant body, including titanium, titanium alloys, CoCr. More recently, metal implant and zirconia or alumina abutment combinations have been explored, primarily because ceramic abutments provide greater aesthetic appeal (Andersson et al., 2001; Nakamura et al., 2010; Rasperini et al., 1998). Like their implant body analogues, however, concerns about premature fracture persist for ceramic abutments.

Today, several hundred different implant body and abutment design combinations are available from over 30 different manufacturers, most notably Nobel BioCare, Astra Tech, Straumann, Zimmer, and Lifecore. The choice of implant body and abutment design, material, and size is governed primarily by four patient-related factors: (1) the width and depth of the underlying bone, (2) density of the alveolar bone, (3) health

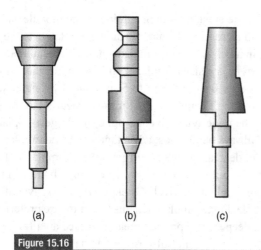

Figure 15.16

Abutment geometries for prosthesis retained through (a) a screw, (b) cement, or (c) attachment.

of the mucosal tissues, (4) position of the missing teeth within the oral cavity and in relation to adjacent teeth.

Since the introduction of the Branemark system opened the flood-gates of dental implantology in the 1980s, over 7 million root-form dental implants have been implanted with very high success rates. Long-term survival rates for titanium implants of over 95% at 10 years for mandible and 90% for maxilla are common. The success of root-form implants varies with the morphology and density of the underlying alveolar bone – implants in the mandible, which has more bone, have traditionally been more successful than those in the maxillary (Albrektsson et al., 1988; Buser et al., 1997). Further, implantation in dense bone is more successful than weaker bone (Bass and Triplett, 1991; Jaffin and Berman, 1991), and consequently, posterior implants, where bone is least dense, have had less success than anteriorly placed implants. Further, the type of restoration may also influence outcomes, with single-root implants less likely to fail than those supporting fixed bridges or overdentures (Brocard et al., 2000; Noack et al., 1999).

The majority of failures of root-form implants occur during healing or in the initial year of function. Failure during healing typically refers to a failure of the implant to sufficiently osseointegrate. The failure during healing may have a biological origin – improper surgical technique, poor oral hygiene, or recurring periodontal disease, all of which promote bacteria-induced inflammation and osteolysis; or, the failure may be mechanical in nature – stresses on the implant from improper placement during surgery or from **parafunctional** oral activity such as clenching, teeth grinding, or tongue-thrusting may cause micromotion and the subsequent formation of a thick, fibrous tissue which does not effectively transmit loads to the implant. In both cases, the implant remains mobile.

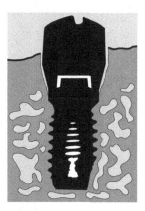

Figure 15.17

Typical v-shaped resorption of crestal bone that occurs around a root-form implant.

Even after sufficient rigid fixation has been achieved, the implant may still become mobile after the restoration is attached and the prosthesis is loaded (Esposito *et al.*, 1998). Root-form implants are particularly susceptible to early bone loss around the crestal region of the alveolar bone, especially in weaker bone. The bone resorbs in a U- or V-shaped pattern, starting at the gum margin (where bone loss is greatest) and culminating at the first few threads of the implant body, irrespective of the design and length of crest module and implant body (Figure 15.17). The **crestal** bone loss may be marginal, or it may be sufficient to cause implant mobility and failure. Following crestal bone loss, failure of the implant due to bacterial invasion is also an increased concern (Wyatt and Zarb, 2002).

Bone atrophy in that region has been attributed largely to the supra-physiologic loading of the crestal bone. Photoelasticity experiments and finite element models of the stress contours around the bone-implant interface reveal that both stresses and stress gradients are greatest near the crestal region of the implant, and, unsurprisingly, form a U- or V-shaped pattern (Figure 15.18). Though the stresses are typically well below those necessary to cause bone fracture, they result in pathological overload and excessive tissue damage. The crestal bone loss is most severe during the first year of loading, often 1–3 mm, after which the atrophy normally subsides to less than 0.2 mm/year. The loss does not continue unabated, since the bone in contact with the inferior regions of the implant typically experiences **anabolic** functional loads that result in remodeling, accommodating the loads initially borne by the crestal bone. Similarly, progressive implant loading has also been found to inhibit crestal bone loss (Misch, 2008).

After the first year, biological failure may still result from poor oral hygiene or infection. Late-stage mechanical failure may be associated with fracture of the implant body, abutment screw, or abutment post, or with abutment screw loosening. Abutment screw loosening is typically a consequence of lateral loading on the prosthesis, which imparts a static and dynamic torque around the abutment screw axis. These loads may

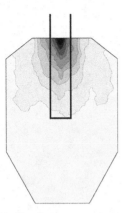

Figure 15.18

Stresses in the alveolar bone surrounding a loaded root-form implant. The stresses increase in the vicinity of the implant and are highest immediately surrounding the crestal region. (After Misch, 2008.)

cause rotation and loss of function of the restoration, or fatigue fracture of the implant-abutment connection. Such fatigue fractures of the abutment connection are common, particularly in implants with external hex crest module geometries (Steinebrunner *et al.*, 2008), and in single-tooth implants where the restoration has been **splinted** (attached) to a healthy adjacent tooth. Since the implant-supported prosthesis deforms and shifts less under occlusal loads than the natural tooth to which it is splinted, the natural tooth essentially becomes a cantilever applying moments on the implant (Shackleton *et al.*, 1994). There is also greater marginal bone loss around a cantilevered implant. Fatigue fracture of the implant body, though rare, may also be associated with malocclusion or parafunctional oral activity, as discussed in greater detail below. Last, blunt fracture of the different implant components may also occur due to trauma or injury.

Up until 1986, blade-form implants were the most popular type of dental implants. Since the area of the bone-implant interface was much greater in blade-form than root-form implants, occlusal loads could be distributed over a greater surface, making blade-form implants viable in moderately and even severely atrophied bone, so long as sufficient bone height was present. However, root-form implants overtook blade-form implants in popularity for several reasons, some of which remain controversial. Though high survival rates were noted in a few animal studies (Natiella *et al.*, 1973), very few osseointegrated, and retrospective studies report a less-than-favorable performance of blade-implants (Cranin *et al.*, 1977; Smithloff and Fritz, 1987). In addition, many dentists and surgeons claimed that single-piece blade-form implants never achieved true osseointegration, and that their success was dependent upon an unforgiving surgical technique that required creating a precise vertical channel in the bone. Further, since fewer blade-form implants were necessary to rehabilitate completely edentulous patients than root-form implants, neither dentists, surgeons, nor device manufacturers had financial incentives to promote their use. Finally, only root-form implants were reclassified by the FDA as Class II devices in 2002 (FDA, 2004).

Though the FDA had been tasked with regulating medical devices starting in 1976, only in 1987 were both subperiosteal and endosteal dental implants officially classified as Class III devices. Since both types of implants had been in use prior to 1976, new dental implants were eligible for approval through the 510(k) notification process if they could prove substantial equivalence to pre-amendment devices. Based on the ubiquitous use and long-term survival rates of root-form implants, the FDA decided to reclassify them as Class II devices. Blade-form endosseous and subperiosteal implants both remained Class III implants. In 2009, the FDA finally required a PMA submission for all pre-amendment blade-form devices as well.

15.4.7 Structural aspects

Healthy teeth experience both static and dynamic loads during mastication, swallowing, and during parafunctional activities such as teeth clenching and grinding. Occlusal forces depend on the type of tooth, and are heavily influenced by age, gender, genetics, and number of teeth present. Though biting forces are significantly diminished in individuals with removable and fixed full or partial dentures, dental implants experience over 90% of the occlusal loads measured in the natural dentition (Misch, 2008).

Most of the forces to which teeth are subject are due to mastication – approximately 1000 chewing cycles/day, experienced in short bouts. Masticatory loads, which are concentrated in the posterior region, are primarily compressive and applied along the long axis of the tooth, orthogonal to the occlusal plane. These forces vary with type of food, and, depending upon the device used to measure them, typically range between 100–250 N in the molars, 40–150 N in the premolars, and less than 50 N in the incisors (Bates *et al.*, 1976; Howell and Brudevold, 1950; Richter, 1995). The maximum bite force (MBF), measured as the maximum force generated during direct contact between the maxillary and mandibular dentition, is significantly greater. Like masticatory forces, the MBF is highest in the posterior molar region and decreases progressively the further anterior the teeth. However, it is much more variable than masticatory forces, and has been measured in the range of 300–1000 N for the molars, with only 30% of those loads observed in the incisal region (Braun *et al.*, 1996; Raadsheer *et al.*, 1999; van Eijden, 1991). Supra-physiologic loads as high as the MBF occur rarely, and account for negligible percentage of the loading experienced by teeth, both natural and artificial.

On the other hand, the occlusal overloads from involuntary or psychological parafunctional activities, particularly clenching and grinding, pose a much greater risk of causing implant damage and failure. **Bruxism** is a cyclic, nonfunctional gnashing, or grinding, of the teeth. Typically, the shear component of masticatory forces is much less than the axial component (de Las Casas *et al.*, 2007). However, bruxism exerts excessive shear force on teeth, comparable in magnitude to vertical forces experienced while chewing. Further, while masticatory loads are only present for about 20–30 minutes/day, forces from bruxism may persist for several hours, particularly at night. These repetitive high-magnitude shear forces are potentially detrimental to both early- and late-stage dental implants. The shear forces experienced by teeth adjacent to the implant are also transferred to the bone-implant interface. The resulting micromotion may delay or inhibit effective osseointegration, or initiate bone loss, all of which preclude implant function. Even after an implant has become rigidly fixed and a prosthesis is in place, bruxism may induce bending moments on the prosthesis that cause implant mobility (e.g., abutment screw loosening) or cause fatigue failure of the implant body or the abutment screw.

Whereas bruxism largely exerts shear forces, clenching essentially applies a static compressive load several times greater than normal masticatory forces for long durations. Clenching is particularly problematic for early implant healing when removable complete or partial dentures are also present over the implant – clenching forces also induce motion and stress around the bone-implant interface. Overloads from clenching may also enhance existing fatigue damage from normal occlusal activity.

In addition to biting forces, teeth experience minor loads from the masticatory muscles and the tongue during swallowing and talking. These are primarily shear forces, and are much lower in magnitude than occlusal forces, reaching only as much as 5–10 N. Teeth experience fewer than 500 cycles of these minor shear loads daily.

15.4.8 Functional requirements

The first official set of guidelines establishing criteria for success in dental implants were documented at the Toronto Conference for Osseointegration in Clinical Dentistry in 1982 (Albrektsson and Wennerberg, 2005) and are summarized in Table 15.1. Like other load-bearing orthopedic implants, dental implants must meet several functional requirements to be deemed successful.

15.4.9 Medical indications and contraindications

Medical indications for subperiosteal implants include edentulism in one or both arches coupled with marked alveolar bone atrophy, and complete or partial edentulism with mild or moderate bone atrophy for endosteal implants. Contraindications for mandibular implants include insufficient alveolar height above the mandibular canal and the inferior alveolar nerves. Contraindications for both maxillary and mandibular implants include decay in the teeth and infection in the periodontal tissues surrounding the edentulous region, which may enhance inflammation around the implant, recent myocardial infarction or stroke, valvular prosthesis surgery, **immunosuppressant** drug use, active infection, or allergies to any implant components. Though not definitively deemed contraindications, bruxism, high blood pressure, type II diabetes, smoking, and intravenous **bisphosphonates** use are considered risk factors that may severely impair implant integration and soft-tissue healing.

Table 15.1 Functional requirements for success in dental implants (Albrektsson and Wennerberg, 2005)

General objective	Description
Restore function	Occlusional loads should be restored to at least 80% of those generated by natural, healthy teeth during mastication. Specifically: Incisors – 10–50 N Canine – 50–100 N Molars – 100–200 N Fit comfortably and naturally in the patient's mouth The implant should be available in many different sizes to accommodate a wide range of patient anatomy
Restore oral aesthetic	Implants should be able to receive natural-looking crowns and bridges
Maintain structural integrity	The implants should be able to withstand maximum bite forces in compression and bruxal shear forces without fracturing or deforming excessively The implants should be able to withstand over 300,000 chewing cycles per year for at least 20 years, with 99% of the load cycles between 5 N and the normal chewing force of the tooth and 1% of load cycles between 5 N and the MBF of the tooth The implant must also be able to withstand over 100,000 cycles/year of dynamic shear loading between 0–10 N for 20 years.
Minimize damage to surrounding tissues	The devices should be implantable without requiring excessive removal of or damage to surviving teeth, surrounding periodontal tissues, and the mandibular canal (alveolar nerves) or mental foreman Heat generation from drilling and tapping should be limited to minimize tissue necrosis Transmit sufficient occlusional loads to the alveolar bone to prevent resorption Guidelines created at the Toronto Congress dictate that less than 1 mm of vertical bone must be lost in the first year, and no more than 0.2 mm each year thereafter
Be biocompatible	Components, in both bulk and particulate form, should not elicit a chronic inflammatory response, local or systemic Components should not be toxic
Have adequate fixation	There should be no movement of the implant at the bone-implant interface
Permit peri-operative visualization	Component should be radiolucent to facilitate periodic imaging

15.5 Temporomandibular joint replacements

15.5.1 Clinical reasons for TMJ replacement

Working in concert with the temporomandibular ligament and several facial muscles, the TMJ permits translational and rotational motion of the jaw. In particular, the articular disk provides a stable platform for movement of the TMJ – articulation of the mandibular condyle with the posterior surface of the disk permits rotation, while gliding of the disk and condyle against the fossa enables translation. Proper articulation of the TMJ is critical for effective mastication and unimpeded speech.

Like other load-bearing joints, including the hip, knee, and shoulder joints, the TMJ is susceptible to many different disorders which can cause dysfunction. An estimated 5–15% of the population is affected by TMJ disorder (NIDCR). Though dysfunction is mostly associated with inflammation of muscles and ligaments around the joint or with displacement/dislocation of the interpositional disk (internal derangement), the primary disorders that necessitate TMJ replacement include both rheumatoid and osteoarthritis, **ankylosis**, congenital deformities, and trauma (Kiehn et al., 1979; Mehra et al., 2009; Moriconi et al., 1986; Worrall and Christensen, 2006).

Rheumatoid arthritis, which is largely considered a genetic disorder, is a systemic, autoimmune condition that causes chronic inflammation of the joints and the surrounding tissues. It preferentially attacks the synovium, causing thickening of the tissue and an overproduction of viscous synovial fluid. This reduction in lubricity hinders joint motion, while the tissue inflammation stimulates osteolysis, both of which can contribute to joint pain and loss of function. Analogous to the cartilage degeneration that characterizes osteoarthritis (OA) in the knee and the hip, OA in the TMJ is accompanied by a loss of the aggregated proteoglycans, an increase in fluid content, and reorientation and fibrillation of the collagen fibers, all of which hinder its load-bearing capacity. The damaged cartilage is eventually worn away by repetitive loading, resulting in painful bone-to-bone contact between the mandibular condyle and the temporal fossa. Most OA is simply age-related, but may also be associated with trauma or be hereditary.

Ankylosis is abnormal joint immobility caused by consolidation of the joint, and is typically bony or fibrous in nature. Bony ankylosis is a fusion of the mandibular condyle and the temporal fossa due to an irregular proliferation of osteoblasts in the joint, while fibrous ankylosis is the excessive growth and consolidation of the fibrous tissues in the joint. Ankylosis may result from a congenital defect, from the prolonged inflammation associated with rheumatoid arthritis, trauma-induced internal bleeding, or a previous surgery (Bumann and Lotzman, 2002).

Though tooth decay is continuous, tooth loss is binary, with all surgical and non-surgical treatment options focused on providing some form of tooth replacement. On the other hand, TMJ disorders are typically progressive, and may require multiple non-surgical or surgical treatments, each with a distinct purpose or targeted tissue. Early TMJ degeneration associated with mild rheumatoid or osteoarthritis, or early-stage ankylosis, may be treated with non-surgical intervention, including a soft food diet, drugs to alleviate inflammation, muscle relaxers, or splints to reposition the condyles within the joint space. For more severe TMJ decay, surgical treatment may be required to restore function. However, the small size and anatomic complexity of the joint complicates surgery, and only 1% of people with symptomatic TMJ disorders undergo surgery.

Primary surgical interventions include arthroscopy, **menisectomy**, and partial and total joint arthroplasty. During arthroscopy, an **arthroscope** is inserted via a small incision into the joint cavity. The arthroscope is not only used to visualize tissues, but has precision surgical tools fitted at the end that may used to irrigate the joint, to inject anti-inflammatory medications into the joint space, to perform **debridement** to clean away damaged segments of cartilage, or to **resect** unhealthy fibrous tissues such as the synovium. Typically, arthroscopy is preferred over open surgery since it is minimally invasive, is associated with fewer complications, and requires a shorter hospital stay. However, the failure of arthroscopy to alleviate pain and restore function may necessitate the more invasive open surgery. **Menisectomy** is surgical removal of the articular disk which may or may not be accompanied by interpositional arthroplasty, or the implantation of a permanent disk implant. Placement of the implant helps to maintain functional ramus height and limit excessive mandibular mobility (Mayo Clinic). Joint arthroplasty is the placement of a partial or total TMJ replacement. Despite difficulties and potential complications, hundreds of TMJ replacements are implanted each year. A modern TMJ replacement typically consists of two components: a temporal component that replaces both the natural glenoid fossa and the interpositional disk as an articular bearing, and a mandibular component, which attaches to the lateral side of the ramus, and replaces the natural condyle as the counter-bearing. The fossa component may also be comprised of two parts – a piece that fits over the fossa and attaches to the cheekbone, and a bearing insert, similar to the interpositional disk, that attaches to the fossa piece and serves as the articular surface.

15.5.2 Historical development of TMJ replacements

From the earliest recorded placement of a TMJ implant in the late 19th century, until the late 1960s, the primary focus of TMJ surgery was on the treatment of ankylosis, largely due to its high rate of recurrence. This early era of TMJ prosthesis development took place primarily in France and Germany. Though records of resectioning of the TMJ can be traced back to the early 19th century, the earliest alloplastic TMJ implant dates back to 1898, when Rosner placed a gold plate into the TMJ to prevent reankylosis after resectioning of the condyle. Orlow later modified Rosner's method by placing gold-coated aluminum plates on resectioned bone surfaces in 1903 (Driemel *et al.*, 2009). Orlow's short-term post-operative success encouraged further experimentation with alloplastic interpositional materials, including gutta-percha, ivory, and magnesium, to find materials with a more favorable host response (Blair, 1914; Risdon, 1934). In 1934, Risdon attempted to use gold foil in the fossa (Risdon, 1934), while Eggers reported successfully placing tantalum foil over the fossa and condylar head in 1946 (Eggers, 1946). As in other implants, the use of these metals was soon abandoned, largely due to their unsuitable mechanical performance. In 1955, Gordon inserted a polyethylene cap onto the condyles of seven patients, which all performed without complications during a three-year follow-up period (Gordon, 1958). Despite its success, it was only in the 1960s, with the introduction of **Silastic**, or silicone rubber, that an alloplastic material achieved ubiquitous clinical use for articular disks. Silastic, manufactured by Dow Corning, was first used to replace torn or displaced disks, in the form of a block carved to the shape of natural tissue (Hansen and Deshazo, 1969). After its highly successful short-term performance, Dow also began marketing Silastic HP sheeting, composed of thin Dacron-reinforced Silastic sheets, in 1976. This was followed, in 1985, by the development of a temporary Silastic TMJ disk based on the design of Dr. Clyde Wilkes (The TMJ Association). Around this time however, reports of widespread failure of Silastic implants began to surface (Bronstein, 1987; Eriksson and Westesson, 1986; Eriksson and Westesson, 1992; Ryan, 1989; Westesson *et al.*, 1987; Dolwick and Aufdemorte, 1985). The implants experienced mechanical failure, undergoing marked wear, with the thinned disks more susceptible to tearing and fracture (Bronstein, 1987; Eriksson and Westesson, 1992; Ryan, 1994). The particulate debris and fragments released from the implants also engendered a local and systemic inflammatory response that led to joint degeneration, bony ankylosis, and lymph node swelling (Dolwick and Aufdemorte, 1985). Dow was aware of these deficiencies in performance for more than 20 years before warnings from the FDA and a statement from the AAOMS compelled the company to pull all of its Silastic TMJ products from the market in 1993 (The TMJ Association; AAOMS, 1993). Contemporaneous interpositional implants made of **Proplast**, a composite of carbon and polytetrafluoroethylene, also met a similar fate, as discussed in greater detail in Section 15.7.1.

Though removable devices were also used in the early 20th century as temporary prostheses to enhance wound healing and provide anatomical support, the first permanent TMJ replacement was developed by Konigh and Roloff in 1908. They anchored an ivory prosthesis onto the spongy bone of the mandible with a spike; poor fixation of the implant eventually led to instability and failure. In the 1910s, both Mohring and Partsch also unsuccessfully attempted to fabricate an ivory prosthesis (Driemel *et al.*, 2009).

It was only in 1960s that the first modern TMJ implant design, comparable to those present today, was conceived. Robinson created a box-like stainless steel fossa for relieving TMJ ankylosis (Robinson, 1960). Although it was later replaced by more versatile designs, his idea for a fossa prosthesis inspired many other investigators. In 1963, Christensen designed the first anatomically fitted fossa-eminence implant made of cast CoCr, which could be manufactured in many sizes and shapes (Christensen, 1963). The implant, which attached to the skull with two or three screws, allowed the meniscal disk to remain intact. Christensen later created a condylar prosthesis – an acrylic (PMMA) head fixed to a CoCr plate – to create the first total joint replacement (Figure 15.19) (Vanloon *et al.*, 1995). The rationale for using the PMMA condylar head was to prevent penetration of the head into the cranial fossa, since the acrylic would

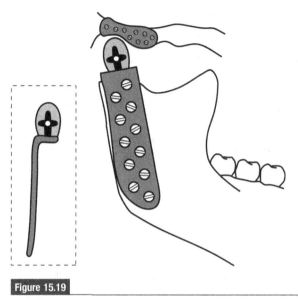

Figure 15.19

Lateral (right) and frontal (left) view of Christensen's early TMJ prosthesis.

wear preferentially over the metal fossa component. Kiehn was the first to utilize PMMA for fixation, which he used to glue both components of a CoCr TMJ prosthesis to the surrounding bone in 1974 (Kiehn *et al.*, 1974). However, the implant provided little joint mobility, in part due the pterygoid muscle loss necessitated by implantation.

In 1971, Morgan modified the design of the Christensen fossa-eminence component to create an implant that only sat on the articular eminence (Morgan, 1971). Soon after, in 1976, he created a condylar component similar to that introduced earlier by Christensen (Morgan, 1988). In the same year, Spiessl also conceived of a new design for a Ti condylar component, known as the AO/ASIF implant, which was comprised of a narrow, L-shaped plate that attached to the ramus and a spherical condylar head that articulated with the natural fossa (Driemel *et al.*, 2009). This design resulted in marked bone resorption of the fossa, leading Lindqvist *et al.* (Lindqvist *et al.*, 1992) to later conclude that such fossa atrophy could not be prevented without implanting a mating glenoid fossa component. Condylar prostheses meant to be implanted using the Charley technique were also designed by Silver in 1977 (Silver *et al.*, 1977) and Flot in 1984 (Figure 15.20). These prostheses had metallic heads and stems that were fixed into the mandibular intramedullary canal after condylar resectioning. Flots's condyle articulated against a metal-backed polyethylene glenoid component, which remained unfixed to the skull to maximize joint mobility (Driemel *et al.*, 2009; Vanloon *et al.*, 1995).

In 1978, Kummoona designed a unique CoCr total TMJ replacement in which the condylar part was cemented between the cortical walls rather than to the exterior of the ramus. Kummoona reported the formation of a thin, fibrous interpositional layer between the two metal surfaces that functioned like a meniscus (Kummoona, 1978).

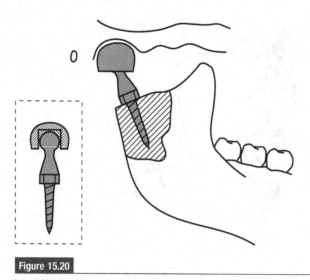

Figure 15.20

Lateral (right) and frontal (left) view of Flot's prosthesis, in which the stem is screwed into the ramus of the mandible. (After Driemel *et al.*, 2009.)

However, the extent to which it bore load was not determined. In 1985, Sonnenburg and Sonnenburg designed a new device which used a polyethylene fossa with a reduced anteroposterior size that was fixed with a single screw and bone cement (Sonnenburg and Sonnenburg, 1985). However, the design met with limited success even in the few patients in whom it was implanted. In 1964, Hahn constructed a ramus prosthesis consisting of an acrylic condyle and a CoCr mesh. The prosthetic mesh was designed to promote the proliferation of fibroblasts that would serve to stabilize the implant following tumor surgery (Hahn, 1964). Similarly, in 1987, Boyne attached a Delrin head onto a titanium mesh condylar prosthesis that was screwed into the mandibular ramus (Boyne *et al.*, 1987).

The 1980s and early 1990s also saw the widespread adoption and rejection of Proplast and Silastic as TMJ materials, which had a lasting impact on regulatory testing, approval, and post-market surveillance of subsequent TMJ replacements. A pilot study by Kent first detailed the use of a porous Proplast coating on a condylar CoCr prosthesis in 1972 (Kent *et al.*, 1972). Proplast, which was initially a composite of graphite and PTFE and later of alumina and PTFE, had been used since 1960 as a filler material in non-loading bearing areas of the body (Wagner and Mosby, 1990). Its porous surface promoted rapid ingrowth of both hard and soft tissues, piquing the interest of the maxillofacial community for its unique fixation properties. Bolstered by its early success, Kent partnered with the company Dr. Charles Homsy of Vitek to develop four different Proplast-Teflon (PT) based TMJ implants. These included the Proplast Interpositional Implant, approved by the FDA in 1983 after proving substantial equivalence to the existing Silastic implants, and the VK glenoid fossa component. The VK fossa was initially a three-layer PT component – a layer of porous Proplast I, which faced the fossa side, laminated to a layer of Teflon-FEP with a polyamide or metal mesh, and a thin, Teflon coating, which faced the condyle. It was modified in 1982 to a two-layer part of Proplast II and Teflon-FEP, at which time the condylar component was also flattened and elongated laterally (Kent *et al.*, 1983). The VK-I and VK-II total TMJ replacements consisted of a CoCr condylar piece articulating again a VK fossa (VK-I) or a Proplast-hydroxyapatite and UHWMPE fossa (VK-II) (Kent *et al.*, 1986; Kent *et al.*, 1993a). With the introduction of Vitek-Kent PT prostheses, TMJ replacement surgery started being performed regularly for a variety of symptoms. Although the early results seemed promising, with a reported 91% success rate (Kent *et al.*, 1986; Vitek, 1986), the continued clinical and radiographic follow-up reported that patients suffered from pain, malocclusion, and foreign body giant cell reaction (FBGCR), resulting in severe degeneration of the tissue surrounding the implant (Feinerman and Piecuch 1993; Lagrotteria *et al.*, 1986; Timmis *et al.*, 1986). The PT prostheses were eventually recalled by the FDA, and the large-scale failure led to the reclassification of TMJ implants as Class III devices in 1993 (FDA, 2001).

Since then, only a few new TMJ devices emerged on the U.S. market. Two of the early models, which were manufactured by Osteomed (Addison, Texas) and Techmedica, adopted many of the standard materials being used in hip and knee joint replacements. The Osteomed model had a polyethylene fossa implant fixed to the zygomatic arch using PMMA and screws, and an adjustable condylar implant modeled after radiographic images. However, the model never received FDA clearance after the reclassification. The Techmedica device was developed for patients that have undergone multiple operations. It consists of a custom-built implant where a 3D CT scan of the patient is used to construct components that conform to the bone anatomy for both the condyle and the fossa. The fossa is composed of titanium mesh layers that permit bone ingrowth and stability, and an articular surface of UHMWPE. The mandibular component is a modular piece with a CoCr head and a Ti-6Al-4V ramus body (Wolford and Mehra, 2000). This device, now marketed as the Patent-Fitted TMJ Reconstruction Prosthesis by TMJ Concepts, Inc. (Ventura, California) received 510(k) clearance in 1997 (TMJ Concepts Inc.) and full approval by the FDA in 1999 (FDA, 1999). A slightly modified Christensen TMJ device, made entirely of CoCr and manufactured by TMJ Implants (Golden, Colorado), has also been on the market since 2001(TMJ Implants Inc.). A prosthesis by Endotec (Orlando, Florida) which uses titanium nitride to harden both the condylar and fossa surfaces has been available since 1995 (Butow *et al.*, 2001), though the device is still classified as an investigational device and is awaiting full FDA approval. Lastl, the Lorentz/Biomet (Warsaw, Indiana) device, composed of UHMWPE fossas component and a CoCr condylar component with a titanium surface, has also received recent market approval. All three of the existing FDA devices have a spherical condylar head (Guarda-Nardini *et al.*, 2008).

Despite stricter regulations and more stringent test guidelines for TMJ replacements enforced by the FDA since the Silastic and PT recalls, the market of TMJ implants has not yet stabilized in the same way that the market for hip and knee replacement has matured. The number of TMJ replacements dropped markedly after the recall. Though the number of annual TMJ arthroplasties is not clear, a much lower failure rate is now observed in TMJ replacements out to 10 years (Guarda-Nardini *et al.*, 2008).

15.5.3 Structural aspects

Despite early controversy about whether the mandible functioned as a link or a lever, and thus whether the TMJ was a load-bearing joint (Gingerich, 1979; Hylander, 1975), it is now commonly accepted that the jaw joint experiences load, acting as a fulcrum for force generation during mastication. The TMJ, and the articular disk in particular, are subject to a combination of different loading regimes during mastication, speech, and clenching and bruxism, all of which generate tensile, compressive, and shearing forces (Tanaka *et al.*, 1994). An appreciation of these complex loads is critical in the design of new TMJ implants.

Though there is disagreement on the ratio of the load transfer, the resultant forces on the TMJ from masticatory loads vary depending on the state of the dentition and the origin of the forces within the oral cavity. For example, 90% or more of occlusal forces from the incisal region are transferred to the TMJ. On the other hand, less than 70% of molar forces may be transferred, with forces directed primarily toward the central part of the intermediate region of the interpositional disk (Abe *et al.*, 2006). Similar bite-force to TMJ load ratios and loading directions exist for clenching.

Since the articular surfaces of the TMJ are highly incongruent, joint contact areas are small, which may lead to significant stress during joint loading. Deformation of the disk during loading is believed to absorb and spread the load over a larger contact area. Though the principal stresses in the condylar and the fossa articular surfaces are compressive, finite element analysis reveals that the disk experiences both tensile and compressive stresses, with principal stresses not necessarily corresponding to regions with the greatest von Mises stresses (Tanaka *et al.*, 1994). Since most jaw loading is dynamic, the TMJ primarily undergoes cyclic loading, with the duration and number of cycles comparable to those on experienced by teeth. Since the age of TMJ implant recipients varies considerably, with many patients in their thirties, TMJ implants must be able to withstand several decades of loading cycles.

15.5.4 Functional requirements

Successful TMJ prostheses must satisfy several functional requirements, which are summarized in Table 15.2.

15.5.5 Medical indications and contraindications

Medical indications for TMJ replacements include rheumatoid and osteoarthritis, malocclusion, internal disk derangement that had not responded to other forms of treatment,

Table 15.2 Functional requirements for TMJ replacements

Requirement	Description
Restores function	Kinematics of joint should be restored to at least 50% of healthy kinematics in translational and rotational directions to allow for mastication and speech. Based on anatomical estimates: Rotation in the sagittal plane should allow patient to open mouth Rotation in the transverse plane Sufficient medial-lateral translation of the jaw Anterior-posterior translation of the jaw Device must not dislocate under normal loads
Maintains/restores anatomic height	Device must maintain or restore anatomic height of ramus and condyle anatomy
Encourages patient fit	Device must fit a wide range of patient anatomy, including unusual features caused by previous revision surgery
Is implanted through a single surgery for reduced medical risk	Parts must be immediately available to surgeon in operating room Part must be able to be implanted immediately after extraction of a failed device during revision surgery
Minimizes damage to surrounding tissue	Design must be implantable without damage to oratory canal, nerves, brain tissue, or surrounding bone Temperatures due to any exothermic reactions must avoid permanent damage leading to necrosis of the tissues Bone must experience 80% of normal stresses to minimize stress shielding
Has structural integrity	Components must not fracture under compressive joint loads of 800 N and shear loads of about 200 N. Components must survive a minimum of 400,000 cycles/year under assumed cyclic loading conditions 20–400 N, with occasional cycles of overload between 20–800 N
Is biocompatible	Components must not be toxic to the patient Components must not trigger a chronic foreign body response
Maintains fixation	Components must remain in place, i.e., no rotation or translation at the bone-implant interface

recurrent fibrous and/or bony ankylosis, malocclusion, loss of alveolar bone due to bone resorption, and atrophy of masticatory muscles. Medical contraindications may include parafunctional activity, such as teeth clenching and/or bruxism, punctures through the glenoid fossa, active or suspected infections, and known allergies to any of the constituent materials.

15.6 Case studies

15.6.1 Failure of vitreous carbon implants (material failure)

Vitreous carbon is formed when a crosslinked polymeric resin, typically a phenol formaldehyde or furfural based polymer, is subjected to temperatures of 1000–3000°C in the presence of a gaseous phase with a controlled composition. The process results in systematic thermal degradation of the non-carbon components, leaving behind a material that is 99.9% pure elemental carbon. Unlike its diamond and graphite counterparts, vitreous carbon has a completely amorphous lattice structure (Cowlard and Lewis, 1967).

Vitreous carbon was originally developed for the aerospace industry, and, unlike other ceramics, it is a dark, glassy material that does not visually mimic teeth or bone (Schnitman and Shulman, 1980). However, vitreous carbon does possess several attributes that could be tailored to make it attractive as a potential dental material, including biological compatibility, corrosion resistance, low fluid permeability, high strength and hardness, and an elastic modulus comparable to mineralized tissues (Cowlard and Lewis, 1967; Fraunhof et al., 1971; Hucke et al., 1973; Pierson, 1993).

Such modulus-matching was not only thought to protect alveolar bone from stress-shielding induced resorption, but also to mitigate shear stresses at the bone-implant interface. Since the bony interface is weakest in shear, Vitredent, the now bankrupt company that manufactured the vitreous carbon dental implants, believed the decreased shear stresses would limit implant failures associated with lack or loss of osseointegration (Vitredent Corporation, 1973).

After preliminary experiments established the biocompatibility of vitreous carbon implant materials in the late 1960s (Benson, 1972), several animal studies with dogs, and later primates, were conducted to investigate its *in vivo* performance (Grenoble et al., 1975; Stallard and El Geneidy, 1975). The implants utilized for the early canine trials were comprised of a hollow stainless steel sleeve that was cemented into the center of a conical-shaped root-form vitreous carbon implant with rounded or triangular annular grooves. Except for the most coronal 2 mm of the crest module, the surface of the implant was roughened via sandblasting. The steel sleeve accepted a stainless steel abutment post, onto which a prosthesis could be cemented. Histological studies reported an absence of inflammation and infection in the peri-implant tissues of both unloaded implants and implanted loaded after several weeks of healing. Though bone grew in close proximity to the implants, a thin, fibrous membrane often encapsulated the implants (Grenoble et al., 1975; Markle et al., 1975). Further, the success of the loaded implants varied, with several of stainless steel abutment posts fracturing through the vitreous carbon bodies. Subsequent animal experiments with baboons cited a 70% implant survival rate (Schnitman et al., 1980). Despite the 30% failure, primarily due to post-healing loss of osseointegration, the survival was highest that had been reported to date for dental implants in baboons.

Figure 15.21

Modified design of the Vitredent vitreous carbon root-form dental implant.

Encouraged by these early animal studies, several groups commenced clinical trials using Vitredent's vitreous carbon Tooth Replacement System (Grenoble and Voss, 1977; Meffert, 1977). Unfortunately, all the major human studies encountered several *in vivo* problems, leading to unacceptably high failure rates. Between 1971 and 1975, Lemons and colleagues inspected several hundred Vitredent implants that were meant for clinical use. They discovered that almost 20% of these implants were manufactured with surface defects – multidirectional fissures, or microcracks, that quickly propagated from the implant surface to the stainless steel core (Lemons, 1975; Lemons *et al.*, 1977; Lemons, 1988). They realized that, *in vivo*, oral fluids could invade these microcracks, forming a galvanic couple between the vitreous carbon and steel. Since stainless steel is more electrochemically active than carbon, it would corrode preferentially in the presence of the fluids (Thompson *et al.*, 1979), and the release of metal ions could then diffuse to the surface of the implant and cause severe inflammation and bone resorption. The group also examined nine implants retrievals for similar defects. Of these, two were found to have surface cracks, one of which had a severely corroded metallic core (Lemons *et al.*, 1977). Once Vitredent was notified of the surface irregularities, the company modified its manufacturing and quality control practices, mitigating the problem in subsequent designs.

Surprisingly, none of the other clinics contemporaneously implanting the Vitredent tooth root system observed these surface-related complications. However, all of them still reported unfavorable *in vivo* results. In the early human trials conducted by Grenoble and co-workers, over 65% of the implants either failed to osseointegrate during healing or lost integration upon the application of occlusal loads. Most of these were freestanding (unsplinted), single-tooth implants. Early implants that had been splinted fared better, but still had a failure rate greater than 25% (Grenoble and Voss, 1977).

In an effort to provide greater mechanical interlock at the bone-implant interface, the vitreous carbon implant body was modified in 1973 to incorporate one or two ridges which were 1 mm deep and 2 mm wide (Figure 15.21) (NIH, 1978). Though these improvements in design and refinements in surgical technique promoted greater bone attachment, more than 30% of subsequently placed implants still failed.

Meffert and colleagues, who implanted 133 vitreous carbon dental implants from 1972 to 1978, met with similar results – by 1982, only 25% of the implants remained in place (Meffert, 1983). Negative survival rates in several clinical studies also led to early suspension of the trials. Most of the failures in these clinical trials occurred within the first year of placement, either due to exposure of the implant into the oral cavity, or excessive implant mobility accompanied by bone resorption. A small percentage of the later-stage failures were also associated with implant body fracture.

Due to the strict self-regulatory practices adopted by dentists and oral surgeons, the Vitredent Tooth Replacement System was short-lived in the clinic. Despite the absence of formal regulatory oversight and related guidance documents, the dental community strove to establish robust guidelines for improving surgical technique, for conducting preclinical and clinical investigations, and for assessing implant success. Hence, once the dismal performance of vitreous carbon implants became clear during a 1978 meeting discussing their risks and benefits, their use was essentially discontinued (NIH, 1978). And although more than 3000 vitreous carbon implants were placed from 1971 to 1978, most of the failures occurred early in treatment, prior to severe bone loss, permitting revision surgeries to be performed without serious complications.

The failure of the vitreous carbon dental implant emphasizes several important issues. It highlights the iterative nature of design and the fact that trade-offs are inherent to the optimization process. The adoption of vitreous carbon as an implant material compromised osseointegrative properties for better modulus-matching to bone. Though the trade-off proved detrimental to device performance, it still informed the hierarchy of design requirements for future dental implant development. The failure also highlighted the importance of thorough clinical testing – given its favorable *in vitro* biocompatibility and its success as a dental implant material in animal studies, the finding that vitreous carbon implants did not translate well to human subjects could not have been surmised without systematic clinical trials.

15.6.2 FDA recall of Proplast-Teflon temporomandibular joint replacements (regulatory and materials failure)

The massive failure of the Vitek-Kent TMJ prosthesis in the mid-1980s has been dubbed by many as one the greatest American medical disasters in history, brought on by the communal failure of Vitek, the manufacturer, the FDA, and the maxillofacial clinical community.

For many decades, controversy surrounded the question of whether the TMJ was a weight-bearing joint (Barbenel, 1972; Hylander, 1975; Roberts, 1974). Despite the uncertainty, clinicians failed to demand, or to conduct, studies that rigorously quantified the multiaxial loads to which the TMJ was subjected. Most simply assumed the TMJ was a non-weight-bearing joint, despite a lack of experimental evidence from animal, cadaveric, or human subjects. At the time, only a few clinicians recognized its anatomic similarity to other loaded joints in the body, and advocated caution before adopting any materials for TMJ implants, especially those known to have poor wear properties *in vitro* and *in vivo* (Gordon, 1958). Only after tens of thousands of Vitek-Kent TMJ prostheses had already been implanted, and, in many cases, failed, were more complete analyses of the biomechanics of the TMJ published (Abe *et al.*, 2006; Faulkner *et al.*, 1987; Throckmorton, 1985).

In light of its low coefficient of friction, there was a growing interest in using Teflon, Dupont's trademark for polytetrafluoroethyelene (PTFE), in biomedical applications in the 1960s. Dr. Charles Homsey sought to use PTFE-based materials for orthopedic implants, and in 1968, he formulated Proplast I. After launching Vitek in 1970, Homsey began working in conjunction with Dr. Kent, who helped him confirm Proplast's ability to promote hard and soft tissue ingrowth (Kent *et al.*, 1972). In an effort to make a device with Proplast that also possessed a low coefficient of friction and a smooth articulating surface suitable for the TMJ, Homsey decided to couple the material with a layer of Teflon-FEP. However, Dupont issued a warning to device manufacturers in 1968, stating that its Teflon products possessed poor wear properties, and, in 1966, Charnley had written a letter emphasizing the poor performance of Teflon in total hip replacements (Charnley, 1966). Despite these warnings, Vitek pressed on with its Proplast-Teflon material, taking the position that the TMJ was either unloaded or only minimally so, even under degenerative conditions (Tanaka *et al.*, 1994). In 1982, Vitek filed for FDA approval for its Interpositional Proplast Implant and circumvented the full PMA process by citing substantial equivalence to the existing Silastic disks already on the market (FDA). In the decade prior to the FDA submission, numerous studies were published highlighting positive results with both Proplast I and II in maxillofacial applications. However, most of these investigations, largely published by scientists at Vitek, used Proplast primarily in the context of facial reconstruction or augmentation rather than in TMJ applications (Block *et al.*, 1984; Gallagher and Wolford, 1982; Kent *et al.*, 1983; Westfall *et al.*, 1982). Further, many of the clinical studies involved small patient populations with short-term follow-up (Hinds *et al.*, 1974; Homsy *et al.*, 1973). During this time, Vitek did not conduct a single animal study using its PT prosthesis in the TMJ. Further, *in vitro* wear testing was simply performed on Teflon-FEP blocks with uniaxial loads much lower than those experienced *in vivo*. Later testing revealed that the wear rate of the prostheses under more physiological loads was nearly 100-fold larger than those initially determined for the Teflon-FEP blocks, and predicted an average implant life of only three years (Fontenot and Kent, 1992; Kent *et al.*, 1993a; Kent *et al.*, 1993b).

Vitek had failed in its responsibilities to rigorously test the efficacy and safety of its materials and its devices. However, there was a dearth of suitable materials for TMJ replacements at that time, and once an initial report cited a 91% success rate for Proplast TMJ prosthesis surfaces (Kent *et al.*, 1986; Kiersch, 1984; Vitek, 1986), Proplast mania quickly ensued. Despite a few early reports on the performance of Proplast in the TMJ that should have been cause for alarm (Lagrotteria *et al.*, 1986; Wade *et al.*, 1986), tens of thousands of people received Vitek-Kent TMJ prostheses between 1982 and 1990.

Whereas clinicians and Vitek failed in their scientific responsibilities, the FDA did not uphold in its regulatory duties. Though Silastic and Proplast were both polymers, the materials possessed very distinct physicochemical and mechanical properties – Silastic is a rubber, while Proplast is a semi-crystalline polymer composite. Hence, claims of substantial equivalence should have been rejected, and a full PMA should have been requested from Vitek for its TMJ prosthesis. By 1988, numerous studies documenting the clinical and radiographic follow-up on patients with the VK implants showed particulate debris from wear of the Teflon surface was resulting in pain, malocclusion, inflammation, and ankylosis of the implanted tissue (Feinerman and Piecuch, 1993; Spagnoli and Kent, 1992; Valentine *et al.*, 1989; Wagner and Mosby, 1990; Wolford, 1997). More than 50% of Vitek-Proplast I prostheses failed 5–6 years postoperatively (Kent *et al.*, 1993b). The widespread failures finally prompted the FDA to issue a regulatory letter to Vitek for violations in medical device reporting and in good manufacturing practices in 1989. The FDA also issued a Class I recall of all of the Proplast-Teflon Vitek TMJ total joint prosthesis in 1990 (Trumpy and Lyberg, 1993), after which Vitek filed for bankruptcy (Speculand *et al.*, 2000). Furthermore, the FDA released a new 510(k) guidance document for the manufacturing of TMJ implants in 1993. Soon after, the FDA reclassified all types of TMJ implants, some of which were considered Class II devices, as Class III medical devices that required a pre-market approval (PMA) application (FDA). The loss of both Silastic and Proplast-Teflon implants left a gaping void in the TMJ prosthesis market, just at the time when tens of thousands of people, still suffering the painful consequences of failed implants, needed a reliable alternative for revision surgery. The failures made apparent the load-bearing nature of the TMJ, and newer implants incorporated materials that were successful in other load-bearing implants, particularly hip and knee replacements.

The massive Vitek-Kent Proplast-Teflon failure highlights the importance of the many key components requisite for the robust design and manufacture of loading-bearing medical implants – a thorough understanding of *in vivo* service loads, rigorous *in vitro* biological and mechanical testing of all materials and components of the device, and systematic *in vivo* evaluation of device performance in both animals and human subjects.

15.7 Looking forward

15.7.1 Dental implants

Besides attempting to improve implant aesthetics through the adoption of ceramic materials, current research in the design of dental implants is focused on reducing the time required to complete implant therapy. The original Branemark protocol required 3–6 months of healing time for the implant, submerged, and unloaded, to ensure sufficient osseointegration before placement of the abutment or the prosthesis (Albrektsson *et al.*, 1981). With such a surgical schedule, the time required to complete implant therapy could exceed 7 or 8 months. Present efforts are aimed at reducing treatment time, either by decreasing the time necessary for osseointegration and soft tissue healing, or through more immediate loading of the implant through earlier attachment of a prosthesis.

Prior efforts to enhance osseointegration have led to the adoption of more osseoconductive implant materials (titanium) and surface coatings (hydroxyapatite), and to designs with greater surface areas (threaded) and rough surface topologies. On the other hand, recent attempts to hasten osseointegration involve optimizing the nanotopology of the implant surface through the fabrication of different nanofeatures, altering the chemistry and thickness of the thin titanium oxide layer, or impregnating surface coatings with bone growth factors such as bone morphogenic proteins (Albrektsson et al., 2008; Elias and Meirelles, 2010; Kim et al., 2008).

Concerns over immediate and early implant loading, defined as loading of the implant within 48 hours or 14 days of placement, respectively, stem from the formation of fibrous, rather than osseous, peri-implant tissues as a result of micromotion during initial healing. However, several recent studies of early and immediate loading of a variety of implant designs in different anatomical locations have reported favorable clinical outcomes, with over 95% survival rates (Ioannidou and Doufexi, 2005). Since the prevalence of failure is greatest in the first year of function, most of these studies simply report the two- or three-year survival rate, though some evidence of long-term success also exists (Proussaefs and Lozada, 2002). Due to these encouraging results, which have been largely attributed to more precise surgical techniques, better pre-operative treatment planning, and improved implant designs, immediate and early loading of dental implants is rapidly becoming a standard procedure in dental implantology.

15.7.2 Temporomandibular joint replacements

Even though they were recalled more than 15 years ago, patients, surgeons, and device manufacturers are still contending with the consequences of the failed Silastic and Proplast-Teflon TMJ devices. Many of the patients that required TMJ replacements in the 1990s were individuals who had had failed PT or Silastic implants. Hence, new designs focused on providing devices for revision surgeries (TMJ Concepts Inc.). With the severe bone resorption and anatomical destruction that accompanied many of the TMJ implant failures, many newer devices were meant to be custom-fit to each patient.

Despite the need for more versatile implants, most of the current research efforts in temporomandibular joint replacements are not focused on implant design but are more basic in nature; refinements in surgical placement of TMJ devices, more accurate characterization of TMJ loading, more robust *in vitro* wear testing of implant materials, more controlled, randomized clinical trials (Guarda-Nardini et al., 2008; Kashi et al., 2006), better tracking of device failures, and greater multidisciplinary cross-talk between TMJ researchers (NIDCR) through the recent creation of a TMJ registry that monitors device performance and failures. Though still at a nascent phase, tissue-engineering solutions for TMJ disorders are also being investigated (Maenpaa et al., 2010; Yuan et al.,

2010), and may ultimately provide a viable alternative to alloplastic TMJ replacements in the future.

15.8 Summary

This chapter discussed the history and evolution of dental implants and temporomandibular joint replacements. Though the first primitive dental implants were placed over 4000 years ago, most of the innovation in dental implant design has occurred following the discovery of osseointegrative properties of CoCr and titanium after the 1950s. Since then, three major types of dental implants – subperiosteal, blade-form, and root-form – have emerged. Though each of these was originally designed for patients with distinct alveolar bone quantity and quality, advancements in root-form implant, such as modular designs and roughened surfaces, have made subperiosteal and blade-form implants largely obsolete. Today, roughened titanium screw-form implants are the most widely used dental implants. These devices are meeting with very high rates of success in completely and partially edentulous jaws, and as single tooth replacements. Current research efforts in dental implants are now focused on reducing the time for treatment.

On the other hand, the development of successful TMJ replacement devices has been problematic. The 1930s–1970s saw a period where clinicians experimented with many different materials and designs, primarily for the treatment of ankylosis. However, these attempts met with little or no success. In the 1980s, Proplast-Teflon and Silastic became very popular, but proper biomechanical testing of the materials and devices was not conducted prior to their adoption, and in the early 1990s, widespread device failures prompted the FDA to recall the implants. TMJ implants were eventually classified as Class III medical devices, and since then, only a few have been granted regulatory approval. Though newer TMJ implants utilize more anatomically fitted designs and materials common to total joint arthroplasty, such as CoCr, titanium, and UHMWPE, they may not provide options for all patients and have been adopted with caution.

15.9 Problems for consideration

Checking for understanding

1. What are the advantages and disadvantages of removable dentures and fixed bridges?
2. Under what clinical circumstances would subperiosteal implants, blade-form implants, and root-form implants be utilized?
3. What are the advantages of titanium as a dental implant material? What about alumina or zirconia?
4. Besides joint arthroplasty, what are other non-surgical and surgical treatments for temporomandibular joint disease?

5. Why has the development of suitable temporomandibular joint implants been plagued with failure?
6. What are the components in modern temporomandibular joint replacements? Of what materials are these components typically made?

For further exploration

7. Several manufacturers of blade-form implants are upset at the FDA's decision to reclassify root-form, but not blade-form and subperiosteal implants, from Class III to Class II. Defend and rebut the agency's actions.
8. What is the current status of tissue-engineered solutions for teeth? Are they worth pursuing?
9. The FDA issued a new guidance document for temporomandibular joint replacements after the disastrous failures of Proplast and Silastic implants. You've just joined the FDA as an engineer, and based on new information regarding TMJ mechanics from the last decade, you are asked to write a revised guidance document for new TMJ implant submissions. Provide a detailed outline of the new and modified requirements you would propose.

15.10 References

(1978). Dental implants – benefits and risk. In *NIH-Harvard Consensus Development Conference*, ed., P. Schnitman and L. Shulman. Cambridge, MA: Harvard Dental School.

AAOMS. (1993). Recommendations for management of patients with temporomandibular joint implants. *Journal of Oral Maxillofacial Surgery*, 51: 1164–71.

Abe, M., Medina-Martinez, R.U., Itoh, K., and Kohno, S. (2006). Temporomandibular joint loading generated during bilateral static bites at molars and premolars. *Medical and Biological Engineering and Computing*, 44(11): 1017–30.

Albrektsson, T., Branemark, P.I., Hansson, H.A., and Lindstrom, J. (1981). Osseointegrated titanium implants – requirements for ensuring a long-lasting, direct bone-to-implant anchorage in man. *Acta Orthopaedica Scandinavica*, 52(2): 155–70.

Albrektsson, T., Dahl, E., Enbom, L., Engevall, S., Engquist, B., Eriksson, A.R., . . . Kjellman, O. (1988). Osseointegrated oral implants. A Swedish multicenter study of 8139 consecutively inserted Nobelpharma implants. *Journal of Periodontology*, 59(5): 287–96.

Albrektsson, T., and Sennerby, L. (1991). State-of-the-art in oral implants. *Journal of Clinical Periodontology*, 18(6): 474–81.

Albrektsson, T., Sennerby, L., and Wennerberg, A. (2008). State of the art of oral implants. *Periodontology 2000*, 47: 15–26.

Albrektsson, T., and Wennerberg, A. (2005). The impact of oral implants – past and future, 1966–2042. *Journal of the Canadian Dental Association*, 71(5): 327.

Andersson, B., Taylor, A., Lang, B.R., Scheller, H., Scharer, P., Sorensen, J.A., and Tarnow, D. (2001). Alumina ceramic implant abutments used for single-tooth replacement: A prospective 1- to 3-year multicenter study. *International Journal of Prosthodontics*, 14(5): 432–38.

Andreiotelli, M., Wenz, H.J., and Kohal, R.J. (2009). Are ceramic implants a viable alternative to titanium implants? A systematic literature review. *Clinical Oral Implants Research*, 20: 32–47.

Anitua, E., and Orive, G. (2010). Short implants in maxillae and mandibles: a retrospective study with 1 to 8 years of follow-up. *Journal of Periodontology*, 81(6): 819–26.

Atkinson, H.F., and Ralph, W.J. (1973). Tooth loss and biting force in man. *Journal of Dental Research*, 52(2): 225–28.

Babbush, C.A. (1972). Endosseous blade-vent implants – research review. *Journal of Oral Surgery*, 30(3): 168.

Bailey, J.H., Yanase, R.T., and Bodine, R.L. (1988). The mandibular subperiosteal implant denture – a 14-year study. *Journal of Prosthetic Dentistry*, 60(3): 358–61.

Barbenel, J.C. (1972). Biomechanics of temporomandibular joint – theoretical study. *Journal of Biomechanics*, 5(3): 251.

Bass, S.L., and Triplett, R.G. (1991). The effects of preoperative resorption and jaw anatomy on implant success. A report of 303 cases. *Clinical Oral Implants Research*, 2(4): 193–98.

Bates, J.F., Stafford, G.D., and Harrison, A. (1976). Masticatory function – a review of the literature. III. Masticatory performance and efficiency. *Journal of Oral Rehabilitation*, 3(1): 57–67.

Benson, J. (1972). Elemental carbon as an implant material. In *Bioceramics – Engineering in Medicine, Journal of Biomedical Materials Research Symposium*, New York: Interscience Publishers, 41–47.

Berge, T.I., and Gronningsaeter, A.G. (2000). Survival of single crystal sapphire implants supporting mandibular overdentures. *Clinical Oral Implants Research*, 11(2): 154–62.

Blair, V.P. (1914). Operative treatment of ankylosis of the mandible. *Surgery Gynecology and Obstetrics*, 19: 436–51.

Block, M.S., Zide, M.F., and Kent, J.N. (1984). Proplast augmentation for posttraumatic zygomatic deficiency. *Oral Surgery Oral Medicine Oral Pathology Oral Radiology and Endodontics*, 57(2): 123–31.

Bodine, R.L. (1963). Implant dentures – follow-up after 7 to 10 years. *Journal of the American Dental Association*, 67(3): 352.

Bodine, R.L. (1974). Evaluation of 27 mandibular subperiosteal implant dentures after 15–22 years. *Journal of Prosthetic Dentistry*, 32(2): 188–97.

Bodine, R.L., Yanase, R.T., and Bodine, A. (1996). Forty years of experience with subperiosteal implant dentures in 41 edentulous patients. *Journal of Prosthetic Dentistry*, 75(1): 33–44.

Boyne, P.J., Mathews, F.R., and Stringer, D.E. (1987). TMJ bone remodeling after polyoxymethylene condylar replacement. *International Journal of Oral Maxillofacial Implants*, 2(1): 29–33.

Branemark, P.I. (1983). Osseointegration and its experimental background. *Journal of Prosthetic Dentistry*, 50(3): 399–410.

Branemark, P.I., Adell, R., and Breine, U. (1969). Intra-osseous anchor of dental prosthesis. I. Experimental studies. *Scandinavian Journal of Plastic Reconstructive Surgery*, 3(2): 81–100.

Branemark, P.I., Hansson, B.O., Adell, R., Breine, U., Lindstrom, J., Hallen, O., and Ohman, A. (1977). Osseointegrated implants in the treatment of the edentulous jaw. Experience from a 10-year period. *Scandinavian Journal of Plastic Reconstructive Surgery Supplement*, 16: 1–132.

Braun, S., Hnat, W.P., Freudenthaler, J.W., Marcotte, M.R., Honigle, K., and Johnson, B.E. (1996). A study of maximum bite force during growth and development. *Angle Orthodontist*, 66(4): 261–64.

Brink, J., Meraw, S.J., and Sarment, D.P. (2007). Influence of implant diameter on surrounding bone. *Clinical Oral Implants Research*, 18(5): 563–68.

Brocard, D., Barthet, P., Baysse, E., Duffort, J.F., Eller, P., Justumus, P., . . . Brunel, G. (2000). A multicenter report on 1,022 consecutively placed ITI implants: A 7-year longitudinal study. *International Journal of Oral and Maxillofacial Implants*, 15(5): 691–700.

Bronstein, S.L. (1987). Retained alloplastic temporomandibular-joint disk implants – a retrospective study. *Oral Surgery Oral Medicine Oral Pathology Oral Radiology and Endodontics*, 64(2): 135–45.

Brose, M.O., Reiger, M., Avers, R.J., and Hassler, C.R. (1988). Eight-year analysis of alumina dental root implants in human subjects. *Journal of Oral Implantology*, 14(1): 9–22.

Bumann, A., and Lotzman, U. (2002). *TMJ Disorders and Orofacial Pain: The Role of Dentistry in a Multidisciplinary Diagnostic Approach*. Stuttgart, Germany: Georg Thieme Verlag.

Buser, D., Mericske-Stern, R., Bernard, J.P., Behneke, A., Behneke, N., Hirt, H.P., . . . Lang, N.P. (1997). Long-term evaluation of non-submerged ITI implants. *Part 1*: 8-year life table analysis of a prospective multi-center study with 2359 implants. *Clinical Oral Implants Research*, 8(3): 161–72.

Buser, D., Schenk, R.K., Steinemann, S., Fiorellini, J.P., Fox, C.H., and Stich, H. (1991). Influence of surface characteristics on bone integration of titanium implants – a histomorphometric in miniature pigs. *Journal of Biomedical Materials Research*, 25(7): 889–902.

Butow, K.W., Blackbeard, G.A., and Van Der Merwe, A.E. (2001). Titanium/titanium nitride temporomandibular joint prosthesis: historical background and a six-year clinical review. *Journal of the South African Dental Association*, 56(8): 370–76.

Cameron, H.U., Pilliar, R.M., and Macnab, I. (1973). Effect of movement on bonding of porous metal to bone. *Journal of Biomedical Materials Research*, 7(4): 301–11.

Cappuccilli, M., Conte, M., and Praiss, S.T. (2004). Placement and postmortem retrieval of a 28-year-old implant – A clinical and histologic report. *Journal of the American Dental Association*, 135(3): 324–29.

Charnley, J. (1966). Letter to the Editor. *Journal of Bone and Joint Surgery – American Volume*, A 48(4): 819.

Christensen, R.W. (1963). The correction of mandibular ankylosis by arthroplasty and the insertion of a cast vitallium glenoid fossa: a new technique. A preliminary report of three cases. *American Journal of Orthodontics and Dentofacial Orthopedics*, 5: 16–24.

Cooley, D.R., Van Dellen, A.F., Burgess, J.O., and Windeler, A.S. (1992). The advantages of coated titanium implants prepared by radiofrequency sputtering from hydroxyapatite. *Journal of Prosthetic Dentistry*, 67(1): 93–100.

Cowlard, F.C., and Lewis, J.C. (1967). Vitreous carbon – a new form of carbon. *Journal of Materials Science*, 2(6): 507–12.

Cranin, A.N., Rabkin, M.F., and Garfinkel, L. (1977). Statistical evaluation of 952 endosteal implants in humans. *Journal of the American Dental Association*, 94(2): 315–20.

Cranin, A.N., Schnitman, P.A., Rabkin, M., Dennison, T., and Onesto, E.J. (1975). Alumina and zirconia coated vitallium oral endosteal implants in beagles. *Journal of Biomedical Materials Research*, 9(4): 257–62.

de Las Casas, E.B., de Almeida, A.F., Cimini Junior, C.A., Gomes Pde, T., Cornacchia, T.P., and Saffar, J.M. (2007). Determination of tangential and normal components of oral forces. *Journal of Applied Oral Science*, 15(1): 70–76.

Depprich, R., Zipprich, H., Ommerborn, M., Naujoks, C., Wiesmann, H.P., Kiattavorncharoen, S., ... Handschel, J. (2008). Osseointegration of zirconia implants compared with titanium: an in vivo study. *Head Face Medicine*, 4: 30.

Diedrich, P.R., Fuhrmann, R.A., Wehrbein, H., and Erpenstein, H. (1996). Distal movement of premolars to provide posterior abutments for missing molars. *American Journal of Orthodontics and Dentofacial Orthopedics*, 109(4): 355–60.

Dolwick, M.F., and Aufdemorte, T.B. (1985). Silicone-induced foreign-body reaction and lymphadenopathy after temporomandibular-joint arthroplasty. *Oral Surgery Oral Medicine Oral Pathology Oral Radiology and Endodontics*, 59(5): 449–52.

Driemel, O., Ach, T., Muller-Richter, U.D.A., Behr, M., Reichert, T.E., Kunkel, M., and Reich, R. (2009). Historical development of alloplastic temporomandibular joint replacement before 1945. *International Journal of Oral and Maxillofacial Surgery*, 38(4): 301–7.

Dubruille, J.H., Viguier, E., Le Naour, G., Dubruille, M.T., Auriol, M., and Le Charpentier, Y. (1999). Evaluation of combinations of titanium, zirconia, and alumina implants with 2 bone fillers in the dog. *International Journal of Oral and Maxillofacial Implants*, 14(2): 271–77.

Eggers, G.W.N. (1946). Arthroplasty of the temporomandibular joint in children with interposition of a tantalum foil – a preliminary report. *Journal of Bone and Joint Surgery*, 28(3): 603–6.

Elias, C.N., and Meirelles, L. (2010). Improving osseointegration of dental implants. *Expert Review of Medical Devices*, 7(2): 241–56.

Eriksson, L., and Westesson, P.L. (1986). Deterioration of temporary silicone implant in the temporomandibular joint – a clinical and arthroscopic follow-up study. *Oral Surgery Oral Medicine Oral Pathology Oral Radiology and Endodontics*, 62(1): 2–6.

Eriksson, L., and Westesson, P.L. (1992). Temporomandibular-joint diskectomy – no positive effect of temporary silicone implant in a 5-year follow-up. *Oral Surgery Oral Medicine Oral Pathology Oral Radiology and Endodontics*, 74(3): 259–72.

Esposito, M., Hirsch, J.M., Lekholm, U., and Thomsen, P. (1998). Biological factors contributing to failures of osseointegrated oral implants – (ll). *Etiopathogenesis. European Journal of Oral Sciences*, 105(3): 721–64.

Faulkner, M.G., Hatcher, D.C., and Hay, A. (1987). A 3-dimensional investigation of temporomandibular-joint loading. *Journal of Biomechanics*, 20(10): 997–1002.

Food and Drug Administration. (1999). *TMJ Concepts Patient-Fitted TMJ Reconstruction Prothesis System – P980052* [Online]. Available: http://www.accessdata.fda.gov/cdrh_docs/pdf/P980052b.pdf [Accessed].

Food and Drug Administration. (2001). *TMJ implants: A consumer information update* [Online]. Available: http://www.fda.gov/downloads/MedicalDevices/Safety/AlertsandNotices/PatientAlerts/ucm108042.pdf [Accessed].

Food and Drug Administration. (2004). *Class II Special Controls Guidance Document: Root-form Endosseous Dental Implants and Endosseous Dental Implant Abutments* [Online]. Available: http://www.fda.gov/downloads/MedicalDevices/DeviceRegulationandGuidance/GuidanceDocuments/ucm072444.pdf [Accessed].

Feinerman, D.M., and Piecuch, J.F. (1993). Long-term retrospective analysis of twenty-three Proplast-Teflon temporomandibular joint interpositional implants. *International Journal of Oral Maxillofacial Surgery*, 22(1): 11–16.

Fettig, R.H., and Kay, J.F. (1994). A seven-year clinical evaluation of soft-tissue effects of hydroxylapatite-coated vs. uncoated subperiosteal implants. *Journal of Oral Implantology*, 20(1): 42–48.

Fontenot, M.G., and Kent, J.N. (1992). In vitro wear performance of Proplast TMJ disk implants. *Journal of Oral and Maxillofacial Surgery*, 50(2): 133–39.

Fraunhof, J.A., Lestrang, P., and Mack, A.O. (1971). Materials science in dental implantation and a promising new material – vitreous carbon. *Biomedical Engineering*, 6(3): 114.

Fritz, M.E., Braswell, L.D., Koth, D., Jeffcoat, M., Reddy, M., Brogan, D., . . . Lemons, J.E. (1996). Analysis of consecutively placed loaded root-form and plate-form implants in adult Macaca mulatta monkeys. *Journal of Periodontology*, 67(12): 1322–28.

Fugazzotto, P.A., Beagle, J.R., Ganeles, J., Jaffin, R., Vlassis, J., and Kumar, A. (2004). Success and failure rates of 9 mm or shorter implants in the replacement of missing maxillary molars when restored with individual crowns: Preliminary results 0 to 84 months in function. A retrospective study. *Journal of Periodontology*, 75(2): 327–32.

Gallagher, D.M., and Wolford, L.M. (1982). Comparison of Silastic and Proplast implants in the temporomandibular joint after condylectomy for osteoarthritis. *Journal of Oral and Maxillofacial Surgery*, 40(10): 627–30.

Gingerich, P.D. (1979). Human mandible – lever, link or both. *American Journal of Physical Anthropology*, 51(1): 135–38.

Goldberg, N.I., and Gershkoff, A. (1952). Fundamentals of the implant denture. *Journal of Prosthetic Dentistry*, 2(1): 40–48.

Gordon, S.D. (1958). Surgery of the temporomandibular joint. *American Journal of Surgery*, 95(2): 263–66.

Greenfield, E.J. (1913). Implantation of artificial crown and bridge abutments. *Dental Cosmos*, 55: 364–69, 430–39.

Grenoble, D.E., Melrose, R.J., and Markle, D.H. (1975). Histologic evaluation of vitreous carbon endosteal implants in occlusion in dogs. *Biomaterials Medical Devices and Artificial Organs*, 3(2): 245–58.

Grenoble, D.E., and Voss, R. (1977). Analysis of five years of study of vitreous carbon endosseous implants in humans. *Oral Implantology*, 6(4): 509–25.

Guarda-Nardini, L., Manfredini, D., and Ferronato, G. (2008). Temporomandibular joint total replacement prosthesis: current knowledge and considerations for the future. *International Journal of Oral and Maxillofacial Surgery*, 37(2): 103–10.

Hahn, G.W. (1964). Vitallium mesh mandibular prothesis. *Journal of Prosthetic Dentistry*, 14(4): 777.

Hansen, W.C., and Deshazo, B.W. (1969). Silastic reconstruction of temporo-mandibular joint meniscus. *Plastic Reconstructive Surgery*, 43(4): 388–91.

Haraldson, T., Karlsson, U., and Carlsson, G.E. (1979). Bite force and oral function in complete denture wearers. *Journal of Oral Rehabilitation*, 6(1): 41–48.

Hayashi, K., Matsuguchi, N., Uenoyama, K., and Sugioka, Y. (1992). Re-evaluation of the biocompatibility of bioinert ceramics in vivo. *Biomaterials*, 13(4): 195–200.

Himmlova, L., Dostalova, T., Kacovsky, A., and Konvickova, S. (2004). Influence of implant length and diameter on stress distribution: a finite element analysis. *Journal of Prosthetic Dentistry*, 91(1): 20–25.

Hinds, E.C., Homsy, C.A., and Kent, J.N. (1974). Use of a biocompatible interface for binding tissues and prostheses in temporomandibular-joint surgery – follow-up report. *Oral Surgery Oral Medicine Oral Pathology Oral Radiology and Endodontics*, 38(4): 512–19.

Homsy, C.A., Kent, J.N., and Hinds, E.C. (1973). Materials for oral implantation – biological and functional criteria. *Journal of the American Dental Association*, 86(4): 817–32.

Howell, A.H., and Brudevold, F. (1950). Vertical forces used during chewing of food. *Journal of Dental Research*, 29(2): 133–36.

Hucke, E.E., Fuys, R.A., and Craig, R.G. (1973). Glassy carbon – potential dental implant material. *Journal of Biomedical Materials Research*, 7(3): 263–74.

Hylander, W.L. (1975). Human mandible – lever or link. *American Journal of Physical Anthropology*, 43(2): 227–42.

Ioannidou, E., and Doufexi, A. (2005). Does loading time affect implant survival? A meta-analysis of 1,266 implants. *Journal of Periodontology*, 76(8): 1252–58.

Ivanoff, C.J., Sennerby, L., Johansson, C., Rangert, B., and Lekholm, U. (1997). Influence of implant diameters on the integration of screw implants. An experimental study in rabbits. *International Journal of Oral and Maxillofacial Implants*, 26(2): 141–48.

Jaffin, R.A., and Berman, C.L. (1991). The excessive loss of Branemark fixtures in the type-IV bone – a 5-year analysis. *Journal of Periodontology*, 62(1): 2–4.

Kapur, K.K., and Garrett, N.R. (1984). Studies of biologic parameters for denture design. Part II: Comparison of masseter muscle activity, masticatory performance, and salivary secretion rates between denture and natural dentition groups. *Journal of Prosthetic Dentistry*, 52(3): 408–13.

Kashi, A., Saha, S., and Christensen, R.W. (2006). Temporomandibular joint disorders: artificial joint replacements and future research needs. *Journal of Long-Term Effects of Medical Implants*, 16(6): 459–744.

Kay, J.F., Golec, T.S., and Riley, R.L. (1987). Hydroxyapatite-coated subperiosteal dental implants – design rationale and clinical experience. *Journal of Prosthetic Dentistry*, 58(3): 339–43.

Kearns, G.J., Perrott, D.H., and Kaban, L.B. (1995). A protocol for the management of failed alloplastic temporomandibular-joint disc implnats. *Journal of Oral and Maxillofacial Surgery*, 53(11): 1240–47.

Kent, J.N., Block, M.S., Halpern, J., and Fontenot, M.G. (1993a). Long-term results on VK partial and total temporomandibular-joint systems. *Journal of Long-Term Effects of Medical Implants*, 3(1): 29–40.

Kent, J.N., Block, M.S., Halpern, J., and Fontenot, M.G. (1993b). Update on the Vitek partial and total temporomandibular-joint systems. *Journal of Oral and Maxillofacial Surgery*, 51(4): 408–15.

Kent, J.N., Block, M.S., Homsy, C.A., Prewitt, J.M., and Reid, R. (1986). Experience with a polymer glenoid fossa prosthesis for partial or total temporomandibular-joint reconstruction. *Journal of Oral and Maxillofacial Surgery*, 44(7): 520–33.

Kent, J.N., Cook, S.D., Weinstein, A.M., and Klawitter, J.J. (1982). A clinical comparison of LTI carbon, alumina, and carbon-coated alumina blade-type implants in baboons. *Journal of Biomedical Materials Research*, 16(6): 887–99.

Kent, J.N., Gross, B.D., Homsy, C.A., and Hinds, E.C. (1972). Pilot-studies of a porous implant in dentistry and oral surgery. *Journal of Oral Surgery*, 30(8): 608.

Kent, J.N., Misiek, D.J., Akin, R.K., Hinds, E.C., and Homsy, C.A. (1983). Temporomandibular-joint condylar porsthesis – a 10-year report. *Journal of Oral and Maxillofacial Surgery*, 41(4): 245–54.

Kiehn, C.L., Desprez, J.D., and Converse, C.F. (1974). New procedure for total temporomandibular-joint replacement – case report. *Plastic and Reconstructive Surgery*, 53(2): 221–26.

Kiehn, C.L., Desprez, J.D., and Converse, C.F. (1979). Total prosthetic replacement of the temporomandibular joint. *Annals of Plastic Surgery*, 2(1): 5–15.

Kiersch, T.A. (1984). The use of Teflon-Proplast implants meniscectomy and disc repair in the temporomandibular joint. *AAOMS Clinical Congress on Reconstruction with Biomaterials*. San Diego, CA.

Kim, T.I., Jang, J.H., Kim, H.W., Knowles, J.C., and Ku, Y. (2008). Biomimetic approach to dental implants. *Current Pharmaceutical Design*, 14(22): 2201–11.

Kohal, R.J., Klaus, G., and Strub, J.R. (2006). Zirconia-implant-supported all-ceramic crowns withstand long-term load: a pilot investigation. *Clinical Oral Implants Research*, 17(5): 565–71.

Koth, D.L., Mckinney, R.V., Steflik, D.E., and Davis, Q.B. (1988). Clinical and statistical-analyses of human clinical trials with the single-crystal aluminum-oxide endosteal dental implant – 5-year results. *Journal of Prosthetic Dentistry*, 60(2): 226–34.

Kummoona, R. (1978). Functional rehabilitation of ankylosed temporomandibular joints. *Oral Surgery Oral Medicine Oral Pathology Oral Radiology and Endodontics*, 46(4): 495–505.

Lagrotteria, L., Scapino, R., Granston, A.S., and Felgenhauer, D. (1986). Patient with lymphadenopathy following temporomandibular joint arthroplasty with Proplast. *Journal of Craniomandibular Practice*, 4(2): 172–78.

Lemons, J.E. (1975). Biomaterials science protocols for clinical investigations on porous alumina ceramic and vitreous carbon implants. *Journal of Biomedical Materials Research*, 9(4): 9–16.

Lemons, J.E. (1988). Dental implant retrieval analyses. *Journal of Dental Education*, 52(12): 748–56.

Lemons, J.E., Baswell, I.L., and Fischer, T.E. (1977). Laboratory and clinical studies of vitreous carbon tooth root replacement system. *3rd Annual Meeting of the Society for Biomaterials and 9th Annual International Biomaterials Symposium*. New Orleans, LA.

Lindqvist, C., Soderholm, A.L., Hallikainen, D., and Sjovall, L. (1992). Erosion and heterotopic bone-formation after alloplastic temporomandibular-joint reconstruction. *Journal of Oral and Maxillofacial Surgery*, 50(9): 942–49.

Linkow, L.I. (1970). Endosseous blade-vent implants – 2-year report. *Journal of Prosthetic Dentistry*, 23(4): 441.

Linkow, L.I. (1990). *Implant Dentistry Today: A Multidisciplinary Approach*. Padua: Piccin.

Maenpaa, K., Ella, V., Mauno, J., Kellomaki, M., Suuronen, R., Ylikomi, T., and Miettinen, S. (2010). Use of adipose stem cells and polylactide discs for tissue engineering of the temporomandibular joint disc. *Journal of the Royal Society Interface*, 7(42): 177–88.

Maniatopoulos, C., Pilliar, R.M., and Smith, D.C. (1986). Threaded versus porous-surfaced designs for implant stabilization in bone-endodontic implant model. *Journal of Biomedical Materials Research*, 20(9): 1309–33.

Markle, D.H., Grenoble, D.E., and Melrose, R.J. (1975). Histologic evaluation of vitreous carbon endosteal implants in dogs. *Biomaterials Medical Devices and Artificial Organs*, 3(1): 97–114.

Mayo Clinic. *TMJ Disease Treatment – Mayo Clinic* [Online]. Available: http://www.mayoclinic.org/tmj/treatment.html [Accessed].

Meenaghan, M.A., Natiella, J.R., Armitage, J.E., Greene, G.W., and Lipani, C.S. (1974). Crypt surface of blade-vent implants in clinical failure – electron-microscopic study. *Journal of Prosthetic Dentistry*, 31(6): 681–90.

Meffert, R.M. (1977). Vitreous carbon–where are we now? *CDS Review*, 70(3): 22–26.

Meffert, R.M. (1983). Vitreous carbon root-replacement system: final report. *Journal of Oral Implantology*, 11(2): 268–72.

Mehra, P., Wolford, L.M., Baran, S., and Cassano, D.S. (2009). Single-stage comprehensive surgical treatment of rheumatoid arthritis temporomandibular joint patient. *Journal of Oral and Maxillofacial Surgery*, 67(9): 1859–72.

Mercier, P., Cholewa, J., and Djokovic, S. (1981). Mandibular subperiosteal implants: Retrospective analysis in light of Harvard consensus. *Journal of the Canadian Dental Association*, 47: 46–51.

Mercuri, L.G. (2000). The use of alloplastic prostheses for temporomandibular joint reconstruction. *Journal of Oral and Maxillofacial Surgery*, 58(1): 70–75.

Michael, C.G., Javid, N.S., Colaizzi, F.A., and Gibbs, C.H. (1990). Biting strength and chewing forces in complete denture wearers. *Journal of Prosthetic Dentistry*, 63(5): 549–53.

Misch, C.E. (2005). *Dental Implant Prosthetics*. St. Louis, MO: Elsevier Mosby.

Misch, C.E. (2008). *Contemporary Implant Dentistry*. St. Louis, MO: Elsevier Mosby.

Moore, D.J., and Hansen, P.A. (2004). A descriptive 18-year retrospective review of subperiosteal implants for patients with severely atrophied edentulous mandibles. *Journal of Prosthetic Dentistry*, 92(2): 145–50.

Morgan, D.H. (1971). Dysfunction, pain, tinnitus, vertigo corrected by mandibular joint surgery. *Journal of the Southern California Dental Association*, 39(7): 505–34.

Morgan, D.H. (1988). Evaluation of alloplastic TMJ implants. *Journal of Craniomandibular Practice*, 6(3): 224–38.

Moriconi, E.S., Popowich, L.D., and Guernsey, L.H. (1986). Allloplastic reconstruction of the temporomandibular joint. *Dental Clinics of North America*, 30(2): 307–25.

Morimoto, K., Kihara, A., Takeshita, F., Akedo, H., and Suetsugu, T. (1988). Differences between the bony interfaces of titanium and hydroxyapatite-alumina plasma-sprayed titanium blade implants. *Journal of Oral Implantology*, 14(3): 314–24.

Nakamura, K., Kanno, T., Milleding, P., and Ortengren, U. (2010). Zirconia as a dental implant abutment material: a systematic review. *International Journal of Prosthodontics*, 23(4): 299–309.

Natiella, J.R., Armitage, J.E., Meenaghan, M.A., Lipani, C.S., and Greene, G.W. (1973). Failing blade-vent implant. *Oral Surgery Oral Medicine Oral Pathology Oral Radiology and Endodontics*, 36(3): 336–42.

NIDCR. National Institute of Dental and Craniofacial Research Launches Important Study on Temporomandibular Joint and Muscle Disorders. Press release on given by Bob Kuska of NIDCR December 5, 2005.NIDCR. *TMJ Implant Registry and Repository* [Online]. Available: http://tmjregistry.org/ [Accessed].

Noack, N., Willer, J., and Hoffman, J. (1999). Long-term results after placement of dental implants: longitudinal study of 1,964 implants over 16 years. *International Journal of Oral and Maxillofacial Implants*, 14(5): 748–55.

Ogata, K., and Satoh, M. (1995). Centre and magnitude of vertical forces in complete denture wearers. *Journal of Oral Rehabilitation*, 22(2): 113–19.

Oliva, J., Oliva, X., and Oliva, J.D. (2007). One-year follow-up of first consecutive 100 zirconia dental implants in humans: a comparison of 2 different rough surfaces. *International Journal of Oral and Maxillofacial Implants*, 22(3): 430–35.

Pierson, H.O. (1993). *Handbook of Carbon, Graphite, Diamonds, and Fullerenes: Processing, Properties, and Applications*. Park Ridge, NJ: Noyas Publication.

Pilliar, R.M., Lee, J.M., and Maniatopoulos, C. (1986). Observations on the effect of movement on bone ingrowth into porous-surfaced implants. *Clinical Orthopaedics and Related Research*, (208): 108–13.

Proussaefs, P., and Lozada, J. (2002). Evaluation of two Vitallium blade-form implants retrieved after 13 to 21 years of function: A clinical report. *Journal of Prosthetic Dentistry*, 87(4): 412–15.

Raadsheer, M.C., van Eijden, T.M., van Ginkel, F.C., and Prahl-Andersen, B. (1999). Contribution of jaw muscle size and craniofacial morphology to human bite force magnitude. *Journal of Dental Research*, 78(1): 31–42.

Rasperini, G., Maglione, M., Cocconcelli, P., and Simion, M. (1998). In vivo early plaque formation on pure titanium and ceramic abutments: a comparative microbiological and SEM analysis. *Clinical Oral Implants Research*, 9(6): 357–64.

Richter, E.J. (1995). In vivo vertical forces on implants. *International Journal of Oral and Maxillofacial Implants*, 10(1): 99–108.

Ring, M.E. (1995a). A thousand years of dental implants: a definitive history –part 1. *Compendium of Continuous Education in Dentistry*, 16(10): 1060, 1062, 1064 passim.

Ring, M.E. (1995b). A thousand years of dental implants: a definitive history –part 2. *Compendium of Continuous Education in Dentistry*, 16(11): 1132, 1134, 1136 passim.

Risdon, F.E. (1934). Ankylosis of the temporomandibular joint. *Journal of the American Dental Association*, 21: 1933–37.

Roberts, D. (1974). Etiology of temporomandibular-joint dysfunction syndrome. *American Journal of Orthodontics and Dentofacial Orthopedics*, 66(5): 498–515.

Robinson, M. (1960). Temporomandibular ankylosis corrected by creating a falsa stainless steel fossa. *Journal of the Southern California Dental Association*, 28: 186.

Romeo, E., Bivio, A., Mosca, D., Scanferla, M., Ghisolfi, M., and Storelli, S. (2010). The use of short dental implants in clinical practice: literative review. *Minerva Stomatologica*, 59(1–2): 23–31.

Rudy, R.J., Levi, P.A., Bonacci, F.J., Weisgold, A.S., and Engler-Hamm, D. (2008). Intraosseous anchorage of dental prostheses: an early 20th century contribution. *Compendium of Continuous Education in Dentistry*, 29(4): 220–22, 224, 226–28 passim.

Ryan, D.E. (1989). The PROPLAST Teflon dilemma. *Journal of Oral and Maxillofacial Surgery*, 47(3): 222.

Ryan, D.E. (1994). *Alloplastic Disc Replacement. Oral Maxillofacial Surgery Clinics North America*, 6: 307–21.

Scheid, R.C. (2007). *Woelfel's Dental Anatomy: Its Relevance to Dentistry*. Philadelphia, PA: Lippincott Williams & Wilkins.

Schnitman, P.A., and Shulman, L.B. (1980). Vitreous carbon implants. *Dental Clinics of North America*, 24(3): 441–63.

Schnitman, P.A., Woolfson, M.W., Feingold, R.M., Gettleman, L., Freedman, H.M., Kalis, P.J., . . . Shulman, L.B. (1980). Vitreous carbon implants – a 5-year study in baboons. *Journal of Prosthetic Dentistry*, 44(2): 190–200.

Shackleton, J.L., Carr, L., Slabbert, J.C.G., and Becker, P.J. (1994). Survival of fixed implant-supported protheses related to cantilever lengths. *Journal of Prosthetic Dentistry*, 71(1): 23–26.

Shulman, L., Feingold, R., Freedman, H., Kalis, P., Woolfson, M., and Schnitman, P. (1980). Vitreous carbon dental implants – prospective clinical trial. *Journal of Dental Research*, 59: 281.

Silver, C.M., Motamed, M., and Carlotti, A.E. (1977). Arthroplasty of temporomandibular-joint with use of a Vitallium condyle prostheses – report of 3 cases. *Journal of Oral Surgery*, 35(11): 909–14.

Smithloff, M., and Fritz, M.E. (1987). The use of blade implants in a selected population of partially edentulous adults – a 15-year report. *Journal of Periodontology*, 58(9): 589–93.

Smithloff, M., Fritz, M.E., and Giansanti, J.S. (1975). Clinical and histologic evaluation of a single blade implant and surrounding bone. *Journal of Prosthetic Dentistry*, 33(4): 427–32.

Sonnenburg, I., and Sonnenburg, M. (1985). Total condylar prosthesis for alloplastic jaw articulation replacement. *Journal of Maxillofacial Surgery*, 13(3): 131–35.

Spagnoli, D., and Kent, J.N. (1992). Multicenter evaluation of temporomandibular-joint Proplast-Teflon disk implant. *Oral Surgery Oral Medicine Oral Pathology Oral Radiology and Endodontics*, 74(4): 411–21.

Speculand, B., Hensher, R., and Powell, D. (2000). Total prosthetic replacement of the TMJ: experience with two systems 1988–1997. *British Journal of Oral and Maxillofacial Surgery*, 38(4): 360–69.

Stallard, R.E., El Geneidy, A.K.S., and Skerman, H.J. (1975). Vitreous carbon implants – An aid to alveolar bone maintenance. *Oral Implantology*, 6(2): 286–308.

Stamfelj, I., Vidmar, G., Cvetko, E., and Gaspersic, D. (2008). Cementum thickness in multirooted human molars: A histometric study by light microscopy. *Annals of Anatomy – Anatomischer Anzeiger*, 190(2): 129–39.

Steflik, D.E., Noel, C., McBrayer, C., Lake, F.T., Parr, G.R., Sisk, A.L., and Hanes, P.J. (1995). Histologic observations of bone remodeling adjacent to endosteal dental implants. *Journal of Oral Implantology*, 21(2): 96–106.

Steinebrunner, L., Wolfart, S., Ludwig, K., and Kern, M. (2008). Implant-abutment interface design affects fatigue and fracture strength of implants. *Clinical Oral Implants Research*, 19: 1276–84.

Strock, A.E. (1939). Experimental work on a method for the replacement of missing teeth by direct implantation of a metal support into the alveolus. *American Journal of Orthodontics and Dentofacial Orthopedics*, 25: 467–77.

Takeshita, F., Ayukawa, Y., Iyama, S., Suetsugu, T., and Kido, M.A. (1996). A histologic evaluation of retrieved hydroxyapatite-coated blade-form implants using scanning electron, light, and confocal laser scanning microscopies. *Journal of Periodontology*, 67(10): 1034–40.

Tanaka, E., Tanne, K., and Sakuda, M. (1994). A three-dimensional finite element model of the mandible including the TMJ and its application to stress analysis in the TMJ during clenching. *Medical Engineering Physics*, 16(4): 316–22.

Taylor, J.C., Driscoll, C.F., and Cunningham, M.D. (1996). Failure of a hydroxyapatite-coated endosteal dental implant: a clinical report. *Journal of Prosthetic Dentistry*, 75(4): 353–55.

TMJ Association. *TMJArchive. TMJ Disorders – Implants* [Online]. Available: http://www.tmjarchive.org/implants5.asp [Accessed].

Thompson, N.G., Buchanan, R.A., and Lemons, J.E. (1979). In vitro corrosion of a Ti-6Al-4V and Type 316L stainless steel when galvanically coupled with carbon. *Journal of Biomedical Materials Research*, 13(1): 35–44.

Throckmorton, G.S. (1985). Quantitative calculations of temporomandibular joint reaction forces. 2. The importance of the direction of the jaw muscle forces. *Journal of Biomechanics*, 18(6): 453–61.

Timmis, D.P., Aragon, S.B., Van Sickels, J.E., and Aufdemorte, T.B. (1986). Comparative study of alloplastic materials for temporomandibular joint disc replacement in rabbits. *Journal of Oral and Maxillofacial Surgery*, 44(7): 541–54.

TMJ Concepts Inc. Available: www.tmjconcepts.com [Accessed].

TMJ Implants Inc. Available: www.tmj.com [Accessed].

Trisi, P., Quaranta, M., Emanuelli, M., and Piattelli, A. (1993). A light-microscopy, scanning electronic-microscopy, and laser scanning microscopy analysis of retrieved blade implants after 7 to 20 years of clinical function – a report of 3 cases. *Journal of Periodontology*, 64(5): 374–78.

Trumpy, I.G., and Lyberg, T. (1993). In vivo deterioration of proplast-teflon temporomandibular joint interpositional implants: a scanning electron microscopic and energy-dispersive X-ray analysis. *Journal of Oral and Maxillofacial Surgery*, 51(6): 624–29.

Valentine, J.D., Jr., Reiman, B.E., Beuttenmuller, E.A., and Donovan, M.G. (1989). Light and electron microscopic evaluation of Proplast II TMJ disc implants. *Journal of Oral and Maxillofacial Surgery*, 47(7): 689–96.

van Eijden, T.M. (1991). Three-dimensional analyses of human bite-force magnitude and moment. *Archives of Oral Biology*, 36(7): 535–39.

Vanloon, J.P., Debont, L.G.M., and Boering, G. (1995). Evaluation of temporomandibular-joint prostheses – review of the literature from 1946 to 1994 and implications for future prosthesis designs. *Journal of Oral and Maxillofacial Surgery*, 53(9): 984–96.

Venable, C.S., and Stuck, W.G. (1941). Three years' experience with Vitallium in bone surgery. *Annals of Surgery*, 114(2): 309–15.

Venable, C.S., Stuck, W.G., and Beach, A. (1937). The effects on bone of the presence of metals based upon electrolysis: an experimental study. *Annals of Surgery*, 105(6): 917–38.

Vitek. (1986). Vitek Survey of TMJ Surgical Procedures. Houston, TX: Vitek, Inc.

Vitek Inc. (1984). Vitek Teflon Proplast Product Information Brochure. Houston, TX: Vitek, Inc.

Vitredent Corporation. (1973). The Vitredent Tooth Rooth Replacement System: Biological, Surgical and Prosthetic Considerations. Los Angeles, CA: Vitredent Corporation.

Wade, M.L., Gatto, D., and Florine, B. (1986). Assessment of proplast implants in meniscoplasties as temporomandibular joint surgerical procedures. *Annual Meeting of the American Association of Oral and Maxillofacial Surgeons*. New Orleans, LA.

Wagner, J.D., and Mosby, E.L. (1990). Assessment of Proplast-Teflon disk replacements. *Journal of Oral and Maxillofacial Surgery*, 48(11): 1140–44.

Westesson, P. L., Eriksson, L., and Lindstrom, C. (1987). Destructive lesions of the mandibular condyle following diskectomy with temporary silicone implant. *Oral Surgery Oral Medicine Oral Pathology Oral Radiology and Endodontics*, 63(2): 143–50.

Westfall, R.L., Homsy, C.A., and Kent, J.N. (1982). A comparison of porous composite PTFE/graphite and PTFE/aluminum oxide facial implants in primates. *Journal of Oral and Maxillofacial Surgery*, 40(12): 771–75.

Wolford, L.M. (1997). Temporomandibular joint devices: treatment factors and outcomes. *Oral Surgery Oral Medicine Oral Pathology Oral Radiology and Endodontics*, 83(1): 143–49.

Wolford, L.M., and Mehra, P. (2000). Custom-made total joint prostheses for temporomandibular joint reconstruction. *Proceedings of the Baylor University Medical Center*, 13(2): 135–38.

Worrall, S.F., and Christensen, R.W. (2006). Alloplastic reconstruction of the temporomandibular joint in treatment of craniofacial development or congenital anomalies. *Surgical Technology International*, 15: 291–301.

Wyatt, C.C.L., and Zarb, G.A. (2002). Bone level changes proximal to oral implants supporting fixed partial prostheses. *Clinical Oral Implants Research*, 13(2): 162–68.

Yanase, R.T., Bodine, R.L., Tom, J., and White, S.N. (1994). The mandibular subperiosteal implant denture – a prospective survival study. *Journal of Prosthetic Dentistry*, 71(4): 369–74.

Yuan, K., Lee, T.M., and Huang, J.S. (2010). Temporomandibular joint reconstruction: from alloplastic prosthesis to bioengineering tissue. *Journal of Medical and Biological Engineering*, 30(2): 65–72.

16 Soft tissue replacements

Inquiry

What methods can be used to keep soft tissue replacements fixed in the body?

A key challenge of developing soft tissue replacements is selecting a *method* for keeping them in place. Whether the implant is highly load-bearing (such as an artificial ligament) or only minimally (as in artificial cheekbones), implant **migration** is a serious issue. Because soft tissue replacements are typically made of compliant materials, implant fixation is particularly difficult. Tools such as screws or wires, commonly used in orthopedic implants, will only serve to damage these more delicate materials. Some typical methods for soft tissue replacements include **suturing** and encouraging tissue ingrowth into porous meshes. Are there ways to incorporate adhesives or less compliant materials for improved fixation?

16.1 Historical perspective and overview

Previous chapters in the clinical section of this textbook have described devices used to replace hard tissues (such as bone or teeth) and blood-interfacing implants (including vascular grafts and stents). Hard tissue replacements have their special challenges: for example, the need for a hip implant to be strong enough to withstand loading but have a low stiffness so as to reduce stress shielding. Blood-interfacing devices are required to provide structural support while preventing adhesion of platelets and without damaging blood cells. A third category of implant is soft tissue replacements. These can be further subdivided into mechanical supports (demanding a specific stiffness or strength as in sutures and synthetic ligaments), space fillers (primarily cosmetic implants, but also including artificial skin, which require compliance match and conformability), and highly specialized ophthalmic implants, which have further specifications related to their optical properties.

Sutures are among the earliest recorded medical devices. Linen fibers were used as sutures nearly 4000 years ago (Roby and Kennedy, 2004). Catgut and silk were the most commonly used suture materials until synthetic polymers were developed starting in the 1940s (Roby and Kennedy, 2004). The ancient Greek physician, Galen, used catgut sutures in 150 CE, making them among the few current medical devices that have been used in the same form for millennia. In contrast to the development of sutures, artificial skin is a more modern construct. Beginning in the late 20th century, research techniques combining advanced materials with tissue engineering concepts resulted in layered implants that are able to mimic the form and function of natural skin.

The development of cosmetic breast implants goes back to 1895, when physicians transplanted a **lipoma** (a benign tumor) to a breast to fill a tissue defect (Curtis and Colas, 2004). In the 1940s, it was common practice to inject substances into the breast for tissue augmentation. Injected substances included paraffin wax and petroleum jellies. The drawbacks of this technique were myriad: pain, skin discoloration, infection, pulmonary embolism, and death (IOM, 2000). In the late 1950s and early 1960s, implantation of the Ivalon sponge, made from PVA and manufactured by Ivano, Inc. (Chicago, Illinois) was popularized and PU sponges followed (Curtis and Colas, 2004). These sponges would sometimes become calcified, a condition known as "marble breast" (Park and Lakes, 2007). In 1961, the first silicone gel-filled implants were marketed, and while designs have evolved since that point, the materials have remained largely the same.

The search for a suitable synthetic ligament replacement is documented in the early 20th century. Scientists experimented with wires made of silver or stainless steel, and tried fibers such as silk and nylon with little success in animals. Because the wires and fibers failed in these studies, there were no recorded clinical trials (Legnani *et al.*, 2010). Current designs rely on polymers such as PP, PTFE, and PET.

Today's ophthalmic implants are based upon the experiences of a Second World War pilot, "Mouse" Cleaver. This pilot's aircraft was hit by a cannon shell, with the result that the PMMA canopy of his plane exploded, causing numerous fragments to become embedded in both eyes. He managed to bail out of the aircraft and survived. Dr. Harold Ridley performed 17 operations over a year and managed to return vision to one eye. During this time, Dr. Ridley recognized that this material was inert in the body and would not cause further damage to the tissue unless it had a sharp edge. Based on this information, he developed the intraocular lens in 1947 using material from the same manufacturer that produced the canopies of the WWII Hurricane and Spitfire aircrafts (Arnott, 2007).

The materials used in modern soft tissue replacement can be subdivided in numerous ways. One method is by chemical makeup. Materials can be classified as natural or synthetic, and synthetic can be further subdivided into resorbable and nonresorbable. Some standard materials and their uses in soft tissue replacements are given in Table 16.1.

These materials can also be classified by the structure used in various devices. Soft tissue implants can be made from solid blocks, porous components, woven or knitted materials, and monofilament or bundled fibers, illustrated in Figure 16.1. Some examples of these usages are given in Table 16.2. Further descriptions of these materials will be given in the implant-specific sections below.

Table 16.1 Materials used in soft tissue replacements, classified by chemical makeup and device area

			Sutures	Cosmetic	Skin	Ligament	Ophthalmic
Natural		Cartilage	1				
		Collagen	4		2		
		Fat		1			
		Silk	3				
		Skin			2		
		Submucosa	4				
Synthetic	Non-resorbable	High-density polyethylene (HDPE)		5,6			
		Polyamide (nylon)	3		2		
		Polyethylene terephthalate (PET)	3			2	
		Poly (2hydroxyethyl methacrylate) (pHEMA)					7
		Polymethyl methacrylate (PMMA)					7
		Polypropylene (PP)	3			2	
		Polytetrafluoroethylene (PTFE)	9	8		2	
		Polyurethane (PU)		4			
		Silicone		4		2	
		Stainless steel	10				
		Ultra-high molecular weight polyethylene (UHMWPE)	11	5			
	Resorbable	Polycaprolactone (PCL)	12				
		Polyglycolic acid (PGA)	12			2	
		Polylactic acid (PLA)	12			2	
		Polydioxanone	12				

[1] Stucker and Lian, 2009, [2] Ventre et al., 2009, [3] Bhat, 2005, [4] Park and Lakes, 2007, [5] Boahene, 2009, [6] Binder et al., 2009, [7] Refojo, 2004, [8] LaFerriere and Castellano, 2009, [9] Cross and Ryan, 2006, [10] Chu, 2002, [11] Kurtz, 2009, [12] Roby and Kennedy, 2004.

In the design of soft tissue replacements, material property mismatch and migration are the primary clinical concerns; that is, the surgeon seeks the best possible compliance match and fixation method to soft tissue so that the replacement feels "natural" and offers long-term stability. Particularly in cosmetic implants used in facial reconstruction, such as the chin, cheek, or nose, or other replacements for tissue defects, component migration is problematic. Also, for cosmetic implants, the "feel" of the replacement is very important, so a material that is too rigid will, while maintaining the correct geometry, still not be a desired implant due to its foreign tactility. In some soft tissue devices, other aspects of the material, such as mechanical strength or optical properties, are the primary concerns, but fixation remains a challenge. The success of modern soft tissue implants is due mainly to the development of polymers and the many available

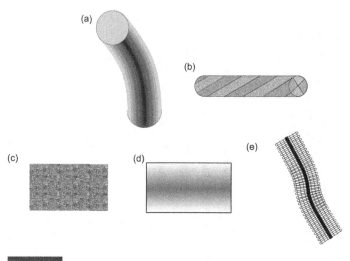

Figure 16.1

Schematics of material structures used in soft tissue replacement. (a) monofilament, (b) multifilament, (c) porous, (d) solid, (e) woven.

forms in which they can be manufactured: solid, gel-like, film, porous, fibrous, knitted, and others.

16.2 Learning objectives

This chapter describes selected soft tissue implants made from synthetic and natural tissues. Mechanical performance requirements will be discussed and additional challenges in device design will be established within the context of the FDA approval process. By the end of this chapter, students will be able to:

1. compare and contrast natural and synthetic materials used in soft tissue replacement
2. succinctly describe the clinical need for sutures, synthetic ligament, skin replacement, ophthalmic implants, and cosmetic implants as well as their historical origins
3. explain the anatomy of the tissues that are being replaced or supplemented with these implants
4. summarize currently available sutures, synthetic ligaments, skin replacements, ophthalmic implants, and cosmetic implants
5. discuss the functional requirements for sutures, synthetic ligaments, skin replacements, ophthalmic implants, and cosmetic implants and predict if a material would be grossly unsuitable
6. cite some challenges inherent to the design and testing of sutures, synthetic ligaments, skin replacements, ophthalmic implants, and cosmetic implants

Table 16.2 Materials used in soft tissue replacement, classified by structure and device area

Material	Solid	Porous	Woven	Monofilament	Multifilament
Collagen		skin[1]		sutures[2]	
High-density polyethylene (HDPE)		cosmetic[3,4]			
Poly (2hydroxyethyl methacrylate) (pHEMA)	ophthalmic[5]				
Polyamide			skin[1]	sutures[6]	sutures[6]
Polyethylene terephthalate (PET)			ligament[1]	sutures[6]	sutures[6]
Polyglycolic acid (PGA)		skin[1]		sutures[6]	sutures[6]
Polylactic acid (PLA)		skin[1]		sutures[6]	sutures[6]
Polymethyl methacrylate (PMMA)	ophthalmic[5]				
Polypropylene (PP)			ligament[1]	sutures[6]	
Polytetrafluoroethylene (PTFE)		cosmetic[7], sutures[8]	ligament[1]		
Polyurethane (PU)	cosmetic[2]				
Silicone	cosmetic[2]	skin[1]			
Ultra-high molecular weight polyethylene (UHMWPE)		cosmetic[3]			sutures[9]

[1] Ventre et al., 2009, [2] Park and Lakes, 2007, [3] Boahene, 2009, [4] Binder et al., 2009, [5] Refojo, 2004, [6] Bhat, 2005, [7] LaFerriere and Castellano, 2009, [8] Cross and Ryan, 2006, [9] Kurtz, 2009

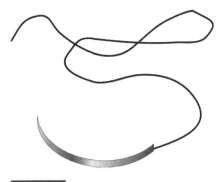

Figure 16.2

A suture with attached needle.

7. analyze a situation in soft tissue replacement history where a device has failed prematurely due to corrosion, fatigue, wear, clinical complication, poor material selection, or inappropriate design

16.3 Sutures

In 2001, the suture market in the United States was approximately $632 million and was divided as follows: 13% natural materials, 41% synthetic non-resorbable, 42% synthetic resorbable (Roby and Kennedy, 2004). The natural sutures include waxed silk and catgut sutures, actually made of bovine or murine submucosa. Catgut sutures can be treated with chromic salts to slow their rate of absorption. Synthetic sutures are made from a wide variety of polymers, discussed in more detail below.

16.3.1 Clinical uses of sutures

Sutures are used to hold tissue together when it has been torn or cut due to injury or surgery. They are typically monofilament or multifilament lengths that are packed as sterilized units with an attached needle, as shown in Figure 16.2. Sutures are Class II devices and are subject to special controls. Some sutures are only approved for specific types of surgery or areas of the anatomy.

16.3.2 Current suture designs

Synthetic non-resorbable sutures can be made from polyamide (nylon), polyethylene terephthalate (PET), polypropylene (PP), ultra-high molecular weight polyethyelene

Table 16.3 Representative non-resorbable sutures currently marketed in the United States, organized by material

Material	Manufacturer	Name	Structure (monofilament or braided)	Coating	Indications
316L	Covidien Syneture (Mansfield, MA)	Steel	M		abdominal wound, intestinal anastomoses, hernial repair, sternal closure
316L	Ethicon, Inc. (Somerville, NJ)	Surgical stainless steel	M		abdominal wound, orthopedic, hernial repair, sternal closure
ePTFE	W. L. Gore & Associates (Flagstaff, AZ)	Gore-Tex	M		anastomoses, valve repair, carotid endarterectomy, ventral hernia
Nylon		Dermalon	M		general soft tissue and ligation, ophthalmic, cardiovascular, neurological
Nylon		Monosof	M		general soft tissue and ligation, ophthalmic, cardiovascular, neurological
Nylon	Covidien Syneture	Surgilon	B	silicone	general soft tissue and ligation, ophthalmic, cardiovascular, neurological
Nylon		Ethilon	B		general soft tissue and ligation, ophthalmic, cardiovascular, neurological
Nylon	Ethicon, Inc.	Nurolon	B		general soft tissue and ligation, ophthalmic, cardiovascular, neurological
Nylon	S. Jackson, Inc. (Alexandria, VA)	Supramid	B		
Nylon		Supramid Extra	B	nylon shell	
Nylon	Teleflex Medical (Research Triangle Park, NC)	Nylon	M		general soft tissue and ligation, ophthalmic, cardiovascular, neurological

(cont.)

Table 16.3 (cont.)

	Manufacturer	Name	Structure (monofilament or braided)	Coating	Indications
PET	Covidien Syneture	Surgidac	B		ophthalmic
		TiCron	B	silicone	cardiovascular
	Ethicon, Inc.	Ethibon Excel	B	polybutilate	general soft tissue and ligation, ophthalmic, cardiovascular, neurological
		Mersilene	B		general soft tissue and ligation, ophthalmic, cardiovascular, neurological
	Teleflex Medical	Cottony II	B		general soft tissue and ligation, ophthalmic, cardiovascular, neurological, orthopedic
		Polydek	B	light PTFE	general soft tissue and ligation, ophthalmic, cardiovascular, neurological, orthopedic
		Tevdek	B	heavy PTFE	general soft tissue and ligation, ophthalmic, cardiovascular, neurological, orthopedic
PP	Covidien Syneture	SurgiPro	M		general soft tissue and ligation, some cardiovascular
	Ethicon, Inc.	Prolene	M		general soft tissue and ligation, ophthalmic, cardiovascular, neurological
	Teleflex Medical	Deklene II	M		general soft tissue and ligation, cardiovascular
UHMWPE	Teleflex Medical	Force Fiber	B		general soft tissue and ligation, cardiovascular, some orthopedic

(UHMWPE), expanded polytetrafluoroethylene (ePTFE), and stainless steel. The processing, chemistry, and morphology of these materials are described in detail in Chapters 2 and 4. Some representative sutures, their structures, and suggested indications are given in Table 16.3.

Resorbable sutures are made from polymers such as polydioxanone (PDO), poly-L-lactic acid (PLLA), polyglycolic acid (PGA), polyglactin 910 (a 90/10 PGA-PLLA random copolymer), and polyglytone (a PGA-PLA-polycaprolactone-trimethylene carbonate random copolymer). These sutures have a characteristic total absorption time in addition to strength estimates at various time intervals post-surgery. For

Table 16.4 Resorbable sutures, materials, structures, and absorption times

Manufacturer	Trade name	Structure (monofilament or braided)	Material	Details
Covidien Syneture (Mansfield, MA)	Biosyn	M	polyglytone	90–110 day absorption, wound support for 21 days
	Caprosyn	M	polyglytone	<56-day absorption, wound support for 10 days
	Maxon	M	polyglytone	180-day absorption, wound support for 42 days
	Polysorb	B	coated PGA/PLLA copolymer	56–70 day absorption, wound support for 21 days
Ethicon (Somerville, NJ)	Monocryl	M	poliglecaprone 25	91–119 day absorption, 50–70% strength at 7 days
	PDS II	M	polydioxanone	183–238 day absorption, 40–70% strength at 28 days
	Vicryl Rapide	M or B	coated polyglactin 910	42 or 56–70 day absorption, 50% strength at 5 days
Teleflex Medical (Research Triangle Park, NC)	Bondek	B	coated PGA	
	Monodek	M	polydioxanone	50% strength at 28 days

example, Coated Vicryl sutures (Ethicon, Somerville, New Jersey) will have 70% strength at 2 weeks and are fully absorbed in 56–70 days. A summary of some resorbable sutures, their materials, structures, and total absorption times are given in Table 16.4.

16.3.3 Functional requirements of sutures

There are many important physical characteristics of sutures. Some have been discussed in other areas of this textbook, such as biocompatibility (Chapter 1), elastic modulus and

Table 16.5 USP sizes, tensile strengths, and needle attachment strengths (USP-NF 2006)

USP size	Limits on average diameter (mm) USP <861>		Knot-pull tensile strength (N) USP <881>	Needle attachment strength (N) USP <871>	
	Min	Max		Avg of 5 Min	Individual Min
10–0	0.020	0.029	0.24*	0.137	0.098
9–0	0.030	0.039	0.49*	0.206	0.147
8–0	0.040	0.049	0.69	0.490	0.245
7–0	0.050	0.069	1.37	0.784	0.392
6–0	0.070	0.099	2.45	1.67	0.784
5–0	0.10	0.149	6.67	2.25	1.08
4–0	0.15	0.199	9.32	4.41	2.25
3–0	0.20	0.029	17.4	6.67	3.33
2–0	0.30	0.349	26.3	10.8	4.41
0	0.35	0.399	38.2	14.7	4.41
1	0.40	0.499	49.8	17.6	5.88
2	0.50	0.599	62.3	17.6	6.86
3 and 4	0.60	0.699	71.5	17.6	6.86

* Measured by straight pull

Figure 16.3

One method for tying a surgeon's knot.

tensile strength (Chapter 6), creep (Chapter 7), and coefficient of friction (Chapter 11). Others, such as knot-pull strength (because the knot acts as a stress concentrator and reduces the tensile load that the suture can hold), **needle attachment force**, and **knot security** (the stability of a knot once it is tied) are specific to sutures. The tests for knot-pull strength and needle attachment force are standardized in USP <881> and <871>, respectively. USP <861>, <871>, and <881> are summarized in Table 16.5. The US Pharmacopeia is a published set of specifications and references, generally used in the pharmaceutical field but which also maintains definitions for measurements related to sutures. For knot-pull strength, a surgeon's knot (illustrated in Figure 16.3) is tied around rubber tubing and the failure load is recorded. The strain rate depends on the suture gauge length. The needle attachment force is a simple tensile measurement of the force required to remove the needle from the suture. Finally, the dimensions and the behavior of the material *in vivo*, such as swelling or degrading, must be quantified. Clinicians who use this product will also be concerned with a suture's handling, which includes properties such as **knot tie-down** (ability to slip a knot down the suture), **first throw hold** (load that a knot can hold before slipping), **tissue drag** (added force needed to pull the suture through tissue due to friction), and **package memory** (a measure of the permanent set of the suture in the packaged configuration). Multifilament designs frequently are used to increase the suppleness of a suture made out of a stiffer material.

The FDA guidance for sutures focuses on biocompatibility, sterility, physical characteristics (such as diameter, needle attachment, and tensile strength), resorption profile, and labeling (kind of surgery, site, and patient population). It does not suggest clinical trials unless there has been a change in technology (such as a new material) or usage (such as a different type of surgery) (FDA, 2003). The main competition for sutures is the development of surgical adhesives and tapes (Smith, 2004).

16.4 Synthetic ligament

Although the geometry of a ligament seems quite simple, the search for a replacement has not yet managed to combine resistance to torsion, abrasion, and cyclic loading. Early failed replacements include a PTFE/carbon composite called Proplast (Vitex Inc., Houston, Texas) and the PP Polyflex (Richard, Memphis, Tennessee), both of which were withdrawn from the market due to rupture problems and inflammatory reactions (Legnani *et al.*, 2010). A further discussion of the use of carbon fiber in ligament replacements can be found in the case studies section of this chapter.

In 2004, there were 200,000 ACL reconstructive surgeries in the United States (Altman and Horan, 2006). The traditional method for replacement is to use a donor graft from another location, such as the patellar tendon as shown in Figure 16.4. This technique has certain drawbacks, including limited graft size, **morbidity** at the donor site, and a long rehabilitation period for the patient after surgery (Legnani *et al.*, 2010; Ventre *et al.*, 2009).

16.4.1 Ligament anatomy and clinical reasons for use

Ligament anatomy and hierarchical organization were described in Chapter 5. Ligaments attach bone to bone and are generally responsible for restraining motion and for transferring load. As illustrated in Figure 16.5, there are four main ligaments in the knee,

16.4 Synthetic ligament

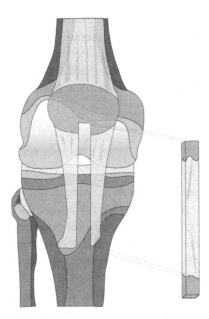

Figure 16.4

The use of the patellar tendon as a donor site for an ACL replacement autograft.

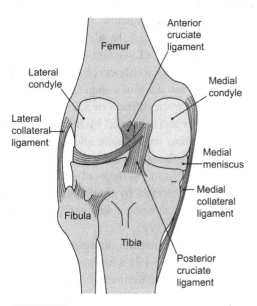

Figure 16.5

The ligaments of the knee.

whose combined placements and strengths are really the only restraints on the motion of that joint. The majority of ligament reconstructive surgeries are performed on the knee, with the most commonly injured ligament being the anterior cruciate ligament (ACL) (Amis, 2006; Ventre et al., 2009). The ACL is the ligament connecting the distal femur to the medial side of the lateral femoral condyle and is thought to rupture due to lateral rotational motion (Altman and Horan, 2006).

16.4.2 Synthetic ACL designs

There are few current options for replacing the ACL. Autografts, as described earlier, have limited infection risk, but are restrained by graft size and donor site morbidity. There is also evidence that these grafts can weaken post-operatively (Ventre et al., 2009). Cadaveric allografts are another alternative. In this case, there is no donor site morbidity (most likely the donor is a cadaver) and limitation to the graft size is less of a concern. Sterilization of allografts, commonly achieved through gamma irradiation or EtO gas exposure (described in detail in Chapter 1), can change their mechanical properties, and these implants are known to weaken 2 weeks to 4 months after implantation (Scheffler et al., 2008; Ventre et al., 2009). Generally there is a risk of infection and/or disease transmission because terminal sterilization would damage the tissue so severely. Synthetic replacements can be classified by use (permanent replacement, permanent augmentation, and scaffold) or by structure (multifilament, woven, and knitted). A discussion of specific implants follows below.

Permanent replacements require high-strength, highly fatigue-resistant materials. Two examples of these are the Gore-Tex Cruciate Ligament Prosthesis (W. L. Gore & Associates, Flagstaff, Arizona) and Styker-Dacron (Meadox Medicals, Oakland, New Jersey) implants. The Gore-Tex replacement was made of a single PTFE fiber arranged in loops. It had a failure load of 4800–5300 N and a stiffness of 322 N/m (Legnani et al., 2010). It was introduced in the early 1980s and approved by the FDA in 1986; however, it was withdrawn in 1993 due to problems with rupture and local inflammatory reaction (FDA, 1993). The Stryker-Dacron prosthesis was a woven PET cable with a velour tube covering. It had a failure load of 3631 N and a stiffness of 420 N/m (Dauner and Planck, 1996; Legnani et al., 2010). It was approved by the FDA in 1988 but withdrawn in 1994, with clinical data from one study showing a 40% rupture rate after 18 months of implantation (Arnauw et al., 1991; FDA, 1993).

Some devices are characterized as permanent augmentation devices, meaning that they provide support for the existing ACL while it heals, and in case it does not recover full strength. Examples of these include the Kennedy Ligament Augmentation Device (LAD) (3M, Minneapolis, Minnesota) and the Trevira (Hoescht/Telos, Germany). The Kennedy LAD is an 8 mm diameter ribbon of woven PP, designed to allow for progressive load transfer from the implant to the healed ligament. It has a failure load of 1500–1730 N and a stiffness of 280 N/m (Dauner and Planck, 1996). It was approved by the FDA in 1987 (FDA, 1993). The Trevira prosthesis is shown in Figure 16.6. Made of woven

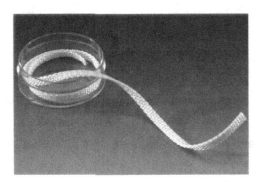

Figure 16.6

The Trevira/Hochfest ligament augmentation device. Image provided courtesy of Telos GmbH.

PET, it is twisted to achieve a specific stress-strain relationship. The untwisted failure strength is 4000 N, but the twisting reduces this to 1866 N. The stiffness is 68 N/m. It has been implanted since 1980 (Dauner and Planck, 1996; Legnani *et al.*, 2010).

Finally, there are the synthetic scaffolds, which provide a framework for tissue ingrowth but are not expected to be the sole source of mechanical strength. These include the Leeds-Keio (Xiros, Leeds, UK) and the Ligament Advanced Reinforcement System (LARS) (Surgical Implants and Devices, Arc Sur Tille, France). The Leeds-Keio prosthesis is a 10 mm diameter tube of woven PET. It has a failure strength of 2000 N and a stiffness of 270 N/m (Dauner and Planck, 1996). It was developed in 1982 but has not shown very strong results in long-term follow-ups (Legnani *et al.*, 2010). The LARS system has a unique design that attempts to capture some of the hierarchical aspects of the natural ligament. At the ends, it consists of bundled PET fibers surrounded by a knitted PET cover, while in the middle it has a more open structure of parallel, twisted PET fibers. The open structure in the articular section is supposed to encourage tissue ingrowth to increase the strength of the implant (Mascarenhas and MacDonald, 2008).

16.4.3 Functional requirements of synthetic ligaments

A healthy ACL will undergo 1–2 million cycles per year with an average peak load of 400 N, although the load in the ACL can be as high as 630 N during activities such as running (Altman and Horan, 2006). Research shows that the ACL can withstand a maximum load of 1716–2160 N and has a stiffness of 182–322 N/m (Altman and Horan, 2006; Legnani *et al.*, 2010). Thus, potential ligament replacement, augmentation, or scaffold should have properties in this range. A summary of the implants described above is shown in Table 16.6.

There are numerous concerns associated with the development of synthetic ligament replacements. The poor performance shown thus far by most synthetic grafts is surprising because they generally have had positive results during *in vitro* testing. *In vivo*, these

Table 16.6 Summary of ACL implants comparing failure load and stiffness (Legnani et al., 2010; Dominkus et al., 2006; Jackson and Evans, 2003)

Name	Use	Material	Structure	Failure load (N)	Stiffness (N/mm)
Healthy ACL				1716–2160	182–322
Patellar tendon				2300	620
Gore-Tex Cruciate Ligament Prosthesis (W.L. Gore & Associates, Flagstaff, AZ)	Replacement	PTFE	woven	4800–5300	322
Stryker-Dacron (Meadox Medicals, Oakland, NJ)	Replacement	PET	woven	3631	420
Kennedy LAD (3M, Minneapolis, MN)	Augmentation	PP	woven ribbon	1500–1730	280
Trevira (Hoescht/Telos, Germany)	Augmentation	PET	woven, twisted	1866	68
Leeds-Keio (Xiros, UK)	Scaffold	PET	woven tube	2000	270
LARS (Surgical Implants and Devices, France)	Scaffold	PET	bundled fibers with knitted cover	1500–4700	

implants have typically failed due to abrasion, fatigue, torsion, and collagen invasion (Ventre et al., 2009). It has been suggested that the inflammatory response due to wear particles could be contributing to osteoarthritis in the knee (Legnani et al., 2010).

Synthetic ligaments are Class III devices and as such, IDE and PMA applications are required. The CDRH does not regulate autografts, but the synthetic permanent and augmentation devices fall under its domain. The FDA guidance for these implants suggests two phases of testing: preclinical and clinical. Preclinical testing should include the following:

- physical/chemical testing (for leachables)
- biological testing (pyrogenicity, hemolytic potential, acute toxicity, intracutaneous irritation, cytotoxicity, genetic toxicity, immunological potential)
- the effect of sterilization on the material stability and mechanical properties
- mechanical testing (tensile, fatigue, bending, fixation, abrasion)
- long-term animal testing (at least 12 months) in which pathology and particulate migration are studied in the tissue and the effect of implantation on the mechanical properties and stability of the material is quantified

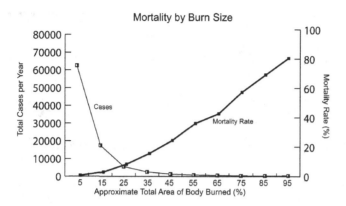

Figure 16.7

Burn victim mortality and number of burns by burn size (ABA, 2007).

In clinical testing, the FDA guidance again recommends two phases: a single-center test with at least 100 devices and a 2-year minimum follow-up, and then a multi-center test with a minimum of six surgeons doing a minimum of 10–15 implants each. It suggests reporting evaluations of the patients before, during, and after the operation as well as follow-up evaluations and general effectiveness data (FDA, 2003).

16.5 Artificial skin

It was not until the 1970s that the value of immediate excision of burned tissue and skin graft was shown (Morgan *et al.*, 2004). Before this, the main treatment for burned tissue was simply to cover it and wait for healing. The last part of the 20th century into the early portion of the 21st century has been marked by rapid advancement in the field of wound healing technologies, with advanced materials and tissue culture techniques leading to fascinating solutions.

The American Burn Association (ABA) estimates 500,000 burns per year requiring medical treatment in the United States, with 40,000 requiring hospitalization (ABA, 2007). The percentages of each burn size from a 2010 report are shown in Figure 16.7.

16.5.1 Anatomy and clinical reasons for use

Artificial skin is used for temporary wound coverage or permanent wound closure. It is used in cosmetic surgery (Becker *et al.*, 2009), to aid in the healing of venous ulcers and diabetic foot ulcers (Zaulyanov and Kirsner, 2007) and in burns or other large-area skin injuries (Bhat, 2005; Park and Lakes, 2007). The layered structure of the skin is discussed in detail in Chapter 5 and is illustrated in Figure 16.8. Human skin is the largest organ in the body, covering approximately 2 m^2 of surface area in an adult human, and

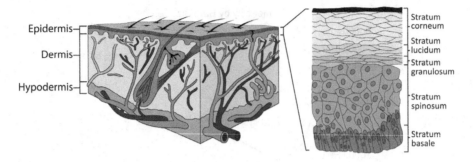

Figure 16.8

The layers of the skin.

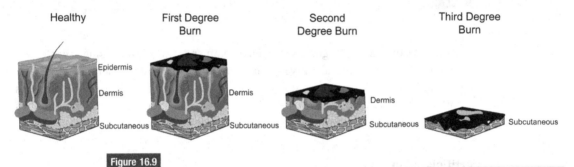

Figure 16.9

The depth of damage corresponding to first-, second-, and third-degree burns.

it comprises 16% of a person's body weight (Bhat, 2005; Silver, 1994). It ranges in thickness from 0.2 mm on the eyelid to 6 mm on the bottom of the foot (Bhat, 2005). Skin is involved in temperature regulation, wound healing, protection from disease, removal of waste, and synthesis of vitamins (Bhat, 2005; Silver, 1994; Ventre *et al.*, 2009).

When a person suffers from a burn, that burn is classified according to its depth. This classification system is shown in Figure 16.9. Burns in which the epidermis is removed are referred to as **first degree burns**. **Second degree burns** include the epidermis and part of the dermis, while **third degree burns** are more severe, extending into the subcutaneous tissue. A person with a large burn is in danger because of the moisture loss that occurs without skin and the potential for infection. Although split-thickness skin grafts can be taken from other parts of the body, sometimes the burn area is simply too extensive. Then, an artificial skin can be used for immediate wound closure.

16.5.2 Current designs for artificial skin

These solutions can be classified as temporary or permanent, natural or synthetic, and immediate (urgent) or requiring lead time (gradual). Another classification scheme divides these implants into epidermal, dermal acellular, dermal cellular and composite.

The current gold standard is a **split-thickness skin graft** (STSG), also known as an autograft. This technique, which uses a **dermatome** to harvest a layer of the epidermis and partial dermis up to 0.03 inches (762 μm) thick, has a low risk of infection and rejection. Autografts are 0.01–0.016 inches (254–406 μm) thick, and although the epidermis regenerates, the dermis does not (FDA, 1997; Morgan et al., 2004). Allografts from cadavers are possible but are usually rejected after approximately three weeks (Ventre et al., 2009). However, this can be enough time for permanent grafts to be generated through other techniques.

One permanent skin replacement is Epicel (Genzyme, Cambridge, Massachusetts). This technique, known as **cultured epidermal autograft** (CEA), was developed by Reinwald and Green in 1975 (Morgan et al., 2004). The implant requires approximately 21 days for cell cultivation from the patient's own **keratinocytes**. The resulting graft is 2–8 cell layers thick and capable of permanently covering large areas. Unfortunately, because it is not multilayered, it can have unstable attachment to the tissue below it and sometimes undergoes scar contraction (Bello et al., 2001; Enoch et al., 2006).

There are three current skin replacements that can be classified as **dermal acellular implants**. The first is Integra (Integra Lifesciences, Plainsboro, New Jersey). Developed by Yannas and co-workers in the early 1980s, Integra is a temporary but immediate skin replacement consisting of a porous crosslinked bovine collagen and **glycosaminoglycan** scaffold with a semipermeable silicone layer on the outermost surface (Morgan et al., 2004). The scaffold is thought to help develop neodermis. Eventually the silicone layer is removed and a split-thickness skin graft is added (Bello et al., 2001; Enoch et al., 2006). Another dermal acellular implant is Alloderm (LifeCell, Branchburg, New Jersey). Alloderm is a human cadaveric skin allograft. It is a non-living matrix with an intact basement membrane. This implant is also temporary, meaning that it will subsequently require a skin graft. It is readily available and is also sometimes used to fill tissue defects in cosmetic surgery (Bello et al., 2001; Enoch et al., 2006). A final example of this group is Biobrane (Smith & Nephew, London, UK). This implant is a silicone film with a partially embedded collagen-coated nylon fabric. This textured surface is thought to help with tissue ingrowth. This is a temporary implant, used to cover extensive burns and/or large donor sites (Enoch et al., 2006; Ventre et al., 2009).

The dermal cellular implants use **neonatal fibroblast** cells to provide a living implant, but do not require donor cells from the patient, which gives these implants a time advantage over cultured epidermal autografts. Transcyte (Advanced BioHealing, Westport, Connecticut) is a silicone membrane with collagen-coated nylon mesh. Neonatal fibroblast cells are cultured onto the nylon fibers. This implant is impermanent and has been shown to be as effective as cadaver skin (Bello et al., 2001). Dermagraft (Advanced BioHealing, Westport, Connecticut) is a polyglactin mesh seeded with neonatal fibroblasts. This permanent graft has good tearing resistance and is only used for diabetic foot ulcers. The mesh disappears after 3–4 weeks (Bello et al., 2001; Enoch et al., 2006).

Apligraf (Organogenesis, Canton, Massachusetts) is an example of a composite implant, or one that has both a dermal-like and epidermal-like layer. It consists of a collagen gel seeded with neonatal fibroblasts, over which keratinocytes are allowed to form a layer similar to stratum corneum. It is a permanent implant and shows expedited healing in excised and full-thickness wounds (Zaulyanov and Kirsner, 2007). One challenge related to its use is that it has relatively short shelf-life and must be stored chilled within a specific temperature range (Bello et al., 2001).

16.5.3 Functional requirements of artificial skin

When developing this type of implant, the primary concerns are limiting fluid loss and achieving good adherence with the tissue. Using a porous material with a pore size of approximately 200 μm is optimal for tissue ingrowth because it allows for capillary formation (Chu, 2002). A match of compliance and mechanical strength to natural tissue should also be a goal so that there is no buckling or peeling of the graft away from the underlying tissue. Furthermore, concerns may arise regarding air bubbles or fluid pooling underneath the implant if there is not good adhesion. Although there is no FDA guidance document for these implants, the submission summaries for Epicel's IDE and Dermagrafts PMA are publicly available and may provide insight. In its submission for Epicel, Genzyme reports results obtained from characterization of the fibroblast cells used to assist in the growth of the donor keratinocytes, the efficacy of the Epicel attachment, and the sterility of the product (FDA, 1997). The PMA for Dermagraft shows the tests conducted during its preclinical and clinical phases. The preclinical testing included characterization of the fibroblasts, small animal feasibility studies, tear strength of the graft, effect of freezing on its mechanical properties, toxicology, stability, and the effect of changing the implant size for commercial scale production. Clinical studies were evaluated for wound closure time and percentage closure at 12 weeks (FDA, 2001).

16.6 Ophthalmic implants

16.6.1 Anatomy and clinical reasons for use

The soft tissues of the eye serve both mechanical and optical purposes. As seen in Figure 16.10, light enters the eye through the cornea, passes through the lens, and is focused on the retina. The average travel distance from cornea to retina is 24.2 mm in a healthy adult eye (Patel, 2004). The adult cornea is about 0.5 mm thick (Meek, 2008) and the curvature of the cornea is responsible for approximately 80% of the focusing of the eye. The tissues of the eye provide both optical properties and mechanical properties, withstanding a normal **intraocular pressure** of 10–20 mm Hg (Snell and Lemp, 1998).

The cornea has five layers: the **epithelium**, **Bowman's layer**, the **stroma**, **Descemet's membrane**, and the **endothelium**. These layers are illustrated in Figure 16.11. The

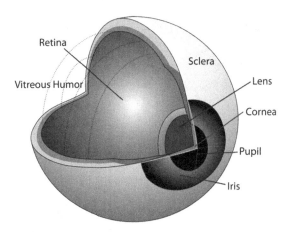

Figure 16.10

The basic anatomy of the eye.

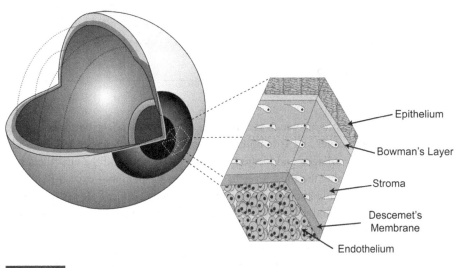

Figure 16.11

The layers of the cornea.

epithelium is the outermost layer and is exposed to the external environment. It is responsible for blocking foreign materials and providing a smooth surface for absorption of oxygen and tears. Because the cornea does not have blood vessels, it is required to access oxygen from the air (Refojo, 2004). The epithelium is approximately 10% of the cornea's thickness and is more densely populated with nerve endings than any other part of the body, which is why the cornea is so sensitive to abrasion (Snell and Lemp, 1998). Below the epithelium is Bowman's layer, composed of layered collagen fibers. It is transparent but can scar if injured, leading to vision loss. The middle layer of the cornea

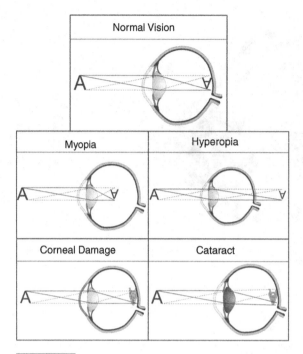

Figure 16.12

Illustrations of various ophthalmologic disorders.

is the stroma, which accounts for around 90% of the cornea's thickness. The stroma is 78% water and 16% collagen. Descemet's membrane is a thin protective barrier made from collagen and located between the stroma and the endothelium. The endothelium maintains the fluid level in the cornea by pumping excess fluid out of the stroma. If the stroma becomes swollen with water, its optical properties are negatively affected. Endothelial cells can not be regenerated, and in the case of injury, the only option is corneal transplant surgery.

The lens is a transparent structure which is approximately 10 mm in diameter and 4 mm thick in adults. It has an elastic capsule, an epithelium, and lens fibers. The capsule is mainly composed of collagen and completely surrounds the lens. The epithelium is responsible for homeostasis. Lens fibers are packed transparent cells that make up the bulk of the lens. The curvature of the lens is controlled by ciliary muscles, which allow the eye to focus at different distances (Snell and Lemp, 1998).

16.6.2 Contact lenses: current designs and functional requirements

Vision can be affected by simple changes in the eye's geometry, as illustrated in Figure 16.12. If the eyeball elongates in the posterior-anterior direction, images will be focused on the interior of the eye instead of on the retina, in a condition known as **myopia**, or nearsightedness. The opposite condition, **hyperopia**, or farsightedness, occurs when the eyeball has become shortened in this direction. These disorders can also occur due to changes in the cornea or lens. Myopia and hyperopia require a relatively common intervention in the form of eyeglasses or contact lenses. Approximately 150 million Americans wear corrective eyewear, and of those, over 38 million wear contact lenses (AAO, 2009). Contact lenses are thin polymer components that are placed over the cornea, reshaping it and improving vision. Contact lenses require a material that has good light transmission, chemical stability, manufacturability, oxygen transmissibility, tear-film wettability, and resistance to protein adsorption (Refojo, 2004). Early contact lens designs could be hard (made of PMMA) or soft (manufactured from hydrogels), but currently most contact lenses are soft. These are generally made from pHEMA, which has about 40% water when swollen in physiological saline. In an effort to increase oxygen transmission, pHEMA can be used in copolymers with materials such as methacrylic acid, acetone acrylamide, or polyvinyl alcohol, which brings the water content as high as 78%. Another method is to use siloxane hydrogels, which have lower water content but higher oxygen permeability. These hydrogels must be surface treated to improve their hydrophilicity (Refojo, 2004).

The FDA Guidance Document for contact lenses suggests that a manufacturer submit the following for a 510(k) (FDA, 1994):

- Chemical composition and purity of monomer
- Manufacturing information (method, polymerization and annealing conditions, manufacturing flow chart and sterilization method, other manufacturing conditions, drawings for lens design, packaging materials, and methods)
- Shelf life information
- Leachability of monomers and tints
- Finished lens parameters
- Preservative uptake and release
- Physiochemical properties (color and light transmittance, refractive index, water content, *in vitro* wetting angle, oxygen permeability, mechanical properties)
- Suppliers of lens blanks

16.6.3 Corneal replacements: current designs and functional requirements

If the cornea is severely injured or scarred, it may require replacement. Natural corneal transplant surgeries have among the highest success rates of any transplant surgery, on the order of 70% success at one-year follow-up (Aldredge and Krachmer, 1981; Refojo, 2004). Should a transplant surgery fail, there are currently two corneal replacements approved by the FDA, and they are very different in design. The first is the Dohlman Doane Keratoprosthesis (Mass Eye and Ear, Boston, Massachusetts), also known as the Boston Keratoprosthesis. This implant was given FDA approval in 1992 (Marten *et al.*, 2009). It consists of a PMMA button that is pressed into a two-piece interface

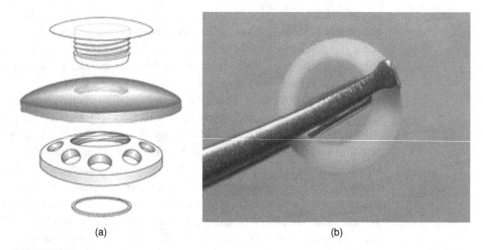

Figure 16.13

Examples of currently marketed keratoprostheses: (a) a schematic of the Boston K-Pro (Image courtesy of the Massachusetts Eye & Ear Infirmary) and (b) the AlphaCor (Image courtesy of Addition Technology, Inc.).

which also grasps a corneal graft, as shown schematically in Figure 16.13(a). The entire assemblage is then implanted in the eye. The second corneal replacement available in the United States is the AlphaCor (Addition Technology, Inc., Des Plains, Illinois). This implant, a one-piece component made of pHEMA, consists of a transparent gel at the center and a porous sponge along the outer ring that promotes tissue integration and provides a location for attachment. The optical area is <30% water by weight and the sponge is 45% water by weight. It requires two surgeries – one for implantation and one to remove scarred tissue after the surgery (Marten *et al.*, 2009). The AlphaCor is illustrated in Figure 16.13(b).

A corneal replacement must be strong enough to maintain the structure and shape of the eyeball and to withstand intraocular pressure. It requires high resistance to tearing and viscoelasticity to accommodate changes to the eye's geometry over time. Human cornea has an ultimate tensile strength of 19.1 MPa as measured in a strip test (Bryant *et al.*, 1994) and an elastic modulus that ranges from 0.1–0.9 MPa (Liu and Roberts, 2005). Intraocular pressure can be modeled *in vitro* by biaxial loading. Concerns with these implants include calcium deposition, shelf life, and preventing scarring of the surrounding tissue. The FDA Guidance Document for keratoprostheses recommends submission of the following information, with all tests performed on the finished device (FDA, 1999):

- Biocompatibility testing (extracts, cytotoxicity, genotoxicity, maximum sensitization per ISO 10993–10, intramuscular animal implantation, ocular implantation, chemical testing, dimensions, and surface quality)
- Optical testing (dioptric power)
- Sterility testing (validation of method)

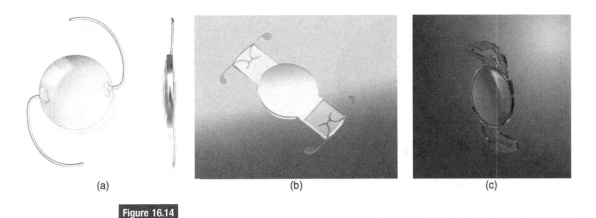

Figure 16.14

Examples of currently marketed intraocular lenses: (a) Tecnis 1-piece Monofocal IOL (Image courtesy of Abbott Medical Optics), (b) Crystalens IOL (Image courtesy of Bausch and Lomb), and (c) 570C IOL (Image courtesy of Rayner Intraocular Lenses Ltd. © Rayner Intraocular Lenses Ltd. 2010).

- Packaging and labeling
- Clinical investigation

16.6.4 Intraocular lenses: current designs and functional requirements

Serious vision problems can result from lens clouding, referred to as a **cataract**. Cataracts are a leading cause of blindness, affecting 22 million Americans over the age of 40 (AAO, 2009). It is expected that they will affect 40 million Americans by 2020 (Patel, 2004). If left untreated, cataracts can lead to **glaucoma** and blindness. When a cataract is sufficiently advanced to require surgery, the lens is removed and replaced by an **intraocular lens** (IOL). Early IOLs were made of PMMA, as described earlier in this chapter. Dr. Harold Ridley first developed an IOL that was 10.5 mm in diameter and weighed 110 mg (current implants are half this size and one-fifth of the weight). Although the Ridley IOLs were not ideal, PMMA IOLs continued to be the standard until the development of foldable IOLs in the early 1980s. Initially made of polysiloxane and pHEMA, foldable IOLs could be inserted through a smaller incision reducing the need for stitching and the risk of surgically-induced **astigmatism**. Approximately 15 years later, the AcrySof IOL (Alcon Laboratories, Inc., Hunenberg, Switzerland) was introduced. This IOL was made of a proprietary copolymer of phenylethyl acrylate and phenylethyl methacrylate. This material was hydrophobic and had a higher refractive index, allowing for a thinner lens. Most modern IOLs are made of soft acrylate or polysiloxane. Their optic diameter ranges from 4–6 mm and they weigh 20 mg on average (Alcon Laboratories, Inc.). They are made by manufacturers such as Alcon Laboratories, Inc., Abbot Medical Optics, Inc. (Santa Ana, California), Bausch and Lomb, Inc. (Rochester, New York), and Rayner Surgical (Bamburg, Germany); some of these are shown in Figure 16.14. STARR Surgical (Monrovia, California) is made

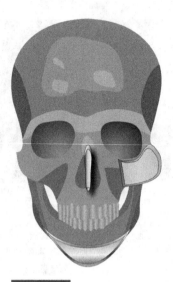

Figure 16.15

Some locations of common facial implants: cheek, nose, and chin augmentation.

from a collagen-pHEMA copolymer which has a high surface water content. Intraocular lenses are Class III devices, whose main functional requirements are optical properties, size, and ability to be folded for implantation.

16.7 Cosmetic implants

Cosmetic implants include both soft tissue reconstruction and augmentation. These implants are used in areas including the face, breast, and buttocks. Estimates for the numbers of invasive cosmetic surgeries performed in the United States in 2009 range from 1.5–4 million (AACS, 2009; ASPS, 2009). The breakdown varies from 799 buttock implants to 375,372 breast augmentations and reconstructions combined. These implants can be necessary due to tissue defects resulting from disease or injury, or they could be used based on the desires of the patient. Locations for some common facial implants are shown in Figure 16.15.

16.7.1 Anatomy and clinical reasons for use

Cosmetic implants can be used to change the perceived shape of the face or other areas of the anatomy through augmentation of the underlying tissue. They are attached to structural tissues, and tissue ingrowth is encouraged to aid in fixation.

Figure 16.16

The evolution of breast implants from the 1960s to the present: (L-R) first generation, second generation, and third generation (Image courtesy of Allergan, Inc.)

16.7.2 General cosmetic implants: current designs and functional requirements

Currently available cosmetic implants are made from ePTFE, silicone, HDPE, PET, and nylon (Fanous et al., 2002). Companies such as Spectrum Designs Medical (Carpinteria, California), Implantech (Ventura, California), and Allied Biomedical (Ventura, California) make a variety of solid silicone and PTFE implants. These solid implants can be further shaped by the surgeon for custom profiles. Some silicones are considered too firm for cosmetic implants because they are readily distinguishable from natural tissue by touch. In thin-skinned individuals, the bright white of PTFE implants can be visible through the skin. Both silicone and PTFE have the disadvantage of being slippery during handling (Fanous et al., 2002). Meshes made from nylon and PET, such as Mersilene (Ethicon, Somerville, New Jersey) and Supramid (S. Jackson, Alexandria, Virginia), can be cut and folded into shape before implantation, and the textures encourage enough tissue ingrowth to maintain implant position (Binder et al., 2009). In addition to being biocompatible and mechanically/chemically stable, ideal cosmetic implants are a good compliance match to tissue, easily shaped, able to maintain form, easily modified or customized before implantation, and have favorable surfaces for positioning.

16.7.3 Breast implants: current designs and functional requirements

The evolution of breast implant design is shown in Figure 16.16. Early designs had a thick, vacuum-molded elastomer shell sealed with a seam, and polyester mesh on the back for improving fixation. These implants had a low rupture rate due to their thick shell, but gel seepage and implant contracture were common (IOM, 2000). In the late 1960s, the first saline-filled implant was designed by Heyer Schulte. Known for their "sloshy" feel, these implants suffered from an enormously high deflation rate of 76% (IOM, 2000).

The late 1970s and early 1980s heralded the development of second-generation breast implants. These implants had thinner shells and double-lumen designs that were added to reduce seepage, although this was never shown to be the case. Another design feature that was added in this time period was an outer covering of PU foam, which helped to reduce **contracture** (Handel et al., 2006). However, the PU foam would begin to disintegrate immediately upon implantation, surrounding the implant with a fibrous capsule filled with foam particles. This made these implants difficult to revise, and they were removed from the market in 1992. These second-generation implants still experienced high rupture and deflation rates.

Third-generation implants were developed beginning in the mid-1980s. These implants featured stronger shells, better deflation rates, and textured surfaces (IOM, 2000). Breast implants were classified as Class III devices in 1988 and PMA certification was required in 1991 (FDA, 2006). Around this time, women with failed breast implants began to sue device manufacturers. The legislative history of breast implants will be further described in one of the case studies presented at the end of this chapter. In the wake of these lawsuits, the FDA took breast implants off the market except as investigational devices in 1991. Numerous studies and committees, including one sponsored by the Institute of Medicine, were commissioned to examine the research on breast implanta and to develop conclusions regarding their safety. Based on these recommendations, in 2000, saline-filled implants were re-approved, and in 2005, silicone gel-filled implants were re-approved (Bruning, 2002).

For breast augmentation and reconstruction, there are only two manufacturers that sell implants: Mentor Corporation (Santa Barbara, California) and Allergan, Inc. (Irvine, California). Both of these companies manufacture silicone gel-filled and saline-filled implants in silicone sacs. An advantage of saline-filled implants is that they can be refilled and adjusted, and they are lower density and easier to handle during surgery (Park and Lakes, 2007). In comparison, silicone gel-filled implants have a more realistic feel and are not prone to collapse the way that saline implants are during a shift of the implant filling. Textured surfaces are associated with a higher rate of deflation in saline-filled implants (Handel et al., 2006).

For breast implants, post-implantation concerns include rupture, deflation, capsular contracture (painful shrinking of fibrous tissue surrounding implant), and pain (IOM, 2000). FDA guidance for submission of PMAs for breast implants suggests the following data for submission (FDA, 2006):

- Chemistry: general chemical information, crosslinking, extractables, volatility, heavy metal content, saline content, silicone gel content, alternative filler content
- Toxicology
- Mechanical testing: fatigue rupture, valve competency, cohesivity, bleed testing, stability testing, shelf-life testing
- Explant analysis
- Clinical trials

Although breast implants show a frequent need for revision surgery due to rupture and other problems, patients consistently rate their satisfaction with the implants very highly (Handel et al., 2006).

16.8 Case studies

16.8.1 Carbon fiber ligament failure (materials failure)

In the late 1970s and early 1980s, the use of carbon fiber composites for ligament replacements was popularized by the Proplast (Vitex Inc., Houston, Texas) and the Intergraft (Osteonics Biomaterials, Livermore, California). Implants designed from carbon fiber alone experienced good results in pre-clinical testing (Jenkins et al., 1977), and in early clinical trials. In a 12–36 month follow-up, no mechanical failure of the carbon-fiber ligament was found (Jenkins and McKibbin, 1980) and optimistically, the authors suggested that none might occur since this material was thought to grow stronger with tissue ingrowth. However, longer-term studies showed that these implants demonstrated extremely poor clinical performance. Carbon fragments were found scattered throughout the knee (Leyshon et al., 1984), and in some cases, the femoral notch was inflamed and stained black (Rushton et al., 1983). These researchers also found evidence that the carbon fiber was abrading the bone in the knee. The poor torsion resistance of the composite implants is credited with their eventual failure, resulting in carbon deposits in the liver and inflammatory problems in the knees (Legnani et al., 2010). The development of these implants demonstrates how there can be a stark difference between *in vivo* and *in vitro* performance, and that pre-clinical mechanical testing should focus on creating testing scenarios that, while not necessarily mimicking *in vivo* conditions, at least test their most challenging modes.

16.8.2 Breast implants (regulatory failure)

The safety of breast implants has been a source of controversy since the mid-1980s, when the wave of legal cases against implant manufacturers began. In 1984, a jury found that a woman's auto-immune disease was caused by her implants, and awarded her $1.7 million (Brunk, 1998). In 1992, breast implants were removed from the market by the FDA and by December 1994, over 19,000 lawsuits had been filed against Dow Corning, a prominent implant manufacturer (Brunk, 1998). Although multiple studies to that point had not shown a link between breast implants and auto-immune diseases or cancer, public opinion played a very strong role. Dow Corning had a particularly difficult case because its own internal records showed that of 329 tests conducted on breast implants, only a few had involved human subjects, and the longest duration was 80 days. In 1999, Dow Corning settled a class action lawsuit with a $3.2 billion settlement (Bruning, 2002).

In parallel with the legal battles, the Institute of Medicine (a division of the National Academy of Sciences), set up a 13-member panel to review scientific literature relating to breast implants, and to study industry reports and hear testimony from the public. In 1999, this panel released its findings (IOM, 2000). Of these, some of the most crucial are:

- Based on current literature, silicone implants do not cause major diseases, including breast cancer, auto-immune or neurological diseases.
- There is no increased silicone in breastmilk due to implants, although breastfeeding itself may be compromised.
- Local complications (such as rupture, contraction, and infection) are frequent with gel-filled silicone implants, but are not life-threatening.

The IOM panel also made a specific recommendation for increased standardization of this industry. Because early breast implants were frequently customized for the patient, there was very little regulatory control over the final product. Before they were taken off the market, it was common practice to implant these devices without any testing. Breast implants were re-introduced to the market in 2000 and 2006 for saline-filled and gel-filled, respectively, and although there are now only two manufacturers who sell devices approved for use in the United States, there used to be many more. Between the customization and the number of manufacturers and models, it is very challenging to track the performance of breast implants sold before 1992. Standardization of testing and designs, and improved tracking of existing implants, could increase the ability of the FDA to evaluate and monitor these implants.

16.8.3 IOL fracture (design failure)

Fracture of the IOL at the optic-haptic junction, while rare, does occur and can have potentially damaging results. Gokhale presents a case study in which a patient develops corneal **edema** one month after IOL implantation. During the initial surgery, one haptic of the foldable lens fractured, and this fragment was removed surgically one month later. The corneal edema was resolved and there was no lasting damage (Gokhale, 2009). Kirkpatrick and Cook describe the case of a cataract surgery in which the dialing hook became stuck in the dialing hole located at the base of the haptic, as shown schematically in Figure 16.17. After the clinician rolled the dialing hook in order to loosen it, the IOL fractured across the dialing hole and the haptic became detached (Kirkpatrick and Cook, 1992). The authors attributed this fracture to the placement of the dialing holes and the possibility that loading of these holes was not properly taken into account during testing. Because these lenses can be made of brittle materials, any stress concentrations induced by the haptic design or other features used for adjustment should be fully evaluated for their effect on the overall integrity of the device.

16.9 Looking forward

The future of soft tissue devices is very much dependent on the invention of novel materials, through new chemistries or tissue engineering. It seems unlikely that the suture market will experience great changes in the future otherwise. One manufacturer,

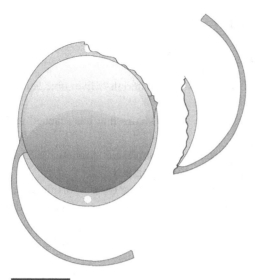

Figure 16.17

An example of intraocular lens haptic fracture.

B. Braun Melsungen AG, held a competition in 2008 encouraging participants to rethink sutures in ways both large (new materials, adaptive sutures, the use of needles) and small (changes to packaging, needle ergonomics) (Braun, 2007). The winners described concepts such as materials that changed color when under excessive force, thus signaling the possibility of tearing during surgery (Braun, 2008). Whether or not this added value is worth applying for pre-market approval through the FDA remains to be seen. In synthetic ligaments, the future is somewhat unclear. Although there has been concentrated research in this area for the past three decades, there is not yet a good solution. Current resorbable materials degrade too quickly for the new ligament to form and have full strength (Amis, 2006). Augmentations through tissue engineering may be a possibility here, and in artificial skin and cosmetic implants.

16.10 Summary

The historical development and current embodiments of selected soft tissue replacements were described in this chapter. Some implants, such as sutures, have seen little change in the modern era, while others, such as artificial skin, have only become possible due to recent scientific advancements. The majority of these implants are polymer-based and promote tissue ingrowth, but perhaps the next generation will be even more bioactive, adapting to their environments and loading in the manner of natural tissues.

16.11 Problems for consideration

Checking for understanding

1. What are the three categories of suture materials? Give some defining characteristics of each.
2. What would be the stress on a size 1 suture that has reached its knot-pull tensile strength?
3. What are the current gold standard methods for synthetic ACL and artificial skin? Why are other options for these being pursued?
4. Name advantages and disadvantages of three types of artificial skin implants.
5. How have the materials used in intraocular lenses evolved over time?
6. What are the main engineering considerations in designing cosmetic implants?

For further exploration

1. Select the most appropriate material for an artificial earlobe: PMMA, UHMWPE, silicone, or PGA. Explain your decision.
2. Find an example in literature of a soft tissue implant failure. Explain how you, as an engineer working for the implant manufacturer, could have prevented this failure.
3. How does LASIK surgery work compared to other vision correction methods described in this chapter?

16.12 References

Aldredge, O.C., and Krachmer, J.H. (1981). Clinical types of corneal transplant rejection. *Archives of Opthalmology*, 99(4): 599–604.

Altman, G.H., and Horan, R.L. (2006). Tissue engineering of ligaments. In *An Introduction to Biomaterials*, ed. S.A. Guelcher and J.O. Hollinger. Boca Raton, FL: CRC Press, pp. 499–524.

American Academy of Cosmetic Surgeons. (2009). 2009 AACS procedural census fact sheet. Chicago, IL.

American Academy of Ophthalmology. (2009). Eye health statistics at a glance. San Francisco, CA.

American Burn Association. (2007). Burn Incidence and Treatment in the US: 2007 Fact Sheet.

American Society of Plastic Surgeons. (2009). 2010 Report of the 2009 Statistics. Arlington Heights, IL.

Amis, A.A. (2006). Artificial ligaments. In *Repair and Regeneration of Ligaments, Tendons, and Joint Capsule*, ed. WR. Walsh. Totowa, NJ: Humana Press, pp. 233–56.

16.12 References

Arnauw, G., Verdonk, R., Harth, A., Moerman, J., Vorlat, P., Bataillie, F., and Claessens, H. (1991). Prosthetic versus tendon allograft replacement of ACL-deficient knees. *Acta Orthopaedica Belgica*, 57(Supplement 2): 67–74.

Arnott, E.J. (2007). *A New Beginning in Sight*. London: Royal Society of Medicine Press.

Becker, S., Saint-Cyr, M., Wong, C., Dauwe, P., Nagarkar, P., Thornton, J.F., and Peng, Y. (2009). AlloDerm versus Dermamatrix in immediate expander-based breast reconstruction: a preliminary comparison of complication profiles and material compliance. *Plastic and Reconstructive Surgery*, 123(1): 1–6.

Bello, Y.M., Falabella, A.F., and Eaglstein, W.F. (2001). Tissue-engineered skin: current status in wound healing. *American Journal of Clinical Dermatology*, 2(5): 305–13.

Bhat, S.V. (2005). *Biomaterials*. Middlesex, UK: Alpha Science.

Binder, W.J., Azzizadeh, B., and Tobias, G.W. (2009) Aesthetic facial implants. In *Facial Plastic and Reconstructive Surgery*, 3d edn., ed. I.D. Papel, J.L. Frodel, G.R. Holt, *et al*. New York: Thieme Medical Publishers, pp. 389–408.

Boahene, K.D.O. (2009). Synthetic implants. In *Facial Plastic and Reconstructive Surgery*, 3d edn., ed. I.D. Papel, J.L. Frodel, G. R. Holt, *et al*. New York: Thieme Medical Publishers, pp. 67–76.

Braun Brochure B16602 (2007). The Future of Sutures. August.

Braun press release (2008). International winners of the ideas competition initiated by B. Braun Melsungen AG receive awards in Berlin. December 9, 2008.

Bruning, N. (2002). *Breast Implants: Everything You Need to Know*, 3d edn. Alameda, CA: Hunter House.

Brunk, C.G. (1998). Managing risks in the restructured corporation: the case of Dow Corning and silicone breast implants. In *The Ethics of the New Economy: Restructuring and Beyond*, ed. L. Groarke. Ontario, Canada: Wilfrid Laurier University Press, pp. 189–202.

Bryant, M.R., Szerenyi, K., Schmotzer, H., and McDonnell, P.J. (1994). Corneal tensile strength in fully healed radial keratotomy wounds. *Investigative Opthalmology and Visual Science*, 35(7): 3022–31.

Chu, C.C. (2002). Textile-based biomaterials for surgical applications. In *Polymeric Biomaterials*, 2d edn., ed. S. Dumitriu. New York: Marcel Dekker, pp. 491–544.

Cross, F.W., and Ryan, J.M. (2006). The severely injured patient. In *General Surgical Operations*, 5th edn., ed R.M. Kirk. Philadelphia, PA: Elsevier, pp. 28–43.

Curtis, J., and Colas, A. (2004). Medical applications of silicones. In *Biomaterials Science*, 2d edn., wd. B.D. Ratner, A.S. Hoffman, F.J. Schoen, and J.E. Lemons. San Diego, CA: Elsevier Academic Press, pp. 698–708.

Dauner, M., and Planck, H. (1996). Ligament replacement polymers (biocompatibility, technology and design). In *Polymeric Materials Encyclopedia*, ed. J.C. Salamone. Boca Raton, FL: CRC Press, pp. 3579–92.

Dominkus, M., Sabeti, M., Toma, C., Abdolvahab, F., Trieb, K., and Kotz, R.I. (2006). Reconstructing the extensor apparatus with a new polyester ligament. *Clinical Orthopaedics and Related Research*, 453: 328–34.

Enoch, S., Grey, J.E., and Harding, K.G. (2006). Recent advances and emerging treatments. *British Medical Journal*, 332(7547): 962–65.

Fanous, N., Samaha, M., and Yoskovitch, A. (2002). Dacron implants in rhinoplasty. *Archives of Facial Plastic Surgery*, 4(3): 149–56.

Food and Drug Administration. (1993). Guidance document for the preparation of investigational device exemptions and premarket approval applications for intra-articular prosthetic knee ligament devices. Rockville, MD: FDA.

Food and Drug Administration. (1994). Amendment 1 to May 12, 1994 Premarket Notification (510(k)) Guidance Document for Daily Wear Contact Lenses. Rockville, MD: FDA.

Food and Drug Administration. (1997). Summary of safety and probable benefit: Epicel. Rockville, MD: FDA.

Food and Drug Administration. (1999). Guidance on 510(k) Submissions for Keratoprostheses. Rockville, MD: FDA.

Food and Drug Administration. (2001). Summary of safety and effectiveness data: Dermagraft. Rockville, MD: FDA.

Food and Drug Administration. (2003). Class II Special Controls Guidance Document: Surgical sutures; guidance for industry and FDA. Rockville, MD: FDA.

Food and Drug Administration. (2006). Saline, silicone gel and alternative breast implants. Rockville, MD: FDA.

Gokhale, N. (2009). Late corneal edema due to retained foldable lens fragment. *Indian Journal of Opthalmology*, 57(3): 230–31.

Handel, N., Cordray, T., Guttierrez, J., and Jensen, J.A. (2006). A long-term study of outcomes, complications, and patient satisfaction with breast implants. *Plastic and Reconstructive Surgery*, 117(3): 757–67.

Institute of Medicine. (2000). *Information for Women About the Safety of Silicone Breast Implants*, ed. M. Griggs, S. Bondurant, V. L. Ernster and R. Herdman. Washington, DC: National Academy Press.

Jackson, D.W., and Evans, N.A. (2003). Arthroscopic treatment of anterior cruciate ligament injuries. In *Operative Arthroscopy, Volume I*, 3d edn., ed. J.B. McGinty, S.S. Burkhart, R.W. Jackson, D.H. Johnson, and J.C. Richmond. Philadelphia, PA: Lippincott, Williams & Wilkins, pp. 347–65.

Jenkins, D.H.R., Forster, I.W., McKibbin, B., and Ralis, Z.A. (1977). Induction of tendon and ligament formation by carbon implants. *Journal of Bone and Joint Surgery (Br)*, 59B(1):53–57.

Jenkins, D.H.R., and McKibbin, B. (1980). The role of flexible carbon-fibre implants as tendon and ligament substitutes in clinical practice. *Journal of Bone and Joint Surgery (Br)*, 62B(4):497–99.

Kirkpatrick, J.N.P., and Cook, S.D. (1992). Broken intraocular lens during cataract surgery. *British Journal of Opthalmology*, 76:509.

Kurtz, S.M. (2009). Composite UHMWPE biomaterials and fibers. In *UHMWPE Biomaterials Handbook*, 2d edn., ed. S.M. Kurtz. Burlington, MA: Academic Press, pp. 249–58.

LaFerriere, K.A., and Castellano, R.D. (2009). Surgical approaches to the midface complex. In *Facial Plastic and Reconstructive Surgery*, 3d edn., ed. I. Papel. New York:Thieme Medical Publishers, pp. 243–59.

Legnani, C., Ventura, A., Terzaghi, C., Borgo, E., and Albisetti, W. (2010). Anterior cruciate ligament reconstruction with synthetic grafts. A review of literature. *International Orthopaedics*, 34(4): 465–71.

Leyshon, R.L., Channon, G.M., Jenkins, D.H.R., and Ralis, Z.A. (1984). Flexible carbon fibre in late ligamentous reconstruction for instability of the knee. *Journal of Bone and Joint Surgery (Br)*, 66B(2):196–200.

Liu, J., and Roberts, C.J. (2005). Influence of corneal biomechanical properties on intraocular pressure measurement: quantitative analysis. *Journal of Cataract and Refractive Surgery*, 31(1): 146–55.

Marten, L., Wang, M.X., Karp, C.L., Selkin, R.P., and Azar, D.T. (2009). Corneal surgery. In *Ophthalmology*, 3d edn., ed. M. Yanof and J.S. Duker. Philadelphia, PA: Elsevier, pp. 351–59.

Mascarenhas, R., and MacDonald, P. (2008). Anterior cruciate ligament reconstruction: a look at prosthetics – past, present and possible future. *McGill Journal of Medicine*, 11(1): 29–37.

Meek, K.M. (2008). The cornea and sclera. In *Collagen: Structure and Mechanics*, ed. P. Fratzl. New York: Springer Science + Business Media, pp. 359–96.

Morgan, J.R., Sheridan, R.L., Tompkins, R.G., Yarmush, M.L., and Burke, J.F. (2004). Burn dressing and skin substitutes. In *Biomaterials Science*, 2d edn., ed. B.D. Ratner, A.S. Hoffman, F.J. Schoen, and J.E. Lemons. San Diego, CA: Elsevier Academic Press, pp. 603–14.

Park, J.B., and Lakes, R.S. (2007). *Biomaterials: An Introduction*, 3d edn. New York: Plenum Press.

Patel, A.S. (2004). Intraocular lens implants: a scientific perspective. In *Biomaterials Science*, 2d edn., ed. B.D. Ratner, A.S. Hoffman, F.J. Schoen, and J.E. Lemons. San Diego, CA: Elsevier Academic Press, pp. 592–602.

Refojo, M.F. (2004). Ophthalmological applications. In *Biomaterials Science*, 2d edn., ed. B.D. Ratner, A.S. Hoffman, F.J. Schoen, and J.E. Lemons. San Diego, CA: Elsevier Academic Press, pp. 584–91.

Roby, M.S., and Kennedy, J. (2004). Sutures. In *Biomaterials Science*, 2d edn., ed. B.D. Ratner, A.S. Hoffman, F.J. Schoen, and J.E. Lemons. San Diego, CA: Elsevier Academic Press, pp. 615–28.

Rushton, N., Dandy, D.J., and Naylor, C.P.E. (1983). The clinical, arthroscopic and histological findings after replacement of the anterior cruciate ligament with carbon-fibre. *Journal of Bone and Joint Surgery (Br)*, 65B(3): 308–9.

Scheffler, S.U., Gonnerman, J., Kamp, J., Przybilla, D., and Pruss, A. (2008). Remodeling of ACL allografts is inhibited by peracetic acid sterilization. *Clinical Orthopaedics and Related Research*, 466:1810–18.

Silver, F.H. (1994). *Biomaterials, Medical Devices, and Tissue Engineering: An Integrated Approach*. London: Chapman and Hall.

Snell, R.S., and Lemp, M.A. (1998). *Clinical Anatomy of the Eye*, 2d edn. Oxford, UK: Blackwell Science, Inc.

Smith, D. C. (2004). Adhesives and sealants, in *Biomaterials Science*, 2nd edn., ed. B.D. Ratner, A.S. Hoffman, F.J. Schoen, and J.E. Lemons. San Diego, CA: Elsevier Academic Press, pp. 572–83.

Stucker, F.J., and Lian, T.S. (2009). Biological tissue implants. In *Facial Plastic and Reconstructive Surgery*, 3d edn., ed. I.D. Papel, J.L. Frodel, G.R. Holt, *et al.* New York: Thieme Medical Publishers, pp. 77–84.

United States Pharmacopeia 29 – National Formulary 24 (2006).

Ventre, M., Netti, P.A., Urciuolo, F., and Ambrosio, L. (2009). Soft tissues characteristics and strategies for their replacement and regeneration. In *Strategies in Regenerative Medicine: Integrating Biology with Materials Design*, ed. M. Santin. New York: Springer, pp. 34–50.

Zaulyanov, L., and Kirsner, R.S. (2007) A review of a bi-layered living cell treatment (Apligraf) in the treatment of venous leg ulcers and diabetic foot ulcers. *Clinical Interventions in Aging*, 2(1): 93–98.

Epilogue

Mechanics of Biomaterials: Fundamental Principles for Implant Design provides the requisite engineering principles needed for the design of load-bearing medical implants with the intention of successfully employing natural or synthetic materials to restore structural function in biological systems. One challenge in the medical device field is the multifactorial nature of the design process. Numerous elements affect device performance, including clinical variables, structural requirements, implant design, materials selection, manufacturing, and sterilization processes. The crucial requirement of any medical device is that it is biocompatible; the implant must restore function without adverse reaction or chronic inflammatory response in the body. In this respect, choice of materials used in the implant is key to the device integrity.

Moreover, the structural requirements of the implant are determined through an assessment of the expected physiological stresses that vary depending upon the patient's anatomy, weight, and physical activity. Analyses of these stresses are key in making certain that the selected material offers the appropriate mechanical properties such as requisite elastic modulus and yield stress, as well as resistance to creep, fracture, fatigue, and wear. This book addresses the complexities that are encountered in physiological loading such as three-dimensional stress states; the complex interplay of dynamic loading, contact mechanics, viscoelastic deformation, and rupture; as well as the combined effects of environmental degradation of the materials owing to biological attack, aqueous environment, and sterilization method employed.

Pre-clinical lab bench or *in vitro* simulator studies are generally used to assess the integrity of the implant and are often necessary in obtaining regulatory (FDA) approval of the medical device. The design process involves the generation of dimensioned drawings for the implant and directly incorporates the necessary geometry that satisfies anatomical constraints and provides the desired function of the device. Engineers often employ an elementary design process to (1) identify the problem, (2) identify criteria and constraints, (3) brainstorm solutions, (4) generate ideas, (5) explore solutions, (6) select the best approach, (7) build protoypes, and (8) refine design. An example of this process is utilized in the hypothetical design of a total hip replacement for a small-framed athletic woman with osteoporosis, illustrated below.

As we look forward in medical device design it is obvious that many emerging technologies will contribute to the long-term restoration of function in the body. These innovations include the emerging technologies of genomics, tissue engineering, imaging, robotic surgery, and the development of minimally invasive surgical methods. Many

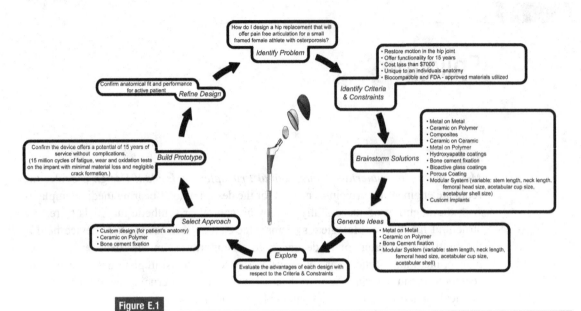

Figure E.1

The design process as utilized in the development of a total hip knee replacement.

tissue-engineered materials are already available for use in bone, cartilage, skin, and ligaments. There are drug-eluting stents on the market that provide pharmaceutical treatment *in vivo* and fully resorbable stents will likely follow in the near future (they are already approved outside of the United States). Novel small-scale device designs that can be passed through a catheter pose endless opportunities for the development of minimally invasive surgery. The possibilities for advancement in biomaterials and medical device technology are vast.

Appendix A
Selected topics from mechanics of materials

A.1 Properties of areas

Circle

$A = \pi R^2$
$I_o = \dfrac{\pi R^4}{4}$

Rectangle

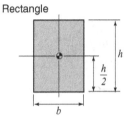

$A = bh$
$I_o = \dfrac{bh^3}{12}$

Triangle

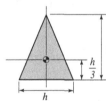

$A = \dfrac{bh}{2}$
$I_o = \dfrac{bh^3}{36}$

Semicircle

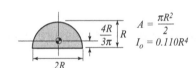

$A = \dfrac{\pi R^2}{2}$
$I_o = 0.110 R^4$

Thin-walled tube

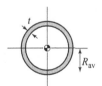

$A = 2\pi R_{av}$
$I_o = \pi R_{av}^3 t$

Half thin-walled tube

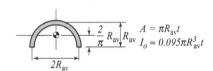

$A = \pi R_{av} t$
$I_o \approx 0.095 \pi R_{av}^3 t$

Figure A.1

Areas and moments of inertia around centroidal axes for basic geometries.

A.2 Thin-walled pressure vessel

For cylindrical pressure vessels, these approximations are within 5% for $t/r_1 < 0.1$ and 10% for $t/r_1 < 0.2$. For spherical pressure vessels, these approximations are within 5% for $t/r_1 < 0.3$ and 10% for $t/r_1 < 0.45$.

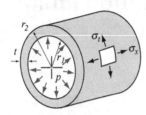

$\sigma_t = \dfrac{pr_1}{t}$, $\sigma_x = \dfrac{pr_1}{2t}$ (closed ends)

$\sigma_t = \dfrac{pr_1}{t}$, $\sigma_x = 0$ (open ends)

$\sigma_r = -p$ (inside)

$\sigma_r = 0$ (outside)

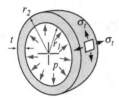

$\sigma_t = \dfrac{pr_1}{2t}$

$\sigma_r = -p$ (inside)

$\sigma_r = 0$ (outside)

Figure A.2

Approximate stresses for thin-walled pressure vessels that are (a) cylindrical and (b) spherical.

A.3 Thick-walled pressure vessel

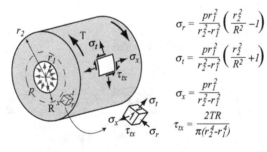

$\sigma_r = \dfrac{pr_1^2}{r_2^2 - r_1^2}\left(\dfrac{r_2^2}{R^2} - 1\right)$

$\sigma_t = \dfrac{pr_1^2}{r_2^2 - r_1^2}\left(\dfrac{r_2^2}{R^2} + 1\right)$

$\sigma_x = \dfrac{pr_1^2}{r_2^2 - r_1^2}$

$\tau_{tx} = \dfrac{2TR}{\pi(r_2^4 - r_1^4)}$

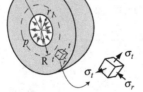

$\sigma_r = \dfrac{pr_1^3}{r_2^3 - r_1^3}\left(\dfrac{r_2^3}{R^3} - 1\right)$

$\sigma_t = \dfrac{pr_1^3}{r_2^3 - r_1^3}\left(\dfrac{r_2^3}{2R^3} + 1\right)$

Figure A.3

Stresses in thick-walled pressure vessels that are (a) cylindrical and (b) spherical.

A.4 Thin-walled tube under torsion and/or bending

These approximations are within 5% for $t/r_1 < 0.1$ and 10% for $t/r_1 < 0.25$.

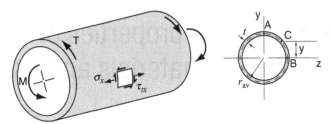

$$\tau_{tx} = \frac{T}{2\pi r_{av}^2 t}, \quad \sigma_{xC} = \frac{My}{\pi r_{av}^3 t}, \quad \sigma_{xA} = \frac{M}{\pi r_{av}^2 t}, \quad \sigma_{xB} = 0$$

Figure A.4

Approximate stresses in a thin-walled tube under torsion and/or bending.

Appendix B
Table of material properties of engineering biomaterials and tissues

Test method & properties measured	Schematic of represented sample & material type	Testing protocol	Summary of test protocol	Equipment needed
Mechanical properties				
Compression – Yield strength – Yield point – Young's modulus – Compressive strength	Metallic materials	Standard Test Methods of Compression Testing of Metallic Materials at Room Temperature E9 – 09	An increasing axial compressive load is applied on the specimen. The load and strain are monitored either continuously or in finite elements, and the specimen's compressive properties are determined.	Compressive testing machine that conforms to requirements of: Standard Practices for Force Verification of Testing Machines
Creep – Cyclic stress-strain deformation response – Cyclic creep deformation response – Cyclic hardening, softening response – Cycles to crack formation	Nominally homogenous materials	Standard Test Method for Creep-Fatigue Testing E2714 – 09	The specimen is subjected to constant-amplitude strain-controlled or constant-amplitude force-controlled tests in the uniaxial direction. This test focuses on creep and fatigue deformation and damage that are generated simultaneously within a given cycle.	Servo-controlled tension-compression fatigue machine in accordance with ISO 7500 – 1 – 2004 or Standard Practices for Force Verification of Testing Machines
Dynamic Testing of Polymers – Transition temperature – Elastic modulus – Loss modulus	Polymers	Standard Practice for Plastics Dynamic Mechanical Properties Determination and Report of Procedures D4065 – 06	Specimen is subjected to mechanical oscillations at either fixed or natural resonant frequencies. The specimen's elastic or loss moduli, or both, are measured while varying time, temperature of the specimen, or both the time and temperature.	Dynamic mechanical or dynamic thermomechanical analyzers

(cont.)

Test method & properties measured	Schematic of represented sample & material type	Testing protocol	Summary of test protocol	Equipment needed
Fatigue Crack Propagation – Crack growth rate	Metallic materials	Standard Test Method for Measurement of Fatigue Crack Growth Rates D4065-06	Precracked notched specimens are cyclically loaded. The crack size is measured as a function of elapsed fatigue cycles. Numerical analysis is used to establish the rate of crack growth.	– Grips and fixtures – Symmetrical force distribution machine
Fractures Toughness – Fracture toughness	Metallic materials	Standard Test Method for Measurement of Fracture Toughness E647 – 08	Fatigue pre-cracked specimen is loaded to induce unstable crack extension and or stable crack extension. This method requires continuous measurement of force versus load-line displacement and crack mouth opening displacement. Fracture instability results in single point-value fracture toughness at the instability point. Stable tearing results in a continuous fracture toughness versus crack-extension curve (R-Curve).	– Apparatus for the measurement of applied force, load-line displacement, and crack-mouth opening displacement. – Displacement gages – Force transducers – Fixtures – Tension testing clevis

Hardness – Indentation hardness	Metallic materials	Standard Test Method for Indentation Hardness of Metallic Materials by Portable Hardness Testers E110 – 02	Specimen is subjected to a force applied by Portable Brinell testers, Rockwell testers, or Vickers testers.	– portable hardness Testers – Hydraulic cylinder with pressure gage – Specimen holder
Impact Testing – Charpy V-notch Toughness – Energy absorbed by materials during fracture	Metallic materials	Standard Test Method for Notched Bar Impact Testing of Metallic Materials E23 – 07	Specimens placed in supports are subjected to a blow from a moving mass. The moving mass must have sufficient energy to break the specimen in its path, and a device is used to measure the energy absorbed by the broken specimen	Pendulum testing machine that conforms to the requirements: Annex A1
J-Integral – J-integral – Crack growth resistance	Polymers	Standard Test Method for Determining J-R Curves of Plastic Materials D6068 – 02	A series of specimens are subjected to different displacements using crosshead or displacement control. The resulting crack fronts are marked and the crack extensions from the fracture surface are measured. The J value is calculated from the indention corrected energy for fracture.	– Displacement control testing machine, which records load versus load-line displacement – Bend test fixture – Grips for specimen

(cont.)

Test method & properties measured	Schematic of represented sample & material type	Testing protocol	Summary of test protocol	Equipment needed
Mixed Mode I-Mode II – Interlaminar fracture toughness G_c	Unidirectional fiber reinforced polymer matrix composites	Standard Test Method for Mixed Mode I-Mode II Interlaminar Fracture Toughness of Unidirectional Fiber Reinforced Polymer Matrix Composites D6671 – 06	Mixed-Mode Bending apparatus load split laminate specimens at various ratios of Mode I to Mode II loading to determine delamination fracture toughness.	– Mixed mode bending fixture – Load indicator – Load point displacement indicator – Load versus load point displacement record – Optical microscope – Micrometer
Punch Test for UHMWPE – Ultimate strength – Yielding – Ductility – Toughness	Ultra-high molecular weight polyethylene	Standard Test Method for Small Punch Testing of Ultra-High Molecular Weight Polyethylene Used in Surgical Implants F2183 – 08	Specimen is subjected to bending by indentation with a hemispherical head steel punch. The hemispherical head punch loads the specimen at a constant displacement rate until failure of the specimen. The load and displacement of the punch is continuously recorded.	– Small punch test apparatus – Guide – Die – Punch – Testing machine described in method D 695 – Compressometer – Compression platen – Micrometers – Thermometer

Static Crack Propagation in Polymers – Slow crack growth resistance	Polyethylene pipes and resins	Standard Test Method for Notch Tensile Test to Measure the Resistance to Slow Crack Growth of Polyethylene Pipes and Resins F1473 – 07	Specimens are notched and exposed to a constant tensile stress at elevated temperatures in air. The time to complete failure is recorded.	– Lever loading machine – Furnace – Temperature controller – Temperature-measuring device – Timer – Alignment jig – Notching machine
Tensile – Elongation	Elastic fabrics	Standard Test Method for Tension and Elongation of Elastic Fabrics (Constant-Rate-of-Extension Type Tensile Testing Machine) D4964 – 08	(A) Loop tension at specific elongation. Specimen is extended at a specified rate to a specified loop tension and returned to a specified rate to zero. The extension-recovery curves are plotted by an automatic recorder. The tension at specific percent elongation is calculated. (B) Elongation at specific loop tension. Specimen is loaded at a specific rate to a specified loop tension, and unloaded at a specified rate to zero loop tension. The tension-recovery curves are plotted by an automatic recorder. The elongation at a specified loop tension is calculated from the plot.	– Tensile testing machine CRE-type conforming to specification D 76 – Band clamps – Sewing machine single-needle

(cont.)

Test method & properties measured	Schematic of represented sample & material type	Testing protocol	Summary of test protocol	Equipment needed
Torsion – Cycles to failure – Location of fractures in relation to the cantilever plane	Metallic stemmed hip arthroplasty femoral components	Standard Practice for Cyclic Fatigue Testing of Metallic Stemmed Hip Arthroplasty Femoral Components with Torsion F1612 – 05	Specimen is subjected to a forcing function at constant periodic amplitude. The amount of horizontal deflection of the head in both the M-L and A-P projections in response to the periodic forcing function.	– Ball or roller bearing, low-friction mechanism – Loading apparatus
Microstructure				
Density – Density	Polymers Any shape	Standard Test Method for Density of Plastics by Density-Gradient Technique D1505 – 03	Specimen is submerged in a liquid column exhibiting a density gradient. The specimen in the liquid column is then compared with standards of known density.	Density-gradient tube Constant-temperature bath Glass floats Pycnometer Liquids Hydrometers Analytical balance Siphon or pipette arrangement
	Polyethylene	Standard Practice for Use of a Melt Index Strand for Determining Density of Polyethylene D2839 – 05	Specimen is placed in boiling water. Specimen is cut into strands and tested according to D1505.	Extrusion plastometer prescribed in test method D1238 Hot plate Beakers

Glass Transition Temperature – Glass transition temperature	Pure materials	Standard Test Method for Assignment of the Glass Transition Temperature By Dynamic Mechanical Analysis E1640 – 09	Specimen is subjected to mechanical oscillation at either fixed or resonant frequencies. Changes in viscoelastic responses of the material are monitored as a function of temperature. The glass transition temperature is determined by extrapolating the onset of the decrease in storage modulus.	Dynamic mechanical analyzer Clamps or calipers – Machine that applies oscillatory stress or strain – Detector – Temperature controller and oven Data collection device Nitrogen, helium, purging gases
Melting Temperature (DSC) – Melting temperature	Pure materials Any shape	Standard Test Method for Melting And Crystallization Temperatures By Thermal Analysis E794 – 06	Specimen is heated at a controlled rate through the region of fusion. The difference in heat flow or temperature and a reference material due to energy changes is recorded. The transition is marked by absorption or release of energy by the specimen resulting in a endothermic (or exothermic) peak in the heating (or cooling) curve.	Differential scanning calorimeter or differential thermal analyzer Furnace Temperature sensor Differential sensor

(cont.)

Test method & properties measured	Schematic of represented sample & material type	Testing protocol	Summary of test protocol	Equipment needed
Molecular weight analysis – Molecular weight averages – Distribution of molecular weights in linear, soluble polystyrene – Molecular weight averages – Molecular weight distributions	Materials of molecular weights from 2000 to 2,000,000 g mol^{-1} Solutions Linear polyolefins Solutions	Standard Test Method for Molecular Weight Averages and Molecular Weight Distribution of Polystyrene by High Performance Size-Exclusion Chromatography D5296 – 05 Standard Test Method for Determining Molecular Weight Averages and Molecular Weight Distribution and Molecular Weight Averages of Polyolefins by High Temperature Gel Permeation Chromatography D6474 – 06	Dilute solution of polystyrene sample is injected into a liquid mobile phase. The mobile phase transports the polymer into and through a chromatographic column. A detector monitors the eluate as a function of elution volume. The elution volumes are converted to molecular weights and other molecular weight parameters. Polyolefin sample is dissolved in a solvent and injected into a chromatographic column, which separates the molecules according to size. The separated molecular are detected and recorded as they elute from the column. The retention times are converted to molecular weights.	– Liquid high-performance size-exclusion chromatography – Solvent reservoir – Solvent pumping system – Sample injector – Columns – Detectors – Recorder or plotter – Data handling system – Chromatography column – Solvent reservoir – Pump – Solvent degasser – Sample injection system – Parked columns – Solute mass detector

Percent Crystallinity (DSC) – Percent crystallinity	Polyetheretherketone PEEK polymers Any shape Must be at least 1–2 mm thick	Standard Test Method for Measurement of Percent Crystallinity of Polyetheretherketone PEEK Polymers by Means of Specular Reflectance Fourier Transform Infrared Spectroscopy (R-FTIR) F2778 – 09	Specimen is evaluated by infrared spectroscopy. The intensity of the absorbance peaks is related to the amount of crystalline regions present in the specimen.	– Infrared spectrometer – Specimen holder – Samples preparation equipment – Software that can use the Ramers-Kronig transformation algorithm

Specific Applications

Environmental Degradation – Stress cracking due to presence of environment for a given time	Ethylene polymers	Standard Test Method for Environmental Stress-Cracking of Ethylene Plastics D1693 – 08	Bent specimens of ethylene plastic, each containing a controlled imperfection on one surface, are exposed to a surface-active agent. The proportion of the total number of specimens that crack for a given time is observed.	– Blanking die-jig – Specimen holders – Test tubes and closures – Aluminum foil – Constant-temperature bath – Test tube rack – Bending clamp – Transfer tool
– Degradation of polymer due to environment	Hydrolytically degradable polymers	Standard Test Method for *in vitro* Degradation Testing of Hydrolytically Degradable Polymer Resins and Fabricated Forms for Surgical Implants F1635 – 04	Samples of polymer resins, semi-finished components, finished surgical implants, or test specimens are placed in buffered saline solution at physiologic temperatures. Samples are periodically removed and tested for various metal and mechanical properties at specified intervals.	– Physiologic soaking solution – Sample container – Constant temperature bath – pH meter – Balance

(cont.)

Test method & properties measured	Schematic of represented sample & material type	Testing protocol	Summary of test protocol	Equipment needed
Stress corrosion cracking – Presence of residual stresses	Copper alloys Any shape	Standard Test Method for Ammonia Vapor Test for Determining Susceptibility to Stress Corrosion Cracking in Copper Alloys B858 – 06	Specimen is placed in a closed container with a specific pH at ambient temperature for 24 hours. After removal, specimen is examined for presence of cracks.	– pH meter – Closed vessel
Wear (Gravimetric weight loss) – Wear rates	Polymers Any shape	Standard Practice for Gravimetric Measurement of Polymeric Components for Wear Assessment- F2025 – 06	Specimen is placed in a soak chamber with optional machines that agitate or cyclically load the specimen. At specified intervals, specimen is removed from chamber and weighed. Wear process characteristics are determined through visual, microscopic, profilometric, or replication techniques.	– Micro-balance – Vacuum jar – Deionized water – Methyl alcohol – Soak chamber – Test lubricant
– Wear rates	Prosthetic hip components	Standard Guide for Gravimetric Wear Assessment of Prosthetic Hip Designs in Simulator Devices F1714 – 96	Specimen is placed in a hip simulator upon completion of the control soak period. At specific intervals, specimen is removed from simulator and weighed. Wear process characteristics are determined through visual, microscopic, profilometric, or replication techniques. Specimen is then replaced, soak controls in fresh lubricant, and wear cycling continues.	– Hip prosthesis components – Hip simulator – Soak chamber – Test lubricant

Appendix C
Teaching methodologies in biomaterials

The field of medical device design is highly multidisciplinary and builds upon a number of specialties including biomaterials science, mechanical engineering, bioengineering, chemical engineering, electrical engineering, integrative biology, public policy, and clinical medicine. With this broad topic comes the challenge of implementing a course that facilitates learning in an interdisciplinary framework. To this end, we have utilized a number of pedagogical techniques to develop an interdisciplinary course entitled *Structural Aspects of Biomaterials*, from which this textbook was developed. Specifically we have utilized tools such as (1) structured learning objectives based on Bloom's taxonomy; (2) active learning practices and inquiry-based lectures; (3) clinical case studies; (4) professional development utilizing interdisciplinary teams with diversified learning styles; (5) outreach teaching (service-based learning) in the K-12 sector; and (6) criteria specified by the Accreditation Board for Engineering and Technology (ABET).

C.1 Structured learning objectives based on Bloom's taxonomy

As there are so many sub-specializations within the field of medical device design it is useful to have distinct learning objectives that are both observable and specific (Anderson *et al.*, 2001). Clearly defined course objectives provide a guide that facilitates the structure of lectures and serves as a study guide for students. In this respect, the application of Bloom's taxonomy (Bloom, 1984) when developing the learning objectives for a course is quite useful. The use of Bloom's taxonomy in the cognitive domain of learning has been updated by Anderson (Anderson *et al.*, 2001) and comprises both lower level thinking skills (remembering, understanding, applying) as well as higher order thinking skills (analyzing, evaluating, creating). Table C.1 shows the ranking (in descending order) of skill levels where Level 6 (L6) represents the highest level of learning (creating) and Level 1 (L1) describes the lowest level of learning (remembering). Creativity is the ultimate outcome for the student and for this reason it is the skill that is listed first in Table C.1. Ideally, each lecture topic addresses all six levels of learning. More important is that the full span of lower-level and upper-level thinking skills is contained within the course content. An example of course learning objectives (http://www.educationoasis.com) that we have employed in past offerings of the *Structural Aspects of Biomaterials course* is given in Table C.2.

Table C.1 Levels of learning and associated skills (listed in descending order) based on Bloom's Taxonomy (Bloom, 1984)

Skill Level	Bloom's Taxonomy Skills	Student skills
L6	Creating	Design (formulate, create, plan)
L5	Evaluating	Make technical judgments (choose, rank, critique)
L4	Analyzing	Explain behavior of a system (interpret, predict)
L3	Applying	Problem-solving (resolve, elucidate)
L2	Understanding	Describe observations (compare, classify)
L1	Remembering	Recall facts (define, list, replicate)

Table C.2 Sample learning objectives for Structural Aspects of Biomaterials

At the completion of the course the student will be able to:

List the specific metals, ceramics, and polymers used in medical devices

Specify the attributes and limitations for each class of materials

Schematically illustrate the stress-strain curve for metals, ceramics, and polymers on one plot

Describe the basic structure of bone, cartilage, dental tissue, vascular tissue, and their constituents

Define Type I, Type II, and Type III devices

Give the requirements for FDA approval of a new device and a "me-too" device

Identify appropriate sterilization protocols for medical devices

Derive the 3D Hooke's Law and use it to solve a 3D loading situation of a medical device

Calculate the effective stress in a component and determine if it will yield or fracture

Determine the critical flaw size in a material and the applied stress intensity in a device

Write the basic equations for the relationship between loading cycles and applied stress or strain

Decide when to use a total life fatigue philosophy versus defect-tolerant approach

Predict the expected life of a device based on loading cycles

Assess a component for wear and corrosion resistance

Explain clinical treatments for osteoarthritis and trauma of articular joints and the spine

Describe the clinical use of dental, soft tissue, and vascular implants

Justify structural and clinical requirements of various medical devices

Design a structural implant used in orthopedics, dental, cardiovascular, or soft tissue

Discuss patent development, device design, and ethical issues with other professionals

C.2 Active learning practices and inquiry-based lectures

Active involvement in the classroom engages students and enhances the learning process (Mooney and Mooney, 2001; Newstetter, 2005). Accordingly, our biomaterials course has evolved from a traditional lecture format to an interactive lecture structure that initiates with a basic inquiry and follows with active learning exercises. To this end, we have utilized a number of engaging activities in the classroom to accompany lecture topics. An example of an inquiry that initiates a lecture topic of orthopedics is "What factors are important for the design of a total hip replacement for an active woman with osteoporosis?" The class is given 1–2 minutes in which individuals list their answers. The class is then quickly divided into groups of two or more and given an additional 2 minutes to discuss their thoughts among themselves. A fun way to engage the class is, for the example of the inquiry given above, to see which group can list the most factors. The question is then repeated to the classroom and answers are solicited from the various groups. Similarly, one might ask which material-design combinations offer the greatest resistance to fatigue failure in a femoral stem. On the topic of fatigue, one simple exercise is to provide small groups of students (2–4) with rubber bands and paper clips to develop a statistical sampling of fatigue life (number of cycles to failure). This is a fun and easy way to engage students with the factors that contribute to failure under cyclic loading conditions.

These types of simple exercises engage the class within the first 5 minutes of the lecture and enable various discussion topics to be emphasized and incorporated into the lecture. This process can be repeated a number of times during the lecture (our lecture time is 90 minutes and we use 2–3 active exercises per lecture). There are numerous mechanisms by which a class can be actively engaged in various subject areas and the reader is referred to the *Engineering Pathway* website for a more detailed presentation of this topic (http://www.engineeringpathway.com/ep/index.jhtml). Typical examples queries that we have used in the course are provided at the start of each chapter in the textbook.

C.3 Clinical case studies

The incorporation of case studies in the lecture framework provides a number of benefits to students (Newstetter, 2005). For example, students get to see a clinical problem and to dissect the multifactorial elements that contributed to an *in vivo* failure. For example, we have used a case study on stress-shielding in femoral stems (given in Chapter 6) to provide essential elements of compliance match, elasticity theory, materials selection, and biocompatibility. Concepts such as intellectual property, patent infringement, and good manufacturing practices; biocompatibility, materials selection, and design evolution; ethical considerations and regulatory issues; as well as failure analysis can be incorporated through the appropriate use of case studies. Over the past decade we have

Table C.3 Learning styles and their characteristic traits

Learning continuum	Learning styles	Characteristics
Perception	Sensing	External inputs (see, hear, touch) and tend to be practical and prefer facts, data, and experimentation
	Intuitive	Internal inputs (thoughts, memories, images) and tend to be imaginative and prefer principles, theories, and models
Input	Visual	Recall what they see (pictures, charts, diagrams)
	Verbal	Remember what they hear and read
Processing	Active	Learn best when doing active experimentation or discussion and prefer group work
	Reflective	Prefer to observe experiments, think about information introspectively, and favor individual assignments
Understanding	Sequential	Like to have information presented in a linear fashion with each new idea building from that previously learned
	Global	Need to understand how all of the information relates to each other before they can understand the details

used case-in-point examples such as failure of Acroflex spinal implant designs, recall of Sulzer hip replacements, litigation involving the Dow-Corning silicone breast implants, fatigue fractures in T-28 femoral stems, welding failures in the Bjork-Shiley heart valves, and material failures in the Poly II knee implants using carbon reinforced UHMWPE as well as TMJ designs using Proplast®. Relevant case studies on the topics of structural biomaterials and medical device design are provided at the end of each chapter in the textbook.

C.4 Professional development using learning styles and interdisciplinary teams

The implementation of an interdisciplinary course can be challenging owing to the difference in technical backgrounds and the range of learning styles expected for students engaged in multidisciplinary topics (Felder and Brent, 2005; Felder and Spurlin, 2005). Students who are aware of their own learning style and know how to communicate effectively with people of different learning styles tend to work more efficiently on interdisciplinary teams (Halstead and Martin, 2002; Patten et al., 2010). Felder and co-workers (Felder and Silverman, 1988) developed a well-known classification of learning styles along a continuum within four categories: perception (sensory or intuitive), input (visual or verbal), processing (active or reflective), and understanding (sequential or global). Table C.3 summarizes the traits of these learning styles. Over the evolution of this course we have found that interdisciplinary teams with balanced gender, majors, and learning styles have worked best (Patten *et al.*, 2010). In particular, one of our student groups has stated "at first the differences in our learning styles made communication more challenging but in the long run the diversity enabled us to work more creatively as a team." Since implementing assessments of learning styles into the group teamwork, we have had an improved success rate, with 19 out of 20 groups reporting no project-limiting conflicts in their teams. This is a vast improvement over years past, where the success rate has been closer to 80–85% on average. It appears that teams with a range of learning styles are able to work through communication barriers and in fact may have more creative outcomes than teams who have no sense of their individual learning styles.

Students enrolled in our course have reported that using learning styles to assign groups brought in new and creative ideas for their design projects (Patten *et al.*, 2010). Wilde discovered similar results: "in the long run teams do better when they are composed of people with the widest possible range of personalities, even though it takes longer for such psychologically diverse teams to achieve smooth communications and good cooperation" (Wilde, 2010). We have found that the use of learning style assessments has enriched the class learning and effectiveness of teams. Furthermore, courses using lecture techniques that implicitly take into account the range of learning styles can be helpful to ensure that students from a variety of backgrounds learn as much of the material as possible (http://www4.ncsu.edu/unity/lockers/users/f/felder/public/ILSdir/styles.htm).

C.5 Outreach teaching (service-learning) in the K-12 sector

There is strong evidence that outreach to the K-12 sector is a vital part of maintaining and improving the numbers of current and potential students who study engineering at the university level (Atwood *et al.*, 2010; Poole *et al.*, 2001). The exposure of children to young, diverse role models reinforces the notion that engineering is a career path that is accessible to people regardless of gender or ethnicity. Moreover, undergraduate students are engaged in an active learning process that develops their communication and collaborative skills while serving society. These undergraduate students also express an increased desire to continue teaching and outreach after experiencing these activities through a structured process in their coursework (Atwood *et al.*, 2010; Chakravartula *et al.*, 2006).

For the past ten years, we have utilized an outreach-teaching project as part of an interdisciplinary team project for *Structural Aspects of Biomaterials*. The course project entails the development of a collection of interactive tabletop exhibits collectively entitled "Body by Design" and is hosted at the Lawrence Hall of Science (LHS), which is our local children's museum, and is offered in a one-day format to children in local elementary and middle schools. As part of this collaboration, the staff at LHS provide 2 hours of lecture on the topic of K-12 education and interactive exhibit design in the formative stages of project development.

Table C.4 Technical concepts associated with medical implants

Medical implant	Design consideration
Stents	Shape memory vs. deformable
Breast implants	Silicone vs. saline
Heart valves	Caged valve vs. tilting disk vs. bileaflet designs
Intervertebral disk	Interbody fusion vs. total disk replacement
Total knee arthroplasty	Cruciate ligament retaining vs. unretained
Dental implant	Thermals stresses and fracture
Total hip replacement	Ceramics vs. polymers

In designing this project, several objectives were kept in mind within a general framework of asking the undergraduates to explain medical device design concepts to children as part of an exhibit at LHS. First, each team exhibit must demonstrate the students' knowledge of course content. The undergraduates were assessed in their project delivery and teaching methodology by the professor, graduate student instructors, and the science museum education staff on the day of the exhibit. Second, the students were also asked to demonstrate outreach teaching to the target age group of third through sixth grade. Each interactive exhibit delivered technical concepts associated with a medical device (Table C.4) while addressing the different learning styles as defined by Felder (Felder and Silverman, 1988). The undergraduates were assessed in their project delivery and teaching methodology by the professor, graduate student instructors, and the science museum education staff on the day of the exhibit.

The outreach-teaching projects used in our course have been highly structured in order to facilitate success of the group work. Benchmarks were assigned to assist students in developing their project over the course of the semester, and the following items were used to evaluate progress and learning:

- Personal assessment of learning styles
- A written report which included the history of the device and its current state-of-the-art configuration
- An exhibit development plan in a specified format
- A two-minute "elevator speech" or demonstration for peers, teachers, and science center staff
- Feedback on the project day from the elementary school students and science center staff as well as from course instructor and graduate student instructors
- A written report that detailed the project development, evaluation, and lessons learned
- Team members' evaluations

Teams were asked to develop their learning objectives, to explain how they planned to draw in the children (also known as their lesson "hook"), to provide details about their interactive demonstrations, and to speak to mechanisms for addressing the various learning styles. The "Body by Design" exhibit has been presented in a large enclosed room within a children's science center for the past several years. One hundred students from local elementary schools are invited annually to participate in the Exhibit Day of this project. In our latest offering, there were 11 different interactive exhibits ranging from preventing stagnant flow in heart valve designs to comparing fixed and mobile total knee bearing designs. One team presented "cemented versus uncemented hip replacements" and developed models with play-dough along with wire meshes of varying hole size to demonstrate the effect of bone porosity on cement interpenetration. This group also utilized changeable geometries and surface finishes between metal and wood to achieve different interface strengths. The children measured strengths in an inquiry-based format using varying weights. The exhibit offered *active* elements that were *sequential* and *global*; *visual* and *verbal* tools; *intuitive* and *reflective* learning through inquiries; and facts for *sensing* learners as well. The entire exhibit was documented with video camera that captured individual projects, interactions between undergraduate students and children, and the diversity within the exhibit hall (Pruitt *et al.*, 2010).

The projects were not graded with a formal rubric but were evaluated with the following elements in mind: correct technical vocabulary (knowledge), organization of information and comparisons to relevant concepts (comprehension), identifying key elements of the problem and simplifying the problem (application), assumptions and trade-offs in the design or solution (analysis), methodology and conclusions (synthesis), and assessment of limitations or consequences of their solutions (evaluation). Projects were assigned letter grades rather than numerical scores and prior to the project, submission definitions were given to the class as to what constituted excellent work (A), good work (B), and acceptable work (C).

The long-term goals of the outreach teaching mission are to disseminate science, engineering, and technology information informally to the public; to create cognizant engineers skilled in teamwork, teaching, communication, and outreach; and to develop a model that can be implemented in general engineering curricula to provide informal science education and accessible outreach teaching modules. In summation, K-12 outreach projects provide a unique opportunity for undergraduate students to demonstrate their grasp of the subject matter while inspiring children to develop their interest in math, science, and technology with the goal of becoming engineers.

C.6 Criteria specified by Accreditation Board for Engineering and Technology (ABET)

An interdisciplinary course on the topic of structural biomaterials and medical device design that utilizes inquiry-based learning, active teaching, case studies, teamwork, professional development, and outreach facilitates achievement of program outcomes specified by the Accreditation Board for Engineering and Technology (Shuman *et al.*,

2005). The Accreditation Board for Engineering and Technology (ABET) criteria for a graduating engineer are given as:

(a) ability to apply knowledge of mathematics, science, and engineering
(b) ability to design and conduct experiments and interpret data
(c) ability to design a system, component, or process to meet desired needs within realistic constraints such as economic, environmental, social, political, ethical, health and safety, manufacturability, and sustainability
(d) ability to function on multidisciplinary teams
(e) ability to identify, formulate, and solve engineering problems
(f) understanding of professional and ethical responsibility
(g) ability to communicate effectively
(h) broad education necessary to understand the impact of engineering solutions in global, economic, environmental, and societal context
(i) recognition of the need for and an ability to engage in lifelong learning
(j) knowledge of contemporary issues
(k) ability to use the techniques, skills, and modern engineering tools necessary for engineering practice.

The Structural Aspects of Biomaterials course (as described above) meets all of the ABET criteria. This has been demonstrated for the past three years in end-of-the-semester teaching evaluations that specifically question whether a course meets the ABET criteria as listed above. Similarly, we believe that if this textbook is coupled with interdisciplinary teamwork in a biomaterials course, then the framework for ABET criteria can be readily implemented.

The pedagogical techniques presented in this Appendix can be applied specifically to enhance students' learning experiences in multidisciplinary courses. Examples of how to appropriately incorporate interdisciplinary teamwork, diverse learning styles, and outreach teaching into an engineering course are given in this Appendix. Several case studies that can act as jumping-off points for inquiry-based lectures are also presented throughout the rest of this textbook. These concepts can be easily modified to fit the objectives of other engineering courses. Deliberate use of Bloom's taxonomy in developing course objectives will lead to a deeper educational experience and engage undergraduates in critical thinking. The integration of the methodologies described in this Appendix can increase student engagement and learning and will facilitate meeting the ABET criteria for a graduating engineer.

C.7 References

Anderson, L.W., Krathwohl, D.R., *et al.* (eds.) (2001). *A Taxonomy for Learning, Teaching, and Assessing: A Revision of Bloom's Taxonomy of Educational Objectives*. Boston, MA: Allyn & Bacon (Pearson Education Group).

Atwood, S., Patten, E. and Pruitt, L. (2010). Outreach teaching, communication, and interpersonal skills encourage women and may facilitate their recruitment and retention in the engineering curriculum. *Proceedings of the Annual Meeting of the American Society for Engineering Education*: Louisville, KY.

Bloom, B.S. (1984). *Taxonomy of Educational Objectives*. Boston: Allyn and Bacon.

Chakravartula, A.M., Li, C., Gupta, S. Ando, B., and Pruitt, L. (2006). Undergraduate students teaching children: K-8 outreach within the core engineering curriculum. *Proceedings of the Annual Meeting of the American Society for Engineering Education*: Chicago, IL, p. 1310.

Felder, R.M., and Brent, R. (2005). Understanding student differences. *Journal of Engineering Education*, 94(1), 57–72.

Felder, R.M., and Silverman, L.K. (1988). Learning and teaching styles in education. *Journal of Engineering Education*, 78(7), 674–81.

Felder, R.M., and Spurlin, J.E. (2005). Applications, reliability, and validity of the index of learning styles. *International Journal of Engineering Education*, 21(1), 103–12.
http://www.educationoasis.com/curriculum/LP/LP_resources/lesson_objectives.htm
http://www.engineeringpathway.com/ep/index.jhtml.

Halstead, A., and Martin, L. (2002). Learning styles: a tool for selecting students for group work. *International Journal of Electrical Engineering Education*, 39(9), 245–52.

Mooney, M.A., and Mooney, P.J. (2001). A student teaching-based instructional model. *International Journal of Engineering Education*, 17(1), 10–16.

Newstetter, W. (2005). Designing cognitive approaches for biomedical engineering. *Journal of Engineering Education*, 90(2), 207–13.

Patten, E., Atwood, S., and Pruitt, L. (2010). Use of learning styles for teamwork and professional development in a multidisciplinary course. *Proceedings of the Annual Meeting of the American Society for Engineering Education*: Louisville, KY.

Poole, S.J., DeGrazia, J.L., and Sullivan, J.F. (2001). Assessing K-12 pre-engineering outreach programs. *Journal of Engineering Education*, 90(1), 43–48.

Pruitt, L. Atwood, S., and Patten, E. (2010). Body by design: A model for K-12 outreach in engineering education. *Proceedings of the Annual Meeting of the American Society for Engineering Education*: Louisville, KY.

Shuman, L.J., Besterfield-Sacre, M., and McGourty, J. (2005). The ABET professional skills – can they be taught? can they be assessed? *Journal of Engineering Education*, 94(1), 41–55.

Wilde, D. (2010). Personalities into teams. *Mechanical Engineering*, February, 22–25.

Glossary

abutment: part of a structure in a dental implant that receives pressure, typically a natural tooth or implant that retains the dental prosthesis.

acetabular cup: portion of a total hip replacement shaped like a cup (inserted in the acetabulum).

acetabulum: cavity (shaped like a cup) at the intersection of the pelvis and femur.

activation volume: rate of decrease of activation enthalpy with respect to flow or yield stress at fixed temperature in the Eyring model. The premise of this model is that the segments of the polymer chain must overcome activation barriers to move from one position to another in the structure.

addition reaction: polymer synthesis mechanism in which free radicals are used in the initiation and propagation stages of chain growth.

adulterated: rendered poorer in quality through the addition of an inferior substance.

adventitia: outer layer of the artery.

age hardening: strengthening mechanism that makes use of a decreasing solubility with decreasing temperature such that precipitates can be formed in the matrix with appropriate heat treatment. This treatment is also termed precipitation hardening.

alloplastic: derived from a non-living source.

alloy: metallic system with two or more elements typically combined to improve mechanical properties or corrosion resistance.

American Academy of Orthopaedic Surgeons (AAOS): medical organization for orthopedic surgeons with nearly 36,000 members.

amorphous polymer: polymer without any long-range order of molecules (absence of crystallinity).

anabolic: tending to promote growth or "buildup" of organs or tissues.

aneurysm: bulge in a blood vessel due to weakening of the vessel wall.

angioplasty: surgical procedure that removes an occlusion and restores flow in a blood vessel.

anion: negatively charged ion created from an atom that accepts electrons.

anisotropic: material behavior in which properties vary along crystallographic axes or directions within the material.

ankylosing spondilosis: systemic disease resulting in inflammation between the vertebrae and the joints between the pelvis and spine.

ankylosis: fusion or consolidation of tissues.

annealing: heat treatment in which the metal system is heated below its melting point for a period of time. This thermal treatment is used to relieve stresses, activate diffusion processes, annihilate dislocations, and to reform grain structure (a process known as recrystallization).

annulus: the ring-like structure of the intervertebral disk.

apex: terminal end of a tooth or implant.

appendicular: of, relating to, or consisting of an appendage or appendages, especially the limbs.

area moment of inertia: also known as the second moment of area, used to predict the stress in a beam due to bending.

artery: blood vessel that carries blood away from the heart.

arthrodesis: immobilization of a joint (surgical).

arthroplasty: surgery to realign or reconstruct a joint.

arthroscope: instrument used to illuminate, examine, and in some cases, repair the inside of a joint that is typically inserted through a small incision.

articular cartilage: tough, resilient tissue that lines joints and selected other surfaces in the body, also known as hyaline cartilage.

articulation: movement between two contacting components.

asperities: small peaks or irregularities on a surface.

astigmatism: condition in which the eye has difficulty focusing due to irregular curvature or scarring of the cornea or lens.

ASTM International: formerly known as the American Society for Testing and Materials, this organization establishes standard protocols as well as specifications for the composition and properties of alloys.

atherosclerosis: inflammatory disease in which the lumen of an artery is narrowed due to plaque accumulation.

atlas: first cervical vertebrae (C1).

atria: two thin-walled upper chambers of the heart, responsible for holding blood that enters the heart.

austenite: FCC solid solution of iron and carbon (steel) that is in its equilibrium form above the eutectoid temperature.

autoclave: sterilization system that employs high-pressure steam.

autogenous: produced from within or originating from one's own body.

autografts: tissue removed from one part of the body and implanted in the same individual.

autologous: from the same organism. In grafts, this refers to the practice of taking tissue from one part of the patient's own body to replace damaged tissue elsewhere.

axis: second cervical vertebrae (C2).

balloon catheter: catheter with an enlarged portion at its tip that can be inflated to enlarge vessels or implant stents.

Basquin equation: equation used in the total life philosophy that relates the number of cycles to failure for given stress amplitude ranges.

bicuspid: having two points or cusps.

bioactive: relating to a substance that has an effect on living tissue.

bioactive glass: silicate glass coatings that encourage bone growth and integration.
biocompatibility: measure of a body's tolerance to a foreign material, such as an implant.
biocompatible: does not elicit an acute inflammatory response in the body.
biodegradable: capable of being fully or partially metabolized by the body. The word is used synonymously with the term resorbable or bioresorbable.
bioreactivity: degree to which a material triggers a chemical or physical response in the body.
bioresorbable: see "biodegradable."
bisphosphonates: class of drugs the slow the resorption of bone by suppressing osteoclast activity.
body-centered cubic (BCC): lattice with one point in the center of the cube and eight corner points. Contains two full lattice points per unit cell.
bone cement: grouting agent (usually polymethylmethacrylate) used to secure an implant to adjacent bone tissue.
bone graft: replacement for bone tissue.
bore: hole (female part) of the femoral head that mounts onto the trunnion of the Morse taper.
Bowman's layer: layer of the cornea made up of aligned collagen fibers.
Bravais space lattices: fourteen configurations of atoms that represent the seven crystal systems: triclinic (simple); monoclinic (simple, base-centered); orthorhombic (simple, base-centered, body-centered, face-centered); tetragonal (simple, body-centered); cubic (simple, body-centered, face-centered); trigonal (simple); and hexagonal (simple).
brittle: material behavior where there is very little strain or deformation before fracture.
bruxism: cyclic grinding of teeth.
Burger's vector: vector representing the direction and magnitude of distortion owing to the presence of a line defect (dislocation) in the atomic lattice of the crystal.
butterfly fracture: center fragment of a broken bone contained by two cracks forming a triangle.
canaliculi: small vessels that allow for transport between osteocytes.
cannulated: containing an opening, such as in the hollow shaft of a cannulated screw.
cardiovascular disease: any disease that affects the heart and blood vessels, including coronary heart disease, congestive heart failure, valvular disease, septal defects, and atherosclerosis.
cardiovascular system: the heart and blood vessels.
cartilage: lubricious soft tissue. One type is articular cartilage which is found at the end of long bones and facilitates lubrication in the articulation process of synovial joints.
casting: manufacturing process whereby a melted material is poured into a mold such that the material retains the shape of the mold upon cooling.
cataract: clouding of the lens, resulting in vision impairment.

catheter: tube that is inserted into the body for a wide variety of purposes such as basic drainage, or transport (delivery) of a balloon or stent in angioplasty procedures.

cation: positively charged ion created from an atom that donates electrons.

cement line: incompletely mineralized tissue between osteons.

ceramic: material comprising a metal and non-metal (e.g., Al_2O_3) that generally offers high melting temperatures, elastic modulus, and wear resistance.

cervical spine: portion of the vertebrae making up the neck.

chondrocyte: cell that develops and maintains cartilage.

Class I device: device that is low-risk, short term, and/or used to treat illnesses that are not life-threatening.

Class II device: moderate-risk device.

Class III device: device that is high-risk and/or used for life-threatening diseases.

coefficient of friction: ratio of the force that maintains contact between an object and a surface and the frictional force (shear force) that resists the motion of the object.

coherent flaw: defect or crack that is large enough to grow under operating stresses.

cold rolling: reduction of metal thickness through rollers that is performed below the recrystallization temperature.

comminuted fracture: broken bone that has splintered.

complex modulus (G^* or E^*): effective material deformation response for a viscoelastic material, represented as a complex number, with the real part indicating the storage modulus and the imaginary part indicating the loss modulus. This quantity is analogous to elastic modulus in linear elastic materials, and so is commonly referred to as a "modulus."

compliance: measure of flexibility that is quantified by the amount of deformation per given load.

compliant: denotes material components that are less stiff.

composite: material comprising two or more structural constituents.

compression fracture: fracture in which the bone gives way (often associated with a fall).

compressive strength: amount of force per area that can be sustained in a material when it is compressed along its loading axis.

condensation reaction: polymer synthesis mechanism in which two initial molecular species are combined and another species (usually water) is released as part of this reaction.

condylar: relating to a condyle.

condyle: rounded feature at the end of a bone used for articulation.

conforming: displaying geometric matching of surfaces.

congenital: existing at or before birth.

congestive heart failure: condition in which the heart can not pump enough blood through the body.

contact stresses: stresses distributed to materials owing to loaded articulation (sliding, rolling, rotation).

contracture: condition in which scar tissue forms around a breast implant, squeezing it and causing it to feel hard.

coordination number: number of nearest neighbors (atoms) in a crystalline lattice.

coral: hard, stony skeleton secreted by certain marine polyps and often deposited in extensive masses forming reefs in tropical seas.

coronal: any vertical plane that divides the body into front and back (anterior and posterior).

coronary artery bypass graft (CABG): procedure in which arteries or veins from another part of a patient's body are used to bypass narrowed coronary arteries in order to increase the circulation of blood to the heart.

coronary heart disease: lack of circulation to the heart and surrounding tissue due to plaque buildup in the coronary arteries.

corrosion: degradation of a metal owing to chemical reactions with its surrounding environment.

cortical bone: dense, compact bone that is the principal structural material of the skeleton.

covalent bonding: chemical bond generated by the sharing of valence electrons.

crack bridging: toughening mechanism that utilizes intact fibers or particles that provide cohesive forces behind the crack tip.

crack initiation time (t_i): amount of time that it takes for a material to nucleate a flaw under quasi-static or monotonic loading conditions.

crack opening displacement (δ): quantitative measure of the vertical gap at the crack tip. In ductile materials this is found by using the 45 degree intercepts in the wake of the crack tip.

craniofacial: pertaining to the cranium and the face.

craniomaxillofacial: pertaining to the head, neck, face, jaws, and the hard and soft tissues of the oral region.

craze: fibrillated polymer with its fibrils oriented in the principal stress direction. Requires a cavitational stress for void formation.

creep: time-dependent strain in a material (usually associated with viscoelastic materials).

crest: a projection or ridge, especially of bone.

crest module: portion of the implant designed to create a transition zone between the prosthesis and load-bearing implant.

crevice corrosion: process where a differential in environments owing to a geometric feature (the presence of a gap, crack, or fissure) results in corrosion of the metal or alloy.

critical flaw size: size of a flaw in a material that results in fast fracture for a given stress state and geometry.

crosslinking: presence of covalent bonds between macromolecules.

crown, natural: part of the tooth that sits above the gumline and is covered by enamel.

crown, synthetic: cap or structural covering placed over a tooth and used to restore strength, shape, or appearance.

crystalline: solid structure of atoms that utilizes a repeated pattern of unit cells.

crystallinity: amount of material that has short-range and long-range order in a polymeric structure. Most polymers are incapable of achieving 100% crystallinity.

cultured epidermal autograft (CEA): skin replacement in which cells are cultured from the donor's own keratinocytes.

cyclic strain hardening: refers to the phenomenon of an increase in yield stress or increase in stress amplitude to maintain a specified plastic strain under repetitive loading conditions.

cyclic strain softening: refers to the phenomenon of a decrease in yield stress or stress amplitude to maintain a specified plastic strain under repetitive loading conditions.

Dacron: trademark for polyethylene terepthalate (PET) or polyester fiber. Used in fabric form for soft tissue grafting and repair.

Deborah number: a dimensionless parameter that captures how solid-like or fluid-like a material is and is defined as the normalized value of relaxation time relative to the time duration of a given experiment or loading event.

debridement: process of removing non-living tissue from wounds.

deciduous: that which will be shed or fall off.

defect tolerant: refers to materials or systems that are able to function structurally in the presence of a crack or flaw.

deformation plasticity: permanent deformation of a material that is related to the second stress invariant or the von Mises stress.

delamination: material failure mode resulting in loss of a sheet-like structure of the material.

dental caries: regions of demineralized enamel and/or dentin.

dentin: tough, mineralized material that makes up the bulk of the tooth.

dentition: type, number, and arrangement of teeth.

deoxyribonucleic acid (DNA): generic framework for the development of organisms.

depressed fracture: fracture where the bone is pushed in, common in skull.

dermal acellular implant: skin replacement that uses a scaffold to encourage tissue ingrowth, but does not have living cells.

dermatome: tool used to harvest skin grafts.

dermis: layer of skin between epidermis and subcutaneous tissue.

Descemet's membrane: basement membrane between the stroma and the endothelium.

deviatoric: component of the stress tensor responsible for shape distortion.

diaphyseal: referring to the shaft of a long bone.

diaphysis: shaft of long bone.

differential scanning calorimetry (DSC): heating technique used to measure the crystallinity of a polymer. An endothermic reaction occurs as the sample absorbs heat to melt the crystals and the peak of the melting curve represents the melting temperature of the polymer. The method can also be used to measure the glass transition temperature.

dilatational: component of the stress tensor responsible for volume change.

disk arthrosis: degeneration of the spinal disks.

dislocation: second-order imperfections that are also known as line defects. Primary dislocations are edge, screw, or mixed in character. The control of movement of defects is key to strengthening metals. The primary dislocations are edge or screw dislocations.

distal: anatomic term referring to a feature located further from the axial skeleton than another; e.g., the knee is found at the distal end of the femur.

distortional energy: portion of strain energy that results in shape change rather than volume change.

driving force: amount of energy available to advance the crack in a material.

ductile: material behavior where there is plastic strain or deformation before fracture.

dynamic mechanical analysis (DMA): experimental approach for probing the response spectrum of a material by varying a sinusoidal stress or strain input over a range of frequencies. This can be coupled with altering the temperature and using time-temperature superposition.

edema: swelling of tissue caused by excess fluid.

edentulism: complete or partial toothlessness.

edentulous: toothless.

edge dislocation: line defect characterized by an extra half plane of atoms into an otherwise ideal crystal lattice that results in elastic displacement, strain, and stress fields around the defect. The displacement (slip) of this type of dislocation is orthogonal to the extra half plane of atoms.

effective stress: scalar representation of a complex stress state.

elastic modulus (E): also known as Young's modulus, the mechanical property that describes the ratio between stress and strain.

elastic-plastic fracture mechanics (EPFM): field of fracture mechanics that accommodates plasticity (beyond small-scale yielding) in the fracture process (at the crack tip).

electronegativity: measure of the tendency of an atom to attract a bonding pair of electrons.

embolus: floating blood clot.

emulsion: suspension of small globules of one liquid in a second liquid.

enamel: extremely hard and durable outer surface of the tooth.

endocardium: inner layer of the heart which lines the chambers of the heart.

endosteal: within bone.

endotenon: tissue that binds the collagen fibers in tendon and contains the nervous, lymphatic, and vascular elements of the tissue.

endothelium: single layer of cells between the cornea and the iris.

endurance limit: stress range below which a material has infinite life.

energetic toughness: amount of energy that can be absorbed by a material without fracturing. This is often simply defined as the area under the stress-strain curve.

energy release rate (G or J): amount of strain energy that is released in the system as the crack grows incrementally.

epicardium: serous membrane that forms the outer layer of the heart.

epidermis: outer layer of skin.

epiphysis: round end of long bone.

epithelium: outer layer of the cornea.

erosion: polymer disintegration through the absorption of water.

ethylene oxide gas: C_2H_4O gas used to sterilize bacterial contamination in medical implants (often used in polymers).

expanded polytetrafluoroethylene (e-PTFE): porous form of a fully fluorinated polymer. The porous structure facilitates tissue ingrowth and fixation.

extension: action at the knee when the leg is straightened.

extracellular matrix: structural support of connective tissue, made up of proteins, proteoglycans, and in some cases, mineral deposits.

extrinsic toughening: mechanisms that are activated during the propagation of a crack and include crack deflection, contact shielding, and zone shielding.

extrusion: manufacturing process that is used to create materials with a fixed cross-sectional shape (such as a rod).

Eyring model: general model that describes how the rate of degradation varies with stress or, equivalently, how time to failure varies with stress. The model includes temperature and is similar to the Arrhenius empirical model in establishing the connection between the enthalpy parameter and the activation energy needed to cross an energy barrier and initiate a reaction.

face-centered cubic (FCC): lattice with a point in the center of each face as well as the corner points, giving it four full lattice points per unit cell.

factor of safety: scalar measure that is obtained from the normalization of the failure stress with respect to the actual stress on the structure.

failure strength (S_f): the point at which failure occurs.

fascicles: substructures of the tendon that are able to move freely past one another.

fatigue: cyclic loading that results in the loss of mechanical or structural integrity of a material.

fatigue strength: structural integrity remaining for a material or component after a specified number of loading cycles.

femoral head: highest portion of the femur (or the portion of the total hip replacement that articulates within the acetabular cup).

femoral neck: proximal end of the total hip replacement that mates into the femoral head.

femoral stem: long portion of the total hip replacement below the head and implanted in the femur.

fenestration: opening in a surface.

fibrillin: glycoprotein that forms a scaffold for elastin deposition.

fibroblast: cell that synthesizes the extracellular matrix and collagen.

fibrocartilage: mixture of fibrous tissue and cartilaginous tissue.

fibrosa: dense connective tissue layer of the heart valves.

filling: material embedded in a dental cavity (can be metallic or polymeric resin).

first degree burn: burn in which the epidermis is removed.

first throw hold: load that a single knot can hold before slipping.

flexion: the action at the knee when the leg is bent.

fluorocarbon polymers: macromolecule that comprises fluorine side groups on the backbone of the chain.

fluoroscopy: medical imaging technique that gives a continuous, real-time X-ray projection.

Food and Drug Administration (FDA): federal agency that oversees the safety, classification, and effectiveness of medical devices in the United States.

forging: mechanical working of a metal to increase strength.

forming: process of working metal that keeps a constant cross-section.

fractional recovery: non-dimensional fractional strain recovery (recovered strain/total creep strain). Used in nonlinear viscoelastic relaxation analysis.

fracture fixation: any method or system used to stabilize a broken bone for healing.

fracture mechanics: discipline of mechanics that studies the nucleation and growth of flaws in structural materials.

fracture plate: thin metal structure used to repair a broken bone.

fracture strength: amount of stress that is sustained in the material at the time of fracture.

fracture toughness: material property that provides a quantitative measure of a material's resistance to fracture via crack advance. It is often used synonymously with plane strain fracture toughness.

fretting: loss of material owing to rubbing of surfaces.

friction, kinetic: friction between moving surfaces.

friction, static: friction between non-moving surfaces.

frontal: pertaining to the front (forward portion) of a feature.

fully reversed loading: cyclic conditions that span between an equal tensile and compressive stress (or strain).

functional requirements: the structural job of a device or biological system.

galvanic corrosion: corrosion process that occurs when two materials of dissimilar electrochemical potential are placed in contact in the presence of an electrolyte in such a way that an electrochemical cell is created.

gamma radiation: shortest wavelength of electromagnetic spectrum, often generated from a Co60 high-energy source, used as a sterilization method in medical devices.

gas plasma: ionized gaseous mixture that is often used as a sterilization method in medical devices.

gel permeation chromatography (GPC): technique that measures the molecular weight distribution of a polymer system. An appropriate solvent is used and molecules are passed through a series of packed columns and separated from one another based on their differences in size and elution time (time to pass through the packed columns).

generalized Kelvin model: arbitrarily large group of Kelvin models connected in series.

generalized linear viscoelasticity: linear viscoelastic modeling of arbitrarily high order, used for precisely fitting material behavior when many transitions occur over a range of time scales of interest.

generalized Maxwell model: arbitrarily large group of Maxwell models connected in parallel.

gingiva: gums, or soft mucous membrane covering the tooth-bearing part of the jaw.

gingivitis: inflammation of the gums.

glass transition temperature (T_g): temperature where an amorphous polymer transitions from glassy behavior to rubbery behavior and represents the transition between large chain motions and small chain motions in a polymer.

glasses: amorphous (non-crystalline) solids that are brittle in their mechanical behavior.

glassy plateau: high stiffness regime of behavior occurring at temperatures and time scales below the viscoelastic transition, where the modulus is approximately constant over a wide range of temperatures and times.

glassy polymers: amorphous (non-crystalline) polymers that are below their glass transition temperatures and are brittle in their mechanical behavior.

glaucoma: disease that is manifested in optical nerve damage, associated with increased intraocular pressure.

glenoid: socket joint of the shoulder.

glenoid fossa: hollow, depressed area in the mandible.

glycosaminoglycan (GAG): long, unbranched polysaccharide that is a key component of connective tissues.

grafts: replacement of tissue that mimics and restores the exact features of the original tissue (e.g., bone grafts, vascular grafts, skin grafts).

greenstick fracture: fracture in which one side of a bone is broken while the other is bent.

hardenability: ability of a metal to be strengthened with improved surface hardness with a combination of working and heat treatment. In steels this is also controlled with carbon content.

haversian canal: main conduit for blood and lymphatic vessels in cortical bone.

haversian system: synonymous with the osteon.

heart valve: structural feature of the heart that controls blood flow. Synthetic versions can be made from biological materials such as porcine valves or engineered systems such as a leaflet design comprised of pyrolytic carbon.

heel region: higher-slope region of stress-strain curve of soft tissues.

hematopoiesis: production of red blood cells.

herniated disk: bulging of the intervetebral disk.

heterograft: also known as a xenograft, refers to a tissue graft from another species.

heteroplastic: derived from a different living species.

hexagonal close packed (HCP): lattice in which atoms are in three layers: the top and bottom forming a hexagon with three atoms in the middle layer.

hierarchical: system of tissue constituents ranked one above the other.

high cycle fatigue: sustained loading is dominated by elastic strain and results in extensive fatigue life.

high-density polyethylene (HDPE): linear form of polyethylene (C_2H_4) that is highly crystalline and has a density in the range of 0.94–0.98 g/cc.

high noble metals: metal systems that contain more than 60% gold, palladium, or platinum with at least 40% of the metal being comprised of gold.
homoplastic: derived from the same species.
Hooke's Law: linear relationship between stress and strain.
horizontal plane: plane that divides the body into superior (top) and inferior (bottom) parts.
hot rolling: reduction of metal thickness through rollers that is performed above the recrystallization temperature. Often used to reduce grain size of a metal system.
HRR field: near-tip stress fields developed by Hutchinson, Rice, and Rosengren (HRR) that capture the elastic-plastic or nonlinear behavior of power-hardening (strain-hardening) materials.
hydrogen bonding: attractive force between the hydrogen atom attached to an electronegative atom of one molecule and an electronegative atom of a different molecule that has a partial negative charge, such as oxygen.
hydrophilic: attracted to water.
hydrostatic stress: stress state where all the normal stresses are equal; also termed hydrostatic pressure.
hydroxyapatite or hydroxylapatite: naturally occurring mineral form of calcium apatite with the formula $Ca_5(PO_4)_3(OH)$ that is chemically analogous to the mineral component of bones and hard tissues in mammals and supports bone ingrowth.
hyperopia: vision defect in which images are focused outside of the eye due to the eyeball being too short in the anterior-proximal direction. Also known as farsightedness.
hyperplasia: increase in the number of cells, beyond normal levels.
hysteresis loops: phenomenon seen in materials that have a memory effect, manifested as a lag in stress-strain behavior under cyclic loading conditions.
immunosuppressant: an agent that suppresses the immune system.
impacted fracture: fracture in which one broken end of a bone is forced into another.
inert: chemically non-reactive.
inferior: anatomic term referring to something that is below or more distant from the head than another; e.g., the foot is inferior to the hip.
inflammatory response: body's immune response to a foreign material or implant.
injection molding: high-volume manufacturing process that forms objects by heating the material to a fluid state and then forcing (injecting) the fluid into a mold of desired shape.
inkjet method: process in which a structure is built layer by layer to achieve the final geometry.
intergranular corrosion: corrosion process that occurs along grain boundaries.
intermetallics: materials comprising two or more metallic elements and whose properties are between those of a metal and a ceramic.
interpositional: placed or inserted between parts, an intervening position.
interstitial sites: spaces between packing of atoms in the FCC, HCP, and BCC structures.

interstitial solid solution: strengthened solid solution in which the geometric mismatch between the lattice (solvent) atoms and solute atoms is greater than 15% and the alloying element is substituted in the interstitial positions. The local strain fields around the interstitial atoms can interact with dislocation strain fields and impede dislocation motion.

intima: innermost, blood-contacting layer of the artery.

intramedullary: refers to the inside of the bone.

intraocular lens (IOL): implant which can replace damaged lenses in the eye.

intraocular pressure: pressure in the eyeball due to the aqueous humor.

intrinsic toughening: mechanisms involving fundamental material properties that enhance the plastic deformation or ductility of the material and are controlled by the microstructure.

investigational device exemption (IDE): approval to use a device in humans for clinical testing before FDA approval.

investigational review board (IRB): local body which the FDA uses to monitor clinical testing.

in vivo: Latin phrase meaning "in the body."

ionic bonding: chemical bond that is formed by electrostatic attraction (typically between a metal and non-metal); occurs in atoms with differences in electronegativity where one atom donates one or more valence electron(s) to another atom resulting in filled energy shells for both atoms involved in the interaction.

irradiation: process by which a component or material is subjected to radiation.

isotropic: deformation in response to load is invariant with respect to direction.

***J*-integral:** path-independent fracture parameter developed by Jim Rice that facilitates nonlinear elastic deformation or deformation-theory plasticity at the crack tip.

Kelvin model: first-order, two-element linear viscoelastic model, with a spring and damper in parallel. Used to model solid-like materials.

keratinocytes: cells responsible for the formation of keratin, a substance that protects the skin.

knot security: measure of the stability of a knot in a suture.

knot tie-down: ability to slip a knot down the suture.

knot-pull strength: strength required to fracture a suture tied in a single knot.

Knowles pin: fracture fixation pin used for intertrochanteric fractures.

kyphosis: curving of the spine that causes the back to bow.

lacunae: pores in the bone containing osteocytes.

lamellae: crystalline plates within a polymer that make use of chain folding to achieve order of molecules, or layers of mineralized tissue that form the osteonal structure.

lamina: tightly woven meshwork that lines the nucleus of the intervertebral disk.

laser etching: method of labeling medical devices.

lattice substitution: an alloying process in which a metallic atom is substituted for one of the primary lattice atom positions. This usually requires that the atomic radii of the alloying elements are within 15% of each other.

ligament: connecting tissue between two bones.

linear elastic fracture mechanics (LEFM): field of fracture mechanics that is based on linear elastic material behavior (no plasticity) or conditions of small-scale yielding such that the region of plastic deformation is small compared to the annular region ahead of a crack where the K fields describe the stresses.

linear superposition: principle that allows for a summation of responses to individual stimuli.

linear viscoelasticity: simplified mathematical treatment of viscoelastic behavior where the time-dependent material behavior obeys the principle of linear superposition.

lipoma: benign tumor.

loss factor (tanδ): ratio of the loss and storage moduli.

loss modulus (G'' or E''): imaginary part of the complex modulus. This corresponds directly to the energy lost during a load cycle, resulting in no stored energy.

low-cycle fatigue: sustained loading involves plastic strain and results in a shortened fatigue life.

lubrication: use of a film or layer to separate opposing surfaces during relative motion.

lubrication, boundary: lubrication scheme in which the lubricant is a solid or gel coating on one or more of the surfaces.

lubrication, elastohydrodynamic: lubrication scheme in which the normal load is transmitted through the lubricant, causing elastic deformation of at least one of the surfaces.

lubrication, hydrodynamic: lubrication scheme in which lubricant is drawn between opposing surfaces through relative motion.

lubrication, mixed: displaying hydrodynamic or elastohydrodynamic lubrication with boundary lubrication.

lubrication, squeeze film: lubrication scheme in which the lubricant demonstrates an elastic response to loading.

lumen: opening of a blood vessel.

machining: manufacturing process in which material is removed with the aid of a power-assisted system such as a lathe or mill.

macrophage: white blood cell that ingests foreign material and aids the immunological response of the body.

malocclusion: improper alignment of the contacting surfaces of opposing teeth.

mandible: lower jaw bone.

martensite: metastable supersaturated microstructure that is created through a displacement transformation by quenching steel rapidly through its eutectoid temperature. The result of excess carbon in the system results in a distorted body-centered tetragonal crystal structure. This form of steel is very hard but has little ductility.

mastication: chewing.

maxilla: upper jaw bone.

maxillofacial: relating to the face or the upper jaw.

maximum shear stress criterion: theory used to predict yield in ductile metals that is based on planes of maximum shear in the material.

Maxwell model: first-order, two-element linear viscoelastic model, with a spring and damper in series. Used to model fluid-like materials.

mean stress (σ_m): average stress sustained in a loading application.

media: middle, muscular layer of the artery.

median: middle plane.

medical device: instrument, apparatus, machine, implant, or related article which is intended for use in the diagnosis of disease; or in the cure, mitigation, treatment, or prevention of disease; or intended to affect the structure or function of the body of man or other animals and which does not achieve its primary intended purpose through chemical action.

melt viscosity: a parameter that describes the rate of extrusion of a polymer melt through an orifice for a given load and temperature.

menisectomy: removal of the meniscus.

metallic bonding: chemical attachment of metal atoms using a "sea of electrons" that enables sharing of free electrons among positively charged metal ions.

metaphyseal: related to the cartilage between the end of the bone and the shaft that later becomes bone.

microcracking: toughening mechanism that uses numerous cracks to dissipate energy in front of an advancing crack tip.

micromechanisms: processes of deformation or fracture associated with the microstructure of a material.

migration: implant motion away from the original location.

Miller indices: convention used to describe crystallographic planes, denoted by the reciprocals of the intersections in the x, y, and z directions in a Cartesian coordinate frame.

mixed-mode: condition of loading that combines Mode I (opening), Mode II (shear), and Mode III (torsion) fracture processes.

modulus of resilience: energy that can be absorbed per unit volume in the material without creating permanent distortion.

Mohr's circle: graphical method for determining stresses after coordinate axis rotation.

molding: manufacturing process in which a melted material is poured into a geometric container whose shape is retained upon cooling (solidification).

molecular weight: mass of one molecule. Because of the synthesis processes involved in polymer chemistry, the chains within a bulk polymer will be distributed in their size. These can be number averaged or weight averaged to represent the polymer system.

morbidity: poor tissue health.

Morse taper: mechanical junction used to connect the femoral stem and femoral head.

myocardium: middle layer of the heart, composed of cardiac muscle.

myopia: vision defect in which the image is focused inside of the eyeball, due to it being too long in the anterior-posterior direction. Also known as nearsightedness.

necking: reduction in cross-sectional area during tensile testing that leads to increased local stress.

needle attachment force: force required to separate a needle from the suture.

neoformation: process of regeneration.

neonatal fibroblast: cells that create extracellular matrix and collagen, cultured from newborn foreskin.

neutral axis: line of zero stress in a beam under a bending moment.

Nitinol: titanium alloy utilizing 50% titanium (atomic) and 50% (atomic) nickel and exhibiting both superelastic (extreme elastic shape recovery) and shape memory (thermal shape recovery) behavior.

non-destructive evaluation: techniques used to assess internal flaws in materials or components that do not cause structural damage, such as ultrasound or X-ray analysis.

non-thrombogenic: substance that is without a tendency to promote formation of clots in the presence of blood.

normal strain: extension or contraction in the direction that a load is applied, normalized by the unit length of fibers in that direction.

notch: geometric cut-out in a structure that is often used in fastening (causes local stress concentration).

nucleus or nucleus pulposa: central portion of the intervertebral disk.

number average molecular weight (M_n): arithmetic mean of individual macromolecule molecular weights for a polymer system.

nylon: generic designation for the polyamide class of polymers.

oblique fracture: slanted breakage of the bone.

occipito-cervical: complex interface between the cranium and the cervical spine.

occlusal: of, or relating to, the contacting surfaces of opposing teeth.

octahedral shear stress: shear stress acting on the octahedral shear plane (the plane that makes equal angles with the principal planes).

octahedral site: space contained within a close-packed octahedron of atoms with a coordination number of 6.

ophthalmologic: pertaining to the eye.

oral muscosa: thin, moist tissue lining the end of the jaw bones and the gums.

orthotropic: material symmetry about three orthogonal planes.

osseointegration: process in which bone tissue bonds with a metal surface at the bone-implant interface and enables fixation.

ossification: formation of bone.

osteoarthritis: arthritis that is caused by the breakdown and loss of cartilage in the articulating joint.

osteoblasts: cells that build bone.

osteoclasts: cells that break down bone.

osteoconduction: process by which growth of new bone occurs.

osteocytes: principal bone cells.

osteogenesis: formation or growth of bone.

osteoid: organic matrix phase of bone microstructure.

osteoinduction: process of stimulating osteogenesis.

osteolysis: resorption of bone by osteoclasts as part of a disease mechanism (often associated with immune response to debris in biomaterials).

osteon: basic unit of cortical bone, also known as the Haversian system.

osteoporosis: loss of bone density usually associated with aging.

oxidation: reaction in a material that involves the uptake of oxidation (usually accompanied by an embrittlement process).

pacemaker: small electronic device that is used to control abnormal heart rhythms (often implanted into the chest).

package memory: measure of the permanent set of the suture in the packaged configuration.

parafunctional: outside or beyond the normal function of.

passivation: protective surface layer or film that forms on a metal and prevents corrosion.

patella: kneecap.

patency: fraction of blood vessels that allow blood flow past the procedure location, sometimes more generally used as a description of an open vessel with free blood flow.

Pauling's rules: chemical rules pertaining primarily to charge neutrality (valence balance) and geometric size differences in atoms that are used to predict where cations and anions can pack in space relative to each other in a crystalline structure.

pearlite: lamellar microstructure of steel comprising layers of cementite and alpha-ferrite upon slow cooling through the eutectoid temperature.

percutaneous: performed through the skin, as opposed to through a larger opening that exposes tissue and organs.

pericardium: double-walled sac that encloses the heart, responsible for keeping the heart in the correct location and preventing it from over-filling with blood.

peri-implant: surrounding an implant.

periodontal: of, or relating to, tissues and structures surrounding teeth.

periodontal disease: deterioration of gum tissue.

periodontitis: infection and/or infection of the tissues and structures surrounding teeth.

periosteum: fibrous sheath (connective tissue) that covers bone.

periprosthetic osteolysis: loss of bone around implants.

phase shift (δ): lag angle that exists between force and displacement behavior in a viscoelastic solid.

phase transformation toughening: crystal lattice conversion ahead of the crack tip that results in a volume change. Constraint by the surrounding material results in compressive stresses that retard crack growth.

physiological stresses: local normalized forces (stresses) that exist in the body or anatomical location.

pile-ups: dislocations that are not oriented for slip or do not meet critical shear stress criteria for the crystal are blocked from further motion. Slip cannot occur until critical stress levels or slip systems are activated.

pin: small thin rod used for securing fractured bones.

pitting: type of crevice corrosion where pits (low points in the surface) may preferentially grow deep from the surface (becoming more like crevices) and thus serve as initiators for cracks.

plane strain fracture toughness (K_{IC}): material property that provides a quantitative measure of a materials resistance to fracture via crack advance. This material property is measured under plane strain conditions (according to ASTM specifications).

plaque, cardiovascular: buildup of fibrous tissue, fatty lipids, and calcifications that occurs on the intimal side of the vascular wall.

plaque, dental: film of mucus and bacteria on the tooth surface.

plaster of Paris: ceramic system comprising calcium sulphate hemihydrate ($CaSO_4 \bullet H_2O$) that is employed as a bone substitute and is used as a structural support for fractured bones.

plastic zone: region directly ahead of the crack tip where damage mechanisms or plasticity processes occur.

plates: smooth, thin piece of metal with uniform thickness that is used to immobilize bone fractures.

Poisson's ratio (v): ratio between the extension and contraction in orthogonal directions of a material under loading.

polyamide: polymer containing an amide link (R_2N) in its chain chemistry.

polydisperity index (PDI): ratio of the weight average molecular weight to the number average molecular weight.

polyester: condensation polymer that contains an ester functional group ($RCOOR'$). Often used synonymously with polyethylene terepthalate.

polyetheretherketone (PEEK): semicrystalline polymer with backbone chemistry $-C_6H_4-O-C_6H_4-O-C_6H_4-CO-$ created by a step-growth reaction, used in spinal cages and composites.

polyethylene: simplest polymer comprised of a repeating ethylene mer (C_2H_4).

polyethylene terephthalate (PET): polyester polymer $(C_{10}H_8O_4)_n$.

polyglycolic acid (PGA): resorbable polymer with chemical formula $(C_2H_2O_2)_n$.

polylactic acid (PLA): resorbable polyester polymer that is often made from an agricultural source such as cornstarch and with a chemical formula $C_3H_6O_3$.

polymer: macromolecule or chain with many repeat units (mers).

polymethylmethacrylate (PMMA): transparent thermoplastic with optical, orthopaedic, and dental applications and formula $(C_5O_2H_8)_n$.

polyp: coelenterate, such as a hydra or coral, having a cylindrical body and an oral opening usually surrounded by tentacles.

polypropylene (PP): vinyl polymer (CH_2CH-R) where R is a methyl group (CH_3). The polymer is semicrystalline and behaves very similarly to polyethylene.

polytetrafluoroethylene (PTFE): polymer that contains fluorine elements along its backbone chain structure (C_2F_4).

polyurethane: polymer that contains urethane ($RO(CO)NHR'$) links in its backbone structure.

pontic: artificial tooth.

poroelastic: mechanical behavior of fluid-saturated porous tissue such as articular cartilage.

powder sintering: fusion of small particles using heat and pressure.

precipitation hardening: see "age hardening."

predicate device: currently approved device whose equivalence to a newer device is used as an argument for that newer device's approval.

Pre-Market Approval (PMA): approval given to devices that are high-risk or novel and post-1976.

Pre-Market Notification (510(k)): approval given to devices that are low-risk and substantially equivalent to a predicate device that must be either pre-1976 or a 510(k)-approved device.

principal directions: three orthogonal directions that correspond to the principal stresses.

principal stresses: stresses on a component that has been rotated to eliminate all shear stresses.

Proplast: composite of polytetrafluoroethylene (PTFE) and carbon or PTFE and alumina.

proteoglycans: macromolecules consisting of a protein core and attached glycosaminoglycan chains.

proximal: anatomic term referring to something that is located nearer to the axial skeleton than another, e.g., the hip is found at the proximal end of the femur.

pterygoid: muscle that descends from the sphenoid bone to the mandible.

pyrogenicity: ability to induce fever.

pyrolytic carbon: crystalline form of carbon that is similar to graphite but is disordered in its covalent bonding between graphine sheets.

quenching: cooling treatment for the metal that is performed rapidly enough to create a metastable structure. Quenching can be used to create non-equilibrium systems such as supersaturated solid solutions and martensite microstructures in steel.

rapid prototyping: automated construction of a singular object that is fabricated free form from data, such as 3D printing. The term is also used to describe products that are brought from concept to market quickly and inexpensively.

real area of contact (A_r): actual asperity-level contact area between two imperfect surfaces.

recrystallization: temperature that results in the nucleation of a fine equiaxed grain structure.

reduced time: non-dimensional recovery time (recovery time/total creep time). Used in nonlinear viscoelastic relaxation analysis.

regurgitation: leakiness in heart valves.

relaxation time (τ): characteristic time of the viscoelastic transition, typically given as the geometric mean of the viscoelastic transition window.

remodeling: bone adaptation in response to loading activity.

resect: surgically remove.

resorbable: capable of being broken down and assimilated back into the body, used synonymously with the term biodegradable.

restenosis: re-closing of the vessel after stenting and/or balloon angioplasty, usually defined as a reduction in cross-sectional area of >50%.

retrodiscal: posterior to the articular disk.

rheopectic: displays lower viscosity during shear motion.

rheumatoid arthritis: autoimmune disease that causes chronic inflammation of the joints.

rod: cylindrical component used to stabilize and align fractures.

root: part of the tooth that sits below the gumline.

roughness, average (R_a): the average value of the distance between the profile and the mean line.

roughness, root mean squared (R_{RMS}): root mean squared value of the distance between the profile and the mean line.

R-ratio: see "stress ratio."

rubbery plateau: low-stiffness regime of behavior occurring at temperatures and time scales above the viscoelastic transition (when the material is a solid), where the modulus is approximately constant over a wide range of temperatures and times.

rubbery polymer: long chain molecule that is capable of coiling and uncoiling to provide large, reversible deformation. At room temperature, these systems are well above their glass transition temperature.

saggital plane: longitudinal mid-line plane that divides the body into right and left.

saphenous vein: large vein of the leg and thigh.

scission: process of breaking bonds, usually in polymers.

scoliosis: abnormal curve of the spine.

screw: fastener with external thread used for fixation and bone repair.

screw dislocation: line defect characterized by slip that occurs parallel to the dislocation. (The Burgers vector is parallel to the dislocation the imperfection.)

secant modulus: elastic modulus defined as the slope of a line connecting the origin to a point on the stress-strain curve.

second degree burn: burn in which the epidermis and part of the dermis are removed.

second-phase additions: particles capable of deformation that are used to dissipate energy ahead of the crack tip.

self-similarity: concept that the stress intensity value in a laboratory specimen is predictive of the stress fields and fracture behavior in a full size structure that contains a flaw with that same value of stress intensity. This is also termed "similitude."

semi-brittle: material behavior where there is a small amount of strain or deformation before fracture.

septal defects: congenital heart defects that allow for unrestricted flow between the right and left sides of the heart.

shear modulus (G): mechanical property that describes the ratio between shear stress and shear strain.

shift factor (a_T): shift in frequency or time scale corresponding to the time-temperature equivalent state. This is often derived empirically from data, or predicted from the WLF equation.

silastic: silicone rubber.

silicone: polymer that incorporates silicon, carbon, hydrogen, and oxygen in its backbone structures.

singularity-dominated zone: region ahead of the crack tip where the K-fields are highly predictive of the near-tip stresses, strains, and displacements.

sintering: manufacturing processing combining heat and pressure that results in local fusion of particles (can be used in ceramics, metals, or polymers).

slip: primary plastic deformation mode when the atoms are loaded in shear and move atop of closely packed planes, generally occurring in the direction of closest packing.

slip system: combination of closest packed planes and directions in which resolved shear stresses enable atom (dislocation) movement.

small-scale plasticity: condition of plastic deformation that is contained in a region that is small compared to the region of K-dominance (singularity dominated zone). This assumption enables materials with limited plasticity to be modeled with linear elastic fracture mechanics. This term is used synonymously with small-scale yielding.

solid-solution strengthening: mechanism of strengthening that works through the addition of an element that has a geometric mismatch to the lattice atoms.

spectrum loading: refers to variable amplitude loading on a component.

spherulites: spherical semi-crystalline regions within a linear (non-branched) matrix of polymer chains.

spinal canal: region where spinal cord passes through the vertebrae.

spinal rods: metallic tubes or cylinders used to provide mechanical support to the vertebrae (spine).

spinous processes: protrusion of structural bone along the backside of the vertebrae.

spiral fracture: helical break in long axis of bone.

splint: appliance for fixation of movable parts.

split-thickness skin graft (STSG): layer of the epidermis and partial dermis harvested from a donor site to cover a wound.

spongiosa: loose tissue layer in the heart valves.

standard engineering materials: commonly used synthetic materials such as metals, ceramics, and hard polymers.

standard linear solid (SLS): first-order, three-element linear viscoelastic model, consisting of a spring element in parallel with a Maxwell model. This is the simplest linear model that can effectively capture both creep and stress relaxation for a solid material.

standards development organization (SDO): body whose members create and formalize testing standards.

stenosis: blockage of an artery or heart valve.

stent: temporary or permanent scaffold used to maintain or increase the arterial lumen.

sterilization: process by which bacterial contamination is removed from a material.

storage modulus (E' or G'): real part of the complex modulus. This corresponds directly to the maximum energy stored in the material during a cyclic loading event.

strain, normal (ε): quantity describing dimensional change owing to a displacement in the direction of loading. Defined as length change normalized by the initial length.

strain, shear (γ): change in angle between two originally orthogonal directions.

strain hardening: strengthening of a metal owing to the increased density of dislocations associated with deformation. The added dislocations cause lattice strain that impedes the motion of other dislocations and increases the required stress necessary to move dislocations (increased yield strength). This term is often used synonymously with cold working.

strain rate: speed at which something is deformed.

strain to failure (ε_f): strain at which failure occurs.

strength of materials: field of mechanics that studies bulk structural properties in a material.

stress (σ): transmission of force through deformable materials.

stress amplitude: half the value of the stress range in a fatigue cycle.

stress concentration factor: value of the localized stress normalized by the far-field stress.

stress corrosion cracking: combined failure mechanism in which a crack initiates or grows under a corrosive environment in the presence of a stress.

stress intensity factor (K): parameter that describes the strength of the stresses, strains, and displacements at a crack tip.

stress range: The span of minimum stress to the maximum stress in a fatigue cycle.

stress ratio: proportion of the minimum stress to the maximum stress in a fatigue cycle. Also known as the R-ratio.

stress relaxation: phenomena by which polymers recover or relieve stress under the application of constant strain.

stress shielding: bone loss in an area where the load normally carried by the bone is being taken up by something else, normally an implant.

stress tensor: representation of the stress at any point in a material body.

stress transformation: change in stresses made by changing from one coordinate system to another.

stroma: middle layer of the cornea composed of layered collagen fibrils.

submergible: capable of being completely immersed below the gums.

subperiosteal: beneath the periosteum.

substantially equivalent: defined in the Safe Medical Devices Act as having the same intended use, having the same technological characteristics OR having different technological characteristics but not raising new questions about safety and effectiveness.

substitutional solid solution: form of solid solution strengthening in which the geometric mismatch between the lattice (solvent) atoms and solute atoms is less than 15% and the alloying element is substituted in the lattice positions. The local strain fields around the substitution atoms can interact with dislocation strain fields and impede dislocation motion.

sulcus: deep, narrow groove or furrow.

superior: anatomic term referring to something that is above or closer to the head end of the body than another; e.g., the neck is superior to the abdomen.

supersaturated solid solution: alloy that has been heat treated to contain a solute content that exceeds its equilibrium solubility. The supersaturated solid solution is thermodynamically unstable and will nucleate precipitates upon moderate heating or extended time (age hardening).

surface energy: amount of work (energy) needed to break bonds and create a new surface.

surface finish: roughness that exists on a machined or manufactured component.

suture: fiber-shaped implant used to hold tissue together after surgery or injury.

suture anchors: implantable devices used for the suture attachment and fixation of tendons and ligaments to bone.

synovial joint: moveable joint surrounded by a capsule that contains synovial fluid.

tan(δ): see "loss factor."

tangent modulus: elastic modulus defined as the slope of a line tangent to the stress-strain curve.

tartar: calcified plaque.

tempering: thermal treatment that provides sufficient energy to facilitate precipitation, grain growth, and relief of residual stresses.

temporomandibular joint: articulating joint of the jaw where the temporal bone and the condyle of the mandible meet.

tendon: connecting tissue between muscle and bone.

tensile strength: amount of force per area that can be sustained in a material when it is pulled apart along its loading axis.

tetrahedral site: space created by the three atoms in the closest packed plane and an atom sitting in a valley above. This atom site has a coordination number of 4.

theoretical cohesive strength: theoretical tensile stress that causes fracture in a material that exhibits no plastic deformation and is free of defects.

thermal expansion coefficient: material parameter that describes how the material volume changes with temperature.

thermoplastic: polymer that can be remelted and processed continuously. The molecules are usually one-dimensional chains.

thermosetting polymers: polymer that cannot be remelted once it is reacted. The molecules are usually crosslinked network systems.

third degree burn: burn in which the epidermis and dermis are removed.

thoracic spine: twelve vertebrae that form the middle segment of the spine (above lumbar spine and below cervical spine).

thrombus: fixed clot.

tibial: pertaining to the tibia (bone beneath the knee).

time-dependent: property that is not constant, but rather varies with time.

time-dependent fracture mechanics (TDFM): discipline of fracture mechanics that studies the time-dependent fracture behavior in materials. This is often used in high-temperature behavior in metals and ceramics that are prone to creep processes (time-dependent strain).

time-temperature superposition: technique for equating a change in temperature in a viscoelastic material to a change in time scale. This can be used for probing a relatively small range of spectral frequency response in DMA, repeated at multiple temperatures to obtain an overall master spectral response over a large range of frequencies.

tissue drag: force imposed on a suture by friction between the suture and tissue.

toe region: initial part of loading curve of many soft tissues, which displays a lower elastic modulus than the higher-strain region.

tolerance: refers to final finish and geometric conciseness in a machined component.

total joint replacements: medical devices used to restore articulation and function to synovial joints such as the hip, knee, shoulder, elbow, wrist, fingers, and big toe.

total life philosophy: refers to the design methodology that employs the belief that the life of a component subjected to sustained loading is owed to both the initiation and propagation of a flaw to a critical length.

trabecular bone: also known as cancellous bone, the less dense bone found at the ends of long bones and in the interior of vertebrae.

transmucosal: entering through, or across, a mucous membrane, such as the gums.

transverse isotropy: material that is symmetric about an axis that is normal to a plane of isotropy.

transverse plane: divides the body into head and tail portions.

transverse processes: protrusion of structural bone along the left and right sides of the vertebrae.

trauma: physical injury that damages the cartilage or fractures the hip joint.

trephine: surgical instrument for cutting circular sections; a cylindrical blade.

tribology: study of friction, lubrication, and wear.

tropocollagen: superhelix of collagen molecules that forms the basis of collagen fibrils.

true strain ($\bar{\varepsilon}$): instantaneous strain, defined as the integral of the incremental strain, dl/l.

true stress ($\bar{\sigma}$): ratio between stress and actual cross-section area, as opposed to initial cross-sectional area.

trunnion: cylindrical protrusion of the Morse taper (male portion) used as the mounting point for the femoral head.

turbostratic carbon: crystalline form of carbon where the basal planes have slipped relative to each other.

ultimate tensile strength (UTS): strength that provides a measure of the peak force (stress) that is sustained in a tensile test.

ultra-high molecular weight polyethylene (UHMWPE): linear polyethylene that has a molecular weight on the order of 2–6 million g/mol.

umbilical vein: vein that delivers oxygenated blood from the placenta to the developing fetus.

unicondylar: single condyle (refers to specific implant design).

uniform attack: corrosion process that occurs when the whole surface of the metal is corroded in an evenly distributed manner.

valgus: distal ends of the bone turn outward (bow-legged).

valvular disease: condition in which heart valves exhibit regurgitation or incompetency.

vanadium steel: steel alloyed with vanadium for enhanced strength.

van der Waals forces: weak attractive interactions that occur between atoms or nonpolar molecules owing to local dipole moments.

varus: distal ends of the bone turn inward (knock-kneed).

vascular: pertaining to the blood vessels in the body.

vascularization: formation of blood vessels.

vein: blood vessel that carries blood toward the heart.

venous ulcers: wounds, usually on the leg, due to improper function of venous valves.

ventricles: two lower chambers of the heart, responsible for pumping blood.

viscoelastic fracture mechanics (VEFM): field of fracture mechanics that studies the processes of fracture in viscoelastic materials.

viscoelastic: displaying both fluid- and solid-like behaviors.

viscoelastic transition: regime of behavior where the material is neither glassy nor rubbery, but displays some aspects of each along with substantial time-dependence of the material properties.

viscoelasticity: material behavior that has both viscous (fluid) and elastic (solid) characteristics.

Volkmann's canals: conduits for blood and lymphatic vessels perpendicular to the long axis of the osteon.

vulcanize: term used to describe crosslinking in rubber materials.

wear: loss of material due to contact, sliding, or articulation.

wear, abrasive: process by which hard asperities remove material from a softer surface through plowing, cutting, or cracking.

wear, adhesive: process by which fragments from one surface are pulled off through adherence to a countersurface during moving contact.

wear, contact fatigue: process by which material removal occurs due to near-surface alternating stresses induced through rolling contact.

wear, corrosive: process by which a corrosive layer is removed through moving contact with a countersurface.

wear, delamination: process characterized by large, thin wear particles which are developed when a void and sub-surface crack grow in response to loading, rolling, or sliding.

wear, erosive: process by which material is removed during bombardment by liquid or solid particles.

wear, fretting: process by which material is removed when two contacts under load experience minute reciprocating or vibrating motion.

weight average molecular weight (M_w): Weight fraction mean based on the weight (mass) of individual molecular weights for a polymer system.

wires: string of metal used for bone fixation.

WLF equation: equation for computing the shift factor for amorphous or molten polymers developed by Williams, Landel, and Ferry.

wrought steel: steel alloy whose mechanical properties can be controlled through composition and heat treatment.

xenograft: see "heterograft."

yield strength (S_y): material property representing the stress at which plastic (permanent) deformation occurs. In metals this is associated with the movement or slip of dislocations.

yield surface: envelope within the space of principal stresses that defines the boundary between elastic and plastic behavior for a material.

Ziegler-Natta catalyst: catalyst that provides steric hindrance while a chain is growing (polymerizing), used to develop highly linear polymers.

Index

(110) plane, schematic illustration, 36
(111) plane, schematic illustration, 35
<1 1 1> family of directions, 35
18 carat gold implant, 515
2-D stress state, 259
316L stainless steel alloy, 61
3D CT scans, 537
3-D problem, simplifying, 261
510(k), *see also* Pre-Market Notification (510(k))

AAOS (American Academy of Orthopaedic Surgeons), 620
AbioCor Replacement Heart, 495
abrasive wear, 382–3
abrasive wear particles, 383
abutment(s), 513, 620
  designs, 525
  geometries, 526
  head, 525
  of same materials as the implant body, 525
  screw, 525, 527
  teeth, 512
acetabular cup, 620
  delamination of highly crosslinked UHMWPE, 462–3
  femoral head penetration into, 237
  highly resistant to wear, 424
  socked portion of the joint, 424
  total hip replacement utilizing an UHMWPE, 117
    as polymer of choice for, 432
acetabulum, 620
ACL, *see* anterior cruciate ligament (ACL)
AcroFlex design, 455
AcroFlex lumbar disk
  replacements, 455, 469–70
acrylic bone cement (PMMA), 431, 433
AcrySof IOL, 583
activation volume, 110, 620
active materials, 71
addition reaction, 106, 620
adhesion, 375
adhesive wear, 381–2
adulterated, 620
adventitia, 155, 481, 620
age hardening, 48, 620
aged dentin, 139
aggregated proteoglycans, loss of, 532
Albee, Fred, 453

Alcmaeon of Crotona, 129
all-ceramic implant bodies, 523
Allergan, Inc., 586
Alloderm dermal implant, 577
allografts, 572, 577
alloplastic, 620
alloplastic interpositional materials, 533
alloplastic material, for dental implants, 505
alloys, 17, 620, *see also* specific alloys, *see also* medical alloys
alloys
  altering structure and properties of, 42
  imperfections in, 38–42
AlphaCor implant, 582
alumina, 80, 86
alveolar bone, 528
alveolar crest, 508
alveolar processes, 508
American Academy of Orthopaedic Surgeons (AAOS), 620
American Burn Association, 575
American National Standards Institute (ANSI), 409
American Society for Testing and Materials, *see also* ASTM International
American Society for Testing Methods (ASTM), standard for surgical grade stainless steel, 417
Amonton's Law, 375
amorphous ceramics, forming, 83
  amorphous polymers, 98, 620
  lacking order, 98
  randomly ordered or entangled chains, 103
  T_g characterization plot, 103
  uses of, 106
amorphous silicate glasses, 71
amplitude loading, 341, 350
anabolic, 620
anabolic functional loads, 527
anatomical study, 129
ANCURE Endograft System, 497–8
aneurysms, 489, 620
angina, 489
angioplasty, 620
angioplasty catheter, 11
anion coordination number, 72
anions, 72, 620
anisotropic, 620
anisotropic behavior, of an arterial graft, 11

anisotropic composite systems, 19
anisotropic material, Hooke's Law for, 184
ankylosing spondilosis, 458, 620
ankylosis, 532, 533, 620
annealing, 46, 48, 621
annulus, 621
    composed of concentric sheets of collagen, 457
    of each disk, 457
    preserving, 454
annulus fibrosus, 457
anode, 55
anodic materials, 55
Anonymous Seven from Guidant
anterior cruciate ligament (ACL), 572
    1–2 million cycles per year, 573
    designs, synthetic, 572–3
    implants, comparing failure load and stiffness, 574
    patellar tendon as donor site, 571
    permanent augmentation devices, 572
    permanent replacements, 572
    reconstructive surgeries, 570
anti-coagulant medications, 487
AO/ASIF implant, 535
aorta, 481
    aneurysm in, 489
    tensile strength, 156
aortic arch, 156
aortic valve, 479
apatite form, of calcium phosphate, 86
apex, 514, 621
apical foreman, 508
Apligraf, composite implant, 578
apparent modulus, 224
appendicular, 621
applied crack driving force, 319
applied mechanics, 130
applied stresses, 36, 183
arbitrary plane, intersecting the y and z axes, 34
area moment of inertia, 196, 621
Arrenhius equation, 110
Arrhenius empirical model, 627
arterial lumen, reduction of, 477
arterial tissue, cross-section of, 481
arterial wall, multi-layered structure of, 155
arteries, 481, 621
arthoplasty, early, 416
arthrodesis, 454, 459, 621
arthroplasty, 621
arthroscope, 533, 621
arthroscopy, 533
artia, 480
articular cartilage, 151–4, 621
    collagen fiber network of, 153
    composition of, 152, 214, 422
    limited capacity for self-repair, 151
    microstructure and ultrastructure of, 152
    non-equilibrium properties, 154
    providing lubrication and smooth articulation, 435
    surgical repair difficult, 152

    time-dependent strain under deformation, 215
    viscoelastic behavior of, 214
articular disks, 533, 534
articular eminence, 509
articular surfaces, of the TMJ, 538
articulation, 276, 621
artificial intervertebral disk, 180
artificial skin, 575–6, *see also* skin
artificial skin
    as a more modern construct, 561
    current designs for, 576–8
    functional requirements of, 578
    uses for, 575
asperities, 371, 621
astigmatism, 583, 621
ASTM D638
    Standard Test Method for Tensile Properties of Plastics, 409
ASTM F562 alloy, 62
ASTM F75, cast form of cobalt-chromium, 62
ASTM F799 alloy, improved mechanical properties, 62
ASTM F90 alloy, utilizing addition of tungsten and nickel, 62
ASTM G115–04
    Standard Guide for Measuring and Reporting Friction Coefficients, 389
ASTM International, 61, 408, 621
ASTM specifications, for wrought Co–Cr alloys used in medical implants, 62
ASTM Standards E466–E468 (American Society for Testing and Materials), 335
atherosclerosis, 477, 483, 621
atlas vertebra, 456
atomic force microscopy (AFM), 134, 387, 388
atomically bonded solid, 285
atoms, extra half plane of, 39
atria, 479, 621
austenite, 49, 621
austenitic stainless steel, 61
Australia, Therapeutic Goods Association (TGA), 411
autoclave, 621
autoclaving, 8
autogenous, 621
autogenous bone grafts, 89
autografts, 487, 572, 577
auto-immune disease, caused by breast implants, 587
autologous, 621
autologous tibial grafts, 453
autologous transplants, for vascular grafts, 478
average roughness, 371
axial load
    stress due to, 204
    stressing a part cross-section, 346
axis vertebra, 456

Braun Melsungen AG, 589
bacteria, housed in the human mouth, 510
bacterial contamination, elimination of, 8

ball and socket articulation mechanism, of the hip, 425
balloon angioplasty catheter, properties of, 11
balloon angioplasty, compressing plaque, 483
    balloon catheter, 483, 621
    polymers in, 94
    stent shipped over outside of, 483
    used in percutaneous transluminal angioplasty, 493
balloon-expanded stent delivery, 483, 484
Barenblatt, 305, 307
basement membrane, Type IV
    collagen in, 133
Basquin equation, 337, 338, 340
Basquin exponent, 337
BCC, see also body-centred cubic (BCC)
beam equation, development of, 193–7
beam theory, 169
    basics of, 193–7
    derivation for, 194
    model for certain loading situations *in vivo*, 205
    solving stresses, 267
beam, under bending moment, 194
bearing pair, for the hip, 426
behavior regimes, for viscoelastic material, 210
bending load, stress due to, 204
bending modulus, for hydrated elastic fibers, 134
bending moment, 193, 194
Bernoulli, Daniel, 194
biaxial stress states, 256
bicuspid, 621
bicuspid pre-molars, 507
bileaflet heart valve
    design, 78, 79
    inherent stress concentrations, 78
bileaflet valves, 487
bioactive, 621
bioactive ceramics, 71, 85
    in the form of coatings, 86
    relying on porosity, 84
    use of, 433
bioactive glass, 86, 425, 622
bioactive polymers, 113
Biobrane, 577
Bioceram, recent withdrawal of, 524
biocompatibility, 7–8, 622
    analysis of, 7
    assessment of, 7
    definition of, 7
    necessity for, 60
    of any implantable device, 4
    of ceramics, 85
    of cobalt-chromium-molybdenum alloys, 506
    of polymers, 113
    studies, 417
biocompatible, 622
biocompatible materials, limited number of, 7
biodegradable, 622
biodegradable ceramics
    composed of calcium phosphate, 86
    used as bone substitutes, 86

biodegradable materials, 71
bioglass, 71, 87
Biolimus, 485
biological attack, long-term resistance to, 7
biological inflammatory response, 65
biological materials, prone to immunorejection or normal resorption processes, 15
biological system, interactions with mechanically damaged synthetic biomaterial, 8
biological tissues, 13
biological valves, reducing risk of clotting, 487
biomaterials
    as hard or soft, 13
    classifying, 13–16
    complicated mechanical behaviors of, 208
    exhibiting viscous or fluid-like characteristics, 318
    experiencing forces or loads, 167
    overview of, 4
    restoration of function of a living being, 13
    specimen configuration requirements for, 303
    surface contact in, 386–7
    testing for friction and wear properties, 370
biomaterials science, 4
BioMatrix stents, 485
biomechanics, 130
bioprosthetics, 486
bioreactivity, 622
    customizable, 72
    illustrating spectrum of, 85
    polymers range of, 94, 113
    schematic illustration showing, 85
    tailoring from inert to fully resorbable, 85
bioresorbability, of certain polymers, 19
bioresorbable, *see also* biodegradable
bioresorbable ceramics 85
bioresorbable polymer sutures, 124
bioresorbable polymers, uses of, 94, 113
bioresorbable suture anchor, 92
biotribology, study of, 387
bisphosphonates, 530, 622
bi-unicondylar knee implants, 437
Bjork-Shiley Convexo-Concave Heart Valve, 400, 496–7
blade-form endosteal implants, 513, 518–19
blockage, 486
blood cells
    emerging from a blood vessel, 482
    types of, 481
blood circulation, working model of, 130
blood vessels, 481
    dimensions and pressures in, 482
    replacements of small, 156
    trilaminate structure of, 155
blood-contacting implants, 478
blunting region, of the JR curve, 316
body-centred cubic (BCC), 622
body-centred cubic (BCC) crystal structure, 32
body-centred cubic (BCC) crystals, 40

body-centred cubic (BCC) lattice
   body-centred atom in, 35
   close-packed direction in, 35
   layering of closest packed planes to create, 32
   tetrahedral site larger than octahedral site, 33
body-centred cubic (BCC) structure, 32
body-centred cubic (BCC) systems, secondary slip systems, 40
bond energy $U(r)$, general form of, 11, 12
bond, stiffness of, 11
bonding and crystal structure, of metals and alloys, 28–32
bonding mechanisms, 16
bone absorption, 203–5
bone architecture, remodeling of, 88
bone cement, 383, 425
bone drilling, 522
bone grafts, 622
   from bone banks, 88
   need for, 87
   obtaining, 70
   physiology of successful, 88
   stabilizing adjacent vertebrae, 454
bone healing, 448
bone marrow, damage to, 447
bone periosteum, tearing of, 447
bone resorption, 168, 520
bone substitutes, 70, 87–9
bone tissue, 140, 144
bone(s)
   arresting cracks, 323
   building up bone density under higher loading, 167
   compliance match to adjacent, 11
   mineral responsible for stiffening, 135
   nanoscale ceramic reinforcement, 286
   resorbing, 527
   taking from elsewhere in the body, 87
   types of, 140
   with fracture and intramedullary rod, 177
bony ingrowth, at the bone-implant interface, 520
bore, of the femoral head, 429
Boston Keratoprosthesis, 581, 582
Boston K-Pro
boundary lubrication, 379
boundary work, 310
bovine serum, 389
Bowman's layer, 579, 622
Branemark, Per Ingvar, 522
Bravais space lattices, 28, 29, 622
breast augmentations and reconstructions, 584
breast implants
   auto-immune disease caused by, 587
   classified as Class III devices in 1988, 586
   current designs and functional requirements, 585–6
   design evolution of, 585
   development of, 561
   evolution of, 585
   legislative history of, 586

post-implantation concerns, 586
regulatory failure, 587–8
second-generation, 585
third-generation, 586
breasts, injecting substances into, 561
brittle, 622
brittle materials crack tip in, 356
   fracture associated with, 284
   fracturing before yield, 241, 269
   highly sensitive to tensile stress, 265
   reducing size to improve strength, 286
   relying on damage mechanisms, 322
   stress singularity at the crack tip, 307
brittle solids, failing, 270
brittle systems, as weak in tension, 244
bruxism, 529, 622
building blocks, of tissue, 131–6
bulk deformation processes, 51, 52
bulk erosion, 124
bulk properties, 371
bulky side groups, 99
bundled fibers, 561
Bureau of Chemistry, 399
Burger's vector, 39, 40, 622
burn victim mortality, 575
burns, 575, 576
butterfly fracture, 622
buttock implants, 584

cadaver autografts, not providing number of needed valve replacements, 487
cadaveric allografts, 572, 577
calcium carbonate, 86
calcium phosphate, 71
Canada, Therapeutic Products Directorate (TPD), 411
canaliculi, 141, 622
cancellous bone, *see* trabecular bone
cancellous screws, 450
canine aorta, mechanical properties of, 156
cannulated, 622
cannulated screws
   for femoral neck repair, 451, 452
   sizes, 450
capsular contracture, 586
carbon fiber
   abrading bone in the knee, 587
   ligament failure, 587
carbon fiber-reinforced UHMWPE (Poly II), 444
carbonate content, of bone, 135
carbonated hydroxyapatite, 86
cardiopulmonary bypass machines, 495
cardiovascular anatomy, 479–83
cardiovascular applications, of e-PTFE, 94
cardiovascular biomaterials, 13
cardiovascular devices, on the horizon, 500
cardiovascular disease, 477, 478, 622
cardiovascular implants, 478
cardiovascular system, 477, 480, 622
cartilage, 622
   ability to regenerate *in vivo*, 15

Index

layer of partially calcified, 153
   normal compared to osteoarthritic, 154
   optimized for compressive loading, 151
   surrounding a synovial joint, 422
cartilage-on-cartilage contact, 386
casting, 51, 52, 622
cataracts, 583, 622
catgut, 124, 560
catgut sutures, 565
catheters, 493, 623
cathode, 55
cations, 72, 623
Cauchy stress (true stress), 245
cavitating ductile material, fracture surface, 271
cavities, 510
CEA (cultured epidermal autograft), 577, 625
cells, in human blood, 481
cement line, 623
cemented hip arthroplasties, 433
cementum, thin layer, 508
Center for Devices and Radiological Health
    (CDRH), 397, 403, 404–8
Centerpulse Orthopedics, 467
ceramic abutments, 525
ceramic coating, surface modifications with, 523
ceramic heart valve, critical crack length in, 78–9
ceramic materials, 72, 623
ceramic systems, 330
ceramic total hip replacement femoral heads, 284
ceramics
   as implant materials, 70, 71
   bonding and crystal structure, 71–5
   characteristics of, 241
   classifying, 85
   complex crystal structures of, 72
   drawback of using, 85
   failure mechanisms in, 272
   fatigue behavior of, 359
   forms of, 71
   fracture mechanisms in, 322
   high elastic modulus values, 16
   in a variety of applications, 70
   in medical implants, 85–7
   in structural applications, 70
   ionic bonding, 72
   low fracture toughness of, 322
   mechanical behavior of, 75–82
   processing of, 82–4
   properties of, 18
   resistance to fatigue crack growth, 330
   sensitive to tensile normal stresses, 77
   stronger in compression than in tension, 273
   structure and mechanical behavior of, 75
   susceptibility to fracture, 71
   toughening mechanisms, 79, 80
   wear resistance, 433
Ceravital, 71, 87
cervical spine, 454, 456, 623
chain branching, in polymers, 98
chain length, impeding long-range ordering, 99
chain scission, 9

chain structure, one-dimensional, 97
chains, entanglement of, 107
characteristic test frequency, inverse of, 227
Charite implant, design evolution of, 461
Charite intervertebral disk, 460
Charnley hip, innovative low-friction, 431
Charnley, Sir John, 390, 417
chemical bonds, 16
Chercheve, Raphael, 522
chewing, mechanical demands of, 136
chondrocytes, 151, 623
chordae tendineae, 481
Christensen TMJ device, slightly modified, 537
chromium carbide, precipitation of, 57
chromium precipitates, 48
chromium-depleted zones, 58
ciliary muscles, 580
Class I devices, 9, 403, 405, 623
Class I implants
Class I recall, of all Proplast-Teflon Vitek TMJ total
    joint prosthesis, 544
Class II devices, 10, 403, 623
Class II implants
Class III devices, 10, 403, 623
Class III implants
classifying, biomaterials, 13–16
cleaning protocols, 7
Cleaver, Mouse, 561
clenching, 530
clinical application, of biomaterials, 13
clinical concerns, in design of soft tissue
    replacements, 562
clinical issues, as paramount, 4
clinical performance, 10
clinical trials, 10
close-packed planes, 30, 31, 34, 75
close-packed spheres, planar section of, 30
close-packed structures, 30
clot formation, reducing blood loss, 483
cobalt-chrome alloys, replaced stainless steel,
    443
cobalt-chromium (Co-Cr) alloys
   blade implants, 518
   cast form limitations, 62
   elastic modulus, 61, 63
   femoral head exhibiting surface damage, 463
   high strength and fatigue resistance, 62
   in hip implants, 432
   in the fracture fixation industry, 453
   intergranular attack of, 66
   providing enhanced corrosion resistance, 27
   uses of, 61
CoCr total TMJ replacement, Kummoona designed,
    535
COD, *see* crack opening displacement (COD)
coefficient K, specifying relative strength of the
    stress singularity, 288
coefficient of friction, 373, 623
   defined, 375
   for common materials pairs, 375
   low for UHMWPE, 117

Coffin-Manson equation, 344
coherent flaw, 334, 623
cohesive stress, 307
cohesive zone, length of, 319
cold forged stainless steels, 62
cold rolling, 47, 623
cold working, 40, 46, 640
cold-rolled brass, 46
collagen
    abundance of, 131
    known types of, 132
    variants of, 133
collagen fibers appearing in a variety of conformations, 132
    composition of, 133
    crimped form of, 151
        in articular cartilage, 153
        in dentin, 139
        schematic of, 132
        structural hierarchy inside, 140
collagen fibrils, 133
collagen molecules, 133, 140, 141
collagen sub-fiber, hole zones in, 141
collagenous osteoid, organization of, 141
comminuted fracture, 448, 623
compact bone, *see* cortical bone
compact tension geometry, 290, 349
compact tension specimen, illustration of, 349
complex modulus, 228, 230, 623
compliance, 11, 623
compliance match to bone, 450
compliance matrix, 205
compliant, 623
compliant materials, stretching at a crack tip, 323
complications, arising from corrosion, wear, fatigue or fracture, 8
composite beam
    bending of, 199
    finding the neutral axis of, 197–9
    for a custom implant, 199
composite materials, 169, 623
composite systems, 19
composite(s)
    characteristics of, 241
    composed of unidirectional fibers, 200
    undergoing longitudinal transverse loading, 200
    with matrices of polymers, metals, or ceramics, 19
compression fracture, 448, 623
compression loading, flaws closing, 77
compression, described by negative values for strain, 171
compressive portion, of variable amplitude loading, 276
compressive strength, 623
    of ceramics, 18, 75
    of enamel, 138
    of human dentin, 139
compressive stresses, 80, 264
compressive yield strength, 264
condensation reaction, 106, 623

condylar, 623
condylar prosthesis, 534
condylar type tibial component, 276
condyles, 10, 623
conforming, 623
conforming ball-and-socket joint, 421
congenital, 623
congenital defect, 486
congenital heart defects, blocking, 495
congestive heart failure, 477, 623
constant bending moment, 335
constant stress (creep) response, 217
constitutive behavior, 181–2
constrained dislocation loop, around particles, 48
contact fatigue, 384
contact lenses
    as common intervention, 581
    composition of, 581
    designs and functional requirements, 580–1
    tearing during ordinary use, 284
contact loading, between tibial insert and femoral component, 278
contact shielding, 320, 355
contact stresses, 66, 277, 623
contracture, 586, 624
contraindications
    for implants, 530
    for maxillary and mandibular implants, 530
    for temporomandibular joint (TMJ) replacements, 538
cooling, very rapid, 49
coordinate axes, changing, 173
coordinate rotation, 260
coordinate transformation, applying, 175
coordination number, 30, 624
    for octahedral site, 33
    for tetrahedral site, 33
co-polymerization, with PLA, 125
co-polymers, 95
coral, 71, 624
    as a bone substitute, 70, 87–9
    as bioresorabable ceramic, 86
    attributes for bone grafting, 88
    porosity of, 86
    structure of, 88, 89
coral-derived bone substitute
    widespread clinical success, 89
coral polyps, hard mineral deposit built up by, 88
coral secant modulus, 188, 639
cornea
    curvature of, 578
    layers of, 578, 579
    ultimate tensile strength, 582
corneal edema, one month after IOL implantation, 588
corneal replacements, 581–3
coronal plane, 624
coronary artery bypass graft (CABG), 489, 624
coronary heart disease, 477, 624
coronary stent designs, 485
corrodibility, scale of, 55

corrosion, 624
    at the mechanical junction, 429
    in metals, 18, 26
    in modular orthopedic implants, 64–7
    pathways in metallic systems, 55
corrosion debris, 53, *see also* wear debris
corrosion processes, 53–60
    at the taper connection, 66
    occurring in a modular total hip replacement, 66
    resulting from reacting with an aqueous environment, 54
corrosion resistance designing for, 54
    of alloys, 27
    of cobalt-chromium alloys, 62
    of metals, 60
    stainless steel known for, 62
corrosion-induced degradation, 417
corrosion-resistant materials (cathodic), 55
corrosive attack, in crevices, 65
corrosive wear, 383
cortical bone, 140–3, 624
    as an example of a hierarchical material, 323
    composition of, 140
    coral as a substitute for, 89
    deformation behavior, 142
    finding strains in, 186
    majority of body's calcium in, 144
    organization into osteons, 142
    stress-strain curve for, 192
    tension and compression response of, 143
cortical screws, 450
cosmetic breast implants, *see* breast implants
cosmetic implants, 584–6
cosmetic surgeries, numbers of invasive, 584
costs, dictating manufacturing method for an implant, 53
covalent bonding, 71, 95, 624
covalent-ionic ceramic structures, 13
covalently bonded materials, 13
crack advance correlated to peak value of stress intensity, 356
    in polymers, 361
    mechanisms, illustration of, 356, 357
    rate of, 348
crack bridging, 81, 320, 624
crack deflection, 320, 355
crack driving force coming from stored energy from, 295
    overcoming, 294
    work in the system providing, 320
crack equilibrium, 294
crack growth
    ahead of the notch, 268
    as stable or unstable, 298
    correlation to stress intensity factor, 356
    energy dissipated by, 295
    in a material, 320
    initiation of, 316
    presence of a notch or small crack, 272
    primary regimes of, 351
    rate, 351

    resistance and, 294–6
    resistance curves, 295, 316–17
    under application of tensile stress, 291
crack initiation criteria, 316
    designing against, 317
    failure criterion, 317
    in Regime II, 316
    time, 319, 624
crack loading, Modes I, II, and III, 292
crack nucleation
    from a sharp notch, 268
    in the component, 266
crack opening displacement (COD), 304–5, 624
crack propagation
    in ceramics, 359
    in most metals, 357
    intrinsic resistance to, 348
    value for the onset of, 59
crack shielding mechanisms, 355
crack surface, 294
crack tip
    brittle materials accommodating stress singularity at, 307
    compliant materials stretching extensively at, 323
    compressive stresses retarding growth of the flaw at, 80
    effective, 301, 304
    elastic-plastic singularity, 311–14
    in a ductile metal, 356
    in brittle materials (ceramics), 356
    inelasticity, 305
    physical as blunted, 304
    plasticity, 304–5
    stress becoming singular near, 289
    stresses, 77, 311
    studying extreme stresses at, 284
    viscous dissipation at, 318
crack velocity, 22, 60
crack wake
    in a ductile metal, 356
    inelastic region around, 320
    mechanisms, 357
crack wedging, 320
crack(s)
    as extreme stress concentrators, 287–8
    behavior under moderate and severe crack tip plasticity, 307
    distance from, 288
    forming during loading, 283
    nucleated, 269
    subjected to tensile stress in a ceramic, 77
    under complicated global stress states, 291
    under complicated load conditions, 296
cracked plate, thickness approaching plastic zone size, 301
crack-opening mode, 291
craniofacial, 624
craniofacial features, 506
craniomaxillofacial, 624
craze, 624
crazing, 274, 275, 276

creep, 19, 208, 210, 624
creep (time-dependent) strain response for the
    Kelvin model, 220
  of the Maxwell model, 218
creep and cycle strain softening behavior, of
    UHWMPE, 225
creep behavior, over a range of applied stresses, 237
creep compliance of the SLS, 223
  over time for UHMWPE, 225
creep response
  of the SLS, 222
  vital to assessment of performance, 212
creep strain, for the Kelvin Model, 220, 221
creep time constant, for the SLS, 223
crest, 624
crest module, 514, 524, 624
crestal bone
  loss as marginal, 527
  supra-physiologic loading of, 527
  v-shaped resorption around root-form implants, 527
crevice corrosion, 58, 624
  in metal components, 26
  minimizing, 26
  schematic illustrating, 58
  stainless steel susceptible to, 62
crimp, in collagen fibers, 149, 151
crimped collagen bundles, 193
critical crack length, 60
  as an inspection parameter, 78
  form of, 78
  resulting in fracture, 297
  solving for, 353
critical crack opening displacement, 304
  critical flaw size, 624
  defining relative fracture toughness, 297
  determining, 298
  for pyrolytic carbon, 79
  for toughened zirconia ceramic, 79
  in component loaded in shear, 293
  ratio of, 298
  using fracture toughness to determine, 77
critical flaw, initiation or propagation of, 333
critical normal stresses, 258
critical strain, for low cycle fatigue in a tibial plateau, 345–6
crosslinked system, lightly compared to fully, 112
crosslinked UHWMPE, sacrificed resistance to fatigue crack propagation, 463
crosslinking, 9, 624
  at the expense of fatigue and fracture resistance, 428
  effect on glass transition temperature of a polymer, 104
  in the collagen network affected by environment, 213
  of collagen increasing with age, 133
  resistance as a function of, 428
crosslinks
  compared to entanglements, 108
  resulting in permanent covalent bonds, 107
cross-shear articulation, 427
crown, 136, 507
crown, natural, 624
crown, synthetic, 624
cruciate ligaments, retaining in the unicondylar procedure, 438
crystal atoms, loading of, 76
crystal lattices
  distortion to, 33
  orienting in a coordinate system, 34
crystal planes, displacement of atoms across, 36, 37
crystal structures in structural ceramics, 72, 73
  of metals, 27
crystal systems, 622
  classification of, 28, 29
  general, 28
crystal(s)
  different kinds of constituents in, 74
  in enamel, 136
Crystalline Bone Screw (CBS), 523
crystalline ceramics, 72, 330
crystalline domains, polymers forming, 98
crystalline lamellae, 98, 101
crystalline materials, metals and alloys as, 28
crystalline solids, 28
crystalline structure, 625
crystallinity, 625
  facilitating through chain folding, 99
  of polymers, 98, 102
crystallite development, in polymer systems, 101
crystallographic planes and directions, 34–5
cubic structures, more available slip systems, 45
cubic system, close packed plane, 34
cultured epidermal autograft (CEA), 577, 625
curved resorbable collagen scaffold, 411
curves, *see* specific curves
curves
  indicating roughness and waviness
  resulting from friction tests
cycles to failure, for a spinal implant, 338–40
cyclic fatigue exponent, 357
cyclic loading
  effects on fatigue or fracture, 261
  of polymers, 360
cyclic plastic strain amplitude, 345
cyclic plastic strains, 335
cyclic softening, in UHMWPE, 225
cyclic strain
  data representing, 343
  hardening, 625
  softening, 625
cyclic stresses
  resulting from kinematics of the joint, 277
  with tensile stresses, 426
cyclic stress-strain loops, evolution of, 334
cyclic yield stress, 333
cylindrical designs, more easily implanted, 524
cylindrical osteons, composition of, 141

da Vinci, Leonardo, 130
Dacron, 478, 625
    fibers, 121
    grafts, 490
Dahl, Gustav, 515
Dalkon Shield, 400
damage mechanisms, 274, 275
damage processes, utilized in ceramics, 274
damper, 217, 230
dashpot, *see* damper
De humani corporis fabrica, 129
Deborah number, 211
debridement, 533, 625
debris, *see also* wear debris, *see also* particulate debris, *see also* corrosion debris
deciduous, 625
deciduous teeth, 507
deep zone, collagen fibers in, 153
defect tolerant, 625
defect-tolerant approach, to fatigue failure, 333
defect-tolerant method, 297
defect-tolerant philosophy, 348–62
defects, presence of, 75
deformable materials, 171
deformation plasticity, 308, 625
deformation(s), 45, 375
    altering orientation of grains, 45
    categories, 167
    enhancing resistance to crack advance, 320
    important in implant design, 167
    in response to loading, 169
    localized and extreme nature of near a sharp flaw, 286
    micromechanism of, 333
degradable polymers, 94
degradation rate, of biodegradable ceramics, 86
deionized (DI) water, 389
delamination, 625
    in total knee replacements, 22, 279
    on articulating surfaces of knee joint, 277
    prevalent region of, 463
    wear, 386
    with clinical dislocation, 464
Delrin, disk made of, 486
dense bone, implantation in more successful, 526
densification, 147
density, relationship with crystallinity, 102
dental applications, from investment casting, 52
dental biomaterials, 13
dental caries, 625
    as a leading cause of tooth loss, 510
    described, 510
dental enamel, *see* enamel
dental implants, 505, 510–30
    anchoring via osseointegration, 520
    clinical reasons for, 510–12
    cylindrical in shape, 82
    described, 505
    examples of different types of, 513
    filling cavities in tooth enamel, 81
    functional requirements, 530
    future improvements, 544–5
    historical development of, 515
    immediate and early loading of, 545
    innovation in, 515
    integrating into underlying bone, 70
    replacement materials used against enamel, 383
    thermal stresses or strains, 81, 82
    types of, 513–15
dental plaque, 510
dental prosthesis, 511
dentin, 136–9, 625
    as more compliant than enamel, 139
    decay of, 510
    middle layer of, 507
    providing protective cover for pulp, 507
    secondary, 508
    structure of, 136
dentinal tubules, 138
    illustration of, 139
dentino-enamel junction, 138
dentition, 511, 625
denture base, 511
dentures, 512
deoxyribonucleic acid (DNA), 625
depressed fracture, 448, 625
Dermagraft, 577, 578
dermal acellular implants, 577, 625
dermal cellular implants, using neonatal fibroblast cells, 577
dermatome, 577, 625
dermis, 155, 625
Descemet's membrane, 580, 625
design factors, for friction and wear, 389
deviatoric, 625
deviatoric component 173
deviatoric portion of the normal and shear stresses, 250
    of the stress, 245
deviatoric stress tensor, 175
device classifications, 403–4
diabetic foot ulcers, 575, 577
diagnostics biomaterials, 13
dialing hook, in cataract surgery, 588
diamond cubic lattice, utilizing FCC lattice, 72
diamond-like carbon (DLC) coatings, 84
diaphyseal, 625
diaphyseal (cortical) bone, 450
diaphysis, 140, 625
die casting, with steel molds, 52
diethylene glycol, 399
differential scanning calorimetry (DSC), 102, 625
dilatational, 625
    component 173
    stress tensor, 174
    stresses, 249
dilational portion, of stress, 249
directions, within the lattice structure, 34
disk arthrosis, spinal fusion used to treat, 454
disk replacement technology, 460
dislocation, 40, 47, 626
dislocation force fields, causes of, 42

dislocation initiation rates, in polymers, 111
dislocation line, 39
dislocation movement (slip), 243
dislocation pile-ups, 42
dislocation slip, 251
dislocation theory, full treatment of, 39
dislocations
  in ceramic structures, 75
  in shear processes in metals, 40
  types of, 39
disorders, necessitating TMJ replacement, 532
displacement vector, 39
dissipated energy, from a single load cycle, 229
distal, 626
distortional energy, 243, 249, 250, 626
distortional energy criterion, 253
distortional or deviatoric stresses
distortional portion, of stress, 249
distortional strain energy, 250
D-lactide form, of PLA, 123
DNA disablement, 8
Dohlman Doane Deratoprosthesis, 581
double modular hip prosthesis, corrosion-induced fracture of, 463–7
Dow Corning, lawsuits filed against, 587
drawing, 51
drawn glass fibers, experimental testing of, 286
driving force, 294, 320, 626
drug efficacy, demonstrating, 400
drug-eluting stents, reducing restenosis, 483
ductile materials
  behavior, 626
  deforming through shear, 243
  facilitating dissipation of near-tip crack stresses, 322
  undergoing fatigue experience cyclic damage, 356
  yielding before fracture, 241, 269
ductile metals, 250, 357
ductile processes, of polymers
ductile systems, generally utilizing shear deformation, 270
ductility
  in metals, 35
  loss of, 46
Dugdale, 305, 306
Dugdale plastic zone, 306
duocondylar knee, 439
dynamic compression plate, 451, 452
dynamic loading, during downhill skiing, 208
dynamic mechanical analysis (DMA), 227, 626
  finding upper and lower limit of, 200–3
  line slope, 187
  material constant, 181
  measured value for, 181
  range for estimates in uniaxial fiber composites, 202
  range of values for, 181

edema, 626
edentulism (tooth loss), 505, 511, 626
edentulous, 626
edentulous individuals, biting forces of, 511
edge dislocations, 39, 271, 626
  forces between, 41
  parallel repelling each other, 40
  visualized as an extra half plane of atoms, 39
edges, common to two anion polyhedra, 73
effective crack tip, 301
effective creep compliance, of the SLS, 223
effective stress, 626
  calculating, 241
  determining, 260
  independent of coordinate frame, 260
  using 2-D analysis, 260
effective stress range, 338
effective von Mises stress, 345
  for a compressive loading state, 264
  for a loading state, 259
Egyptians, placing hand-shaped prosthesis into deceased, 515
eigenvalues, 179, 180–1
eigenvectors, 179, 180–1
elastic behavior, 109, 245
elastic collagenous tissues, elastin found in, 133
elastic component, of the strain amplitude, 344
elastic deformation, 16, 109, 167
elastic energy, material's ability to store, 243
elastic fibers
  elastin and fibrillin-based microfibrils, 133
  mechanical properties of, 134
  recoil in, 134
  schematic of, 134
elastic J, dominating in LEFM, 314
elastic modulus, 15, 181
  alternative methods for calculating, 188
  described, 16
  estimating, 11
  finding from a stress-strain curve, 189
  following rule of mixtures, 202
  for ligaments, 150
  for skin and blood vessels, 156
  for stainless steel, 61
  for viscoelastic material, 211
  formula for, 13
  in a fiber-reinforced composite, 202
  lower reducing complication of stress shielding, 63
  materials with high, 16
  materials with lower, 16
  of articular cartilage, 154
  of C-Cr alloys, 63
  of dentin, 139
  of enamel, 136
  of human enamel, 138
  of human trabecular bone, 148
  of hydrated collagen fibers, 133
  of materials, 11, 81, 187
  of tendons, 149
  of the Maxwell model, 219
  physical foundation for, 11

range associated with each material class, 13
uses of, 205
values, comparison of, 17
elastic modulus (E), 626
elastic strain, 13, 109
elastic strain energy, storing, 109
elastic stress, 287
elastic-plastic behavior, 308
elastic-plastic crack tip singularity, 311–14
elastic-plastic fracture handbook, 314
elastic-plastic fracture mechanics (EPFM), 303–5, 626
   admitting inelastic behavior, 308
   deformation plasticity, 308
   fracture instability in, 316
   fracture toughness measure, 316
   J for many engineering problems of interest, 316
   summary of, 318
   true conditions, 314
   valid for moderate yielding, 312
elastic-plastic generalization, of crack driving force, 303
elastin, 133–4
   described, 131
   in arterial wall, 156
   in ligament, 148, 151
   in tendon, 151
   responsible for stiffness of soft tissues, 134
elastohydrodynamic lubrication, 377
elastomeric material, under mixed mode conditions, 470
elastomeric polymer, time temperature superposition behavior for, 233
elastomeric-viscoelastic properties, of silicones, 123
Electric Power Research Institute (EPRI), at General Electric, 314
electron beam method, 9
electron disruptions, damaging living cells, 8
electronegativity, 72, 626
electrospinning, 157
electrostatic attraction, 28
electrostatic valency, 73
ellipse, equation for, 253
ellipsoidal pores, in bone, 141
elongation at break, 187
elution time, 107
embedded ellipse, in an infinite plate, 266
embolus, 483, 626
embrittlement, caused by cold working, 46
emulsion, 626
emulsion methods, used for ceramics, 83
enamel, 136–9, 626
   acids in the plaque reacting with, 510
   brittleness of, 138
   composition of, 136
   crystals larger than bone, 135
   elastic modulus of, 136, 138
   energetic toughness of, 138
   energy of fracture of, 138
   most mineralized tissue in the body, 136
   outer layer of, 507
   tensile strength of, 138
   thermal expansion coefficient for, 82
endocardium, 479, 626
endosteal, 626
endosteal dental implants, 513, 514, 529
endotenon, 149, 626
endothelial cells, 155, 157
endothelium, 580, 626
endovascular stent-graft, alternative conduit for blood flow, 489
endurance limit, 626
   at zero mean stress, 340
   between test specimens and components, 348
   compared to the cyclic driving force, 340
   defined, 336
   error, 347–8
   factors affecting, 337
   for a notched bar, 267
   for an unnotched bar, 267
   for certain metals, 336
   for the material, 329
   of metals, 330
endurance limiting factors, 346
energetic description, of crack growth, 295
energetic methods, in fracture mechanics, 293–4
energetic toughness, 49, 626
   of enamel, 138
   of UHMWPE, 117
energy dissipation, 208
energy of fracture for human dentin, 139
   of enamel, 138
energy release rate, 294, 626
   crack as a growth driving force, 294
   critical value of, 296
   describing an energetic condition, 296
   for an unbounded center-cracked plate, 296
   form of the negative of the derivative of a potential, 294
   J as, 309
energy storage component, 229
energy, supplying to sustain crack growth to failure, 295
engineering attributes, of ceramics used in medical implants, 72
engineering design analysis, for total hip replacement, 435
engineering materials defined, 14
   properties remaining unchanged over large durations of time, 209
   selecting for an implant design, 15
   used in the body, 13
   utilizing for medical implants, 4
engineering mechanics, in tissues, 130
engineering polymers, overlap with hard and soft tissues, 17
engineering strain, 191, 205
engineering stress, 191, 205
engineering stress-strain
engineering stress-strain behavior, 242

engineering stress-strain curve, 191
entanglements, compared to crosslinks, 107, 108
enzymes, serving as a solubilizing agent to the polymer, 125
EPFM, *see* elastic-plastic fracture mechanics (EPFM)
epicardium, 479, 626
Epicel, 577
epidermis, 154, 627
epiphysis, 627
epitenon, 149
epithelium, 579, 580, 627
EPRI estimation procedure, 315
e-PTFE, *see also* expanded polytetrafluoroethylene (e-PTFE)
equal biaxial stress, 256
equal yet opposite edge dislocations, 40
equibiaxial stress state, 256
equilibrium atomic separation, 75
equilibrium condition, 294
equilibrium elastic modulus, for articular cartilage, 153, 154
erosion, 124, 627
erosive wear, 384
ethylene oxide (EtO) gas sterilization, 9
ethylene oxide gas, 627
Euler, Leonhard, 194
European Union (EU), Directorate General for Health and Consumers, 410
eutectoid reaction, microstructure development through, 49
Elixir Sulfanilimide, for sore throats, 399
expanded polytetrafluoroethylene (e-PTFE), 113, 490, 627
  in facial reconstruction implants, 120
  introduced in the 1970s, 478
  microporous structure, 120
  reconstructing ligaments, 94
expendable mold, 52
extension, 171, 627
extension flexion, of the joint, 277
extracellular matrix, 627
extrinsic crack-tip shielding effect, 355
extrinsic mechanisms, altering effective driving force, 320
extrinsic toughening, 320, 355, 627
extrusions, 51, 330, 627
eye
  anatomy of, 578, 579
  changes in geometry of, 580
eyeglasses, 581
Eyring model, 109, 111, 620, 627

fabrics, for synthetic grafts, 490
face-centred cubic (FCC), 30
face-centred cubic (FCC) lattice, 627
  close packed direction in, 35
  metals flowing under stress, 40
  packing density, 30
  packing density of, 30–2

transformation into a HCP structure, 62
yielding, 30
faces, common to two anion polyhedra, 73
facet screws, for spinal stabilization, 453
facial changes, associated with aging, 511
facial implants, locations of common, 584
facial reconstruction implants, 120
  uses of e-PTFE in, 94
factor of safety (FS), 627
  against failure, 241
  against plastic deformation, 255
  against yielding, 249, 255, 259
  determining, 264, 339
failure assessment diagrams, two-parameter, 305, 317
failure criterion, basic form of, 265
failure envelope, 265
failure mechanisms
  in ceramics, 272
  in metals, 271–2
  in polymers, 274–6
failure strength (S_f), 187, 627
failure stress, 251, 265, 340
failure theories, for ductile materials, 279
family of directions, denoting, 35
1 1 1 family of planes, 34
far field, 289
fascicles, 149, 151, 627
fast fracture criterion, of the implant, 330
fast fracture regime, 351
Father of Bioengineering, 130
fatigue, 331–3, 627
fatigue behavior
  difference between ductile and brittle solids, 355
  of an engineering biomaterial, 333
  of ceramics, 359
  of metals, 357–9
  of polymers, 360–2
  of structural materials, 354–7
  under cyclic compressive loading, 277
fatigue characterization, of a material, 334
fatigue crack(s)
  growth rate, 359, 360
  nucleating under cyclic compression loading, 268
  propagation, 349–54, 362
  propagation resistance, 350
  under cyclic tensile loading in UHMWPE, 277
  velocity of a moving, 350
fatigue cycles
  for a total hip replacement, 329
  mean stress of, 338
fatigue damage, mitigating, 330
fatigue data, from accelerated and idealized tests, 346
fatigue failures
  as a consequence of crack nucleation, 333
  in medical devices, 329
  occurring well below yield strength of material, 330
fatigue fracture resistance, loss in, 21
fatigue fractures

in trapezoidal hip stems, 362–4
of the abutment connection, 528
fatigue life
  based on propagation of an initial flaw, 348
  compared, 329
  of flawed components, 297
fatigue loading, 363
fatigue resistance design centred upon stress-based tests, 335
  designing for, 333
  for bearing surfaces in knee arthroplasty, 443
  in ball-and-socket joint of the hip, 11
  remaining limited in ceramics, 330
fatigue resistant implant, 329
fatigue sensitivity, to notches, 268
fatigue strength, 336, 337, 627
fatigue stress concentration factor, 267, 268
fatigue striations, 358
fatigue tests, in fully reversed loading, 341
fatigue threshold, 351
FCC, *see* face-centred cubic (FCC)
FDA, *see* Food and Drug Administration (FDA)
feathered fusion, 453
Fe-C steel, phase diagram for, 50
Fe-C system, eutectoid point, 49
Federal Food, Drug and Cosmetic Act, 399
femoral head, 627
  articulating junction with acetabulum, 426
  penetration into an acetabular cup, 237
  properties of, 424
  providing ball portion of the joint articulation, 424
  utilizing ceramics, 433
femoral neck, design requirements, 424
femoral stem, 627
  design evolving away from sharp corners, 364
  design requirements, 424
fractures, 362
  multiaxial loading, 259–61
  side crack in, 365
  total life approximation for, 340–1
femoral trabeculae, stress lines in, 146
femoral trabecular bone, stress trajectories in, 146
femur, compliance matrix for dry, 186
fenestrations, 516, 627
ferrous alloys (steel), endurance limit, 336
fiber recruitment, 151
fibers, thinner as stronger, 286
fibril alignment, 149
fibrillin, 133, 627
fibrils, in skin and arterial wall, 155
fibroblasts, 627
  embedded among collagen fibers, 149
fibrocartilage, 627
fibrocartilage callus, 448
fibrosa, 479, 627
fibrous peri-implant tissues, formation of, 545
fibrous pseudo-periodontal membrane, 520
filling, 627
filling material, within tooth enamel, 81
first degree burns, 576, 627

first moment of molecular weight, 107
first-order imperfections, 38
first throw hold, 570, 627
first-order viscoelastic material models, 217–21
fixation materials, properties of, 433
fixation, of knee components, 442
fixed bridges, 512
fixed partial bridges, 512
fixed-hinge-knee design, 437
flat bone, 140
flat plane *X*-ray, to catch strut fracture, 499
flaw size, reaching critical value, 59
flaws
  evaluating in ceramic components, 78
  in ceramics, 75
  making cast form of Co-Cr prone to fracture or fatigue failures, 62
  subjected to a tensile stress in a ceramic, 77
flexibility, of an arterial graft, 11
Flexicore device, 461
flexion, 628
fluid
  bearing pressure, 214
  constituent behavior, 214
fluid flow, regulating without damaging red blood cells, 477
fluorocarbon polymers, 119, 628
fluoroscopy, 499, 628
fold length, measure of, 102
food additives, testing of, 398
Food and Drug Administration (FDA), 399, 628
  510(k) guidance document for TMJ implants, 544
  approval, 7
  avoiding giving specific test requirements, 408
  classifying implants, 9
  final word in medical device regulation, 401
  gaining approval from, 397
  guidance contact lenses, 581
  guidance for keratoprostheses, 582
  guidance for submission of PMAs for breast implants, 586
  guidance for sutures, 570
  guidance for synthetic ligament implants, 574
  guidance in the submission process, 408
  history of, 401
  legislative history, 398–401
  medical devices regulated by, 92
  medical devices subject to pre-market approval by, 93
  not upholding regulatory duties, 544
  products regulated by, 397
  reclassified all types of TMJ implants, 544
  recognizing international testing standards, 408
  risks characterized for grafts, 491
  schematic representation of the organization, 405
  timeline of legislative history, 402
  types of characterization for mechanical heart valves, 489
Food and Drug Administration Modernization Act (1997), 400
Food and Drug and Insecticide Administration, 399

food and drugs, adulterated and misbranded, 398
Foods and Food Adulterants experiment, 398
force vector, resolving, 171
force(s)
  for activating yield, 40
  for the separation of atoms, 11
  physical definition of, 294
  translating between a plate, screw and bone, 450
Ford, Gerald, 400
foreign body giant cell reaction (FBGCR), 537
forging, 51, 628
forming, 51, 628
fossa prosthesis, 534
fossa-eminence implant, 534
Fourier Transform infrared (FTIR) spectroscopy, 21
Fox and Flory equation, 104
fractional lifetime, 341, 342
fractional recovery, 236, 628
fractography, 290
  estimating fracture toughness, 290–1
  insight into mechanism of failure, 270
  of fractured trapezoidal T-28 stem, 363
fracture behavior, 303
fracture callus, 448
fracture fixation, 446–53, 628
  broadly defined, 446
  clinical reasons for, 448
  designs, 450–1
  in a spine, 451
  in a spine using plate and screws, 451
  materials used in, 451
  plate, alloy used in, 60
fracture fixation devices
  design concerns in, 453
  functional requirements, 449
  properties, 450
fracture healing anatomy of, 447
  improvement in understanding of, 447
fracture instability, in EPFM, 316
fracture mechanics, 628
  concepts, 348–62
  described, 284
  ensuring appropriate structural analysis, 274
  estimating fatigue life, 329
  origin of modern, 283
fracture mechanisms
  in ceramics, 322
  in metals, 322
  in structural materials, 322–3
  in tissues, 323
fracture plate, 628
fracture processes, associated with chronology of failure events, 463
fracture resistance, of ceramics, 85
fracture strength, of ceramics, 272
fracture surface of a cavitating ductile material, 270
  orientation of, 270
  with fiber debonding in the UHMWPE matrix, 444
fracture toughness, 15, 59, 60, 77, 78, 297, 628

  as different from mechanical strength, 284
  associated with stress corrosion cracking, 322
  EPFM measure, 316
  estimate of, 291
  estimating from fractography of a tibial implant, 290–1
  improvement in, 301
  influencing critical flaw size, 79
  low for ceramics, 18, 433
  of ceramics, 75
  of elephant dentin, 139
  wide range in polymers, 322
fracture types, basic classifications associated with, 448
fracture(s) as catastrophic, 283
  characterized by types of loads generating, 449
  compressing, 451
  critical tensile stress for, 324
  failure mode distinct from strength, 284
  governed by localized processes, 284
  of trabecular bone, 148
fractured bones, implants used to properly align, 449
free volume, for polymers
frequency domain analysis, 227–33
fretting, 628
  at the mechanical junction, 429
  in metals, 17
  minimizing, 26
  wear, 384
  wearing away protective oxide film, 464
friction, 373–5
  classic definition, 374
  kinetic, 628
  static, 628
  wear test methods and, 387–9
friction test, standard curve, 375
frictional forces, reducing amount of force applied to teeth, 369
frictionless contact, not existing in the real world, 369
frontal, 628
FTIR spectra, showing oxidation of an UHMWPE component, 21
fully reversed loading, 338, 628
function, restoring to the tissue or biological component, 14
functional requirements, 628
  breast implants, 585–6
  contact lenses, 580–1
  corneal replacements, 581–3
  cosmetic implants, 585
  for dental implants, 530
  for temporomandibular joint (TMJ) replacements, 538
  for TMJ replacements, 538
fracture fixation devices, 449
  hip replacement, 424–9
  intraocular lenses, 583
  of artificial skin, 578
  of components of total knee replacement, 443

of fracture fixation devices, 449
of knee replacements, 442–3
of sutures, 568–70
of synthetic ligaments, 573
spinal implants, 459

G, measured value of, 297
gait analysis, showing loads on the hip, 425
Galen, catgut sutures used by, 560
Galen, Claudius, 129
Galileo, 130
galvanic corrosion, 18, 57, 628
galvanic potential, minimizing, 26
galvanic series, 55, 56
gamma (γ) phase, of iron (austenite), 49
gamma irradiation, generating free radicals in polymers, 20
gamma process, as deeply penetrating, 9
gamma radiation, 3, 628
gamma sterilization process, 8
gap, at the interface, 65
gas plasma, 9, 628
gel permeation chromatography (GPC), 107, 628
general failure criterion, 251
general yield criterion, benefit of developing, 255
generalized Hooke's law, 250, 264
generalized Kelvin model, 226, 628
generalized linear models, strength of, 227
generalized linear viscoelasticity, 226–7, 628
generalized Maxwell model, 226, 629
generic composite beam, 198
geometric parameters, tabulation of, 315
geometry, of the critical flaw size, 297–8
Gershkoff, 515
gingiva, 629
gingival ligaments, 508
gingivitis, 510
glass ceramics, 86
glass fibers, 283, 286
glass hip socket, 417
glass transition temperature, 103, 210, 234, 629
glasses, 83, 629
glassy plateau, 210, 629
glassy polymers, 629
  characteristics of, 103
  undergoing shear banding and crazing, 276
glaucoma, 583, 629
glenoid, 629
glenoid components, shoulders, 432
glenoid fossa, 509, 629
Global Harmonization Task force, 411
global loading, stored energy from, 295
Gluck, Thoeodorus, 436
Gluck, Thomas, 416
glycoprotein, forming a scaffold for elastin deposition, 134
glycosaminoglycan (GAG), 629
glycosaminoglycan scaffold, with a semipermeable silicone layer, 577
glycosylation reactions, 133
gold, as dental material, 26

Goldberg, 515
gold-silver binary system, used in dentistry, 47
Goodman equation, 340, 341–2
Goodman line, 339
Goodman relationship, between endurance limit and mean stress, 339
Gore-Tex Cruciate Ligament Prosthesis implant, 572
grafts, 491, 629
grain boundaries, 42
  as a nucleation site for dislocation pile-ups, 42, 46
  chromium-depleted zones at, 62
  corrosion along, 57
  creating more, 46
  dislocations at, 271
  mismatch modeled with an array of parallel edge dislocations, 42
  modeled as an array of edge dislocations, 43
  serving to impede dislocation motion, 42, 43
grain size
  making smaller, 42
  reduction in, 46
  relating yield strength to, 46
  smaller increasing yield stress, 44
grain(s), 42
  as a source of weakness, 57
  large reducing strength of Co-Cr alloys, 62
  larger facilitating the working process, 46
graphite, structure of, 84
green body, 83
Greenfield, Edward, 520
greenstick fracture, 449, 629
Griffith energy balance, 294
Griffith, A.A., 283, 285, 286
Gruentzig, Andreas, 483
Guidance Documents, with FDA recommendations, 408
Gunston, 437

Hall-Petch equation, 46
Hall-Petch relationship, 272
hard (bony) callus phase, 448
hardenability, 629
  of metals, 35
Harvey, Wiley, 398
Harvey, William, 130
Havers, Clopton, 141
Haversian canals, 141, 629
Haversian system, 141, 629, 635
HCP structure, see hexagonal close packed (HCP) structure
HDPE, see high density polyethylene (HDPE)
head diameter, optimizing, 390
healing abutment, 514
healing, failure during, 526
healthy hip joint, frontal cross-section of, 423
heart
  anatomy of, 479
  chambers, 479
  layers, 479
heart muscle, electrical simulation of, 477

heart valves, 486–9, 629
  design, 78, 487
  replacements currently available, 488
  representative sample of, 488
  schematics of, 480
heat treatments
  of alloyed titanium, 63
  tailoring an alloy with, 42, 43, 44
  thermal processes, 48
heel region, 134, 629
helical polypeptide chains, 132
helical wire implants, 520
hematoma, formation of, 447
hematopoiesis, 144, 629
hemi-osteons, 144
hemocompatibility, of pyrolytic carbon, 79
herniated (ruptured)
  intervertebral disk, 459, 629
Hertzian contact, 376
heterogeneous mixtures, of tissues, 323
heterograft valves, 487
heterografts, 629
  using bovine vasculature, 478
heteroplastic, 629
heteroplastic material, for dental implants, 505
hexagonal close packed (HCP), 629
hexagonal close packed (HCP) structure
  c/a ratio for, 31
  creating, 30
  FCC lattice transformation into, 62
  packing density, 30–2
  titanium, pure existing as, 63
hexagonal closest packed (HCP) lattices, 30
hexagonal manner, packing, 30
hexagonal yield surface, 247
hexene-based elastomer, 469
Hibbs, Russell, 453
hierarchical, 629
hierarchical structure, of articular cartilage, 152
hierarchical toughening mechanisms, 89
high cycle fatigue, 334, 344, 629
high density polyethylene (HDPE), 94, 116, 117, 629
high-frequency loading, 208
high noble metals, 26, 630
high-strength Ti alloy, 303
high-toughness Co-Cr alloy, 303
high-cycle fatigue, withstanding, 477
highly crosslinked acetabular liners, fracture of, 324–5
highly crosslinked UHMWPE, fracture toughness test, 303
hinge mechanism first utilized in total knee reconstruction, 436
  replicating kinematics of the knee, 436
hip, 421
  anatomy of, 422
  forces of everyday activities, 426
hip fractures, number of, 431
hip implants
  contemporary, 430–1
  designs, 431

fatigue crack propagation in a flawed, 353–4
showing signs of moderate-to-severe corrosive attack, 65
success rate of, 341
taking up much of the loading, 167
hip joint, anatomy of a healthy, 422
hip prostheses, bolted design and early ball and socket design, 418
hip replacements
  engineering design process for, 434
  experiencing low contact stresses, 421
  functional requirements, 424–9
  material and design evolution in total, 5
  total, 15
hip stem
  circular and rectangular cross-section for, 197
  cross-section, designing, 196–7
  factors in the design of, 203
  fixed into a femur, 204
  processes involved in machining, 52, 53
hole zones, 132, 140
hollow basket root-form implant, schematic of, 521
homolytic bond cleavage, 20
homoplastic, 630
homoplastic material, for dental implants, 505
homoplastic tooth replacements, utilizing, 515
homopolymers, 95
Homsey, Charles, 543
Hooke's Law, 181, 630
  3D, 254
  applying 3D, 254
  applying to multiaxial loading, 182
  developing three-dimensional, 182
  development of 3D, 182
  for an orthotropic material, 185
  for anisotropic material, 184
  linear portion described by, 187
  versions of, 205
horizontal plane, 630
hot isostatic pressing (HIP), 62
hot rolling, 46, 630
hot working, grain size remaining unaltered, 46
HRR field, 311, 312, 630
HRR stress singularity, 312
Hufnagel, Charles, 486
hyaline cartilage, *see* articular cartilage
hydrated tissues
  in articular cartilage, 214
  multi-phase behavior, 214
hydration, collagen requiring, 133
hydrodynamic lubrication, 377
hydrogels, limited fracture toughness, 284
hydrogen bonding, 95, 630
hydrophilic, 630
hydrophilic ceramics, molding, 83
hydrophilic nature, of a polymer, 124
hydrophilic polymers, 125
hydrophilicity, degree of, 125
hydrostatic component, *see* dilatational component

hydrostatic component of stress, 264
hydrostatic portion, of strain energy, 250
hydrostatic pressure, 258
hydrostatic stress, 247, 257, 630
hydroxyapatite (hydroxlapatite), 71
    carbonated biologically active, 86
    composites used as bone substitutes, 81
    converting coral into, 88
    described, 131, 135–6
    implants with coatings of, 523
    lending itself to substitution, 86
    platelets spreading out between collagen molecules, 141
    serving as a reservoir for mineral exchange, 135
    similar to mineral component of bones and hard tissues, 86
    wet precipitation techniques yielding, 84
hydroxyapatite-coated polymer sleeve, cross-section of, 203
hydroxyl groups, substitutions for, 86
Hylamer polymer system, 445
hyperopia, 581, 630
hyperplasia, 486, 630
hypothetical yield surface, 245
hysteresis loops, 630
hysteretic heating, 333, 361

idealized stress-strain curves, 192
immunosuppressant agent, 630
impacted fractures, 448, 630
imperfections
    in ceramics, 75
    in metals and alloys, 38–42
    presence of, 75
    utilized to strengthen metals, 46
implant body, 513, 514
implant design, 6
implant loading, 545
implant migration, as a serious issue, 560
implant therapy, time required to complete, 544
implant tissue system, 125
implant(s)
    bodies, width and height of, 525
    costs dictating manufacturing methods, 53
    fatigue resistance of, 331
    in vivo degradation of, 7
    mechanical fixation, 424
    monitoring of, 7
    partial loosening of, 427
    restoring function, 462
    structural requirements of, 7
    used to treat coronary heart disease, 478
    utilized in the body, 81
in vivo, 631
in vivo degradation, 4
    of implants, 7
    owing to biological attack or immunological response, 126
incremental damage, 342
inelastic zone, engulfing singularity zone, 299, 303
inert, 630
inert biomaterials, 71
inert ceramics, 71, 85, 86
inert medical polymers, uses of, 113
inferior, 630
inferior vena cava, 481
infinitesimal cube, inside an object under loading, 174
infinitesimal element, undergoing rotation, 176
inflammation, around posterior abutments of subperiosteal implants, 517
inflammatory response, 7, 630
initial flaw size, from non-destructive evaluation (NDE) techniques, 352
initial relaxation response, misinterpreting, 237
initiation, of a critical flaw, 333
injection molding, 83, 630
inkjet method, 83, 630
in-phase component, 229
instability condition, 296
Institute of Medicine, 587
Integra, temporary but immediate skin replacement, 577
integrity, restoring to the tissue or biological component, 14
interatomic force potential, assessment of, 11
interfacial bone tissue growth, 85
Intergraft carbon fiber composites, 587
intergranular corrosion, 57, 62, 630
interlamellar layer, 142
intermediate crack growth, 351
intermetallic compounds, fatigue of, 357
intermetallics, 17, 630
internal fracture fixation development of, 446
    for a fractured tibia, 447
internal tensile stresses, components with, 58
international regulatory bodies, 410–11
International Team of Oral Implantology (ITI), 523
Inter-Op acetabular shells, recall of, 467
interpositional, 630
interpositional disk, 509
Interpositional Proplast Implant, 543
interstitial defect, 38
interstitial sites, 32, 38, 630
interstitial solid solution, 47, 631
intervertebral disks
    afflicted with scoliosis and kyphosis., 458
    between vertebra, 457
    degenerative disease of, 459
    primary components of, 458
    serving as shock absorbers, 457
intima, 155, 481, 631
intra-aortic balloon pump (IABP), 493, 494
intramedullary, 631
intramedullary rods, 417
    for fracture in long bones, 451, 452
    forces in, 176
    repairing a tibial fracture, 176
    stabilizing long bones, 193

intraocular lenses (IOLs), 583, 631
  currently marketed, 583
  designs and functional requirements, 583
  development of, 561
  foldable, 583
  fracture, 588
  haptic fracture, 589
intraocular pressure, 582, 631
intrinsic dissipation mechanisms, 320
intrinsic failure criteria, 297–8
intrinsic mechanisms, inherent to a material, 320
intrinsic microstructural process, 320
intrinsic toughening, 320, 355, 631
intrusions, initiating flaws, 330
investigational device exemption (IDE), 406, 408, 631
Investigational Review Boards (IRBs), 406, 631
investment casting, 52, 54
ionic bonds, 72, 631
ionization (electron disruptions), producing, 8
ionized gas, deactivating biological organisms, 9
ionizing radiation sterilization, deterioration of orthopedic-grade UHMWPE due to, 20–2
iron carbide, precipitating, 49
iron pins, used in the mid-1800s, 447
irradiation, 631
  resulting in chain scission or crosslinking mechanism, 9
  resulting in cross-linking, 9
  using for polymeric materials, 9
irregular bone, 140
Irwin corrected plane stress plastic zone size, 306
Irwin Corrected Plastic Zone Size, 301
Irwin Plastic Zone Size, 300
Irwin, G.I.
Irwin, G.R., 283, 286
ISO (Organization for Standardization), 409
isotropic, 631
isotropic case, defined by two independent constants, 184
isotropic material, 184–6, 255
isotropic pyrolytic carbon, used in mechanical heart valves, 487
Ivalon sponge, implantation of, 561

J, computing for linear elastic solids and fully plastic solids, 314
Japan, Pharmaceutical and Medical Safety Bureau, 411
jaw joint, 509, 538
jaw loading, as dynamic, 538
jaw sizes, accommodating, 525
J-integral, 309–11, 631
  approximating, 314
  as nonlinear energy release rate, 310
  at crack initiation, 316
  integral contour path for, 309
joint arthroplasty, 533
joint replacement, implanted to restore function, 14
joint surface, removal of and replacement with a synthetic component, 152

joints
  lubrication in, 379
  most commonly replaced, 421
J_R curve, 316, 317
Jungle, The (Sinclair), 399
  dependence on global stress and crack length, 289
  dependent on stress triaxiality, 297
  for a constant pressure applied to crack tip, 306

k values, typical, 382
K_c
  as a function of plastic constraint, 300–3
  variations in, 302
K-dominance, region of, 639
Kefauver–Harris Amendments of 1962, 400
Kelvin model, 214, 217, 631
  as solid-like, 217
  chained in series, 226
  complex modulus of, 230
  equal strain in two parallel elements, 219
  generalized form of, 226
  incapable of predicting stress relaxation behavior, 220
  limitations of, 222, 231
  parallel network, 217
  relaxing from a highly stiff response, 220
  with theoretically infinite instantaneous elastic modulus, 220
Kelvin-type SLS, 224
Kennedy Ligament Augmentation Device (LAD), 572
keratinocytes, 577, 631
keratoprostheses, currently marketed, 582
kinetic barriers, preventing or limiting corrosion, 55
kinetic friction, 373
knee
  anatomy of, 152, 435
  articulation of the non-conforming, 439
  complex anatomy with relatively low conformity, 421
  flexion, 437
  joint, 439
  ligaments, 150, 151, 570, 571
knee replacements, functional requirements of, 442–3
knot security, 569, 631
knot tie-down, 570, 631
knot-pull strength, 569, 631
Knowles pin, 451, 452
kurtosis, 372
kyphosis, 458, 631

laboratories, achieving repeatable results, 168
lacunae, 141, 631
LAD (Kennedy Ligament Augmentation Device), 572
lamellae, 631
  composition of, 142
  function of, 457
  mechanical properties changing, 142
  of mineralized tissue, 141

lamina, 457, 631
Lange, Fritz, 453
laser etching, 353, 631
lateral screw rod systems, 454
lattice atoms, element with a geometric mismatch, 47
lattice structure, of ceramic materials, 72
lattice substitution, 45, 47
layered devices, optimal thicknesses in, 202–3
LDPE, *see* low-density polyethylene (LDPE)
lead-coated platinum post, as a dental implant, 515
leaflets (or cusps), with three layers of tissue, 479
leakiness, 486
Leeds–Keio prosthesis, 573
LEFM, *see* linear elastic fracture mechanics (LEFM)
left ventricular assist device (LVAD), 493, 494
lens
  fibers, 580
  of the eye, 580
life-supporting or life sustaining devices, 10
ligament (material ahead of the crack tip), 289
Ligament Advanced Reinforcement System (LARS), 573
ligament reconstructive surgeries, 572
ligament replacements, carbon fiber composites for, 587
ligaments, 148–51, 631
  anatomy of, 570
  as structures of discrete components, 151
  attaching femur and tibia and fibula, 435
  composition of, 148
  in unaltered state, 135
  of the knee, 571
  reconstructing, 94
  showing relative deformation response, 150
  structure of, 149
  transition from compliant to stiff behavior, 151
  tying bones together, 149
likelihood of failure, predicting, 255
line defects, 39–42, 626
linear chains, facilitating efficient packing of molecules, 99
linear crack growth, 350, 352
linear elastic element, 216
linear elastic fracture mechanics (LEFM), 77, 285–303, 349, 632
  based on stress and strain field solutions, 308–9
  COD methods as alternative approach to, 304
  crack tip effective in modified, 304
  elastic J dominating in, 314
  in situations where inelasticity present but limited, 299
  modified methods in, 299–303
  summary and limitations of, 298
linear elastic isotropic materials, 288
linear elastic strains, 311
linear individual model elements, 216
linear low-density polyethylene (LLDPE), 116, 117
linear polyethylene chains, packing, 118
linear portion, described by Hooke's Law, 187
linear region, 134

linear relationship, between stress and strain, 181
linear response, representing with complex numbers, 230
linear spring, dissipating no energy, 230
linear superposition, 632
  invoking, 218
  principle of, 182
  properties of, 215, 292
linear viscoelastic elements, 216–17
linear viscoelastic networks, 214–27
linear viscoelasticity, 632
linear viscous response, 216
linen fibers, used as sutures, 560
Linkow, Leonard, 517, 518
lipoma, 561, 632
liquid phase sintering, 83
L-lactide form, of PLA, 124
LLDPE, *see* linear low-density polyethylene (LLDPE)
load event, duration of, 227
load-bearing devices, 483–96
load-bearing human tissue constituents, values for, 137
load-bearing tissues, 136–56
loading, 332, 350
loading Modes, independent, 291
loading problems, solving variable, 343
loading scenarios, 255
loading situations, as three-dimensional, 168
loads, in the body, 10
local stresses
  controlled by presence of a notch, 267
  exceeding strength of materials under loading, 287
  finding, 267
localized crack phenomena, alternative measure of, 294
long bones
  anatomy of, 140
  epiphyses and a shaft, 140
long neck lengths, resulting in increased stresses, 466
longitudinal direction, as stiffer than transverse, 143
longitudinal testing, of cortical bone, 142
long-term creep tests, on UHWMPE, 225
Lorentz/Biomet (Warsaw, IN) device, 537
loss factor, 232, 632
loss modulus, 228, 229, 230, 232, 632
lost-wax casting, 52
low-cycle fatigue, 335, 344, 345–6, 632
low density polyethylene (LDPE), 116, 117
low frequency, modulus of Kelvin model purely elastic, 230
low-carbon steel, heat treatment of, 49
low-cycle fatigue component failure, 269
lower back pain, incidence of, 458
lower bound elastic modulus, 19
low-friction contact, leading to appreciable wear, 369
low-friction hip arthroplasty, 417
lubricant layer, maximizing, 390
lubricant, choice of, 389

lubricated articulation, in the healthy body, 369
lubrication, 377–81, 632
    boundary, 632
    elastohydrodynamic, 632
    hydrodynamic, 632
    mixed, 632
    squeeze film, 632
    types of, 377–81
lumbar spine, 456
lumen, 632

machine coolant, leaking during manufacturing process, 467
machining, 51, 52, 632
macromolecular structures mechanical behavior of, 274
    of polymeric materials, 94
macrophages, 7, 427, 632
malleability, of metals, 28
malocclusion, 511, 632
mandible, 508, 632
mandibular bone, schematic of, 509
mandibular implants, contraindications for, 530
man-made materials, versus biological materials, 14
manufacturability, of polymers, 112
manufacturing methods for implants, 7
    of ceramic materials, 82
    run cost, 52
    utilized in polymeric materials, 114
marble breast, 561
Marin factors, 346–8
Marin, Joseph, 346
martensite, 49, 632
martensite-austenite phase transformation, 64
mastication, 632
    1000 chewing cycles/day, 529
    efficacy of, 510
    severe mechanical demands of, 136
masticatory loads, 529
material behavior, over a range of loading frequencies, 227
material choices, for orthopedic implants, 203
material development, used in medical devices, 4
material failure, process of, 241
material phase, as a true viscous medium, 214
material properties
    measuring in a uniaxial test, 187
    need for certain, 15
    processing metals for, 45–50
    tailoring through heat treatment, 44
    time-dependent, 209, 210–12
material resistance, to crack growth, 294
material selection, as a multifactorial problem, 60
material symmetry, utilizing, 261
materials
    deforming plastically, 299
    selecting for devices and implants, 7, 14
materials science, of medical alloys, 27
Maverick implant, 461
maxilla (upper jaw), 506, 508, 632

maxillary and mandibular implants, contraindications for, 530
maxillofacial, 632
maxillofacial implants, 506
maximum bite force (MBF), 529
maximum contact pressures, experienced in clinical total knee replacement designs, 278
maximum distortional energy, 249–55
maximum principal stresses, 324
maximum shear stress, 38, 245
    at the center of the contact area, 376
    criterion, 245
    determining using Mohr's circle, 178
    finding, 179
    found using Tresca yield criteria, 260
    localized, 243
    occurrence of, 246
    planes of oriented, 246
    reaching yield stress, 245
    relationship with principal (normal) stress, 246
    Tresca yield criterion, 245–9
maximum stress, finding, 196
Maxwell material, 221
Maxwell model, 217, 633
    as fluid-like, 217
    characteristic time as defined by, 221
    combined in parallel, 226
    generalized form of, 226
    imposed sinusoidal strain in, 231
    loss modulus, 232
    series network, 217
    storage and loss modulus, 232
    storage modulus of, 231
Mayan civilization, teeth implanted into live patients, 515
McKee-Farrar prosthesis, 417
mean stress, 633
    effect on fatigue behavior of a material, 338
    effect on fatigue life of a material, 339
    endurance limit decreasing with increasing, 339
    incorporated into the life calculations, 329
    mechanical behavior of polymers, 109–12
    of loading cycle, 332
    of structural tissues, 129–31
mechanical failure, of Silastic implants, 534
mechanical heart valve replacement, 486
mechanical heart valves, 477
mechanical properties commonly compared, 205
    concomitant with variations in microstructure, 50
    describing linear elastic, isotropic deformation, 168
    describing ratio between stress and strain, 181
    for tissues, 158
    for trabecular bone, 147
    of corals, 89
    of e-PTFE, 120
    of eye tissues, 578
    of inert medical ceramics, 86
    of metals yield strength, 27
    sensitivity to different test parameters, 168
mechanical requirements, on an implant, 14

mechanical supports, 560
mechanical test, for material characterization, 241
mechanical variables, associated with sustained
    (cyclic) loading, 331
media, 155, 481, 633
median, 633
medical alloys
    elastic modulus, yield strength and tensile
        strength for most commonly used, 61
    fatigue crack propagation behavior of a range,
        358
    materials science of, 27
    variety of uses in the body, 60
medical ceramics, summary of, 87
Medical Device Amendments of 1976, 400, 404
Medical Device Report (MDR), with the FDA, 498
Medical Device User Fee and Modernization Act
    (2002), 400
medical devices, 633
    approval, 397
    brought under FDA control, 399
    complexity of, 401
    definition of, 401
    definitions and classifications, 401–4
    design, 4
    early, 3
    for load-bearing applications in the body, 13
    for which failure cannot be tolerated, 297
    fractures, 283
    long-term success of, 4
    multifactorial factors contributing to
        performance, 6
    regulating advertising of, 400
    structural requirements of, 6, 10–13
    subject to pre-market approval, 400
medical implants
    ceramics in, 85–7
    designing, 3
    metals in, 60–4
    polymers in, 113–24
    requiring very specific properties, 11
    utilized in the body, 14
medical indications for subperiosteal implants, 530
    for TMJ replacements, 538
medical polymers, 116
medium density polyethylene (MDPE), 117
Medtronic catheter, bursting during an angioplasty,
    401
melt viscosity, scaling with molecular weight, 107
melting curves, of polymers, 102
melting temperature, relationship with lamellae
    thickness, 103
meniscectomy, 533, 633
Mentor Corporation, 586
mers units, in polymers
meshes, made from nylon and PET, 585
metal backing, added to polyethylene tibial insert,
    443
metal bonds, breaking and reforming, 35
metal corrosion, characteristics determining, 55
metal head, displacement into the polymer cup, 237

metal ions, 54
metal processing, 42–53
metal systems
    alloying, 47
    defect types and sources of microstructural
        features, 272
    utilized as medical alloys, 27
    yield stress saturating under sustained loading,
        333
metal(s)
    altering structure and properties of, 42
    as intrinsically metastable, 53
    characteristics of, 241
    expression for theoretical strength of, 38
    failure mechanisms in, 271–2
    fatigue behavior of, 357–9
    fracture mechanisms in, 322
    high fracture toughness and resistance to crack
        propagation, 330
    imperfections in, 38–42
    in medical implants, 26–64, 67
    limitation of using, 17
    making use of metallic bonding and closely
        packed planes of atoms, 16
    mechanical properties for load-bearing medical
        devices, 17
    plastically deformed, 35
    processes strengthening, 45
    processing for improved material properties, 45–8
    processing for shape-forming, 53
    sensitivity to corrosion, 322
    slip systems in, 36
    strengthening mechanisms in, 45
    theoretical shear strength of, 36–8
    utilizing metallic bonding, 27
    with limited slip systems, 270
metal/metal pairings, failed lubrication in, 382
metallic bonding, 13, 633
    bonding structure in, 53
    described, 34
    material behaviors, 27
metallic femoral stem, 259, 262, 354
metallic implants, degradation of, 53
metallic stent, schematic of, 352
metal-on-metal conical taper connections, to assist
        modularity, 64
metal-on-metal designs, in intervertebral disk
        replacement, 461
metal-on-polymer articulation, implant design
        employing, 439
metal-oxide passive film, on the metal surface, 55
metaphyseal, 633
metaphyseal (trabecular) bone, 450
microcrack propagation, in young and aged human
        teeth, 139
microcracking, 80, 633
microguidewire, in an artery, 178
micromechanisms, 321, 330, 633
micromotion
    during initial healing, 545
    experienced by implants, 519

microscopic flaw, introducing, 284
microscopic sharp cracks, 287
microstructure, of metals, 27
microvoid coalescence, 270, 274
middle zone, collagen fiber orientation, 153
migration, 633
Miller indices, 34, 633
mineral crystals, in dentin, 136
misalignment
  clinical reasons for, 436
  of the knee, 436
mitral valve, 479
mixed lubrication, 380
mixed-mode condition, 633
mixed-mode crack propagation, 349
mixed-mode loading, 292, 293, 297
mobile bearing designs, 440
Mode I
  crack-opening mode, 291
  equation, 292
  loading, 77
  opening mode, 349
  stress intensity factor, 349
  strip-yield zones in thin steel sheets in plane stress, 305
  typically most critical, 297
Mode II, shear mode, 291, 349
Mode III
  out-of plane shear mode, 349
  twisting (or tearing) mode, 291
modeling packages, computing J-integral at blunt or sharp crack tips, 311
modeling techniques, predicting behavior of polymers, 212
modular hip stems, assembly of, 385
modular implant design, trade-off in, 466
modular implants
  conical press-fit fixation, 65
  degradation, 65
  tapered shank, 65
modular neck
  added length, 66
  fracture of, 465
modularity, benefits of, 64, 429
modulus, 623
modulus of resilience, 243, 633
modulus values, for alloy classes used in medical implants, 44
Mohr, Otto, 176
Mohr's circle, 176, 633
  basic equations for, 260
  coordinate system maximizing shear stress, 178
  finding principal stresses, 248
  for uniaxial loading, 246
  plot for uniaxial loading, 246
  representation of a stress state, 178, 180
  representation of a two-dimensional problem, 177
  showing maximum shear stress, 256
  using to find principal stresses, 179
molding, 633
molds, intricacy of, 52

molecular bonding level, of structural materials, 15
molecular characteristics, limiting ability of a polymer to form crystalline structures, 98
molecular effects, in polymers, 105
molecular structures, in ceramics, metals, and materials, 18
molecular variables effect on time-temperature behavior of polymer systems, 112
  strengthening a polymer system, 112
  tailoring in polymers, 19
molecular weight, 633
  affecting glass transition of a polymer system, 104
  determination of, 108–9
  distribution in polymers, 106–9
  effect on structure, 100
  impeding long-range ordering, 99
molecule size distribution, in a typical polymer, 106
moment area of inertia, around the neutral axis, 198
monofilament, 561
monomers, 95
morbidity, 570, 633
Morse taper, 26, 58, 65, 633
  achieving junction between femoral head and neck, 429
  requirements for, 424
  schematic illustration, 65, 430
Morse Taper cone, showing burning, fretting and intergranular attack, 66
Morse Taper junction, burnishing and fretting at, 66
motion, in knee, 435
Mouse Cleaver, 561
multiaxial loading, 182–3, 255–61
multiaxial stress state, 241
multiaxial stresses, sources of, 259
multiaxial testing, of a material, 187
multifilament designs, increasing suppleness of a suture, 570
multiphase polymers, toughness of, 323
muscle, motive action of, 149
mussel shells, toughness of, 287
myocardium, 479, 633
myopia, 581, 633

natural corneal transplant surgeries, 581
natural joints, coefficient of friction in, 386
natural materials, 561
natural options, for vascular grafts, 490
natural sutures, 565
natural tissues, *see* tissues
natural tissues, dramatic increase in understanding of, 129
  enabling energy dissipation, 322
  existing along a continuum, 136
  mechanical properties of, 129
  properties of, 158
  stress-strain curves in, 191–3
Naval Ordinance Laboratory (NOL), 63
nearsightedness, 633
near-tip stresses, maximized, 77
neck-head junction, 424

necking, 189, 633
neck-stem junction, bending stresses, 466
needle attachment force, 569, 634
neoformation, 88, 634
neonatal fibroblast, 577, 634
nerve endings, in the cornea, 579
network polymer, three-dimensional, 97
neutral axis, 194, 197–9, 634
Nitinol, 64, 634
Nitinol Ni-T alloy, 63
nitinol stent
   fatigue design of, 352–3
   strut failure, 498–500
noble alloys, 26
noble metals, in structural dentistry, 26
non-conforming design, 441
non-conforming surfaces, creating large Hertz contact stress, 442
non-destruction evaluation (NDE) techniques, 352
non-enzymatic crosslinks, 133
non-expendable mold, 52
nonlinear (NL) elasticity (E), 308
nonlinear creep functions, 236
nonlinear elastic material, 308
nonlinear elasticity, 308
nonlinear models, common, 235–6
nonlinear relaxation, 236–7
nonlinear viscoelasticity, 235–7
non-resorbable materials, 561
non-resorbable polypropylene suture, used for wound repair, 121
non-resorbable sutures, currently marketed in the US, 566
non-submergible implant, 514
non-thrombogenic properties, of polyesters, 121
non-thrombogenic substance, 634
normal strain, 171, 634
normal stress, 172, 173, 244
normal stress criteria, 244, 265, 279
Not Significantly Equivalent (NSE), 412
notch sensitivity, 267
notches, 634
   materials sensitive to the presence of, 267
   presence of, 266
   serving to locally concentrate stresses, 330
novel device, path to market, 406
nucleation site
   for dislocation pile-ups, 42, 46
   of critical defects, 330
nucleus, 634
   of each disk, 457
nucleus pulposa, 457, 634
number average molecular weight, 106, 109, 634
nylon, 121, 124, 337, 585, 634
nylon balloon, in deployment of a vascular stent, 122
nylon polymers, tensile strength of, 94

OA, *see* osteoarthritis (OA) object, under loading
oblique fractures, 449, 634
observer invariance, 173
occipito-cervical, 634
occipito-cervical junction, 454
occlusal, 634
occlusal forces, 529
occlusal loads, 506, 516
occlusal overloads, 529
octahedral shear plane, 251
   orientation with respect to principal stress, 251
   plotted in Mohr space, 252
   shear stress on, 251
octahedral shear stress, 251, 634
octahedral shear yield criterion, 251
octahedral site, 33, 634
Office of Combination Products, 401
oily residue, on defective implants inhibiting osseointegration, 467
one-dimensional polymers, 113
one-dimensional systems (chains), 95
open cell foam, 147
open pore architecture, of coral, 88
ophthalmic implants, 560, 561, 578–84
ophthalmologic, 634
ophthalmologic disorders, 580
optical micrographs, 428
optical properties, of eye tissues, 578
optic-haptic junction, fracture of IOL at, 588
oral and maxillofacial implants, demand for, 505
oral implants, 505
oral muscosa, 634
Organization for Standardization (ISO), 409
orientation factor, for resolved shear stress, 40
orientation mismatch, 42, 46
orthopedic biomaterials, 13
orthopedic implants challenges remaining in the development, 416
   corrosion in, 64–7
   described, 416
   engineering challenges and design constraints of, 462
   looking forward in, 470
orthotropic, 634
orthotropic material, possessing symmetry, 185
orthotropy, 185
osmosis, recruiting lubricating fluid into tissue, 151
osseointegrated implants, failure of, 520
osseointegration, 634
   at bone-implant interface, 63
   efforts to enhance, 545
   lack of, 519
   of implants, 505
   supporting, 86
osseointegrative ability, of titanium, 506
osseointegrative potential, of ceramics, 524
ossification, 448, 634
ostcoclasts, eating away at surrounding bone tissue, 427
osteoarthritis (OA), 423, 424, 532, 634
osteoblasts, 634
osteoclasts, 88, 634
osteoconduction, 88, 634

osteoconductive properties, of coral, 88
osteocytes, 141, 634
osteogenesis, 88, 634
osteoid, 140, 634
osteoinduction, 88, 634
osteolysis, 416, 427, 635
   complications with, 424
   due to poor wear resistance, 20–2
   loosening associated with, 422
   propensity for, 455
osteonal system, 141
osteons, 635
   as a basic unit, 141
   composition of, 141
   in cortical bone, 323
   organization of cortical bone into, 143
   resulting from remodeling processes, 144
   separated by the cement line, 142
osteoporosis, 70, 448, 635
osteoporosis-related fractures, increased risk, 144
outer scaffold layer, 157
out-of-phase component, 229
overall strain response, of the SLS in creep, 223
overall stress response, for Maxwell model, 231
oxidation, 635
   as preferable energy state for metals, 53
   creating structural degradation of UHMWPE, 21
   long-term, 9
   subsurface peak in, 21
oxidation-induced embrittlement, gamma radiation resulting in, 3
oxidative degradation, enhanced susceptibility of Hylamer, 445
oxidative reaction, 55

pacemaker, 635
package memory, 570, 635
packing density, determining, 30
packing efficiency, 30
Palmgren-Miner accumulated damage model, 341
Palmgren-Miner's rule, 342, 343
parafunctional, 635
parafunctional activities, 529
parafunctional oral activity, 526
Parallel Axis Theorem, 198
parallel network relationship, 217
Paris equation, 358
Paris law, 351, 352
Paris regime, 351
Paris regime slope, 359
particulate debris, see wear debris
particulate debris, increased levels of, 440
Pascals [Pa]
passivated surface, 58
passivation, 55, 635
passive oxide film, 55
patella, 635
patellar tendon, 570, 571
patency, 500, 635
Patent-Fitted TMJ Reconstruction Prosthesis, 537
pathways, for corrosion in metallic systems, 55

patient factors, 5, 525
Pauling's rules, 72–5, 635
peak energy, 229
peak stored energy, 229
peak stress, 288
peak stress intensity, 356
peak-to-valley height, 372
pearlite, 49, 635
pedicle screw spinal fixation method, 469
pedicle screws, 467, 468, 469
pedicles as fixation points, 453
   spinal fixation through, 454
pelvis, foundation of, 456
percutaneous, 635
percutaneous procedures, implanting stents, 483
percutaneous surgeries, increased, 500
pericardium, 479, 635
peri-implant, 635
peri-implant tissues, 519
periodontal, 635
periodontal disease, 87, 510, 635
periodontal ligaments, 508
periodontal tissues, 508
periodontitis, 510, 635
periosteum, 635
periprosthetic osteolysis, 635
permanent abutment, 514
permanent ACL replacements, 572
permanent augmentation devices, providing support for existing ACL, 572
PET, see polyethylene terephthalate (PET) polymer
Peterson equation, developed for steels, 268
PGA-PLA polymers, increased resorption time, 125
phase diagram, for Fe-C steel, 50
phase shift, 227, 635
phase toughening mechanism, for zirconia, 80
phase transformation toughening, 80, 635
pHEMA Poly(2-hydroxyethyl methacrylate), contact lenses made of, 581
physical crack tip, as blunted, 304
physiological problems, measurement methods in, 130
physiological stresses, 6, 78, 635
pile-ups, 635
pin-on-disk testing, 389
pins, used in internal fracture fixation, 450
pitting, 277, 636
pitting corrosion, 58, 59
planar defects, 42
planar organization, of tissues, 155
planar two-dimensional polymer, 97
plane sections, remaining plane during bending, 195
plane strain fracture toughness, 301, 636
   plastic zone size, 299
planes
   arbitrary intersecting the y and z axes, 34
   describing within the lattice structure, 34
   of maximum principal stress, 270
   of the body, 420

plaque
    cardiovascular, 636
    dental, 636
    formation of, 477
Plaster of Paris, 70, 636
plastic blunting, of the crack tip, 303
plastic component, of the strain amplitude, 344
plastic constraint
    in plane stress and plane strain, 307
    K_c as a function of, 300–3
plastic deformation
    associated with hydrostatic loading, 257
    increasingly severe, 313
    local, 266
    of metals, 35
    onset critically linked to resolved shear stresses, 42
    within crystal systems, 34
plastic flow, region of reversed, 268
plastic strain amplitude, 344
    avoiding sustained, 344
    resulting in permanent deformation, 243
    some portion of the energy lost, 243
plastic wake, 317
plastic zone, 636
    ahead of apparent crack tip, 305
    correction to, 300
    growing singular stress extinguished, 306
    in high toughness materials, 308
    radius, 315
    size, 300
    size affecting distribution of stresses, 304
    size approximation, 300
plasticity, 308, 312
platelets, 482
plates, 450, 636
PMMA, *see* polymethylmethacrylate (PMMA)
point defects, 33, 38
Poison Squad, 398
Poisson's ratio, 181, 205, 636
poly (lactic acid) (PLA), 123
polyamide polymer, 636
polyamides, uses of, 121
polycentric implant design, 438
polycentric knee, invention of, 437
polycentric motion, in the knee, 438
polycentric pathway, rotation about, 437
polycrystalline aluminum oxide (alumina), 523
polycrystalline ceramic glasses, 71
polycrystalline materials, metals as, 42
polycrystalline metal, 42
polycrystalline solids, metals more precisely termed, 42
polydimethylsiloxane (PDMS), 123
polydisperity index (PDI), 107, 109, 636
polyester polymers, tensile strength of, 94
polyesters, 121, 636
polyetheretherketone (PEEK) polymer, 636
    carbon fiber reinforced, 19
    uses of, 123

polyethylene (UHMWPE), as acetabular linear in a modular total hip replacement, 118
polyethylene fossa, 536
polyethylene polymer, 636
polyethylene polymer systems, 99
polyethylene terephthalate (PET) polymer, 121, 636
    dynamic scan of, 231
    flexible tubes of, 478
    mechanical properties for, 229
    meshes made from, 585
    no endurance limit, 337
polyethylenes
    molecular weight of, 116
    uses of, 116–17
polyglycolic acid (PGA), 94, 113, 125
    in sutures, 19
    polymer, 636
polylactic acid (PLA) polymer, 94, 113, 485, 636
polymer chains
    accommodating through chain folding, 102
    foldering into ordered structures, 95
    increased length of, 107
    segments overcoming activation barriers to move, 110
polymer composites, 19
polymer crystals, thickness of, 102
polymer resorption, rate of, 125
polymer structure, density of, 102
polymer synthesis, result of basic, 106
polymer systems
    crystallinity of, 101
    elastic modulus, 16
polymer(s), 636
    as viscoelastic materials, 212
    bonding and crystal structure of, 95–106
    characteristics of, 241
    deforming elastically under small stresses, 109
    ethylene oxide sterilization used in, 9
    failure mechanisms in, 274–6
    fatigue behavior of, 330, 357, 360–2
    fracture mechanisms in, 322
    improvement of, 93
    in load-bearing applications in medical devices, 93
    in medical implants, 92, 113–24
    low elastic modulus values, 16
    mechanical behavior of, 109–12
    melting rather than fracturing, 361
    melting temperature of, 102
    molecular weight distribution in, 106–9
    offering lower elastic modulus, 13
    porous by design, 113
    processing, 83, 112
    properties of, 19
    requiring low temperature methods, 9
    structural properties of, 95
    susceptible to creep and strain rate effects, 360
    tensile loading versus compression loading, 276
    time-dependent crack propagation in, 305
    toughening with addition of a second phase, 323

polymer(s) (cont.)
  tribological behavior, 387
  uses of, 94
  varying greatly in molecular weight, crystallinity, and density, 99
  with high glass transition temperature, 103
polymeric biomaterials, 19
polymeric coatings, 94
polymeric implant, modified yield criteria for, 263–4
polymeric materials highly dependent upon local stress states, 361
  primary advantage of, 94
  softening, 333
  susceptible to irradiation damage, 9
polymeric tibial insert, 263
polymethylmethacrylate (PMMA), 118, 636
  as load transfer medium and grouting agent, 119
  biodegradation in, 19
  button pressed into a two-piece interface grasping a corneal graft, 581
  condylar head, 534
  IOLs, 583
  transparency of, 119
  uses of, 92
polyp, 636
polypropylene (PP) polymer, 120, 636
polytetrafluoroethylene (PTFE) polymer, 636
  as a bearing surface, 93
  cup, 417
  flexible tubes of, 478
  in a low-friction bearing pair, 390
  in dental implants, 93
  materials for orthopedic implants, 543
  slippery during handling, 585
polyurethane (PU) foam disintegration of, 586
  susceptible to chronic inflammatory response, 93
polyurethane polymer, 636
polyurethanes, 94, 122
polyvinyl chloride (PVC) balloon catheters, 483
pontics, 511, 636
pore spaces mineralization spreading into, 140
  reduction in, 83
pores, 132
poroelastic behavior, 637
porosity
  of cortical bone, 145
  of trabecular bone, 145, 147
porous coatings, 425, 431
porous components, 561
porous meshes, 560
positive values, for stress
post-amendment devices, 400
posterior cruciate ligament-retaining design, 441
posterior cruciate ligament-substituting design, 441
posterior implants, less success, 526
post-market surveillance, of a device, 498
potential energy, 310
pounds per square inch [psi], 172

powder metallurgy, 83
powder sintering, 637
power-law correlation, 236
power-law deformation plasticity material, 312
power-law empirical correlation scheme, 235
power-law plasticity, near the crack tip, 312
power-law time-dependent compliance, 319
power-law viscoplasticity, 236
PP Polyflex, 570
precipitates, 48
precipitation hardening, 47, 620, see also age hardening
predicate creep, 412
predicate device, 404, 637
Pre-Market Approval (PMA), 404, 637
Pre-Market Notification (510(k)), 637
  device path to market, 406
  examining approval of and subsequent FDA internal review, 411–13
  for a device, 404
  using surgical meshes as predicate devices, 412
primary body planes (cross-sections), describing anatomical positioning, 420
primary slip systems, 35
principal directions, 179, 637
principal material coordinate system, 145
principal stresses, 179, 637
processing of the implant, 7
  polymers, 112
Pro-Disk intervertebral implant, 460
ProFemur Z hip stem, 463
profilometer, ID curve taken using, 372
profilometry measurements, 388
propagation process, of a critical flaw, 333
Proplants, interpositional implants made of, 534
Proplast, 536, 570, 637
  carbon fiber composites, 587
  coating, 536
Proplast I, 543
Proplast Interpositional Implant, 536
Proplast-Teflon (PT) as TMJ implant material, 506
  TMJ implants, 536
Proplast-Teflon Temporomandibular Joint Replacements, 542–4
prosthesis, see specific types
prosthesis, assessing amount of wear, 237
  views of Flot's, 536
prosthetic heart valves, need for, 486
proteins, in the body's structural tissues, 131
proteoglycans, 148, 152, 637
proximal, 637
proximal-distal slipping, 65
pterygoid muscles, 509, 637
PU foam, see polyurethane (PU) foam
pulmonary valve, 479
pulp, interior of, 507
pulsatile fatigue, 500
Pure Food and Drug Act of 1906, 399
pure shear loading, 256
pyrogenicity, 637

pyrolytic carbon, 83, 637
    fatigue crack propagation rates, 360
    properties of, 86
    replacement of Delrin by, 487
    structural arrangement of carbon in, 84
pyrolytic carbon coating, 83
pyrolytic carbon heart valves, fracture of, 284

quenching, 49, 637

R curve shape, sensitive to stress triaxiality, 296
R curves, 295
    for monotonic (quasi-static) loading, 320
radians, measuring shear strain, 170
radiation chemistry, of specific polymer systems, 9
radicular pulp, 508
radiopacity, for post-surgical monitoring, 489
radius ratio, 72
ramus blade, 519
ramus blade-form implant, schematic of, 519
ramus frame, 519
Rankine, 244
rapid crack growth, 351
rapid prototyping, 83, 637
rate sensitivity, of polymer mechanical behavior, 212
ratio of maximum stresses, 298
R-curve (resistance curve) behavior, resistance curve behavior, 360
reactions, when metals corrode, 55
real area of contact, 371, 637
recrystallization, 48, 621, 637
recrystallization temperature, 46
rectangular and circular cross-sections, area moments of inertia for, 196
red blood cells, 481
red bone marrow, 144
reduced time, 236, 637
reduction reaction, 55
reference temperature, for WLF equation, 234
ReGen Menaflex Collagen Scaffold (MCS), 411
region of material breakdown, 289
regulatory affairs, 397
regulatory issues, 7, 9
regurgitation, 486, 637
rehabilitation period, after surgery, 570
relative strength, of viscous component of the spectral response, 229
relative time scale, of relaxation and loading event, 211
relaxation spectrum, 227, 233
relaxation time, 208, 210, 637
    for Silly Putty, 212
relaxation transitions, in polymers, 233
remodeling, 140, 637
removable dentures illustration of, 512
    primary components, 511
repeatable data, accumulating large amounts of, 346
resecting, 533, 637
residual compression, 269
residual stresses, developing at notch tip, 268
residual tensile stress, zone of, 268

residual tension, created upon unloading, 269
resilience
    of polymer structures, 106
    of polymers, 19
resolved shear stress, 40
resorbable, 638
resorbable (biodegradable) ceramics, 84
resorbable materials, 71, 561
resorbable sutures, 11, 124–5
    made from polymers, 567
    materials, structures and absorption times, 568
restenosis, 483, 638
retrieval studies, providing insight into adverse reactions or complex problems, 10
retrodiscal, 638
retrodiscal tissues, 510
rheopectic, 638
    as synovial fluid, 380
rheumatoid arthritis, 532, 638
    as source of damage in articular cartilage, 424
    in the knee, 436
Ridley IOLs, 583
Ridley, Harold, 561, 583
Riegel v. Medtronic (2008), 401
Riegel, Charles, 401
Ringer's solution, 389
RMS roughness, 378
Roberts, Ralph and Harold, 519
rods, used in internal fracture fixation, 450
rolling, 51
rolling contact fatigue, *see* contact fatigue
Roosevelt, Theodore, 399
root, 638
    of a tooth, 136
root canal, 508
root mean squared (RMS) roughness, 371
root-form endosteal dental implants, 513, 520–9
root-form implants
    failures during healing, 526
    geometries for, 524
    overtaking bladeform implants in popularity, 528
    reclassified by the FDA as Class II devices, 528
    regions of, 514
    requiring a healthy, intact vertical column of alveolar bone, 514
    success of, 526
    susceptible to early bone loss, 527
    with surfaces roughened, 522
rough surfaces, coming into contact, 371
roughened surfaces, particularly in cylindrical bodies, 523
roughness, 371
    average, 638
    calculating, 372–3
    root mean squared, 638
roughness measurement methods, 388
R-ratio, 332, *see also* stress ratio rubber disk, embedded between two titanium endplates, 455

rubber particles, nucleating multiple crazes, 277
rubbery plateau, 210, 638
rubbery polymers, 104, 638

S.E. Massengill Co., 399
Sabins, Gary, 467
sacrum, 456
Safe Medical Devices Act, 641
Safe Medical Devices Act (1990), 400, 640
safety factor, as proximity to yield boundary, 253
safety-critical applications, defect-tolerant approach used in, 348
sagittal plane, 638
saline, 389
saline-filled breast implants, 585, 586
Sandhaus, S., 523
saphenous vein, 490, 638
sapphire implant, 524
sapphire, more favorable survival rates, 524
saturation crack length, effect of load ratio on, 279
SB Charité intervertebral disk, 455
scaffold, creating for tissue engineering, 156–8
Schulte, Heyer, 585
scission, 9, 638
scoliosis, 453, 458, 638
screw dislocations, 39, 40, 271, 638
screw type designs, 524
screw(s), 638
  types of, 450
  used in internal fracture fixation, 450
screw-plate construction, for stabilization of a spine, 454
screw-rod system, 454
screw-type implants, 514
screw-type root-form implant, 520
sea coral, *see* coral secant modulus, 188, 638
second degree burns, 576, 638
second moment of molecular weight, 107
second-order imperfections, 39
second phase additions, 80, 638
second phase particle strengthening mechanism, 48
second phase particles, 45, *see also* precipitates
second phase particles, strengthening achieved with, 48
second-phase particles, 45
self-expanding stent delivery, 483, 484
self-interstitial atom, 38
self-interstitial point defect, 38
self-similarity, 288, 638
self-tapping implants designs, 522
semi-brittle, 638
semi-brittle processes, of polymers
semi-brittle solid, cleavage mechanisms in, 273
semi-brittle system, 241, 269
semi-constrained dynamic plates, 454
semi-crystalline polymer, 98
  having crystalline and amorphous domains, 98
  prediction of shift factors, 235
  schematic illustration depicting, 103
  uses of, 106

with glass transition temperatures well below body temperature, 105
semilunar valves, 479
septal defects, 638
septal occluders, 495, 496
series network relationship, 217
series representations, for relaxation behavior, 226
sesamoid bone, 140
severe stress, 287
shape memory behavior, 63, 64, 634
sharp cracks, 273, 288
sharp fatigue crack, in UHMWPE, 279
sharp flaws, central role of, 283
sharp notches, 266
shear banding, 275, 276
shear cavitational fracture processes, 257
shear forces, repetitive high-magnitude, 530
shear modulus, 182, 638
  of human dentin, 139
  relating strain, shear stress, and shear strain, 37
  solute atom locally altering, 47
shear plane, 41
shear processes, deforming metals, 245
shear strain, 169, 171
shear stresses, 172, 173
  common at interfaces, 291
  necessary to force dislocations between particles, 48
  plane of greatest, 247
  relationship with normal stresses, 179
  required for continued deformation, 45
shear, as mechanism for plastic deformation, 257
shearing mode (Mode II), 291
shift factor, 233, 234, 639
short bone, 140
short fibers, 202
short implant survival, 525
sigmoidal fatigue crack propagation plot, 351
Significant Risk Study Protocol, 407
Significantly Equivalent (SE), 510(k) application, 412
Silastic (silicone rubber), 639
  as TMJ implant material, 506
  compared to Proplast, 544
  introduction of, 534
Silastic HP sheeting, 534
Silastic implants, widespread failure, 534
Silastic TMJ disk, temporary, 534
silica glass, 72, 74
silicone breast implants, 561
  more realistic feel and not prone to collapse, 586
  re-approved, 586
  removed from the biomaterials market, 93
  utilizing crosslinked form of silicone, 123
silicone elastomer, in place of a rubber one, 455
silicone implant design, 123
silicone polymer
silicone, slippery during handling, 585
silicones, 93, 122, 639
silk, as suture materials, 124, 560
Silly Putty, 212, 213

siloxane hydrogels, in contact lenses, 581
siloxanes, *see* silicones
similitude, *see* self-similarity
Sinclair, Upton, 399
single craze, in a polymer
single crystal aluminum oxide (sapphire), 523
single crystal orientation, regions of, 42
single-stage surgery, 514
single tooth crowns, cemented to the abutment, 525
single-root implants, less likely to fail, 526
singular region, 289
singularity dominated zone, 289
singularity stresses, extending beyond crack tip, 288
singularity-dominated zone, 639
sintering, 83, 639
sinusoidal deformation, over a wide range of frequencies, 227–30
sinusoidal strain input, resulting stress response, 227
sinusoidal stress displacement curve, 36
skeletal fractures, incidence of, 446
skeletal regeneration, of fractured bone tissue, 447
skeleton, principal structural material of, 140
skewness, 372
skin, *see* artificial skin
skin and blood vessel walls, 154–6
skin
    as largest organ in the body, 575
    cellular structure of the epidermis, 154
    functions of, 576
    layers of, 154, 575, 576
    mechanical properties of, 156
sleep away implant, 522
slip, 639
    atomic movement denoted as, 35
    defined, 28
    due to grain orientation mismatch, 45
slip plane, capturing motion of the dislocation, 39
slip systems, 35, 36, 639
slow crack growth, 351
small-scale plasticity, 299–300, 312, 313, 639
small-scale yielding, *see* small-scale plasticity
small-scale yielding, conditions of 348
S-N curve, generating, 335
S-N fatigue data, for several polymer systems, 337
S-N plots
    for several polymer systems used in medical devices, 337
    on linear-logarithmic scale, 336
sodium chloride (NaCl) structure, as lattice of sodium atoms, 72
soft tissue biomaterials, 13
soft tissue devices, future of, 588
soft tissue repair, polymetric suture anchor providing, 124
soft tissue replacements
    classifying materials used, 561
    in facial augmentation, 506
    materials used classified by chemical makeup, 562
    materials used classified by structure and device area, 564
    method for keeping them in place, 560
    schematics of material structures used in, 563
    types of, 560
soft tissues
    elastically blunting, 323
    exhibiting an initially low stiffness, 323
solid blocks, 561
solid-solution strengthening, 45, 47, 639
solute atom size, 47
solution jet, stretched into a thin fiber, 157
sp^3 hybridization
    facilitating four bonds, 95
    in carbon, 96
space fillers, 560
spatial configuration in polymers, 95
specific volume, as a function of temperature, 104
spectrum loading, 331, 341, 639
spheres, contact between two elastic, 376
spherulite development, illustration of, 100
spherulites, 99, 639
spinal arthroplasty, clinical success of, 454
spinal canal, 457, 639
spinal column, composition of, 456
spinal fixation designs, contemporary, 454
spinal function, restoring, 459
spinal fusion, 459
spinal fusion procedure, performing, 453
spinal implants, 453–62
    clinical reasons for, 458
    cycles to failure for, 338–40
    design concerns in, 462
    designs, 460–1
    functional requirements, 459
    materials used in, 461
    Tresca stress, 248–9
spinal instrumentation, failure of pedicle screws, 467–9
spinal rods, 639
spine, 456
    anatomy of, 456–7
    biomechanics of, 460
    degrees of freedom in kinematics of, 459
    medical devices used in, 453
    post-operative stabilization of, 469
    regions of, 456
    stresses of, 454
spine disorders, 458
spinous processes, 453, 639
spiral fractures, 449, 639
spiral implant, schematic of, 521
splint, 639
splinted restoration, to a healthy adjacent tooth, 528
split-thickness skin graft (STSG), 577, 639
spongiosa, 479, 639
spring, 217
spring and damper elements, 230
spring and dashpot, schematic showing, 216
spring constant, of the bond, 11
squeeze film lubrication, 379
stability, addressing for knee replacement, 440

stabilization, of spine disorders and fractures, 453
stabilized zirconia, fracture toughness test, 303
stable crack propagation, 295, 316
stable equilibrium, 295
stainless steel alloys as material for fracture fixation devices, 59
   used medical devices, 61
stainless steel total hip replacement stem, 347
stainless steel(s) as implant materials, 26
   chromium added to facilitate passivation, 57
   elastic modulus, yield strength and tensile strength for, 61
   in medical implants for corrosion resistance, 61
   strengthening by cold working, 61
standard engineering materials, 169, 639
standard linear solid (SLS), 639
   as an improved Kelvin model solid, 222
   model, 222, 232
standard metal processing, categories of, 51
standards development organizations (SDOs), 409, 639
standards, defined, 408
STARR Surgical IOL, 583
Starr-Edwards Silastic Ball Valve, 486
state of plane stress, 252
static driving force (mean stress), 340
static friction, 373
statics, problems of, 130
steam, high pressure, 8
steel plates, as internal fixation devices, 26
steel rods, in combination with wire to stabilize the spine, 453
steel(s)
   concentration of interstitial carbon, 38
      having near identical elastic modulus values, 44
      heat treatment of low-carbon, 49
Steffee, Arthur, 469
stem-neck design (modularity), contributing to corrosion-induced fracture, 464
stenosis, 486, 639
stent characterization, non-clinical tests for, 486
stent delivery, methods for, 483
stent device, made of Nitinol, 352
stent fracture, 499
stent grafts, 491, 497–8
stent(s), 483–6, 639
   composition of, 485
   fully absorbable, 500
   improved testing, 500
   keeping arteries open better and longer, 483
   marketed in the United States, 485
   representative sample of, 485
   sizing as crucial, 483
   structural requirements of, 485
sterility, 8–9
sterilization, 8, 639
   methods, 3, 9, 10
   options, 8–9
   procedure, 126
   protocols, 7
stiff tissues, 323

stiffness evolution, in a viscoelastic biomaterial, 224–6
storage modulus, 228, 229, 230, 231, 640
stored energy component, 229
strain
   amplitude, 333
   complete equations for, 171
   definition of, 169–71
   definitions, 170
   enabling, 110
   equation, 171
   imposing equal across each element, 217
   in cortical bone, 186
   normal (ε), 640
   parameters used in defining types of, 170
   shear (γ), 640
   types of normal (or extensional) strain, 169
strain energy, 250, 310
   basic form or, 250
   obtaining the total change in, 229
strain hardening, 640
   activated as dislocations interact, 45
   adding line defects, 45
   of metals and alloys, 40
   simple demonstration of, 46
   source of, 42
strain rate, 109, 640
   deceleration with increasing deformation, 236
   effect of increasing on stress-strain behavior, 111
strain response, of the Maxwell model, 218
strain to failure, 27, 187, 640
strain-based loading, 343–6
strain-based tests, 334, 343
strain-based total life philosophy, 344
strength described, 284
   designing for, 284
strength of materials, 284, 640
strengthening behavior, as a measure of strain hardening, 272
stress amplitude, 640
   determining, 338
   increasing or decreasing, 333
   of loading cycle, 332
   relationship with number of cycles to failure, 337
stress amplitude cycle (S-N) fatigue plots, 334
stress and strain, local fluctuations in, 189
stress concentration factor, 266, 267, 287, 640
stress concentrations
   cracks as extreme, 287–8
   in a ceramic system, 273
   locations of occurrence, 266
stress corrosion cracking, 640
   in a medical grade alloy, 59–60
   in device components, 59
   in metals, 17
   metals susceptible to, 330
   stainless steel susceptible to, 62
stress criteria, for crazing, 276
stress distribution, in a total joint replacement, 276–9
stress intensity, 357

calculating values of, 60
for an embedded penny-shaped flaw, 293
reaching critical value for material, 58
stress intensity factor (K), 60, 77, 288–91, 348, 640
for a flaw, 78
formula, 289
of an edge-notched crack, 298
overall, 293
range, 351
rewriting for individual loading modes, 292
stress intensity parameter, 349
stress invariants used to describe the yield surface, 245
utilized in fundamental failure theories, 245
stress matrix, components of, 259, 260
stress power law, governing strain rate, 236
stress range, 332, 640
stress ratios, 279, 640
stress relaxation, 210
accounting for, 19
for the Maxwell model, 219
in a polymeric implant, 221
Maxwell model illustrating, 218
stress relaxation response, of the SLS, 224
stress response, of the SLS, 232
stress shielding, 167, 640
bone loss due to, 416
Co-Cr prone to, 63
defined, 329
leading to bone resorption, 203
stress states
complicated, 291
components of, 173
contained within the ellipse as safe, 253
for pure shear, 258
of an object, 173
reduced to two dimensions, 259
stress tensor, 173–81, 640
stress transformation, 173, 640
stress triaxiality, 301
stress(es), 640
ahead of plastic zone, 301
at which plastic deformation begins, 36
complex states of, 243
decomposition of, 249
definition of, 171–3
distributing in the singularity region, 288
finding the highest, 179
imposing equal across each element, 217
in several different sets of coordinate axes, 175
in the Maxwell arm of the SLS during creep, 223
loads resolved into, 10
localization of, 266
measured in Pascals [Pa] or pounds per square inch [psi],
partitioning into the Maxwell arm of the SLS, 223
peak intensity of, 77
sharing over time, 222–30
upper limit of, 243
versus cycles to failure (N) data, 335
stress-based loading, 334–7

stress-based test, 334
stress-corrosion data, 59
stress-induced transformation, of crystal structure, 80
stress-strain behavior
of several polymers, 109, 110
of the human artery, 193
showing modulus of resilience, 243
stress-strain curves, 187–90
area under, 626
finding elastic, 189
for a material undergoing necking, 189
for brittle and elastomeric materials, 188
for brittle and rubbery materials, 188
for cortical bone, 193
for skin and artery, 155, 194
for tendon and ligament, 149
in natural tissues, 191–3
plotted data given, 190
toe and heel regions of, 323
traditional, 187
stress-strain response, 188, 228
striation markings, 357
Stribeck curve, 380, 384
Stribeck, Richard, 380
strip-yield zone diagram, 306
strip-yield crack tip cohesive zones, 305–8
strip-yield models, 308
strip-yield shape, of the plastic zone, 307
strip-yield zones, 307
Strock brothers, 520
stroma, 580, 640
structural analysis, presuming implant materials to be unflawed, 283
structural aspects
of dental implants, 529–30
of TMJ replacements, 538
structural components, as intrinsically flawed, 348
structural discontinuity, magnitude of stress near, 266
structural factors, determining elastic interaction with a dislocation, 47
structural implants examples of, 14
sustained loading of, 329
structural layers, separated by a soft interspersed phase, 323
structural materials
fatigue behavior of, 354–7
fracture mechanisms, 322–3
under sustained loading, 333
structural medical ceramics, comparison of mechanical properties of, 76
structural properties, for materials used in implants, 14
structural relaxation effects, 210
structural requirements, of medical devices, 6, 10–13
structure-property relationships, 16–17
strut fracture, 487
in a mechanical heart valve, 496
in a stent, 498
Styker-Dacron implant, 572

submergible, 640
submergible implant, 514
submissions, types for medical devices, 404–8
subperiosteal, 640
subperiosteal dental implants, 513, 514, 515–18, 529
subperiosteal framework, design of metal in, 516
subperiosteal implant strut configurations, 517
substantially equivalent, 640
substantially equivalent device, 404
substitutional solid solution, 47, 640
sulcus, 511, 641
Sulzer Inter-Op cup
Sulzer Orthopedics, recall, 467
super saturated solid solution, creating, 48
superelastic behavior, 63, 634
superelastic traits, 64
superficial tangent zone, 153
superior, 641
superposition
    determining stresses individually, 291
    of G, 296–7
supersaturated solid solution, 47, 641
Supramid, 585
surface coatings, for fixation, 433
surface contact mechanics, 375–7
surface contact, in biomaterials, 386–7
surface damage, for crosslinked UHMWPE used in the hip, 463
surface energy, 117, 641
surface factor, as a function of ultimate tensile strength, 347
surface features, relative sizes of, 374
surface finish, 641
surface properties, 371
surface roughness
    controlled through processing, 372
    curve illustrating, 371
    of components, 346
    resulting from conventional machining, 347
surface topography, 371
surface treatments, as a protective mechanism, 55
surgeon, role of a good, 5
surgeon's knot, 569, 570
surgical biomaterials, 13
surgical interventions, for TMJ disorders, 533
surgical meshes, reinforcing natural tissue, 412
surgical tools, sterilizing, 8
sustained loading, 331
sustained loading cycle, mean stress of, 338
suture anchors, 92, 124–5, 221, 641
suture materials employed in biomedical implants, 230
    examples of, 124
sutures, 565–70, 641
    among earliest recorded medical devices, 560
    as Class II devices, 565
    clinical uses of, 565
    current designs, 565–8
    functional requirements of, 568–70
    physical characteristics of, 568
    polymers in, 94
    properties of, 11
    with attached needle, 565
    with structures and suggested indications, 567
suturing, 560
switchboard model, 100
symmetric stress tensor, 249
symmetry, special cases of, 185
synovial fluid, 422
    as example of a lubricant, 380
    drawing into tissue, 153
    penetrating articular cartilage, 379
synovial joint, 422, 641
synovial membrane (capsule), covering the knee, 435
synthetic biomaterials attributes and limitations of, 17–20
    designing an implant using, 130
    in load-bearing medical devices, 14
synthetic grafts, 490
synthetic hydroxyapatites, elastic moduli of, 136
synthetic ligament replacements concerns associated with development of, 573
    search for a suitable, 561
synthetic ligaments, 570–5
    Class III devices, 574
    functional requirements of, 573
    not yet a good solution, 589
synthetic materials, 561
    categories based on structure, bonding, and inherent properties, 13
synthetic non-resorbable sutures, materials used, 565
synthetic scaffolds, 573
synthetic sutures, 565
synthetic vascular grafts, 491

tailbone (coccyx), 456
tan (δ), see loss factor
tangent modulus, 187, 641
tartar, 641
tartar-induced inflammation, of the gums, 510
tearing instability, 316
teeth, see tooth
Teflon, 478, 543
Teflon products, 543
temperatures, methods withstanding high, 9
tempered martensite, 49
tempering, 49, 641
temporomandibular joint (TMJ), 509, 641
    anatomy of, 509
    ankylosis, box-like stainless steel fossa for relieving, 534
    articular surfaces, 538
    as weight-bearing joint, 542
    devices, refinements in surgical placement of, 545
    disorder (NIDCR), 532, 533
    functions of, 533
    implants, reclassification as Class III devices, 537
    in the OA, 532
    loading, 545
    malocclusion generating super-physiologic loads on, 511
    mechanics, 506

prosthesis, Christensen's early, 528
registry, 545
resultant forces on from masticatory loads, 538
temporomandibular joint (TMJ) replacements, 506, 532–9, 545–6
   clinical reasons for, 532–3
   contraindications for, 538
   disorders necessitating, 532
   first permanent, 534
   functional requirements for, 538
   historical development of, 533–8
   history of, 506
   medical indications for, 538
   placement of a partial or total, 533
   structural aspects of, 538
tendon(s), 148–51, 641
   as assemblies of bundled collagen fibers, 129
   as structures of discrete components, 151
   composition of, 148
   hierarchical structure of, 149
   high load-carrying ability, 149
   high proportion of Type I collagen in, 133
   protection from abrasion and wear, 149
   schematic showing relative deformation response, 150
   structurally in parallel with muscles, 151
   structure of, 149
   transition from compliant to stiff behavior, 151
tennis players, bone strength of the dominant arm, 167
tensile force, 201
tensile strength, 641
   for elastic fibers, 134
   of a brittle material, 286
   of ceramics, 18
   of Co-Cr alloys, 63
   of collagen fibers, 133
   of cortical bone, 142
   of enamel, 138
   of human dentin, 139
   of most ceramic structures, 75
tensile stress, 172
tensile stress-strain curve, for articular cartilage, 153
tensile test, metrics derived from, 168
tensile testing configuration, 242
tension-compression stress-strain curve, 192
test methodology, reporting all aspects of, 169
test specimen, dimensions, 301
testing standards anatomy of, 408–9
   development of, 409–10
   from international standards development organizations, 168
tetrahedral site, 33, 641
tetrahedron, geometry of, 31
Thalidomide disaster, 400
theoretical cohesive strength, 75, 641
theoretical shear strength, 36–8, 75
theoretical strength, for small displacements, 75
theoretical value of strength, 75
thermal expansion coefficient, 81, 82, 641
thermal strains, 81

thermal stresses, 81–2
thermal treatments, controlling grain size, 49
thermal variation, experienced by dental implants, 81
thermodynamic driving forces, 55
thermoplastic polymer, 641
thermoplastics, characteristics of, 113
thermosetting plastics, 113
thermosetting polymers, 112, 641
thick specimen, plane strain and plane stress, 302
thin-walled cylindrical pressure vessel, 257
third degree burns, 576, 641
third-order imperfections, 42
thoracic spine, 456, 641
three-dimensional Tresca yield envelope, 248
three-dimensional hydrostatic stress, 257
three-dimensional polymer networks, 95
threshold regime, 351
threshold stress intensity factor, 353
thrombosis, in synthetic grafts, 491
thrombus, 483, 641
tibial, 641
tibial components, in total knee arthroplasty, 241
tibial plateau, 183, 184
time power law, 235
time, under a constant load, 319
time-dependence, associated with a characteristic time scale, 208
time-dependent, 641
time-dependent creep component, 361
time-dependent fracture mechanics (TDFM), 318–20, 641
time-dependent material behavior, 210
time-dependent material properties, 210–12
   describing viscoelastic materials, 208
   making design process more challenging, 209
time-dependent vanishing yield strengths, 318
time-temperature equivalence, 233
time-temperature superposition, 233–4, 642
time-temperature-transformation (TTT) plot, for steel, 50, 51
time-varying stress, in the Maxwell element during creep of SLS, 223
tissue drag, 570, 642
tissue engineering, creating a scaffold, 156–8
tissue ingrowth, assuring valve placement is secure, 488
tissue integration, available in ceramic systems, 85
tissue microstructure, as predictive of material response, 131
tissue valves, 486
   currently marketed in the United States, 489
   vulnerable to calcification, 487
tissue-engineering solutions, for TMJ disorders, 545
tissues, *see* natural tissues
tissues as viscoelastic solids, 213–14
   building blocks of, 131–6
   characteristics of, 241
   fracture mechanisms in, 323
   hydrated to maintain function, 214
   load-bearing, 136–56

titanium
  as a surgical material, 27
  elastic modulus compared to Co–Cr, 63
  forming an outer oxide layer, 18
  natural propensity for integration with bone, 63
  offering exceptional strength, 63
  osseointegrative properties of, 522
  pure existing as HCP structure, 63
  susceptible to corrosive wear, 383
titanium alloys
  elastic modulus, yield strength and tensile strength for, 61
  in the fracture fixation industry, 453
  introduced into orthopedics, 418
  offering exceptional strength, 63
  specials surface treatments with, 18
Ti-6Al-4V, 63, 432
titanium endplate-rubber core interface, failures occurring at, 455
titanium implant, Branemark's original, 522
titanium nitride, in an investigational device, 537
TMJ, *see* temporomandibular joint (TMJ)
toe regime behavior, 151
toe region, 134, 149, 642
tolerance, 642
tooth
  anatomically distinct regions, 507
  diagram indicating enamel and dentin, 138
  mineral responsible for stiffening, 135
  sections of, 136
tooth crown, 507, 508
tooth enamel, *see* enamel
tooth loss, 510, 511
tooth root
  below the gum margin, 508
  schematic of, 508
Toronto Conference for Osseointegration in Clinical Dentistry, in 1982, 522
torsional loading, 256
total artificial heart (TAH), 495
total condylar knee
  illustration of, 440
  in 1974, 439
  knee design founded upon, 439
total damage, 342
total hip arthroplasty, 422–33, 435
total hip replacements
  components of, 424
  constituents of contemporary, 419
  design of, 429
  elements of a contemporary, 418
  endurance limit error, 347–8
  functional properties of components in, 425
  historical evolution, 416
  material properties summarized, 432
  path to market, 407–8
  showing porous coating in, 431
  utilizing an UHMWPE acetabular cup, 117
total joint arthroplasty (TJA), 421
total joint replacements, 421–2, 642
  causes for, 422
  evolution in design and materials, 419
  failure rate, 422
  polymer sleeve assisting with bone ingrowth in, 202
  stress distribution in, 276–9
  use of composites in, 390
total knee arthoplasty, 435–46
  design concerns, 446
  materials used in contemporary, 443–6
total knee reconstruction, in mid-20th century, 437
total knee replacements design of, 443
  functional properties of components, 444
  functional requirements of components, 443
total life approximation for a femoral stem, 340–1
  using the Goodman equation, 341–2
total life design methodology, 333
total life philosophy, 334–7, 642
total life tests, specimens utilized in, 335
total strain amplitude, of the fatigue cycle, 344
toughening ceramics, in engineering applications, 322
toughening mechanisms classifications of, 354
  types of, 320
  utilized in ceramic structures, 80
  utilized in ceramics, 79
toughness, of polymer structures, 106
trabeculae
  arrangement of, 144
  beam or plate-like behavior of, 146
  excessive remodeling of in old age, 144
  in vertebrae, 145
  orientation coordinated, 145
trabecular bone, 143–8, 642
  architecture of, 130
  cellular structure of, 144
  composition of, 144
  coral as a substitute for, 89
  dependence on porosity and trabecular architecture, 147
  dominant orientation of architecture of, 145
  exhibiting different architectures, 145
  increased surface area of, 144
  macroscopic mechanical performance of, 148
  mechanical behavior of, 145
  network structure of, 145
  open network structure of, 146
  relationship between ultimate tensile strength and porosity, 148
  structural performance of, 148
  structure, schematic of, 144
traction vector, defining, 172
Transcyte silicon membrane, 577
transformation toughened ceramic, 80
transition region, 210
transition time, viscoelastic transition around, 233
transmucosal, 642
transmucosal abutments, 513
transparency, of PMMA, 119
transverse fracture, 449
transverse isotrophy, 185, 186, 642
transverse plane, 642

transverse processes, 457, 642
transverse testing, of cortical bone, 142
Trapeziodal-28 (T-28) hip stem design
    failure mechanisms of, 362
    failures of, 426
trauma, 424, 642
trephine, 520, 642
Tresca stress, 248–9
Tresca yield criterion, 293
    determining effective stress, 338
    developed for a plane stress state, 247
    expression for, 248
    for a stress state, 256
    shear stress needed for plastic deformation, 258
Tresca yield surface, 247
Tresca, Henri, 243
Trevira/Hochfest ligament augmentation device, 572, 573
triaxial stress state, 258
tribology, 369, 642
tricuspid valve, 479
tripod blade-form ramus frame implant, 519
tripodal subperiosteal implants, 517
trobocollagen, 642
tropocollagen molecules, 132
true EPFM conditions, 314
true material properties, 320
true strain, 191, 642
true stress, 191, 642
true stress and strain compared to engineering stress and strain, 192
    realistic picture of deformation behavior, 205
true stress–strain behavior, 242
trunnion, of femoral shaft, 429
truss-like frameworks, with strategically placed struts, 516
turbostratic carbon, 84, 642
turbulent flow, reduction of, 488
twisting or tearing mode (Mode III), 291
two-dimensional polymer systems, weak van der Waals forces, 95
two-stage surgeries, adoption of, 522
two-stage surgery, 514
Type I collagen, 132
Type II collagen, 132
Type III collagen, 132
Type IV collagen, 133
Type IX collagen, 132
Type V collagen, 133
Type VII collagen, 132
Type XI collagen, 133
Type XII collagen, 133
Type IX collagen, 133

UHMWPE, *see* ultra-high molecular weight polyethylene (UHMWPE)
ultimate strain
    for elastic fibers, 134
    of tendon and ligament, 150
ultimate strength, of metals, 27
ultimate tensile strength (UTS), 150, 187, 642
ultra-high molecular weight polyethylene (UHMWPE), 117, 642
    acetabular cup, delamination of highly cross-linked, 462–3
    as bearing material, 417
    as polymer of choice for acetabular cups, 432
    bearing components in total hip replacements, 284
    bearings, wear in total joint arthroplasty, 237
    components sterilizing with EtO or gas plasma methods, 21
    delamination wear of, 21
    fatigue crack propagation data for, 361
    good wear properties, 390
    highly crosslinked form of, 428
    in a constraining frame of Co–Cr, 183
    in total joint joint reconstruction, 225
    in total joint replacements, 93
    master curve correction factors for, 225
    master curve for, 236, 237
    material properties, 432
    mechanical characterization of, 361
    medical grade, 117
    poor oxidation resistance of, 20, 22
    subsurface fatigue fracture (delamination) of, 442
    tibial knee component, oxidation of, 21
    uniaxial yield strength of, 254
    MPa, 254
umbilical vein, 478, 490, 642
umbilical vein homograft, 490
unhealthy contact, leading to pain and tissue damage, 369
uniaxial elastic deformation, of an isotropic material, 181
uniaxial fibers, 201
uniaxial load, bar placed under, 171
uniaxial loading analysis of, 175
    as not a typical stress state, 255
uniaxial stress, 12
uniaxial stress state, in an isotropic ductile material, 256
uniaxial tensile test, 241
uniaxial tension, bar under, 172
uniaxial yield strength, 249, 251
uniaxial yield stress, 241, 255
unicompartmental designs, 438
unicondylar, 642
unicondylar knee arthroplasty, 437
unicondylar knee implants, 437
unidirectional composite, example, 19
unidirectional fiber composite, 20
uniform attack, 643
uniform attack mode of corrosion, 55, 56
United Kingdom, Medicines and Healthcare products Regulatory Agency (MHRA), 410
upper bound elastic modulus, 19
upper thoracic spine, 454
US Pharmacopeia, 569

vacancies, 39
vacant lattice site, 38
VADs (ventricular assist devices), 495
valence state, increase of during corrosion, 55
valgus, 643
valgus (knock-kneed) alignment, 436
valleys, as lubricant reservoirs, 372
valve design, requirements in, 488
valves, in the heart, 479
valvular disease, 477, 643
van der Waals forces, 95, 643
vanadium steel, 417, 643
variable amplitude loading, Palmgren–Miners rule for, 343
varus, 643
varus (bow-legged) knees, 436
vascular, 643
vascular grafts, 489–91
   evolution of materials selection for, 478
   options, 490
   representative sample of, 492
   showing a range of diameters, 492
vascular implants, development of, 156
vascular replacement, creating a trilaminate, 157
vascularization, 88, 643
vasculature, interruption of the local, 447
vector components, within the lattice, 34
veins, 481, 643
venous homografts, 490
venous ulcers, 575, 643
vent-plant implants, 522
ventricles, 479, 480, 643
ventricular assist devices (VADs), 495
ventricularis, 479
vertebra
   constituents in individual, 457
   individual illustration of, 457
   stacked upon each other, 456
Vesalius, Andreas, 129
viscoelastic, 233, 643
viscoelastic (glass) transition temperature, 210
viscoelastic behavior, 210, 214
viscoelastic body
viscoelastic fracture mechanics (VEFM), 319–20, 643
viscoelastic materials, 208, 212
viscoelastic model, 214
viscoelastic nature, of most polymers
viscoelastic response, 223
viscoelastic solids, 209, 213–14
viscoelastic transition, 210, 643
viscoelasticity, 109, 643
   described, 209
   introduction to, 209–14
   of polymers, 19
viscous (fluid) behavior, 109
viscous dissipation, 229, 318
viscous loss component, 229
viscous solids, differing from inelastic solids, 318
Vitek, failed in responsibilities to rigorously test, 543
Vitek-Kent PT prostheses, 537, 542

Vitek-Proplast I prostheses, failure rate, 544
Vitredent (company), 540
Vitredent Tooth Replacement System, 542
Vitredent vitreous carbon root-form dental implant, 541
vitreous carbon
   as a potential dental material, 540
   biocompatibility of implant materials, 540
   forming, 540
vitreous carbon implants
   dismal performance of, 542
   failure of, 540–2
vitreous carbon Tooth Replacement System, 541
VK glenoid fossa component, 536
VK implants, showing particulate debris, 544
Volkmann's canals, 141, 643
volumetric expansion, for dental implants, 82
volumetric thermal expansion, estimating, 82
von Helmholtz, 130
von Meyer, 130
von Meyer, Herman, 146
von Mises effective stress
von Mises ellipse yield surface, schematic of, 252
von Mises ellipse, circumscribing the Tresca hexagon, 253
von Mises stress effective, 251, 253–5, 260, 290
   finding, 254
   large sub-surface, 442
von Mises yield criteria, 251, 260
   better matching experimental data, 253
   capturing complete loading, 261
   for stress state, 256, 258
   indicating required shear stress, 258
   modified, 264
   modifying, 263
   writing in general stress component terms, 252
von Mises yield envelope, 253, 254
von Mises yield space
von Mises, Richard, 243
vulcanizing, to the titanium plates, 455

water, in articular cartilage, 151
wax prototype, 52
wear, 381–6, 643
   abrasive, 643
   adhesive, 643
   as situational, 391
   contact fatigue, 643
   corrosive, 643
   delamination, 643
   erosive, 643
   for any long-term implant problem, 370
   fretting, 643
   in a clinical setting, 237
   mechanism resulting in lamellar alignment in UHMWPE, 427
   properties at expense of fracture properties, 429
   rates, 237
   stages of, 381
   types of, 381
wear bearings, highly resistant, 424

wear debris, *see also* particulate debris, *see also* corrosion debris
wear debris
   from fretting wear, 385
   generation of, 369
   particles invoking inflammation response, 381
   setting off inflammatory cascade, 370
wear resistance
   improved by enhanced lubrication schemes, 237
   paramount for knee reconstruction, 11
wear simulators, 389
wear testing, standardization of, 370
   wide variety of test methods, 389
wear-induced osteolysis, associated with loosening, 427
wedge-shaped cohesive zone, for a propagating crack in UHMWPE, 305
wedge-shaped plastic zone crack in tip zone, 305
weight average molecular weight, 107, 109, 643
wet precipitation techniques, 84
Wetherill, Charles M., 398
white blood cells, 482
Wiles, Phillip, 417
Wilkes, Clyde, 534
Williams, Landel, and Ferry (WLF) equation, 234–5
wire methods, of fixation, 453
wires, 450, 644
WLF equation, 234, 644
wound healing technologies, 575
woven or knitted materials, 561
wrought cobalt-chromium based steels, used in orthopedic bearings, 47
wrought Co-Cr, in hip arthroplasty, 432
wrought steel, 47, 644

xenografts, using bovine vasculature, 478
xenograft, *see* heterograft
XL-UHMWPE acetabular liners, fractured
x–y plane, of an infinitesimal element, 175

yield behavior, of polymer systems, 274
yield criteria, 243, 261, 263

yield envelope, for the plane stress case, 253
yield strain, occuring under uniaxial yielding, 345
yield strength, 15, 36, 187, 644
   higher, 42
   information given by, 205
   normalizing with effective stress, 255
   of a material, 243
   of cobalt-chromium, 62
   of Co-Cr alloys, 63
   value for a metal, 38
yield stress, 243
   as a function of grain size, 44
   as failure criteria, 340
   calculated for uniaxial loading, 256
   for 316L stainless steel, 363
   in compression, 263
   in tension, 263
   increasing with increasing strain rate and decreasing temperature, 111
   measured in uniaxial tension, 251
yield surfaces, 244–5, 644
   boundary of, 247
   for a planar (two-dimensional), 245
   overlay of von Mises and Tresca, 253
yield, in multiaxial loading conditions, 255–61
yielding
   criterion in metals based on shear stresses, 42
   occurring with maximum distortional energy, 249
   occurrence of, 249
   onset of, 243
Young, Thomas, 130
Young's modulus, 134, 181
yttria-stabilized zirconium oxide, 523

z-direction, neglecting stresses acting in, 259
Ziegler-Natta catalysts, 99, 644
zirconia
   properties of, 86
   used as a bearing material, 80
zirconia implants, manufacturers of, 524
zirconium oxide, newer devices adopting, 506
zone of HRR singularity, 312
zone shielding, 320, 355

Printed in the United States
By Bookmasters